Nanomaterials Recycling

Nanomaterials Recycling

Edited by

MAHENDRA RAI
Department of Biotechnology, SGB Amravati University, Amravati, Maharashtra, India

TUAN ANH NGUYEN
Institute for Tropical Technology, Vietnam Academy of Science and Technology, Hanoi, Vietnam

Elsevier
Radarweg 29, PO Box 211, 1000 AE Amsterdam, Netherlands
The Boulevard, Langford Lane, Kidlington, Oxford OX5 1GB, United Kingdom
50 Hampshire Street, 5th Floor, Cambridge, MA 02139, United States

Copyright © 2022 Elsevier Inc. All rights reserved.

No part of this publication may be reproduced or transmitted in any form or by any means, electronic or mechanical, including photocopying, recording, or any information storage and retrieval system, without permission in writing from the publisher. Details on how to seek permission, further information about the Publisher's permissions policies and our arrangements with organizations such as the Copyright Clearance Center and the Copyright Licensing Agency, can be found at our website: www.elsevier.com/permissions.

This book and the individual contributions contained in it are protected under copyright by the Publisher (other than as may be noted herein).

Notices

Knowledge and best practice in this field are constantly changing. As new research and experience broaden our understanding, changes in research methods, professional practices, or medical treatment may become necessary.

Practitioners and researchers must always rely on their own experience and knowledge in evaluating and using any information, methods, compounds, or experiments described herein. In using such information or methods they should be mindful of their own safety and the safety of others, including parties for whom they have a professional responsibility.

To the fullest extent of the law, neither the Publisher nor the authors, contributors, or editors, assume any liability for any injury and/or damage to persons or property as a matter of products liability, negligence or otherwise, or from any use or operation of any methods, products, instructions, or ideas contained in the material herein.

British Library Cataloguing-in-Publication Data
A catalogue record for this book is available from the British Library

Library of Congress Cataloging-in-Publication Data
A catalog record for this book is available from the Library of Congress

ISBN: 978-0-323-90982-2

For Information on all Elsevier publications
visit our website at https://www.elsevier.com/books-and-journals

Publisher: Matthew Deans
Acquisitions Editor: Simon Holt
Editorial Project Manager: Gabriela D. Capille
Production Project Manager: Debasish Ghosh
Cover Designer: Greg Harris

Typeset by MPS Limited, Chennai, India

Contents

List of contributors	*xv*
Foreword	*xix*
Preface	*xxi*

Section I Environmental impacts of nanowastes

1. Nanomaterial recycling: an overview   3

Pooja Sharma, Praveen Guleria and Vineet Kumar

1.1	Introduction	3
1.2	Classification of nanowastes	4
1.3	Sources and routes of nanowaste contamination	7
1.4	Toxic effects of nanowastes	9
1.5	Impact of nanowastes on environment	11
1.6	Nanowaste treatment strategies	13
1.7	Regulatory bodies for nanowaste generation and management	15
1.8	Future perspectives and challenges	16
1.9	Conclusion	17
	References	17

2. Nanomaterial waste management   21

Thodhal Yoganandham Suman and De-Sheng Pei

2.1	Introduction	21
2.2	Nanomaterials: definition and trends of the world nanomaterials market	21
2.3	Nanowastes	22
2.4	Carbon-based nanomaterials	23
2.5	Silver nanoparticles	23
2.6	Titanium dioxide nanoparticles	24
2.7	Prospective concerns around nanowastes	24
2.8	Challenge of nanowastes	24
2.9	Classification of nanowastes	25
2.10	Difficulties and concerns about nanowastes management	28
2.11	Incineration of waste that contains nanomaterials	28
	2.11.1 Nanowaste treatment in waste treatment plants	28
	2.11.2 Nanowaste treatment in waste incineration plants	29
	2.11.3 Nanowaste treatment in landfills	30

vi Contents

2.11.4 Recycling of waste containing nanomaterials	31
2.11.5 Nanowaste management problems and issues	32
2.11.6 Legislative framework	33
2.12 Conclusions	33
Acknowledgments	34
Conflicts of interest	34
References	34

3. Classification and sources of nanowastes   **37**

Prashant Gupta and Subhendu Bhandari

3.1 Introduction	37
3.2 Types of nanomaterials	38
3.2.1 Carbon-based nanomaterials	38
3.2.2 Organic nanomaterials	40
3.2.3 Inorganic nanomaterials	40
3.3 Classification of nanowastes	42
3.4 Sources of nanowastes	45
3.4.1 Stationary sources	47
3.4.2 Dynamic sources	48
3.4.3 Miscellaneous sources	51
3.5 Conclusion	52
References	52

4. General regulations for safe handling of manufactured nanomaterials   **61**

Maria Batool, Muhammad Nadeem Zafar and Muhammad Faizan Nazar

4.1 Introduction	61
4.1.1 Precautionary principles	62
4.2 Precautionary measures	63
4.2.1 Technical measures	63
4.2.2 Organizational measures	63
4.2.3 Personal measures	63
4.3 Health hazards	63
4.3.1 Exposure routes	64
4.4 Fire and explosion hazards	66
4.5 Environmental hazards	66
4.6 Risk assessment and safety precautions for nanomaterial use	66
4.6.1 Risk evaluation	66
4.6.2 Controlling exposure	67
4.7 Storage, waste handling and spills	71
4.7.1 Storage	71

Contents vii

	4.7.2	Waste handling	71
	4.7.3	Spills	71
4.8	Regulations		72
	4.8.1	Legislation for suppliers	72
	4.8.2	Legislation for recipients and users of chemicals	73
	4.8.3	Legislation for regulatory authority	73
	4.8.4	Chemicals (hazard information and packaging for supply) regulations	75
4.9	Workplace risk management		76
	4.9.1	Control of substances hazardous to health regulations	76
	4.9.2	Dangerous substances and explosive atmospheres regulations	79
	4.9.3	Existing substances regulation	80
	4.9.4	Biocidal products regulations	80
	4.9.5	Control of major accident hazards regulations	80
4.10	Conclusion		80
References			81

5. Safety and global regulations for application of nanomaterials 83

Md Abdus Subhan and Tahrima Subhan

5.1	Introduction		83
5.2	Risks management for environment and health safety		88
5.3	Approaches of democratic governance to nanotechnology		89
5.4	Nano inventiveness		91
5.5	International law on nanomaterials		91
5.6	Arguments against regulation of nanomaterials		92
5.7	Response from governments all over the world		92
	5.7.1	The United States	92
	5.7.2	The United Kingdom	95
	5.7.3	The European Union	95
	5.7.4	Canadian policy on nanotechnology	99
	5.7.5	Japanese nano policy	100
	5.7.6	South Korean policy on nanotechnology	100
	5.7.7	Application of nanotechnology in Thailand	101
	5.7.8	Response from advocacy groups	102
	5.7.9	Some technical aspects of nanomaterials	103
	5.7.10	The regulation of nanomaterials for clinical application	103
5.8	Conclusion and future perspectives		105
References			105

6. Nanowaste disposal and recycling 109

Sakshi Gupta and Manish Kumar Bharti

| 6.1 | Introduction | 109 |

viii   Contents

6.2	Classifications of nanowaste	112
6.3	Disposal and recycling of nanowaste	114
	6.3.1 Disposal of nanowaste	115
	6.3.2 Recycling of nanowaste	116
6.4	Conclusion	119
	References	120

7. Management of nanomaterial wastes   125

Nakshatra B. Singh, Martin F. Desimone, Ratiram Gomaji Chaudhary and W.B. Gurnule

7.1	Introduction	125
7.2	Types of nanomaterials and nanowaste	125
7.3	Synthesis of nanomaterials	127
7.4	Toxicity of nanomaterials and their release to the environment	127
7.5	Generation of nanowaste	131
7.6	Impact of nanowaste on the environment	131
7.7	Impact of nanowaste on health	133
7.8	Biological treatment of nanowastes	134
7.9	Recycling of nanowastes	135
7.10	Challenges in nanowaste management	140
7.11	Conclusions	142
	References	142

Section II Methods for recycling of nanomaterials

8. General techniques for recovery of nanomaterials from wastes   147

Kuray Dericiler, Ilayda Berktas, Semih Dogan, Yusuf Ziya Menceloglu and Burcu Saner Okan

Abbreviations		147
8.1	Introduction	148
8.2	Types of nanomaterial wastes	149
	8.2.1 Carbon-based nanomaterials	150
	8.2.2 Ceramic-based nanomaterials	151
	8.2.3 Metal-based nanomaterials	152
	8.2.4 Nanomaterial-reinforced composite materials	155
8.3	Types of techniques used for the recovery of nanomaterials from wastes	157
	8.3.1 Magnetic separation technique	157
	8.3.2 Antisolvent technique by using CO_2	160
	8.3.3 Aqueous dispersion techniques	162
	8.3.4 Colloidal solvent technique	163
	8.3.5 Centrifugation/solvent evaporation technique	163

Contents ix

	8.3.6 Other approaches for the recovery of nanomaterials	164
8.4	Conclusions and outlook	165
References		168

9. Procedures for recycling of nanomaterials: a sustainable approach 175

Ajit Behera, Deepak Sahini and Dinesh Pardhi

9.1	Introduction	175
	9.1.1 Nanomaterials posing risks to humans and environment	179
9.2	Classification of nanowaste	183
9.3	Typical safety guidelines for handling nanoparticles	186
9.4	Disposal of nanoparticle waste	188
9.5	Various processes for nanowaste recycling	188
	9.5.1 Physical processes	189
	9.5.2 Chemical processes	190
	9.5.3 Thermal processes	194
	9.5.4 Electrodeposition deposition and electrokinetic process	195
	9.5.5 Sludge treatment process	196
	9.5.6 Microemulsion process	197
	9.5.7 Microbiological process	197
	9.5.8 Coagulation technique	198
	9.5.9 Nanoporous materials and membrane separation	198
	9.5.10 Glucose reduction process	199
	9.5.11 Layer-by-layer assembling	199
9.6	Various nanowaste recycling products	199
	9.6.1 Nanomaterials in concrete production	199
	9.6.2 Nanomaterials applied in suspensions	200
	9.6.3 Low-cost sensors for energy storage applications	201
9.7	Recycling of nanocomposites	202
9.8	Benefits of nanomaterials recycling	202
9.9	Limitations of nanomaterials recycling	202
9.10	Conclusions	203
References		203

10. Recycling of nanomaterials by solvent evaporation and extraction techniques 209

Haleema, Muhammad Usman Munir, Duy-Nam Phan and Muhammad Qamar Khan

10.1	Introduction	209
10.2	The importance of recycling in waste management	209
	10.2.1 Classification of wastes	210
10.3	Nanomaterials in the environment	210
	10.3.1 Recovering nanomaterials from the environment	211

x   Contents

10.3.2	Recovering nanomaterials from products	212
10.4	Nanomaterial recycling techniques	213
10.4.1	Solvent evaporation method	214
10.4.2	Solvent extraction method	215
10.5	Recycling of nanomaterials via solvent evaporation and extraction	216
10.5.1	Recycling of nanomaterials by solvent evaporation method	216
10.5.2	Recycling of waste by solvent extraction method	218
10.5.3	Nanoparticle recovery using a microemulsion	218
10.5.4	Nanoparticle recovery by cloud point extraction	219
10.5.5	Recycling nanoparticles by employing a colloidal solvent	219
10.6	Potential opportunities for the recovery and reuse of nanowaste	220
10.7	Conclusion	220
	References	220

11. Using pH/thermal responsive materials 223

Soheyl Mirzababaei, Kiyana Saeedian, Mona Navaei-Nigjeh and Mohammad Abdollahi

	Abbreviations	223
11.1	Introduction	224
11.2	pH-responsive materials	225
11.3	Thermoresponsive materials	226
11.4	Stimuli-responsive nanostructures	227
11.4.1	Micelle	228
11.4.2	Vesicle	229
11.4.3	Polymer brush	230
11.4.4	Hydrogel	230
11.4.5	Core-shell NP	231
11.5	Applications in nanomaterials recycling	232
11.5.1	Aqueous two-phase system	232
11.5.2	Catalysts	238
11.5.3	Adsorption	240
11.6	Conclusions	242
	References	243

12. Nanomaterials recycling standards 249

Arsalan Ahmed, Muhammad Fahad Arian and Muhammad Qamar Khan

12.1	Introduction	249
12.2	Fundamental of nanoparticles	251
12.3	Classification of nanoparticles	252
12.3.1	Carbon-based nanoparticles	253
12.3.2	Ceramic nanoparticles	253
12.3.3	Metal nanoparticles	253

Contents **xi**

	12.3.4 Semiconductor nanoparticles	253
	12.3.5 Polymeric nanoparticles	254
	12.3.6 Lipid-based nanoparticles	254
12.4	Scope of nanotechnology	254
12.5	Recycling of nanomaterials	255
12.6	The importance of recycling in waste management	256
12.7	Uses of nanomaterials	258
12.8	Uses of nanomaterials in consumer products	258
12.9	Nanomaterials and the industries they are used in	258
12.10	Categories of wastage of nanomaterials	259
12.11	Nanowaste ecotoxicology and treatment	259
12.12	Waste generated during production	260
12.13	Disposal and recycling of nanomaterials in waste	262
12.14	Risks related to nanomaterials in waste	262
12.15	Nanomaterials in recycling operations and potential exposure	263
12.16	Nanotechnology is dangerous for humans	264
12.17	Possible dangers of nanotechnology	264
12.18	Conclusion	265
	References	266

13. Ionic liquids for nanomaterials recycling 269

Hani Nasser Abdelhamid

13.1	Introduction	269
13.2	Scope of ionic liquids	270
13.3	Synthesis of ionic liquids	271
13.4	Types of ILs	272
13.5	Recycling ionic liquid	273
13.6	The applications of ionic liquids for recycling	276
13.7	Conclusions	278
	References	278

Section III Properties of recycled nanomaterials

14. Techniques used to study the physiochemical properties of recycled nanomaterials 291

Vojislav Stanić

14.1	Introduction	291
	14.1.1 X-ray diffraction method	291
14.2	Fourier transform infrared spectroscopy	296
14.3	Raman spectroscopy	298
14.4	X-ray photoelectron spectroscopy	299
14.5	Auger electron spectroscopy	303

xii   Contents

14.6	X-ray fluorescence analysis	305
14.7	Scanning electron microscopy	306
14.8	Transmission electron microscopy	308
14.9	Atomic force microscopy	309
14.10	Electron probe microanalysis	312
	Acknowledgments	314
	References	314

15. Mechanical properties of recycled nanomaterials   317

S. Behnam Hosseini

15.1	Introduction	317
15.2	Recycling of nanomaterials	318
	15.2.1 Recycling definitions	318
15.3	Mechanical properties of recycled nanomaterials	324
	15.3.1 Recycled nanoclay	324
	15.3.2 Recycled nano-$CaCO_3$	324
	15.3.3 Recycled nanosilver	326
	15.3.4 Recycled nano-zinc oxide	326
	15.3.5 Recycled carbon nanotubes	326
	15.3.6 Recycled nano-SiO_2	327
	15.3.7 Recycled lignocellulose nanofibers	328
	15.3.8 Recycled nanoalumina	330
15.4	Conclusions	331
	References	332

Section IV Applications of recycled nanomaterials

16. Industrial scale up applications of nanomaterials recycling   341

Ajit Behera and Suman Chatterjee

16.1	Why industrial-scale recycling is require for nanomaterials	341
16.2	Market scenario of recycled nanomaterials	343
16.3	Nanowaste generation from various industries and practices	344
	16.3.1 Medical industries' nanowaste	345
	16.3.2 Electronic industries' nanowaste	346
	16.3.3 Energy-harvesting industries' nanowaste	347
	16.3.4 Construction industries' nanowaste	347
	16.3.5 Other industries' nanowaste	348
16.4	Potential nanobyproduct materials, their recovery, and recycling	349
	16.4.1 Carbon-based nanobyproducts	349
	16.4.2 Metal nanobyproducts	350
	16.4.3 Metal oxide and nonmetal oxide nanobyproducts	351
	16.4.4 Polymer nanobyproducts	353

Contents xiii

16.5 Nanowaste recycling processes 353
    16.5.1 Mechanical recycling 353
    16.5.2 Chemical recycling 354
    16.5.3 Thermal processing 355
    16.5.4 Landfilling treatment 356
    16.5.5 Microemulsion process 356
    16.5.6 Other methods of recycling 357
16.6 Summary and future perspectives 358
References 358

17. Recycled nanomaterials for construction and building materials   363

P.O. Awoyera, C.O. Nwankwo and D.P. Babagbale

17.1 Introduction 363
17.2 Application of nanomaterials in construction 363
17.3 Nanomaterials in concrete design and development 364
    17.3.1 Concrete 364
    17.3.2 Asphalt concrete 369
    17.3.3 Other building applications 370
    17.3.4 Health implications 370
17.4 Conclusion 371
17.5 Recommendations 371
References 372

18. Nanomaterials recycling in industrial applications   375

Marjan Hezarkhani, Abdulmounem Alchekh Wis, Yusuf Menceloglu and Burcu Saner Okan

Abbreviations 375
18.1 Introduction 376
18.2 Current problems in nanowaste sustainability and its management 377
    18.2.1 Regulations and standards for nanowaste management 377
    18.2.2 Identification of recovered waste nanoparticles by characterization techniques 378
18.3 Recovery of sustainable metals and inorganic nanoparticles from waste for catalytic and magnetic properties 380
18.4 Utilization of recycled nano-scale and micron-scale reinforcements in composite applications 383
18.5 The growth of nanomaterials from waste plastics by an upcycling process 386
18.6 Recovery and reuse of metal nanoparticles from waste electronic components 388
18.7 Selective recovery of metal nanoparticles by using α-cyclodextrin 389
18.8 Nanomaterials for environmental cleanup applications 390
18.9 Conclusions and potential outlook 391
References 393

Index 397

List of contributors

Hani Nasser Abdelhamid
Advanced Multifunctional Materials Laboratory, Department of Chemistry, Faculty of Science, Assiut University, Assiut, Egypt; Proteomics Laboratory for Clinical Research and Materials Science, Department of Chemistry, Assiut University, Assiut, Egypt

Mohammad Abdollahi
Pharmaceutical Sciences Research Center, The Institute of Pharmaceutical Sciences (TIPS), and School of Pharmacy, Tehran University of Medical Sciences, Tehran, Iran

Arsalan Ahmed
Department of Textile and Clothing, Faculty of Engineering & Technology, National Textile University Karachi Campus, Karachi, Pakistan

Muhammad Fahad Arian
Department of Textile and Clothing, Faculty of Engineering & Technology, National Textile University Karachi Campus, Karachi, Pakistan

P.O. Awoyera
Department of Civil Engineering, Covenant University, Ota, Nigeria

D.P. Babagbale
Department of Civil Engineering, Covenant University, Ota, Nigeria

Maria Batool
Department of Chemistry, Faculty of Science, University of Gujrat, Gujrat, Pakistan

Ajit Behera
Department of Metallurgical & Materials Engineering, National Institute of Technology, Rourkela, India

S. Behnam Hosseini
Department of Wood and Paper Science and Technology, Faculty of Natural Resources, University of Tehran, Karaj, Iran

Ilayda Berktas
Sabanci University Integrated Manufacturing Technologies Research and Application Center & Composite Technologies Center of Excellence, Teknopark Istanbul, Istanbul, Turkey

Subhendu Bhandari
Department of Plastic and Polymer Engineering, Maharashtra Institute of Technology, Aurangabad, India

Manish Kumar Bharti
Department of Aerospace Engineering, Amity School of Engineering & Technology, Amity University Haryana, Gurugram, India

Suman Chatterjee
Department of Mechanical Engineering, National Institute of Technology, Rourkela, India

Ratiram Gomaji Chaudhary
P.G. Department of Chemistry, S.K. Porwal College, Kamptee, India

Kuray Dericiler
Sabanci University Integrated Manufacturing Technologies Research and Application Center & Composite Technologies Center of Excellence, Teknopark Istanbul, Istanbul, Turkey

Martin F. Desimone
Universidad de Buenos Aires, Consejo Nacional de Investigaciones, Científicas y Técnicas (CONICET), Instituto de Química y Metabolismo del Fármaco (IQUIMEFA), Facultad de Farmacia y Bioquímica, Buenos Aires, Argentina

Semih Dogan
Sabanci University Integrated Manufacturing Technologies Research and Application Center & Composite Technologies Center of Excellence, Teknopark Istanbul, Istanbul, Turkey

Praveen Guleria
Plant Biotechnology and Nanobiotechnology Lab, Department of Biotechnology, School of Life Sciences, DAV University, Jalandhar, India

Prashant Gupta
Department of Plastic and Polymer Engineering, Maharashtra Institute of Technology, Aurangabad, India

Sakshi Gupta
Department of Civil Engineering, Amity School of Engineering & Technology, Amity University Haryana, Gurugram, India

W.B. Gurnule
Department of Chemistry, Kamla Nehru Mahavidyalaya, Nagpur, India

Haleema
Nanotechnology Research Lab, Department of Textile & Clothing, Faculty of Engineering & Technology, National Textile University, Karachi Campus, Karachi, Pakistan

Marjan Hezarkhani
Sabanci University Integrated Manufacturing Technologies Research and Application Center & Composite Technologies Center of Excellence, Teknopark Istanbul, Istanbul, Turkey

Muhammad Qamar Khan
Nanotechnology Research Lab, Department of Textile & Clothing, Faculty of Engineering & Technology, National Textile University, Karachi Campus, Karachi, Pakistan

Vineet Kumar
Department of Biotechnology, School of Bioengineering and Biosciences, Lovely Professional University, Phagwara, India

Yusuf Ziya Menceloglu
Sabanci University Integrated Manufacturing Technologies Research and Application Center & Composite Technologies Center of Excellence, Teknopark Istanbul, Istanbul, Turkey; Department of Materials Science and Nano Engineering, Faculty of Engineering and Natural Sciences, Sabanci University, Istanbul, Turkey

List of contributors xvii

Soheyl Mirzababaei
Pharmaceutical Sciences Research Center, The Institute of Pharmaceutical Sciences (TIPS), and School of Pharmacy, Tehran University of Medical Sciences, Tehran, Iran

Muhammad Usman Munir
Department of Materials Engineering, Kaunas University of Technology, Kaunas, Lithuania

Mona Navaei-Nigjeh
Pharmaceutical Sciences Research Center, The Institute of Pharmaceutical Sciences (TIPS), and School of Pharmacy, Tehran University of Medical Sciences, Tehran, Iran

Muhammad Faizan Nazar
Department of Chemistry, University of Education, Lahore, Pakistan

C.O. Nwankwo
Department of Civil Engineering, Covenant University, Ota, Nigeria

Burcu Saner Okan
Sabanci University Integrated Manufacturing Technologies Research and Application Center & Composite Technologies Center of Excellence, Teknopark Istanbul, Istanbul, Turkey

Dinesh Pardhi
Faculty of Health Sciences, School of Pharmacy, University of Eastern Finland, Kuopio, Finland

De-Sheng Pei
School of Public Health and Management, Chongqing Medical University, Chongqing, China; Chongqing Institute of Green and Intelligent Technology, Chinese Academy of Sciences, Chongqing, P.R. China

Duy-Nam Phan
School of Textile – Leather and Fashion, Hanoi University of Science and Technology, Hanoi, Vietnam

Kiyana Saeedian
Biotechnology Group, Faculty of Chemical Engineering, Tarbiat Modares University, Tehran, Iran

Deepak Sahini
Department of Production Engineering, Birla Institute of Technology, Ranchi, India

Pooja Sharma
Department of Biotechnology, School of Bioengineering and Biosciences, Lovely Professional University, Phagwara, India

Nakshatra B. Singh
Department of Chemistry and Biochemistry, SBSR, Sharda University, Greater Noida, India

Vojislav Stanić
Laboratory of Radiation and Environmental Protection, Vinča Institute of Nuclear Sciences, University of Belgrade, Belgrade, Serbia

Md Abdus Subhan
Department of Chemistry, Shahjalal University of Science and Technology, Sylhet, Bangladesh

Tahrima Subhan
Department of Social Work, Shahjalal University of Science and Technology, Sylhet, Bangladesh; The Sylhet Khajanchibari International School and College, Sylhet, Bangladesh

Thodhal Yoganandham Suman
School of Public Health and Management, Chongqing Medical University, Chongqing, China; Chongqing Institute of Green and Intelligent Technology, Chinese Academy of Sciences, Chongqing, P.R. China; Ecotoxicology Division, Centre for Ocean Research (DST-FIST SPONSERED), Sathyabama University, Chennai, India

Abdulmounem Alchekh Wis
Sabanci University Integrated Manufacturing Technologies Research and Application Center & Composite Technologies Center of Excellence, Teknopark Istanbul, Istanbul, Turkey

Muhammad Nadeem Zafar
Department of Chemistry, Faculty of Science, University of Gujrat, Gujrat, Pakistan

Foreword

There are many nondissipative applications of nanomaterials that allow for recycling. Nanomaterials are applied as catalysts or adsorbents in reactors and many large-sized nanocomposites belong to this category. Moreover, recycling is possible when there is limited dissipation of nanomaterials, allowing for the recovery of nondissipated material. Textiles coated with antibacterial nanoparticles are examples thereof. Also, nonproduct outputs generated in nanomaterial production and processing may be recycled.

Nanomaterials are applied for their special properties or functionality. Recycling should preferably conserve the functionality of nanomaterials. Currently, the functional recycling of nanomaterials is limited. For instance, in a study regarding Switzerland (a country that is in the vanguard of circular economy policies), Cabbalero-Guzman et al. [1] found that less than 10% of the engineered nanomaterials nano-TiO_2, nano-Ag, and carbon nanotubes present in large-sized end-of-life nanocomposites were used functionally in secondary production. In the case of platinum group nanomaterials applied in car catalysts, end-of-life recycling focuses on metals rather than nanomaterials. In Europe (which has a circular economy policy) the end-of-life recycling rate for platinum group metals from car catalysts is 50%−60% [2]. These studies involving Switzerland and Europe show disappointing recycling percentages.

The environmental burden of nanomaterials should, moreover, not be neglected because nanomaterials are small: per unit of mass, the input of natural resources tends to be much larger than that for large-sized materials with the same composition. Therefore the limited character of the current functional recycling of nanomaterials has significant negative environmental consequences. The production of virgin nanomaterials often requires larger inputs of natural resources than the functional recycling of nanomaterials. Nonproduct outputs associated with the production of virgin nanomaterials can be significant environmental burdens. Also, nanomaterials that are lost from use in the economy may be hazardous.

In view of these issues, the present book is very welcome. Much attention is given in the book to methods for recycling nanomaterials and the properties and applications of recycled nanomaterials. Also, the regulatory context for nanomaterial recycling is covered. Though considerable additional research and development effort will be needed as a solid basis for widespread high-quality recycling of the already very large variety of nondissipated nanomaterials that is currently around, this book can serve as a basis for substantial advances in nanomaterial recycling.

Reijnders Lucas

Faculty of Science, Institute for Biodiversity and Ecosystem Dynamics,
University of Amsterdam, Amsterdam, The Netherlands

References

[1] A. Cabbalero-Guzman, T. Sun, B. Nowak, Flows of engineered nanomaterials through the recycling process in Switzerland, Waste Manag 36 (2015) 33—43.

[2] M. Saidani, A. Kendall, B. Yannou, Y. Leroy, E. Cluzel, Closing the loop on platinum from catalytic converters: contributions from material flow analysis and circularity indicators, J. Ind. Ecol. 23 (2019) 1143—1158.

Preface

Market consumption of nanomaterials is accelerating, mainly in electronics, energy, biomedical, and agricultural applications. Freedonia Group expected demand for nanomaterials to grow to USD $100 billion by 2025. However, the uses of nanomaterials in commercial products have been increasing at a higher speed than the development of appropriate recycling strategies. Currently, only limited recycling and reuse approaches have been developed for nanomaterials. In fact, after recycling, most nanomaterials will go to landfills or incineration plants.

Compared to conventional materials, nanomaterials are more expensive and hazardous. Therefore nanomaterial recycling offers many benefits in both environmental and economic respects. Nanomaterials can be recycled both from new and pure products (from nanomanufacturing) and from used products (nanowaste, which is waste from nanointegrated products). In addition, current recycled construction waste, plastics, and textiles may contain nanomaterials. These nanosources in the waste stream recycling process could be liquid (suspensions containing nanoparticles) or solid (solid matrices integrated with nanoparticles). To recover these nanowastes, various methods have been used for the separation of nanoparticles, such as centrifugation, solvent evaporation, magnetic separation, using pH- or thermal-responsive materials, molecular antisolvents, or nanostructured colloidal solvents. To achieve efficient recycling, the effect of the recycling process on the intrinsic recyclability properties of nanomaterials (thermal, mechanical, chemical, magnetic, and optical properties) must be determined.

This book sheds light on the recycling of nanomaterials. In the first part of the book the chapters focus on the environmental impact of nanowastes, especially for nanowaste management (Chapters 2, 3, 6, 7) and regulations for safe handling, manufacturing, and application of nanomaterials (Chapters 4, 5). The second part emphasizes the methods for recycling of nanomaterials, such as general techniques, procedures, and standards for recycling (Chapters 8, 9, 12), solvent evaporation and extraction techniques (Chapter 10), and using pH- and thermal-responsive materials (Chapter 11). The third part presents the properties of recycled nanomaterials and the techniques used to study their physicochemical properties. In the last part, promising applications of recycled nanomaterials are discussed through their industrial scale-up applications (Chapters 15, 17), especially in construction and building materials (Chapter 16).

This book could be an essential reading for postgraduate students of nanotechnology, biotechnology, chemistry, and environment technology; researchers; nongovernmental organizations; and policymakers who are interested in sustainable nanotechnology.

We thank all the contributors for their generous cooperation and efforts to provide up-to-date chapters. We are thankful to Elsevier and the authors of the chapters whose research work has been cited in the present book. We also thank the entire team at Elsevier for their cooperation, timely help, and patience in the publication of this book.

Mahendra Rai[1] and
Tuan Anh Nguyen[2]

[1]*Department of Biotechnology, SGB Amravati University, Amravati, Maharashtra, India*
[2]*Institute for Tropical Technology, Vietnam Academy of Science and Technology,*
Hanoi, Vietnam

SECTION I

Environmental impacts of nanowastes

CHAPTER 1

Nanomaterial recycling: an overview

Pooja Sharma[1], Praveen Guleria[2] and Vineet Kumar[1]

[1]Department of Biotechnology, School of Bioengineering and Biosciences, Lovely Professional University, Phagwara, India
[2]Plant Biotechnology and Nanobiotechnology Lab, Department of Biotechnology, School of Life Sciences, DAV University, Jalandhar, India

1.1 Introduction

Nanotechnology is a growing field that focuses on control over the synthesis and applications of nanomaterials. The nanomaterials are foreseen as the key components of many day-to-day use materials. Their large surface area, very small size, and ability to react with other compounds make the availability of nanomaterials in waste a safety concern. In addition, the ability of nanoparticles (NPs) to cross cell membranes poses serious concerns over the safety of NPs [1,2]. Therefore information about nanotechnology and its associated risks is a priority in order to avoid health and environment hazards [3].

Nanowaste is a waste originating from materials having constituents with at least one dimension in the range of $1-100$ nm. The major uncertainty associated with nanotechnology arises as a result of nanowastes generated after its applications. There are certain regulations that are meant for bulk materials and their waste. But there is lack of knowledge about nanowaste. Nanomaterials are generally classified on the basis of their composition, nature, and type of synthesis method. Different types of nanomaterials generate different types of nanowaste. The unique properties possessed by nanomaterials and insufficient knowledge to determine the appropriate treatment process for the diverse types of nanowaste makes nanowaste disposal difficult. Nanowaste is an emerging problem that needs to be dealt with and cannot be considered to be a future problem [4].

Nanowaste generally comprises particles in multiple groups released into the surroundings or thrown away without proper treatment. Nanomaterials that are made up of inert metals such as gold also becomes reactive at nanometer scale. The framework for proper monitoring and disposal of waste materials containing nanomaterials needs to be updated. The quantification of the waste generated is initial step in waste management. The size and shape dependent change in properties of the nanomaterials makes the waste quantification difficult. The nano-sized particles occur regularly in the environment due to their widespread use. The regular production and usage of nanomaterials create large pile of nanowaste. During their stay in the environment and

---

Nanomaterials Recycling
DOI: https://doi.org/10.1016/B978-0-323-90982-2.00001-9

© 2022 Elsevier Inc.
All rights reserved.

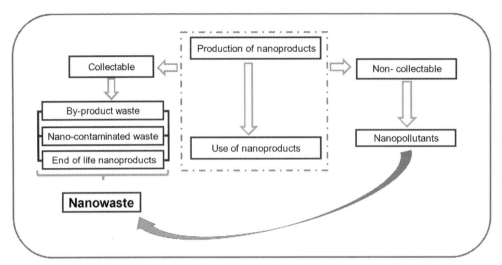

Figure 1.1 Overview of nanowaste production and nanopollution.

other systems, nanomaterials undergo many transformations depending upon their atomic arrangements and sometimes become very hazardous to the environment. At the end of their life cycle, nanoproducts also join the waste circulation and creates problem [3,5,6]. There is contradictory data in the literature about the adverse effects of nanomaterials on health and the environment [7−10].

Nanowaste is considered to be a separate category of waste. The main reason is the uncertainty posed by the extreme small size and shape of nanowaste, its chemical reactivity, and its biocompatibility, which make nanowaste stand out from the normal categories of waste materials [11,12]. Certain nanomaterials can enter the earth from diffuse sources and are termed potential nanopollutants [13,14]. Very few diverse and logical strategies are available for the characterization of nanomaterials as waste in nature. These strategies are dependent on the molecule number, elemental organization, particle size, molecule dispersion, conglomeration state, shape, structure, surface area, surface charge, and surface characteristics of the nanomaterials [12]. The properties of nanowaste are not same as those of general wastes that are discarded in the environment, and it is difficult to recognize these materials with recent technology. Legal guidelines and administrations play an important role in the differentiation and control of nanowaste [15] (Fig. 1.1).

1.2 Classification of nanowastes

The existence of a large diversity of nanomaterials is one of the major challenges for nanowaste segregation. Bulk materials are classified according to their benign or hazardous nature, which depends upon their chemical nature and the dose and the quantity of

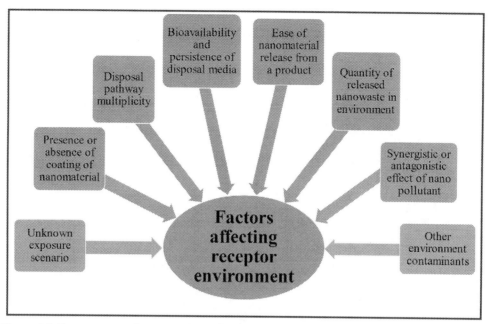

Figure 1.2 Nanowaste production and its different forms.

waste. But it is difficult to classify nanomaterials on this basis. However, there is a lack of international framework for the classification, regulation, and management of nanowaste. Musee [16] has given a qualitative classification of nanowaste depending upon the amount of risk factors that will help in the management of various types of waste streams. Products containing nanomaterials and nanoproducts have the potential to generate nanowaste and can affect the environment, as summarized in Fig. 1.2.

Nanomaterials that are firmly bound to a solid nanoproduct pose no or a very low potential of exposure, whereas free or loosely bound nanomaterials in liquid suspensions have a high to very high degree of exposure. Risk is the function of both hazard and exposure. According to Musee [16], the degree of hazard also needs to be evaluated to fully characterize the risk posed by nanowastes. The potential ease of release of nanomaterials from nanoproducts, leading to exposure and the degree of hazard of released NPs, collectively account for nanotoxicity behavior. The hazard can be derived from the ecotoxicity data of some nanomaterials published in the scientific literature. For the degree of hazard, some well-known nanomaterials were ranked in the following way: fullerenes (high), single-wall carbon nanotubes (CNTs) (high), multiwall CNTs (high), SiO_2 (low), silver (medium) and titanium dioxide (low). Based on the hazards and various risk factors associated with nanowaste exposure, Musee[16] classified them into five classes which are listed in Table 1.1.

Class I is barely to moderately dangerous, while class V is very hazardous. Although this classification is very useful, the worldwide nanowaste risk assessment

Table 1.1 Nanowaste classification on the basis of risk factors.

Type of class	Nature			Remarks	Examples
	Hazard	Exposure	Risk profile		
Class I	Very low to no toxicity	Less to more	No to very low risk	Concern arises when parent bulk material accumulates in the body beyond threshold. Otherwise, no special disposal required	Solar panels, television screens, polishing agents, memory chips
Class II	Harmful to toxic	Low to medium	Low to medium	May cause potential effects that raise concerns for proper management during waste disposal	Paints and coatings, display backplanes of television screens, polishing agents
Class III	Moderate toxic to very toxic	Low to medium	Medium to high	Require standard protocols and research to determine the optimum management and infrastructure for waste disposal	Personal care products, food packaging and additives, polishing agents, pesticides
Class IV	Moderately toxic to very toxic	Moderate to high	High	Waste disposal should be at designated sites, as improper management could lead to threats to surroundings system and humans	Personal care products, paints and coatings, pesticides
Class V	Very toxic to extremely toxic	Medium to high	Very high	Improper waste management could lead to extreme nanopollution, which would be difficult to mitigate and can cause serious ill effects	Sunscreen lotions, pesticides, food and beverages having fullerenes in their colloidal solution

framework is still lacking. Boldrin et al. [13] have provided a framework for the assessment of risk posed by exposure to NPs in environment.

1.3 Sources and routes of nanowaste contamination

Nanowastes can arise from various nanomaterials sources. Nanomaterials are used in various commercial products, such as medicines, diagnostic kits, cosmetics, bioimplants, textiles, electronic devices, paints, and lubricants. Release of nanomaterial during manufacturing and from end-of-life components are some common sources of nanowaste. Such runtime-significant wastes originate from various nano-based products, fabrication, and manufacturing processes such as metal waste/scrap, paper and cardboard, plastics, textiles and leather, electronics, batteries, solar cells, tires, medical devices, construction, and demolition wastes. The nanowastes from such nanomaterials can leach through different routes to contaminate the surrounding environment [16,17] (Fig. 1.3).

The nanomaterials may get dispersed into the wastewater stream and finally reach the treatment plant. Alternatively, nanomaterials may be collected from solid waste when it reaches a landfill. In a wastewater treatment plant, upto 90% of NPs are filtered Organisation for Economic Cooperation and Development [18]. The major portion of NPs would be deposited in waste sludge that will reach a landfill as final destination.

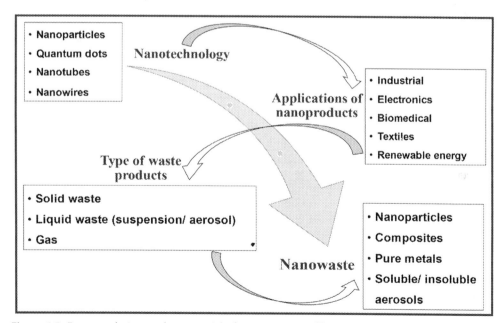

Figure 1.3 Factors relating to the potential of nanowaste to affect environment.

During the manufacturing process, released hazardous vapors of nano-based metal oxides might enter the atmosphere and can cause serious health problems. Even from the landfills, it is possible for certain nanowastes to get oxidized with other waste contaminants and exit through the vent as gas into the atmosphere. Nanomaterials may leach from various disposal sites, such as landfills, incineration units, and wastewater treatment plants, or may remain there for the materials' lifetime [19]. Hence any waste product should be analyzed according to a framework with a hierarchy starting from raw material acquisition to manufacturing, supply chain, and finally the end user.

The rapidly growing nanomaterials and nanoproducts are simultaneously enhancing nanowaste production [20]. NPs have a scale similar to that of cell segments and bigger proteins. This may enable quicker bonding of nanomaterials with toxins because of their enormous surface area, huge quantum impacts, organic reactivity, shape, and particularly size. Nanomaterials may move with high speed through soils and springs, which leads to the quicker movement of these pollutants through air, soil, and water [21]. However, the recovery of important precious metals and other rare earth metals from waste in the presence of nanowaste is a big problem [22,23].

Industries are the main source of nanowaste generation. Cosmetic and personal care—based industries contribute almost half of the total nanowaste [16]. According to the British Standards Guide (BSI), the environment contains four type of nanowastes: unadulterated, material and surface incorporated nanowaste, fluid stage incorporated nanowaste, and solid lattice incorporated nanowastes. There have been a few studies about nanowaste well-being and toxicological impacts, but the data on the control and effective management of nanowaste are very limited [16,24–27] (Fig. 1.4).

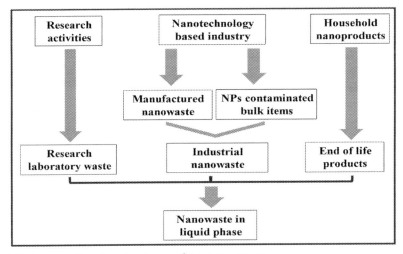

Figure 1.4 Different routes of contamination from nanowastes.

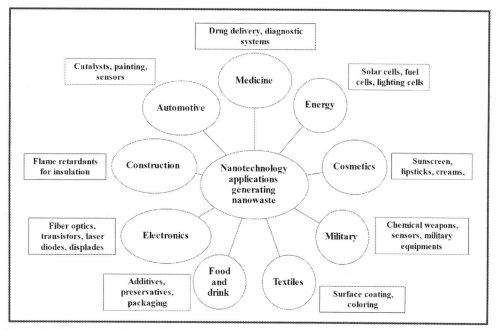

Figure 1.5 Nanotechnological advancements leading to generation of nanowastes.

1.4 Toxic effects of nanowastes

In the last decade, the tremendous development of nanotechnology brought about various new applications in the fields of electronics, medication, agriculture, and food. Enormous amounts of effort and investment are being devoted to nanotechnology. Some economically accessible items are now benefited by nanotechnology. Worldwide nanotechnology market was valued USD 1.75 billion in 2020 and is projected to reach USD 33.63 billion by 2030. Consistent funds are spent on nanotechnology-related exploration, though substantially less consideration is given to facilitating the utilization and removal of nanowaste. Traditional methods for identification of hazardous compounds are not enough. Certain uncommon and customized methods need to be designed. About 4% of the government nanotechnology research expenditure goes into examining the well-being and natural impact of nanomaterials [4] (Fig. 1.5).

The impact of nanotechnology can be seen in two aspects: one in which the potential of nanotechnology advancements in health care is seen and the other in which the potential ill effects of nanomaterials are seen. Owing to their extremely small size, NPs are effectively absorbed into the human body. Direct exposure of human tissues to nanomaterials poses various health concerns. NPs can enter the human body by four different routes: inhalation, swallowing, through the skin, and by

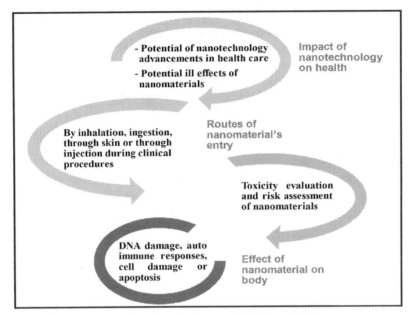

Figure 1.6 Assessment of nanotechnological effects on health and body.

injection. After entry into the body, they become mobile and in certain cases are even able to cross the blood-brain barrier. The chemical nature, surface topography and charge, stability in solutions, and type of function groups are some of the properties of NPs that influence their toxicity [28]. Since the toxicity is influenced by many variables, it is difficult to assess the ill effects related to nanomaterials. Therefore every nanomaterial should be studied so as to have a generalized idea about their toxicity and factors influencing it (Fig. 1.6).

The scanty literature and insufficient scientific data about the toxic effects of various nanomaterials are the biggest concerns. Such drawbacks hamper the management of nanowaste. Further, nanowaste can effectively show different toxicological features in different media and different conditions [16]. The different physical and chemical attributes of nano-scale materials in comparison their bulk materials tend to induce higher toxic effects and high chemical activity. An example is asbestos, which was used in large quantities in its nano-scale form. Very little data about its toxicity and ill nature are available; however, it has been banned in many countries because of several health hazards caused by asbestos usage in the past [4]. Another example is titanium dioxide NPs, which were considered safe owing to their stable chemical nature. However, some recent studies have shown their interaction with cell organelles causing oxidative stress, gene damage, and eventually cell death. Hence the International Center for Cancer Research, an organization of the World Health Organization (WHO) has kept titanium dioxide NPs in Group 2B: Carcinogen class [29]. Silver

NPs, which tend to have antibacterial properties, can cause damage to the microbial cell membrane and to the DNA [30]. However, apart from the rapid growth of the nanomaterial industry there are many concerns left to deal with nanowaste generation and safe disposal in order to minimize their toxic and hazardous effects, as the mechanism of their transport in biological systems is still not understood and the effect of nanowaste in food chain is not yet clear.

1.5 Impact of nanowastes on environment

Various industries produce nanowaste containing functionalized or nonfunctionalized nanomaterials. Such nanowaste is disposed of in water streams, soil, and air in the nearby surroundings. The nanomaterials can either degrade or accumulate. The nanowaste that is not degraded tends to have less toxicity and short-term impact. In contrast, degradable and less soluble nanowaste has higher toxicity. Such nanomaterials also have a higher tendency to bioaccumulate in biological and ecological systems. It is still very difficult to estimate toxicity and biomagnification of nanomaterials in various living systems [17]. According to the literature, risk categories of nanomaterials come under two types: known risks and potential risks. The potential risks category includes those NPs for which it is unknown whether their physical or chemical nature poses danger or some significant destruction during application and waste management [31].

Nanowaste can potentially pollute the environment either in free form or in fixed form. When nanowaste is released in the environment they get accumulated and termed as free form nanowaste. When nanowaste is an integral part of manufactured product, they must be recycled and proper disposal should be there and this type of waste generation is termed as fixed form nanowaste. NPs are released in aerosol forms in the soil and water in environment depending upon their type. They can be released in their bare form or sometimes in their functionalized form as aggregates or matrix embedded form [32]. When NPs undergo bioaccumulation and subsequent biodegradation, they can behave as ecotoxicological threat and prove hazardous [33]. The formation of NPs in diesel exhaust has also been reported [34]. Owing to incomplete combustion, aerosol NPs, generally of carbon based, are emitted by cars. These days, most vehicles are furnished with discharge prevention devices that involve catalysts such as platinum and other noble metals [35,36]. A couple of years ago, it was reported that platinum NPs with measurements in the range 0.8—10 nm were delivered from vehicle catalysts during their lifetime [37]. The automobile industry recently introduced new type of platinum and palladium catalysts in extreme small sizes and works better than traditional catalysts [36]. However, some of the studies have reported that nano-enabled catalysts can efficiently accumulate in living tissues [38]. Potential ecological dangers related to rising nanotechnologies are connected to the bioaccumulation of NPs in characteristic frameworks [38]. NPs contained in products

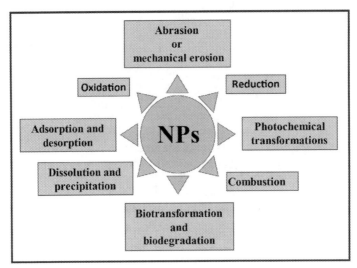

Figure 1.7 Processes affecting engineered nanomaterials.

such as sunscreens can pollute water and soil, adding to their bioaccumulation in the natural course of life [39]. Silver NPs discharged in waste have been observed in sewage and may be spread further on horticultural fields [40]. CNTs are probably the least biodegradable building nanomaterials, since they are lipophilic and insoluble in water (and therefore can possibly enter cell layers) and increase the likelihood of biocollection [41]. Hence interdisciplinary efforts are needed to reveal definite information about various nanomaterial-based hazards [42]; Assessment of ecological exposure must consider the accompanying issues, such as identification and quantification of sources, assurance of the natural delivery design, concentration estimation and quantification in the environment, and assessment of the likely bioaccumulated particles [33]. The destiny of designed NPs in the surroundings could be surveyed by using the data accessible for common NPs [32,43]. The arrival of NPs from nanowaste into the natural compartments—air, water, and soil—depends to a huge degree on the conditions and processes, as shown in Fig. 1.7.

The first step is to identify the possible area in which release takes place, followed by measuring the quantity, frequency, and duration of exposure. The final step is safe disposal of nanowaste. If a quantitative evaluation is difficult, then a subjective methodology could be use, in which the potential for introduction is identified as low, medium, and high. If information for subjective assessment is perceived, then it can be supplemented with quantitative research assessments. Other than proposing a structure for an ecological evaluation of NPs in solid waste, Boldrin et al. [13] proposed routes of exposure in three models: polyester material incorporating nanosilver, nano-scale titanium dioxide in sunscreen moisturizer, and tennis racquets containing CNTs. The silver NPs released in wastewater could

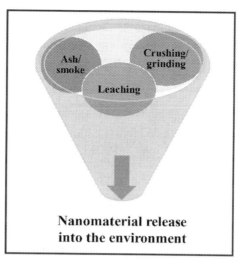

Figure 1.8 Production processes leading to release of nanomaterial into the environment.

affect microbial flora. The wastewater treatment depends on heterotrophic microorganisms for natural and supplemented removals. While autotrophic microorganisms have a significant role in nitrification, nitrifying microbes in sewerage frameworks are particularly defenseless to restraint by silver NPs [4,44]. Titanium dioxide nanowaste present in nature can be harmful to beneficial soil organisms and microorganisms, potentially altering the environmental balance [45,46]. The toxicity of natural chemical compounds, such as polycyclic sweet-smelling hydrocarbons, can be enhanced in the presence of CNTs. CNTs can harm the lung epithelium [47,48]. It has been assessed that roughly 2,000,000 laborers will be exposed to nanowaste by 2027, and around 2000 specialists could be possibly exposed to NPs in the United Kingdom [49] (Fig. 1.8).

1.6 Nanowaste treatment strategies

Nanowaste has great diversity, and its complex nature makes it difficult to deal with. At present, there is no globally acceptable method to tackle the problem of waste production. The general methods of waste management include disposal in waste streams, but this method is not applicable for nanowaste. Exposure to harsh environments is also not an effective method to treat nanowaste. When cerium oxide NPs are exposed to high heat, they do not burn but can be found intact on residual material or sometimes in the incinerator. Two studies were done to test the incineration methods. In one experiment, 10 kg of cerium oxide NPs with a diameter of 80 nm were kept in an incinerator for the process, and the incinerator was equipped with filters; also, separation of fly ash was facilitated with the help of electrostatic filters and scrubber of wet nature. The process

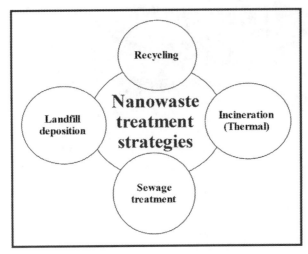

Figure 1.9 Different strategies for nanowaste treatment.

was operated at a speed of 8 tons of waste per hour [50]. In another experiment, a spraying technique was used for direct combustion of NPs in the incineration chamber; this experiment was based on a simulation of future cases in which the particles are released into environment in high amounts [51]. In both experiments, incineration had no effect on NPs, and they were not removed or disintegrated. Other cases for nanowaste generation include the recycling process of metal recovery from batteries, which requires quite high temperature. However, for the NPs contained in lithium batteries enabled with nanotechnology, a heating type of recovery process is inadequate. The smelting process, which is generally applied in nano-enabled cases, requires a very high temperature, which in turn needs higher energy consumption, thereby increasing the overall case load in terms of energy [52]. Some of the studies have reported promising methods of nanowaste treatment. A method involving electroplating waste sludge employing selective crystallization and formation of acid-insoluble nanowires of SnO_2 has been used for the recovery of SnO_2 NPs. Another example is the utilization of uranium-rich nanocrystals prepared by absorption-induced crystallization for uranyl enrichment. Additionally, thermo reversible fluid stage change and cloud point extraction hold promise for the fruitful division and recovery of basic, high-worth, and asset-restricted materials from nanowaste [53] (Fig. 1.9).

As nanotechnology-based items come into routine use, some of the nanowaste will reach wastewater streams. It needs to be removed at this stage; otherwise, it will enter the surroundings. There are four techniques for treatment of nanowaste to change it into less harmful and less complex particles: burning, land filling, reusing, and natural treatment [54] (Table 1.2).

Table 1.2 Nanowaste sources and their treatment strategies.

Type of waste	Nanomaterial in waste	Nanowaste category	Recycling procedure
Scrap from metal waste	CNTs, other metal oxides	Class II or class III	Shredding or smelting
Electronic waste	CNTs, nano-FeO, ZnO, carbon black	Class IV or class V	Dismantling, shredding, fractionation, metal extraction
Textiles	Silver NPs, CNTs	Class II or class III	Shredding
Batteries	Nanophosphate or electrodes containing CNTs	Class I	Mechanical, chemical or thermal process after collection and sorting
Plastic	CNTs, oxides of nano silicon and titanium	Class I or class II	Feedstock recycling after shredding is required
Construction waste	CNTs, iron NPs, silicon NPs, copper NPs	Class I or class II	Shredding and grinding
Paper waste	Ink containing carbon black	Class II or class III	Wet processes, for example, deinking process
Residues from incineration of waste	Nanomaterials that stay in incinerated ash that contains nondestroyed municipal waste	Class II or class III	Separation of bottom ash from waste

1.7 Regulatory bodies for nanowaste generation and management

There is an absence of proper guidelines and strategies for the utilization, safe removal, treatment, and reuse of nanomaterials. Public and research associations, governments, and ecological offices need to thoroughly analyze the advancement of pathways for safe removal of nanowaste with proper arrangements for their disposal in the coming years. The OECD was one of the first global associations to recognize and examine the issue of the possible risks posed by nanomaterials and nanowaste to the environment and human well-being. OECD has delivered five reports that portray their advantages. The International Union for Conservation of Nature (IUCN), the European Commission (EC), the UN Environment Program, the United States Environmental Protection Agency (EPA), and the International Organization for Standardization, among others, are working autonomously to solve this problem [4].

Nanomaterials Recycling

Table 1.3 Nanotechnology safety regulations in some major countries.

Country	Nanotechnology safety regulations
United States	TSCA, FIFRA, FFDCA
European Union	REACH, CLP, CR, RoHS
India	BSI
Japan	MOE-METI
Thailand	NANOTEC
Australia	NICNAS
Korea	Comprehensive plans of nano safety (2011.10)
Turkey	MWCR
Austria	AWG 2002

Likewise, the EPA, the United States Occupational Safety and Health Administration, and WHO are subsidizing concentrations on the well-being and natural dangers presented by nanomaterials. The United Kingdom's Royal Academy of Engineering and the EC are creating rules to shield people and the earth from nanomaterials and nanowaste. The area of nanowaste treatment is full of opportunities for environment engineers, ecological analysts, and nanotechnologists. It is anticipated that in the next several years, research in this field will progress immensely, particularly since this problem was discussed in a G20 forum in 2017 in Germany and it is a piece of Agenda 2030 objectives [4] (Table 1.3).

1.8 Future perspectives and challenges

It is evident from the literature that the reactivity of nanowaste is a major concern. For eliminating the reactivity of nanowaste, extending the life cycle of engineered nanomaterials would be a feasible and ecofriendly solution. By dividing the nanowaste produced into two categories such as intentionally produced (engineered nanomaterials) and incidentally produced, we can have a better framework for their disposal and/or recycle methodologies. Civil background demands certain recycle and reuse options for nanowaste that is produced incidentally. For example, nanowaste that is produced by a factory (dry process) can be collected and stored separately and can be utilized in external coatings by adding required ingredients. Incidental nanowaste can be used in various areas if information about their characteristics and possible disposal behavior are present. A soluble incidental nanowaste can be utilized as a major ($>75\%$) partial replacement ingredient for sand in high-grade concrete mixes. Nanowaste can be mixed with certain appropriate chemicals and can be applied over the surfaces of spillways in dams. It can be utilized to prepare geopolymer boards, which can be used in partition wall panels. Incidental solid nanowaste can be mixed with epoxy injecting materials to evaluate its performance for deteriorated building components. Nanowaste can be used as binder material with asphalt concrete for making pavements, as pothole filler material, as filling material in building foundations, and as a base course component. Nanowaste in the sludge obtained as a leftover of waste treatment plants can be utilized as an ingredient for the preparation of construction blocks.

Nanotechnology is a field that has cost-effective solutions for problems related to cost-effective energy, pollution, and medicine. Hence the production of nanowaste is expected to increase in the future. Researchers have the obligation of finding methods to deal with this problem. Also, international legislation needs to create better policies for nanowaste treatment, especially in industries in which a huge amount of nanowaste is produced. This legislation must have an international classification for nanomaterials to make the development of nanowaste treatment easier. Furthermore, governments must designate more economic resources for nanowaste research.

1.9 Conclusion

There has been an exponential increase in nanowaste generation, but there is a large gap between waste generation and waste management. The recycling and management of these nanocontaminants is not improving, and there is a lack of better disposal practices. Broadly, nanowaste can be categorized into five different classes on the basis of hazard risk. Class I nanowastes are less toxic, while class V nanowaste has a high risk of inducing toxicity. The main concern for nanopollutants is their extremely small size, which makes their effective disposal very difficult. Hence keeping in view the hazardous implications of nanowaste, there is a need to focus on the regulation of nanowaste generation and disposal. More attention needs to be paid to following proper guidelines and frameworks to deal with such types of nanopollutions. There are different organizations, including the EC, OECD, IUCN, EPA, and WHO, that are seriously working to find a solution to challenges posed by nanowastes.

References

[1] S.A. Love, M.A. Maurer-Jones, J.W. Thompson, Y.-S. Lin, C.L. Haynes, Assessing nanoparticle toxicity, Annu. Rev. Anal. Chem. 5 (2012) 181−205.
[2] B.J. Marquis, S.A. Love, K.L. Braun, C.L. Haynes, Analytical methods to assess nanoparticle toxicity, Analyst 134 (2009) 425−439.
[3] A. Caballero-Guzman, T. Sun, B. Nowack, Flows of engineered nanomaterials through the recycling process in Switzerland, Waste Manage. 36 (2015) 33−34.
[4] T. Faunce, B. Kolodziejczyk, Nanowaste: Need for disposal and recycling standards. G20 Insights, Policy Era: Agenda. 27: 2030 (2017).
[5] F. Gottschalk, T. Sun, B. Nowack, Environmental concentrations of engineered nanomaterials: review of modeling and analytical studies, Environ. Pollut. 181 (2013) 287−300.
[6] N.C. Mueller, B. Nowack, Exposure modeling of engineered nanoparticles in the environment, Environ. Sci. Technol. 42 (2008) 4447−4453.

[7] V. Kumar, N. Dasgupta, S. Ranjan (Eds.), Nanotoxicology: Toxicity Evaluation, Risk Assessment and Management, CRC Press, Taylor and Francis, Boca Raton, FL., 2018, p. 684.

[8] V. Kumar, N. Dasgupta, S. Ranjan (Eds.), Environmental Toxicity of Nanomaterials, CRC Press, Taylor and Francis, Boca Raton, FL, 2018, p. 523.

[9] V. Kumar, A. Kumari, P. Guleria, S.K. Yadav, Evaluating the toxicity of selected types of nano-chemicals, Rev. Environ. Contam. Toxicol. 215 (2012) 39−121.

[10] V. Kumar, P. Guleria, Application of DNA-nanosensor for environmental monitoring: recent advances and perspectives, Curr. Pollut. Rep. (2021). Available from: https://doi.org/10.1007/s40726-020-00165-1.

[11] N. Nanoodpadow, Risk of nanowastes, Inzynieria i Ochrona Srodowiska 19 (2016) 469−478.

[12] F. Part, G. Zecha, T. Causon, Current limitations and challenges in nanowaste detection, characterisation and monitoring, Waste Manage. 43 (2015) 407−420.

[13] A. Boldrin, S.F. Hansen, A. Baun, N.I.B. Hartmann, T.F. Astrup, Environmental exposure assessment framework for nanoparticles in solid waste, J. Nanopart. Res. 16 (6) (2014) 2394. Available from: https://doi.org/10.1007/s11051-014-2394-2.

[14] A. Ramakrishnan, L. Blaney, J. Kao, R.D. Tyagi, T.C. Zhang, R.Y. Surampalli, Emerging contaminants in landfill leachate and their sustainable management, Environ. Earth Sci. 73 (2015) 1357−1368.

[15] G. Bystrzejewska-Piotrowska, J. Golimowski, P.L. Urban, Nanoparticles: their potential toxicity, waste and environmental management, Waste Manage. 29 (2009) 2587−2595.

[16] N. Musee, Nanowastes and the environment: potential new waste management paradigm, Environ. Int. 37 (2011) 112−128.

[17] G.B. Piotrowska, J. Golimowski, P.L. Urban, Nanoparticles: Their potential toxicity, waste and environmental management, Waste Manage. 29 (2009) 2587−2595.

[18] Publishing OECD, et al., Extended producer responsibility: Updated guidance for efficient waste management, Extended Producer Responsibility (2016). Available from: https://www.oecd.org/development/extended-producer-responsibility-9789264256385-en.htm.

[19] A.A. Keller, A. Lazareva, Predicted releases of engineered nanomaterials: from global to regional to local, Environ. Sci. Technol. Lett. 1 (2014) 65−70.

[20] C.R. Ratwani, Nanowaste: tiny waste that matters a lot, Int. J. Curr. Res. 10 (2018) 70262−70268.

[21] B. Mrowiec, Risk of nanowastes, Inżynieria i Ochrona Środowiska 19 (4) (2016) 469−478.

[22] H.U. Sverdrup, K.V. Ragnarsdottir, D. Koca, An assessment of metal supply sustainability as an input to policy: security of supply extraction rates, stocks-in-use, recycling, and risk of scarcity, J. Clean. Prod. 140 (1) (2017) 359−372.

[23] S. Zhang, Supply and demand of some critical metal sand present status of their recycling in WEEE, Waste Manage. 65 (2017) 113−127.

[24] V.L. Colvin, The potential environmental impact of engineered nanomaterials, Nat. Biotechnol. 21 (2003) 1166−1170.

[25] R.D. Holbrook, K.E. Murphy, J.B. Morrow, K.D. Cole, Trophic transfer of nanoparticles in a simplified invertebrate food web, Nat. Nanotechnol. 3 (2008) 352−355.

[26] M.N. Moore, Do nanoparticles present toxicological risks for the health of the aquatic environment? Environ. Int. 32 (2006) 967−976.

[27] A. Nel, T. Xia, L. Madler, N. Li, Toxic potential of materials at the nanolevel, Science 311 (2006) 622−627.

[28] E. Herzog, A. Casey, F.M. Lyung, G. Chambers, H.J. Byrne, M. Davoren, A new approach to the toxicity testing of carbonbased nanomaterials-the clonogenic assay, Toxicol. Lett. 174 (2007) 49−60.

[29] R. Baan, K. Straif, Y. Grosse, B. Secretan, F. El Ghissassi, V. Cogliano, et al. WHO International Agency for Research on Cancer Monograph Working Group, Carcinogenicity of carbon black, titanium dioxide, and talc, The Lancelet Oncology 7 (2006) 295−296.

[30] C.N. Lok, C.M. Ho, R. Chen, et al., Proteomic analysis of the mode of antibacterial action of silver nanoparticles, J. Proteome Res. 5 (2006) 916−924.

[31] H.H. Eker, M.S. Bilgili, E. Sekman, S. Top, Evaluation of the regulation changes in medical waste management in Turkey, Waste Manage. Res. 28 (2010) 1034−1038.

[32] B. Nowack, T.D. Bucheli, Occurrence, behavior and effects of nanoparticles in the environment, Environ. Pollut. 150 (2007) 5–22.

[33] A. Dhawan, V. Sharma, Toxicity assessment of nanomaterials: methods and challenges, Anal. Bioanal. Chem. 398 (2010) 589–605.

[34] K. Vaaraslahti, A. Virtanen, J. Ristimaki, J. Keskinen, Nucleation mode formation in heavy-duty diesel exhaust with and without a particulate filter, Environ. Sci. Technol. 38 (2004) 4884–4890.

[35] V. Nischwitz, B. Michalke, A. Kettrup, Speciation of Pt(II) and Pt(IV) in spiked extracts from road dust using on-line liquid chromatography-inductively coupled plasma mass spectrometry, J. Chromatogr. A 1016 (2003) 223–234.

[36] N. Stafford, Catalytic converters go nano, Chem. World. 4 (2007) 16.

[37] S. Artelt, O. Creutzenberg, H. Kock, K. Levsen, D. Nachtigall, U. Heinrich, et al., Bioavailability of fine dispersed platinum as emitted from automotive catalytic converters: a model study, Sci. Total. Environ. 228 (1999) 219–242.

[38] W. Hannah, P.B. Thompson, Nanotechnology, risk and the environment: a review, J. Environ. Monit. 10 (2008) 291–300.

[39] G.J. Nohynek, E.K. Dufour, M.S. Roberts, Nanotechnology, cosmetics and the skin: is there a health risk? Skin. Pharmacol. Physiol. 21 (2008) 136–149.

[40] S.A. Blaser, M. Scheringer, M. MacLeod, K. Hungerbühler, Estimation of cumulative aquatic exposure and risk due to silver: contribution of nano-functionalized plastics and textiles, Sci. Total. Environ. 390 (2008) 396–409.

[41] C. Botta, J. Labille, M. Auffan, D. Borschneck, H. Miche, M. Cabié, et al., TiO_2-based nanoparticles released in water from commercialized sunscreens in a life-cycle perspective: structures and quantities, Environ. Pollut. 159 (2011) 1543–1550.

[42] J.E. Hutchison, Greener nanoscience: a proactive approach to advancing applications and reducing implications of nanotechnology, ACS Nano 2 (2008) 395–402.

[43] M. Hasellöv, J.W. Readman, J.F. Ranville, K. Tiede, Nanoparticle analysis and characterization methodologies in environmental risk assessment of engineered nanoparticles, Ecotoxicology. 17 (2008) 344–361.

[44] N. Lubick, Nanosilver toxicity: ions, nanoparticles – or both? Environ. Sci. Technol. 42 (23) (2008) 8617.

[45] A. Mathur, A. Parashar, N. Chandrasekaran, A. Mukherjee, Nano-TiO_2 enhances biofilm formation in a bacterial isolate from activated sludge of a waste water treatment plant, Int. Biodeter. Biodegr. 116 (2017) 17–25.

[46] H. Shi, et al., Titanium dioxide nanoparticles: a review of current toxicological data, Part. Fibre Toxicol. 10 (2013) 15.

[47] J. Bouillard, Nanosafety by design: risks from nanocomposite/nanowaste combustion, J. Nanopart. Res. 15 (2013) 1519.

[48] A. Magrez, et al., Cellular toxicity of carbon-based nanomaterials, Nano Lett. 6 (6) (2006) 1121–1125.

[49] B.M. Díaz-Soler, M. López-Alonso, M.D. Martínez-Aires, Nanosafety practices: results from a national survey at research facilities, J. Nanopart. Res. 19 (2017) 169.

[50] E.P. Vejerano, et al., Toxicity of particulate matter from incineration of nanowaste, Environ. Sci. Nano 2 (2015) 143–154.

[51] D. Mitrano, K. Mehrabi, Y. Arroyo, C. Nowack, Mobility of metallic (nano)particles in leachates from landfills containing waste incineration residues, Environ. Sci. Nano 4 (2017) 480–492.

[52] S. Olapiriyakul, R.J. Caudill, Thermodynamic analysis to assess the environmental impact of end-of-life recovery processing for nanotechnology products, Environ. Sci. Technol. 43 (21) (2009) 8140–8146.

[53] Z. Zhuang, et al., Treatment of nanowaste via fast crystal growth: with recycling of nano-SnO_2 from electroplating sludge as a study case, J. Hazard. Mater. 211–212 (2012) 414–419.

[54] A.U. Zaman, A comprehensive review of the development of zero waste management: lessons learned and guidelines, Journal of Cleaner Production 91 (2015) 12–25.

CHAPTER 2

Nanomaterial waste management

Thodhal Yoganandham Suman[1,2,3] and De-Sheng Pei[1,2]
[1]School of Public Health and Management, Chongqing Medical University, Chongqing, China
[2]Chongqing Institute of Green and Intelligent Technology, Chinese Academy of Sciences, Chongqing, P.R. China
[3]Ecotoxicology Division, Centre for Ocean Research (DST-FIST SPONSERED), Sathyabama University, Chennai, India

2.1 Introduction

Nanotechnology is broadly described as using technology whose materials have dimensions around 100 nm or less; some authors extend this concept to 1 μm [1]. The use of nano-sized devices has already revolutionized the areas of computing, energy collection, and medicine. Lately, nanotechnology has spread to other fields, including construction technology. It was projected that the global construction sector would reach USD $10.3 trillion in 2020 [2]. Richard Feynman developed the concept of nanotechnology in 1959. According to data reported on the StatNano website (https://statnano.com), about 166,000 papers on nanotechnology were listed in the 2018 Web of Science Database, China (39.47%), and India (8.45%) as countries publishing nanotechnology-related research [3,4]. Increased public interest in nanotechnology has driven the United States government to establish new policies and regulations. Representatives of federal agencies started conducting nanotechnology meetings in 1996 and became an Interagency Nanotechnology Working Group (IWGN) in 1998. The IWGN is regulated through a Science and Technology National Council. Several years later, the National Nanotechnology Program was formed to support nanotechnology research. Nanotechnology has entered common consumer goods over the past two decades, including sports equipment, consumer electronics, and personal care products [5].

2.2 Nanomaterials: definition and trends of the world nanomaterials market

Nanomaterials (NMs) are materials that frequently have specific properties resulting from their small particle size. There are several definitions of nanoparticles (NPs). One of these states that if a given material is composed of particles with dimensions of 100 nm or less, that material may be called a NM. Another definition defines a NM as a material that reveals altered physicochemical properties compared with the bulk products after miniaturizing [6,7]. The modern drivers of the world's NMs market are the

Nanomaterials Recycling
DOI: https://doi.org/10.1016/B978-0-323-90982-2.00002-0

© 2022 Elsevier Inc.
All rights reserved.

growing market penetration of existing materials, reduction in NM prices, improvement in the properties of the materials, R&D expenditure associated with new materials, growing public and private nanotechnology research investment, government agencies' increasing support, rapid production of new materials and technologies [8,9], the successful operation of domestic and international firms and organizations' associations and strategic alliances, and the growing number and partnership between industrial players, such as increased international cooperation in nanotechnology research and development. Currently, the market for NMs includes sporting goods, health care, personal care and cosmetics, automotive, computing, and consumer electronics. The United States National Nanotechnology Initiative has shown that federal support for nanotechnology rose from around $464 million in 2001 to almost $1.9 billion in 2012 [10]. From 2005 to 2010 the European Union and Japan invested nearly $1.5–$1.8 billion, respectively, in nanotechnology. In nanotechnology, China, Korea, and Taiwan spent $300, $250, and $110 million, respectively. The study entitled "The worldwide economy of NMs in 2010−25" offers a valuable view and anticipated growth of the NMs market [11].

2.3 Nanowastes

Residues of NMs that are commonly identified in the waste stream, as shown in Fig. 2.1. NMs can be discharged into the environment by mechanical and/or chemical effects during the product life cycle. According to Boldrin et al. [12], NMs from diffuse sources entering the atmosphere can be categorized as possible nanopollutants, for instance, titanium dioxide (TiO_2) NPs released into surface water by sunscreen lotions. The word *nanowastes* is first used when NMs make contact with solid waste and can be processed separately. Nanoparticulated products can disperse NPs in different ways into the environment. NPs enter the atmosphere in three significant ways. First, raw

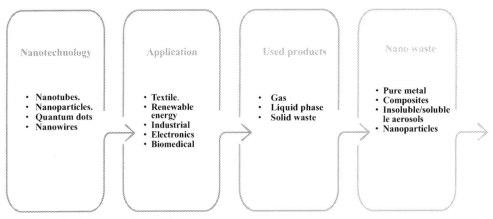

Figure 2.1 Production pathway of nanowastes.

materials, such as those produced through mining and processing, are processed. Second, NPs emitted during use, such as NPs in cosmetics or sunblock, are flushed into the environment. Third, after disposal or use of NPs in the waste management process, NPs can be found in wastewater streams and sewage [13]. Particles of TiO_2 larger than 100 nm diameter are deemed biologically inert for humans and animals [14]. TiO_2 NPs are already commonly utilized in various products, such as white pigments and food coloring [15]. Containers that contain residual material should be disposed of as hazardous waste. NMs can also seep from façades and reach the soil and water bodies through rainwater or the sewer system [13]. Nano-scale waste materials often have deposition as their final destination. At the third level, during the processing of certain items, dangerous nano-based vapors are released. Metal oxides that enter the atmosphere can cause severe health problems. Also, from landfills, nano-proteins may be oxidized and released into the atmosphere as gas by other waste pollutants [16−18].

2.4 Carbon-based nanomaterials

Carbon nanodots, carbon nanotubes (CNTs), graphene, and fullerenes are known as carbon-based NMs. CNTs used for the lightweight reinforcement of building materials such as pavers might cause asbestos-like problems when electric cutters slice the materials. Therefore, special protection or security should be used to deal with these products. CNTs can absorb organic compounds, including polycyclic aromatic hydrocarbons (PAH), which enhance their toxicity. CNTs also create mesotheliomas that weakened the lung epithelium similarly to the effects of asbestos [19]. Since NMs were introduced from nanoproducts, current waste management technologies have been presumed to efficiently extract NMs from the waste streams. However, insufficient data are available to verify this assumption. It is estimated that in the next 10 years, about two million employees will be exposed and that around 2000 workers in the United Kingdom may be exposed to NPs [20].

2.5 Silver nanoparticles

The amount of Ag NPs containing waste has been increasing; this is a disturbing trend, since Ag NPs released into domestic wastewater are transformed into silver ions, for instance, complexing through additional ions and molecular groups agglomerate or stay as NPs. This is the most dangerous illustration of nanowastes. Their potential for environmental toxicity is based on their catalytic and biocidal effects on a wide variety of species, such as earthworms, fungi, and bacteria, and reactions to other lethal agents, a deadly impact on groundwater, and food chain accumulation. Wastewater treatment is based on heterotrophic microorganisms for the organic removal and the reduction of nutrients, whereas autotrophic microorganisms are essential to the process of nitrification. Bacteria that nitrify sewage systems are particularly likely to inhibit Ag NPs [21,22].

2.6 Titanium dioxide nanoparticles

TiO_2 NPs used in sunscreens may enter the skin and bloodstream, causing irreversible effects on the central nervous system by intranasal instillation and brain neuroinflammation. Data indicate that exposure to TiO_2 NPs in the ecosystem can damage desirable bacteria and soil microbes, completely changing the equilibrium of the ecosystem. Furthermore, TiO_2 dust is listed as a Group 2B carcinogen, probably cancerous to humans, according to the International Agency for Research on Cancer. TiO_2 NPs also caused oxidative damage to DNA [23,24].

2.7 Prospective concerns around nanowastes

Usually, with the implementation of nanowastes through nanoproducts, the latest waste management methods are assumed to remove NMs from solid and liquid wastes efficiently and effectively. However, there is no strong evidence to support this statement, and the consequences for the advancement of waste management practices are possibly far-reaching. Two studies have reported for pyre treatment [25,26]. In an initial test, 10 kg of 80-nm-diameter cerium oxide particles were incinerated in a communal waste disposal plant with new filters and a wet scrubber and fly ash filter. It should be noted that 8 tons of waste was destroyed every hour at the site. In the second study, the samples were directly added into combustion chambers, and a possible worse scenario of primary NP emission during combustion was simulated. Regrettably, in both cases, the procedure did not remove NPs. Thus while nanowastes are not considered a possible environmental hazard, present water amenities will face eliminating NMs as the volume rises in the near future. By way of a consequence, NMs are likely to present new challenges to existing waste-processing technology, such as decreasing its operating effectiveness, owing to surface coating. Ironically, the layer is necessary because it makes NMs inert and thus restricts their adverse effects from a toxicological or ecotoxicological point of view until they remain in the ecosystem as much as the coating persists. This applies only if the layer is benign. Therefore this issue must be resolved in the future by the design and production of nonintended surface coating for therapeutic technologies in contrast to their value in shielding probable receptors of NPs in the ecosystem.

2.8 Challenge of nanowastes

The problems of nanowastes involve matters linked to the tracking, safe usage, disposition, and productive use of NMs and recycling of NPs where appropriate. One of these difficulties concerns the extremely small size of the product, which makes it challenging to track and control in both the environment and the human body. Also, preliminary research related to NPs' toxicity and chemical reactivity indicates serious problems requiring new and customized disposal and recycling procedures. Nanotechnology-related research costs hundreds of millions of dollars annually, but much less money and focus are spent on creating devices to promote the use and elimination of nanowastes. While Hallock et al. [27] suggested that numerous approaches are employed in nanowaste organization in the research laboratories of the Massachusetts Institute of Technology; the efficacy of protection of staff and the environment during and after disposal appears technically baseless. This knowledge gap threatens the ongoing, safe, inclusive, and defensible nanotechnology development. However, the current regulatory regime had never foreseen nanowastes sources, which poses concerns about the adequacy of established statutory mechanisms regulating traditional waste management paradigms. At the center of the debate is whether or not the present waste management legislation model will afford effective guidance on the storage, care, and disposal of nanowastes.

Recent research has shown that even under extreme conditions, synthetic nanowastes do not disappear. For example, cerium oxide NPs may never exhaust or modify the temperature of a waste combustion plant. They remain in burning waste or the incineration system. The researcher splashed a total of 10 kg of cerium oxide particles, about 80 nm in diameter, in a waste incinerator plant with modern filters and fly ash filters. At the plant, upto 8 tons of waste is incinerated every hour. The samples were released straight into that combustion chamber in the conducted test to simulate a potential "good possible case" with a complete leak of NPs throughout combustion. In both cases, NPs were not extracted.

The properties of NMs and NPs are not sufficiently understood. Not all NMs are harmful or poisonous, but they differ from bulk (nonnano) artifacts of the same substance. Gold, widely used in jewelry, is not detrimental in its large form, but when it is reduced to nano-size, it becomes chemically reactive, particularly when illuminated with light. The photocatalytic characteristics of gold NPs can promote chemical reactions, such as organic compound degradation and oxidation reactions [28,29].

2.9 Classification of nanowastes

The proposed classification of NM waste streams offered by Musee [30], is presented in Fig. 2.2. The distinction is relevant because the dangerous threat of NMs changes in different parts of their life cycle (production, usage, and disposal). fullerenes, multi-wall CNTs, and single-wall CNTs, used in automotive components, have a high risk of being used, but they pose a medium risk for waste disposal [30].

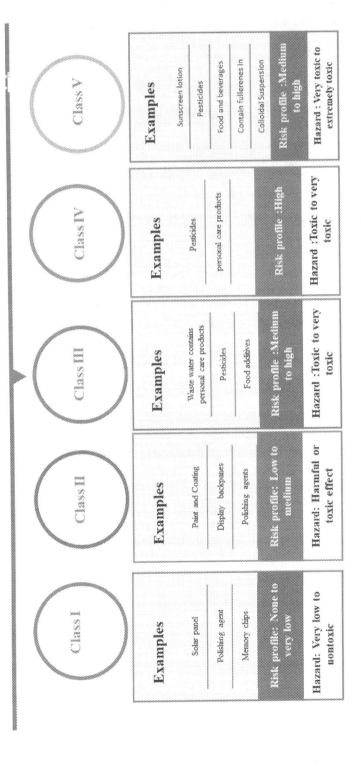

Figure 2.2 Classification of nanowastes.

Class I nanowastes: Class I nanowastes are very low risk or not harmful to people and ecosystem systems, mostly because the NMs are not dangerous materials. Even if the NMs are destined to the surface or within the bulk of the material, the exposure potential for class I nanowastes does not affect the waste stream's risk. Examples are those from the backplane screens in solar panels, memory chips, television screens, and silicon nanowires, for which low to high exposure levels can be seen when the NMs break free or leach out.

Class II nanowastes: Class 2 nanowastes, which range from low to high in risk, can have harmful effects on humans and the environment. The total exposure risk is closely related to the hazard potential as a result of nanostructures on the region or inside the bulk of the material, depending on the outcomes derived from the matrix for the nanowaste groups. If the intensity of exposure is minimal or impossible, such waste can be considered nontoxic, but it contains highly toxic content. Instances are nanowastes produced after backplane displays and storage chips have been discarded. These include mostly single-walled CNTs that are incredibly toxic to organisms.

Class III nanowastes: A stream of nanowastes is rated as class III when its toxic effect is conceivable. The disposal process is classified as toxic to moderately toxic and associated with low to possibly medium exposure. Therefore the resulting waste stream is likely to have a medium hazard effect on an ecological system and should be considered dangerous waste. Nanowaste arising from pesticides, polishing agents, food additives, and food packaging and can be classified as class III nanowastes.

Class IV nanowastes: The NM toxicity risk in class IV varies between toxic and incredibly lethal, and the possible acquaintance ranks from average to high, since nanoproducts are supposed to be freely bound. For organisms as well as the environment, waste streams are considered extremely risky. Therefore such waste streams need specialized care and are adequately handled by immobilizing or neutralizing the NMs. Nanowastes from personal care products, paints and coatings, and pesticides comes under class IV.

Class V nanowastes: Class V nanowastes are highly dangerous, as they are highly exposed and harmful. They range from very harmful to severely toxic. Such sources of waste need skilled handling, efficient treatment, and disposal at well-designated discarding places. Regular inspection of the sites is advised to safeguard proper management of the leachates from the discarding place. The most acceptable waste management methods include immobilization and neutralization processes. Not only is the waste stream highly toxic, but it can also have quite a high potential for contamination when released into the atmosphere, since it is liquid, which can facilitate encounters with environmental organisms. Pesticides, sunscreen lotions, food, and beverages containing colloidal suspensions of fullerenes are highly hazardous and listed as class V nanowastes.

2.10 Difficulties and concerns about nanowastes management

Although new products have been prioritized in nanotechnology, knowledge about the level of nanowastes produced throughout the manufacture of fresh materials is missing. However, nanowastes are recognized as the fundamental cause of environmental NMs [30]. When nanowaste processing has taken the lead in the past, new applications have been considered sufficient to effectively remove these small materials through waste and water resources. However, the latest findings have shown that current wastewater systems are not sufficient to extract NMs [31].

In response to studies from some modeling and estimation studies, no information on industrial nanotechnology or the quantities of nanowastes produced by consumers are available for analysis. Even so, the large variety of nanotechnology applications makes real estimates of the quantities of nanowastes extremely difficult. Their behavior has recently been examined in wastewater treatment plants. However, there are not enough existing data and knowledge to decide the overall designs appropriate for nanowastes treatment. The possibility of nanowastes may be due to the dearth of international classification. The implementation of nanotechnology as green technology is another aspect that delays efforts in some cases.

2.11 Incineration of waste that contains nanomaterials

2.11.1 Nanowaste treatment in waste treatment plants

The 1628 consumer goods based on nanotechnology on the market constitute the most significant and fastest-growing segment of Ag NPs (383 goods) and TiO_2 NPs (179 products) [32]. Hence the deliberate or unintentional release into the environment of new and aged NPs is inevitable. For instance, ZnO and TiO_2 could be emitted by scrubbing consumer products (sunshades and UV lotions) [33,34]. As for most NPs, such as TiO_2, most NMs are likely to be released into sewage systems. Therefore, municipal waste treatment plants (WTPs) act as a gateway to control the discharge to the aquatic environment by treating effluent from domestic and/or industrial sources into surface water [35].

Earlier studies of NM removal (especially Ag NPs) indicated that nearly 90% of spiked NMs are effectively decreased and accumulated in active sludge and biosolids via biological treatment using batch bioreactors in the laboratory [36]. Based upon batch bioreactors in a laboratory, earlier studies on the removal of NM (especially Ag NPs) showed that approximately 90% of spiked NMs are accumulated in active sludge or biosolids via biological treatment.

In a modern simulation model, the concentration of Ag NPs in WTP sludge was calculated to be 1.55 mg/kg [37]. Since they demonstrate significant antimicrobial activity against a range of pathogens, the possible adverse effects of Ag NPs on waste and wastewater treatment are progressively concerned. The approximate concentration varies from 0.7 to 16 μg/L TiO_2 released into water bodies [38]. Despite low levels of released NM from the WTP effluent, chronic toxicity can be caused by the ongoing exposure of aquatic species to NMs.

Numerous research studies on removing and releasing particular engineered NMs (ENMs) from the WTP have been recorded [39–43]. In an analysis of field samples obtained from nine municipal WTPs in Germany, more than 72% of the residual Ag NPs were reduced by biological treatment in wastewater after mechanical treatment [37]. These procedures have decreased 95% of the Ag NPs entering the WTPs to a low level of Ag NPs in effluents ($<$12 ng/L). A current vulnerability analysis in triggered sludge treatment system shows that over 90% of Ag NPs were biomass correlated with an average silver concentration of less than 0.05 mg/L in the effluent wastewater [39,43]. In one WTP sample, in which raw sewage contained from 100 to 300 parts per billion (ppb) Ti, concentrations of 1–6 μg Ti/mg biosolids were collected in established solids [40]. As ZnO and TiO_2 slots may ultimately be deposited in sites, it is stated that roughly 75% of the total TiO_2 in WTPs ends up eventually in landfill sites. A model test for eliminating TiO_2, CeO_2, Fe_2O_3, SiO_2, $Ca_3(PO_4)_2$, and ZnO NPs in the WTP will absorb the maximum of oxide NPs by adhesion to clearing sludge. Most of the NPs produced were removed through a wastewater processing method, and upto 6 wt.% of CeO_2 were found at the plant in the production system [11]. Thus wastewater effluents are predominantly released into surface water and constitute a significant inherent source of NMs that enter the ecosystem. Although a persistent entry of Ag NPs into wastewater did not alter the removal of chemical oxygen demand (COD) substantially, only at the start of the batch experiment the removal of NH_4 was cut [44]. Shortly, surfactant-stabilized ENMs emitted into sewage would probably have significant adverse impacts on the COD and NH_4 removal in urban WTPs of activated sludge processes. Therefore the cytotoxicity of ENMs should be investigated in activated sludge microorganisms. The ENMs reached in WTPs are often extracted and collected in sludge and biosolids. Ultimately, this waste sludge containing ENMs and consumer goods is put into settlements at the end of their lifespan. Meanwhile, the biosolids are generally used in settlements as agricultural land amendments (fertilizers), incinerated, or deposited in oceans; the biosolids thus constitute a possible environmental ENM release source. This source is slightly different from WTP waste, but those biomass exposures and subsequent exposure to ecosystems remain uncertain [40].

2.11.2 Nanowaste treatment in waste incineration plants

Over 100 million tons of urban solid waste is destroyed worldwide each year. However, the outcome for NMs in incineration is little known, although the prevalence of leftover NPs is anticipated to rise [45]. China intended to boost waste incineration to eliminate untreated waste at sites [46]. After combustion or in the furnace's gas stream, the ENMs were contained directly in the trash, and thus waste ash emitted by waste incineration plants (WIPs) may include the ENMs. As was stated earlier, WTPs transfer ENMs from wastewater efficiently into sludge, which is then frequently incinerated. Owing to the complex structures with interactive effects from various surplus matrices, the destiny and efficiency of removing ENMs have not yet been specified for WIPs.

In a research paper on the persistence of ENMs in the WIP [45], Walser and collaborators showed that CeO_2 NPs added in a complete WTP gently bind to the exhaust phase's sludge and could easily be extracted from the flue gas by using existing filters [45]. They indicated that waste could be incinerated without discharging the ENMs into the atmosphere. Instead, the particles bind the ENMs to waste or recycled raw materials. Those are passed to subsequent processing sites and final dumps where wastes from scratch and fly ash are ultimately collected and processed. A document has also been published on the synthesis of PAH and dioxins from the combustion of ENMs, NiO, TiO_2, CeO_2, C_{60}, and Ag NPs [47]. In the waste stream, the existence of ENMs resulted in more significant emissions of specific PAHs, depending on the waste type. Chlorinated furans have been produced with high concentrations of utilizing waste Ag NPs and TiO_2 NPs. They argued that combining NMs with the high specific surface area and catalytic properties could influence toxic pollutants during the entire combustion. Even though WTPs and WIPs are used to remove ENMs effectively, waste sludge and slag or ash bearing the ENMs must be further handled. In other words, the end of the life cycle should include ENMs and consumer goods.

2.11.3 Nanowaste treatment in landfills

It is well understood that Ag NPs and silver ions are often collected in sludge in WTPs. Ag NPs may also be injected into the sludge and mixed into it, creating new items such as Ag_2S [36]. Waste disposal is the ultimate potential spot of the rendered sludge or the ENM waste products. A new design analysis estimates that 4.77 tons of Ag NPs are dumped annually into sites [48]. On the basis of box models and scenario analysis, the Nowack Group showed that garbage processing was the input-output process for ENMs [49]. The bottom ash to the waste disposal for TiO_2, ZnO, and Ag NPs was projected to fill 58%−62% of the main flow. Direct deposition in waste sites, especially with building waste, is also between 23% and 29%. All such flows, including fly ash trade and fly ash deposit, are not very significant. Water and air discharge are also redundant. For CNTs, flows vary significantly, and approximately 94% are burned.

As a consequence, burning could have a profound impact on many ENMs, but the most significant ENM mass can be enclosed in locations. Also, certain toxic ENMs in the deposition may reduce the production of methane by reducing methanogens. Although silver ions did not affect methane production at 10 mg/kg solids from landfills, Ag NPs inhibited methanogenesis at this concentration. This inhibition may be due to the gradual and long-term release of silver ions from the Ag NPs' dissolution in the region [39,43]. In recent years, nanowaste disposal in deposits also posed issues involving the effect of added ENMs on waste degradation and waste sites' leachate processing and potential ENM release into the environment through contact with waste site leachate [50]. The quantity of ENMs in waste leachate is compiled and exists in solid waste form as larger particles, while leachate passes through the site. This means that ENMs continue to be released into soil and underwater through the long-term involvement of ENMs in waste dumping and disposal leachates.

2.11.4 Recycling of waste containing nanomaterials

Recycling usually involves various mechanical, physical, and chemical processes, such as sorting, cleaning, grinding, grinding, pulping, melting, extrusion, desiccation, refining, and reforming. The release of ENMs during recycling from the waste matrix relies on the mechanisms involved, the matrix stiffness; the temperature achieved during the phase; and the affinity of ENMs to the air, solid, and liquid phases. When plastic waste is recycled, processes that minimize sizes (e.g., grinding, shredding, and cutting) might release ENMs into the air. The degree to which this occurs is uncertain, since different studies yield different results. Zhang et al. [51] documented the substantial releases of nano-scale and macro-scale particles by grinding CNT-filled polypropylene, although no free CNTs were identified.

In comparison, in 2017 the National Research Centre for the Working Environment recorded very low emissions of CNT-containing polypropylene and organic pigments in ENM shredding. Although plastic is exposed to high temperatures during deformation, it is not likely to release most inorganic ENMs through evaporation [52]. For CNT, particulate emissions are far greater than the context when polypropylene is melted and molded with CNT [51].

When recycled, glass wastes are either used as inert aggregates or recycled to manufacture new glass goods. Reuse as an inert aggregate may lead to the release of ENMs into water by surface abrasion. This mechanism can be expected to be very sluggish, owing to the resistant nature of glass. A remelting of the glass is carried out at high temperatures (e.g., 1400°C−1600°C), and ENMs can be evaporated and released in the exhaust gas [52]. The boiling and melting points of each ENM and their efficient removal in the flue gas purification system are critical factors in deciding the release of the ENMs into the atmosphere [52]. Similarly, metal waste (e.g., remelted) can be recycled during metal waste processing in high-temperature ovens.

The recycling of paper waste involves the processes of pulping, de-inking, and refining waste paper to create new paper. This method produces a range of pollutants, including solid waste (e.g., sludge) and liquid waste. Both streams need additional treatment, and water pollution may be a concern. Different subprocesses of the recycling facility, such as during chemical processing, require air pollution monitoring systems, which involve chemical air emissions. According to an evaluation of other chemicals carried out by Pivnenko et al. [53], the two key factors that could have consequences for chemical releases during paper recycling are biodegradability (i.e., persistence) and affinity for the air, aqueous, or solid phase. ENMs with a small air phase affinity can end up in recycled materials or in solid and liquid waste. Nonbiodegradable ENMs will accumulate during any paper recycling cycle (such as inorganic ENMs).

Construction and destruction (C&D) scrap is processed and used in various applications, such as a subbase unit for roads, for backfilling excavations, and as a filler in asphalt concrete. Strong materials (i.e., concrete, bricks, and mortar) make up a large proportion of C&D waste, all of which could include ENMs. ENMs can be airborne during grinding, shredding, and milling during the recycling of C&D waste [16–18]. If C&D scraps are used as aggregates for example, road building, ENMs can be discharged into leachate once generated in the presence of water.

ENMs are not released during recycling from the waste matrix and may finish up in recycled goods or materials. Research on recycling particular ENMs in various product categories based on a probabilistic mass flow simulation of ENMs in recycled processes found that less than 10% of the ENMs of the waste containing NMs would return to the supply chain [54]. Destruction of ENMs during recycling is an issue that has not been well studied. It can be hypothesized that ENMs can be thermally destroyed if exposed to sufficiently high temperatures or chemicals. The destruction of organic ENMs in glass and metal smelting, which takes place at very high temperatures, could therefore be expected.

2.11.5 Nanowaste management problems and issues

Although the production of new products has been a concern in nanotechnology, the amount of nano-based waste generated throughout new materials production is uncertain. However, NMs are known as NMs' essential origin throughout the environment [30]. While recent decades have the development of nanowastes on the agenda, current technologies were thought to suffice to efficiently remove such negligible materials through water sources. However, new studies have shown that current wastewater treatment systems are not appropriate for NM removal [31]. Little information on industrial processes linked to nanotechnology or about the quantities of nanowastes produced by end users is available to date except for the findings from specific models and estimating research. However, the widespread applications of nanotechnology make accurate estimates of the quantity of nanowastes challenging. A recent study has been performed on the treatment of water and sludge NMs in wastewater treatment plants. Even so, the data and knowledge available are not adequate to establish general models for treating nanoresidues, which can be included in the international classification to measure nanowastes threats. In certain cases, the adoption of nanotechnology as a green technology has slowed down efforts.

2.11.6 Legislative framework

An analysis of the present regulatory framework reveals that volume-based toxicity is expressed in terms of mass; however, the toxicity of NMs depends on their shape, scale, surface, and reactivity. Therefore, the existing toxicity estimation system is likely to be inadequate to express the real environmental risk levels of NMs in nanowastes. The conventional dose-response curves for macro-scale standard waste streams for toxicity estimation do not have a useful predictive parameter for NMs. The explanation is that NM toxicity does not depend only on aspects such as surface area, size, or transport and uptake rate [55]. Therefore it is suggested that nanoresidues be routinely detected by creating helpful toxicity indices for NMs in aquatic and terrestrial species. To be successful, this project requires techniques to be quickly established in the actual environmental compartments that can detect and track nanoresidues and calculate various NM physico-chemical characteristics (size, shape, surface area, chemical reactivity, etc.). Such methods are central to the creation and enforcement of policies and laws to minimize the spread of nanopollution. Based on the first component, the existing regulatory structure does not include the environmentally admissible limit concentrations of NMs, which can require several indexes and units to be expressed. This knowledge needs immediate attention by conducting and concentrating research on the growing accumulation of NMs in the ecosystem. On the other hand, it is necessary to verify the degree to which the current measuring metrology of contaminants in the environment applies in the sense of nanoresidues NMs and the kinds of changes required.

2.12 Conclusions

NM goods reflect a very developed research area of current technology and promise fresh possibilities across several industries that may alter our reality radically for a healthier life in the future. However, thorough application of nanotechnology and novel NPs to items will generate large quantities of nanowaste in the long term. Further studies should be aimed at discovering novel strategies of nanotoxicology, recognizing the biological impact of NPs throughout the atmosphere, and developing nanobiomonitoring bases. New forms of environmental threats through the launch of ecological NMs must be addressed. Nanotoxicity should be discussed before nanowastes occur throughout the environment and, consequently, before novel nanoproducts are introduced into the market.

Acknowledgments

T.Y.S. (the first author) and D.S.P. (Corresponding author, peids@cigit.ac.cn) contributed equally to this work. The authors are thankful for the supports from the CAS Team Project of the Belt and Road (to D.S.P.), Chongqing Key Program of Basic Research and Advanced Exploration Project (No. cstc2019jcyj-zdxmX0035 to D.S.P.), Three Hundred Leading Talents in Scientific and Technological Innovation Program of Chongqing (No. CSTCCXLJRC201714 to D.S.P.), Program of China—Sri Lanka Joint Center for Water Technology Research and Demonstration by Chinese Academy of Sciences (CAS)/China—Sri Lanka Joint Center for Education and Research by CAS, and International Partnership Program of CAS (No. 121311kysb20190071).

Conflicts of interest

The authors declare no conflict of interest.

References

[1] X. Li, N. Anton, C. Arpagaus, F. Belleteix, T.F. Vandamme, Nanoparticles by spray drying using innovative new technology: the Büchi nano spray dryer B-90, J. Control. Release 147 (2) (2010) 304—310.

[2] PR Newswire, Global construction market worth $10.3 trillion in 2020 (50 largest, most influential markets), 2016.

[3] Statnano, Nanotechnology publications of 2018: an overview. <https://statnano.com/news/65056/Nanotechnology-Publications-of-2018-An-Overview>, 2019.

[4] D. Enescu, M.A. Cerqueira, P. Fucinos, L.M. Pastrana, Recent advances and challenges on applications of nanotechnology in food packaging. A literature review, Food Chem. Toxicol. 134 (2019) 110814.

[5] S. Zanganeh, H.R. Nejad, J.F. Mehrabadi, R. Hosseini, B. Shahi, Z. Tavassoli, et al., Rapid and sensitive detection of staphylococcal enterotoxin B by recombinant nanobody using phage display technology, Appl. Biochem. Biotechnol. 187 (2) (2019) 493—505.

[6] R.W. Kelsal, I.W. Hamley, M. Geoghegan, Nanotechnologie, PWN Warszawa (2008) 19—29.

[7] A. Henglein, Q-particles: size quantization effects in colloidal semiconductors, in: H. Hoffmann (Ed.), New Trends in Colloid Science. Progress in Colloid & Polymer Science, vol. 73, Steinkopff, 1987, pp. 1—3. Available from: https://doi.org/10.1007/3-798-50724-4_55.

[8] E. Inshakova, O. Inshakov, World market for nanomaterials: structure and trends, MATEC Web Conf. 129 (2017) 02013.

[9] Lux Research, Nanotechnology Update: United States Leads in Government Spending Amidst Increased Spending Across Asia, Lux Research Inc, Boston, MA, 2015.

[10] J.F. Sargent, The National Nanotechnology Initiative: Overview, Reauthorization, and Appropriations Issues, Congressional Research Service Report, Washington, DC, 2014.

[11] A.W. Salamon, P. Courtney, I. Shuttler, A primer, Frequently Asked Questions, Nanotechnology and Engineered Material, 2015. Available via Perkin Elmer <http://www.perkinelmer.com/Content/Manuals/GDE_NanotechnologyPrimer.pdf> (accessed 5).

[12] A. Boldrin, S.F. Hansen, A. Baun, N.I.B. Hartmann, T.F. Astrup, Environmental exposure assessment framework for nanoparticles in solid waste, J. Nanopart. Res. 16 (6) (2014) 2394.

[13] M. Bundschuh, J. Filser, S. Lüderwald, M.S. McKee, G. Metreveli, G.E. Schaumann, et al., Nanoparticles in the environment: where do we come from, where do we go to? Environ. Sci. Eur. 30 (1) (2018) 6. Available from: https://doi.org/10.1186/s12302-018-0132-6.

[14] J.R. Gurr, A.S. Wang, C.H. Chen, K.Y. Jan, Ultrafine titanium dioxide particles in the absence of photoactivation can induce oxidative damage to human bronchial epithelial cells, Toxicology 213 (1—2) (2005) 66—73.

[15] K. Donaldson, V. Stone, C.L. Tran, W. Kreyling, P.J.A. Borm, Nanotoxicology, Occup. Environ. Med. 61 (9) (2004) 727−728. Available from: https://doi.org/10.1136/oem.2004.013243.

[16] OECD, FDI in Figures, Organisation for European Economic Cooperation, Paris, 2016.

[17] OECD Publishing, Nanomaterials in Waste Streams-Current Knowledge on Risks and Impacts, OECD Publishing, Paris, 2016.

[18] OECD, OECD Workshop on Recent Scientific Insights into the Fate and Risks of Waste Containing Nanomaterials, November 2016c, Paris France − Agenda.

[19] J.X. Bouillard, B. R'Mili, D. Moranviller, A. Vignes, O. Le Bihan, A. Ustache, et al., Nanosafety by design: risks from nanocomposite/nanowaste combustion, J. Nanopart. Res. 15 (4) (2013) 1519.

[20] B.M. Díaz-Soler, M. López-Alonso, M.D. Martínez-Aires, Nanosafety practices: results from a national survey at research facilities, J. Nanopart. Res. 19 (5) (2017) 169.

[21] T. Faunce, B. Kolodziejczyk, Nanowaste: need for disposal and recycling standards. G20 Insights, Policy Era: Agenda, 2030, 2017.

[22] N. Lubick, Nanosilver toxicity: ions, nanoparticles or both? Environ. Sci. Technol. 42 (23) (2008) 8617.

[23] A. Mathur, A. Parashar, N. Chandrasekaran, A. Mukherjee, Nano-TiO_2 enhances biofilm formation in a bacterial isolate from activated sludge of a waste water treatment plant, Int. Biodeterior. Biodegrad. 116 (2017) 17−25.

[24] H. Shi, R. Magaye, V. Castranova, J. Zhao, Titanium dioxide nanoparticles: a review of current toxicological data, Part Fibre Toxicol. 10 (2013) 15.

[25] D.M. Mitrano, K. Mehrabi, Y.A.R. Dasilva, B. Nowack, Mobility of metallic (nano) particles in leachates from landfills containing waste incineration residues, Environ. Sci. Nano 4 (2) (2017) 480−492.

[26] E.P. Vejerano, Y. Ma, A.L. Holder, A. Pruden, S. Elankumaran, L.C. Marr, Toxicity of particulate matter from incineration of nanowaste, Environ. Sci. Nano 2 (2) (2015) 143−154.

[27] M.F. Hallock, P. Greenley, L. DiBerardinis, D. Kallin, Potential risks of nanomaterials and how to safely handle materials of uncertain toxicity, J. Chem. Health Saf. 16 (1) (2009) 16−23.

[28] T. Jiao, Y. Wang, W. Guo, Q. Zhang, X. Yan, J. Chen, et al., Synthesis and photocatalytic property of gold nanoparticles by using a series of bolaform Schiff base amphiphiles, Mater. Res. Bull. 47 (12) (2012) 4203−4209.

[29] Y. Wen, B. Liu, W. Zeng, Y. Wang, Plasmonic photocatalysis properties of Au nanoparticles precipitated anatase/rutile mixed TiO_2 nanotubes, Nanoscale 5 (20) (2013) 9739−9746.

[30] N. Musee, Nanowastes and the environment: potential new waste management paradigm, Environ. Int. 37 (1) (2011) 112−128.

[31] G.G. Leppard, I.G. Droppo, M.M. West, S.N. Liss, Compartmentalization of metals within the diverse colloidal matrices comprising activated sludge microbial flocs, J. Environ. Qual. 32 (6) (2003) 2100−2108.

[32] W.W. Center, The Project on Emerging Nanotechnologies: Consumer Product Inventory, 2015.

[33] N. von Goetz, C. Lorenz, L. Windler, B. Nowack, M. Heuberger, K. Hungerbuhler, Migration of Ag-and TiO_2-(Nano) particles from textiles into artificial sweat under physical stress: experiments and exposure modeling, Environ. Sci. Technol. 47 (17) (2013) 9979−9987.

[34] A.P. Gondikas, F.V.D. Kammer, R.B. Reed, S. Wagner, J.F. Ranville, T. Hofmann, Release of TiO_2 nanoparticles from sunscreens into surface waters: a one-year survey at the old Danube recreational Lake, Environ. Sci. Technol. 48 (10) (2014) 5415−5422.

[35] L. Li, G. Hartmann, M. Döblinger, M. Schuster, Quantification of nanoscale silver particles removal and release from municipal wastewater treatment plants in Germany, Environ. Sci. Technol. 47 (13) (2013) 7317−7323.

[36] J. Liu, K.G. Pennell, R.H. Hurt, Kinetics and mechanisms of nanosilver oxysulfidation, Environ. Sci. Technol. 45 (17) (2011) 7345−7353.

[37] F. Gottschalk, T. Sonderer, R.W. Scholz, B. Nowack, Modeled environmental concentrations of engineered nanomaterials (TiO_2, ZnO, Ag, CNT, fullerenes) for different regions, Environ. Sci. Technol. 43 (24) (2009) 9216−9222.

[38] N.C. Mueller, B. Nowack, Exposure modeling of engineered nanoparticles in the environment, Environ. Sci. Technol. 42 (12) (2008) 4447−4453.

[39] Y. Yang, C. Zhang, Z. Hu, Impact of metallic and metal oxide nanoparticles on wastewater treatment and anaerobic digestion, Environ. Sci. Process. Impacts 15 (1) (2013) 39−48.

[40] M.A. Kiser, P. Westerhoff, T. Benn, Y. Wang, J. Perez-Rivera, K. Hristovski, Titanium nanomaterial removal and release from wastewater treatment plants, Environ. Sci. Technol. 43 (17) (2009) 6757−6763.

[41] L.K. Limbach, R. Bereiter, E. Müller, R. Krebs, R. Gälli, W.J. Stark, Removal of oxide nanoparticles in a model wastewater treatment plant: influence of agglomeration and surfactants on clearing efficiency, Environ. Sci. Technol. 42 (15) (2008) 5828−5833.

[42] H.P. Jarvie, H. Al-Obaidi, S.M. King, M.J. Bowes, M.J. Lawrence, A.F. Drake, et al., Fate of silica nanoparticles in simulated primary wastewater treatment, Environ. Sci. Technol. 43 (22) (2009) 8622−8628.

[43] Y. Yang, S. Gajaraj, J.D. Wall, Z. Hu, A comparison of nanosilver and silver ion effects on bioreactor landfill operations and methanogenic population dynamics, Water Res. 47 (10) (2013) 3422−3430.

[44] L. Hou, K. Li, Y. Ding, Y. Li, J. Chen, X. Wu, et al., Removal of silver nanoparticles in simulated wastewater treatment processes and its impact on COD and NH(4) reduction, Chemosphere 87 (3) (2012) 248−252.

[45] T. Walser, L.K. Limbach, R. Brogioli, E. Erismann, L. Flamigni, B. Hattendorf, et al., Persistence of engineered nanoparticles in a municipal solid-waste incineration plant, Nat. Nanotechnol. 7 (8) (2012) 520−524.

[46] H. Cheng, Y. Hu, Municipal solid waste (MSW) as a renewable source of energy: current and future practices in China, Bioresour. Technol. 101 (11) (2010) 3816−3824.

[47] E.P. Vejerano, A.L. Holder, L.C. Marr, Emissions of polycyclic aromatic hydrocarbons, polychlorinated dibenzo-p-dioxins, and dibenzofurans from incineration of nanomaterials, Environ. Sci. Technol. 47 (9) (2013) 4866−4874.

[48] F. Gottschalk, T. Sun, B. Nowack, Environmental concentrations of engineered nanomaterials: review of modeling and analytical studies, Environ. Pollut. 181 (2013) 287−300.

[49] N.C. Mueller, J. Buha, J. Wang, A. Ulrich, B. Nowack, Modeling the flows of engineered nanomaterials during waste handling, Environ. Sci. Process. Impacts 15 (1) (2013) 251−259.

[50] S.C. Bolyard, D.R. Reinhart, S. Santra, Behavior of engineered nanoparticles in landfill leachate, Environ. Sci. Technol. 47 (15) (2013) 8114−8122.

[51] J. Zhang, A. Panwar, D. Bello, T. Jozokos, J.A. Isaacs, C. Barry, et al., The effects of recycling on the properties of carbon nanotube-filled polypropylene composites and worker exposures, Environ. Sci. Nano 3 (2) (2016) 409−417.

[52] A. Boldrin, L. Heggelund, N. Hundebøll, S.F. Hansen, Guidelines for safe handling of waste flows containing NOAA. Deliverable report 7.6 for the EU-FP7 project "SUN − Sustainable Nanotechnologies," Grant Agreement Number 604305, October, 2016.

[53] K. Pivnenko, E. Eriksson, T.F. Astrup, Waste paper for recycling: overview and identification of potentially critical substances, Waste Manage. 45 (2015) 134−142.

[54] A. Caballero-Guzman, T. Sun, B. Nowack, Flows of engineered nanomaterials through the recycling process in Switzerland, Waste Manage. 36 (2015) 33−43.

[55] C.M. Sayes, A.M. Gobin, K.D. Ausman, J. Mendez, J.L. West, V.L. Colvin, Nano-C60 cytotoxicity is due to lipid peroxidation, Biomaterials 26 (36) (2005) 7587−7595.

CHAPTER 3

Classification and sources of nanowastes

Prashant Gupta and Subhendu Bhandari
Department of Plastic and Polymer Engineering, Maharashtra Institute of Technology, Aurangabad, India

3.1 Introduction

Nanomaterials have revolutionized technology, offering many benefits for humankind. Nanomaterials are simply the materials that have grained sizes in the order of a billionth of a meter [1]. They exhibit a host of useful and attractive physicochemical properties and are extremely formable, due to which they are employed in structural and nonstructural applications such as microelectronics, insulation, cutting tools, pollutant removal, batteries, magnets, and sensors. They are used in the automotive, healthcare, aerospace, pharmaceutical, marine, and agriculture sectors along with specialized military applications, such as lightweight guns, electromagnetic launchers, and kinetic energy penetrators for projectiles [2].

An increasing influx of funding for research and development of nanomaterials by government and private sector, miniaturization of devices, and strategic alliances between countries have propelled industries to exploit the expanding range of novel properties of nanomaterials and their subsequent use. The efforts to improve existing production technologies, particularly for quality and yield improvement, have been taking it closer to commercial feasibility for producing nanomaterials on a mass scale [3]. The global nanotechnology market is expected to top USD $125 billion by 2027, growing at a compounded annual growth rate (CAGR) of 13% from USD $50 billion in 2020. However, the world is still coming to terms with the novel coronavirus (COVID-19), which has required countries to invest huge amounts of economic wealth in healthcare and public welfare. The business implications of the pandemic and slowdown observed in the second and third quarters of the year 2021 have led to a CAGR readjustment to 11.3% [4]. Some of the examples given earlier for the increased use and applications of nanotechnology have a broad and fundamental impact of all areas of industry along with economy. The three major drivers of nanomaterials, which enjoy an 85% market share in nanotechnology, are electronics, energy applications, and biomedical applications [5].

Nanomaterials Recycling
DOI: https://doi.org/10.1016/B978-0-323-90982-2.00003-2

© 2022 Elsevier Inc.
All rights reserved.

The most popular nanomaterials that have been reported, studied, and used are titanium dioxide [6−9], carbon nanotubes [10−13], graphene and graphene oxide [13−18], silicon dioxide [19−22], bismuth oxide [23−26], copper oxide [27−29], aluminum oxide [30,31], cerium oxide [32−34], antimony tin oxide [35−37], gold [38,39], silver [40,41], zinc [42−44], and clay [45−47]. The increased use of nanomaterials will lead to an increase in nanomaterial waste. The environmental issues, health factors, and safety concerns related to nanomaterials have contributed to hampering the commercial growth of these very useful materials. More work has to be done for a detailed study of these concerns, as dealing with them during their manufacture and handling them at the end of their life cycle are complicated activities. Also, as they can't be seen with the naked eye, they become really hard to track and monitor. Because they are highly reactive, nanomaterials behave in a chemically different manner compared to their ordinary counterparts, making it all the more difficult to predict their action under varying conditions in the environment. Furthermore, disposing of chemicals with such reactivities without proper deactivation and stabilization can potentially have a negative impact on human health in the form of respiratory issues, gastrointestinal issues, and skin damage and may also have deleterious effects on the aquatic environment, soil, and wildlife [48]. The lack of clarity about the usage, disposal, and recycling standards of nanomaterials and the nanowaste generated from their use makes the widespread commercial acceptance of nanomaterials riskier. In the absence of clear standards, organizations such as the Organisation for Economic Co-operation and Development, the World Health Organization, the United Nations Industrial Development Organization, the United States Environmental Protection Agency, the International Union for Conservation of Nature, and the National Institute of Environmental Health Sciences have been pivotal in laying out precautionary principles.

This chapter will present a comprehensive overview of nanomaterial classification along with an explanation of various types and categories of nanowaste, and a discussion of various sources responsible for the generation of nanowastes.

3.2 Types of nanomaterials

3.2.1 Carbon-based nanomaterials

Among all types of nanomaterials, carbon-based nanomaterials have received attention from researchers because of their electrical, electrochemical, and mechanically reinforcing characteristics. Among different carbon-based nanomaterials graphene, carbon nanotube, carbon fiber, and carbon black are noteworthy for their applications in various fields.

Graphene is a two-dimensional arrangement of carbon atoms in a single layer of honeycomb structure [49]. Its hexagonal rings contain both sp^3 and sp^2 hybridized carbon atoms [50]. It can be produced by either a top–down approach or a bottom–up approach. The most widely used methods of graphene synthesis are based on chemical vapor deposition (CVD) over transition metals, micromechanical exfoliation, and thermal exfoliation of graphite oxide to yield graphene oxide, followed by its reduction [51].

Carbon nanotubes are molecular-scale graphitic structured nanotubes consisting of single or multiple layers. It may be considered the intermediate between flat layered grapheme and spherical fullerene in which the single or multiple parallel layers of grapheme sheets are rolled to form tubular nanostructures. The typical aspect ratio (length-to-diameter ratio) of carbon nanotubes is higher than 1000. The electrical, electrochemical, and mechanical properties vary with structural differences.

Carbon nanofiber has emerged as an important reinforcing as well as electrically conductive material. It can be manufactured via catalytic conversion of gases containing carbon in its molecular structure. Its length ranges between 0.1 and 1000 μm, whereas the diameter varies from 3 to 100 nm. Depending on the synthesis procedure, carbon nanofiber may be classified as vapor-grown nanofiber or catalytically grown nanofiber [52]. Among different techniques of carbon fiber manufacturing, CVD, electrospinning, and template process are noteworthy. Carbon nanofiber is widely used in reinforced composites [53] as well as electronic applications [54–56] because of its high electrical conductivity.

Carbon black is a carbon-based small particulate material. Depending on the preparation process, it may be broadly classified as furnace black or thermal black [57]. Furnace carbon blacks are manufactured by controlled decomposition of hydrocarbons in presence of a high-velocity stream of combustion gas, whereas thermal blacks are produced by controlled thermal decomposition of hydrocarbons in the absence of combustion gas or air. Among other methods of carbon black manufacturing, lamp black process, channel black process, gas black process, and acetylene black process are noteworthy. However, carbon black may also be retrieved from polymeric waste materials, such as discarded tires [58] and polyethylene terephthalate bottles [59]. Most furnace blacks exist as agglomerates of nanoparticles, but thermal black contains macro-sized (>100 nm) primary particles. The primary particles exist in agglomerated fractal form [60,61]. The primary particle sizes of different grades of carbon black are listed in Table 3.1 [62].

Carbon black is widely used in different applications, such as tires [63], conveyor belts, hoses [64], extruded products, seals [65], gaskets [66], automotive parts, EMI shielding etc. [67].

Fullerene is a typical allotrope of carbon in which 60–100 carbon atoms form a three-dimensionally spherical nanostructure, sometimes referred to as a buckyball. It was first traced by Smalley et al. in 1980 in carbon soot. A typical C_{60} fullerene contains 12 pentagonal and 20 hexagonal carbon rings. It may be manufactured by arc vaporization of graphite, laser ablation of graphite, or a hydrocarbon combustion

Table 3.1 Different grades and primary particle size of carbon black.

Type	Carbon black	Abbreviation	ASTM D1765 nomenclature	Primary particle size (nm)
Furnace	Superabrasion furnace	SAF	N110	15–18
Furnace	Intermediate superabrasion furnace	ISAF	N220	20–25
Furnace	High abrasion furnace	HAF	N330	28–36
Furnace	Fast extrusion furnace	FEF	N550	39–55
Furnace	General purpose furnace	GPF	N660	56–70
Furnace	Semireinforcing furnace	SRF	N770	71–96
Thermal	Fine thermal	FT	N880	180–200
Thermal	Medium thermal	MT	N990	250–350

Adapted with permission from M. Bera, P. Gupta, P.K. Maji, Structural/load-bearing characteristics of polymer − carbon composites, in: M. Rahaman, K. Dipak, A.K. Aldalbahi (Eds.), Carbon-Containing Polymer Composites. Springer Series on Polymer and Composite Materials, Springer, Singapore, 2019, pp. 457−502. https://doi.org/10.1007/978-981-13-2688-2 [62].

technique. Fullerene finds its application in diversified fields, for example, solar cells [68] and biomedicine [69]. Some fullerene derivatives are highly toxic [70].

3.2.2 Organic nanomaterials

Polymeric nanoparticles are organic nanoparticles that are widely used in drug delivery and therapeutics [71]. Polymeric nanoparticles also find potential use in electronic applications such as supercapacitors [72], fuel cells [73], and sensors [74]. Polymer-based nanoparticles may be synthesized either by using templates or via template-free self-assembly routes. Among different polymeric nanoparticles, polyaniline, polypyrrole, polythiophene, polylactic acid, polyethylene glycol, poly(ε-caprolactone), and poly(lactide-co-glycolide) are noteworthy [75]. Some of the polymers may form nanoparticles themselves, whereas hybrids of the polymers may form nanoparticles. The later type mostly exhibits either spherical or core-shell nanostructures. Polymeric nanoparticles may be synthesized either via the self-assembly route or by using soft or hard templates. Nanostructures formed via the self-assembly or template-free route may be tuned in a wide range by varying different synthesis parameters, whereas the use of templates produces nanostructures of predefined shapes and dimensions as the hollow channels inside the templates.

3.2.3 Inorganic nanomaterials

Inorganic nanomaterials encompass a wide range of metals as well as metal oxides. Owing to their mechanical, electrical, thermal, electrochemical, optical, and magnetic properties as well as responses in biosystems, inorganic nanoparticles find applications in diverse fields.

3.2.3.1 Metallic nanoparticles

Among metallic nanoparticles, gold [76], silver [77], iron [78], and platinum [79] have received the attention of researchers.

Gold nanoparticles are probably the most important type of metallic nanoparticles and find application in biology, optics, and catalysis. The optical properties of gold nanoparticles largely depend on the morphology and aspect ratios [80]. Gold nanoparticles may be prepared by either top-down or bottom-up approaches. The top-down approach is of limited use because of its lesser control for tuning particle shape and dimensions. By contrast, the bottom-up approach involves a biological or chemical reduction step. Reduction involving both nucleation and successive growth simultaneously is called as in situ synthesis, whereas nucleation followed by successive growth in sequence is termed a seed-growth or seed-mediated-growth method [81]. A high chemical supersaturation condition yields spherical gold nanoparticles of diameter 1—5 nm that are generated via the fastest nucleation rates [82].

Metallic silver is not attacked by acids or water but continuously releases small number of ions, which exhibit antibacterial activity. Silver nanoparticles are synthesized by reducing a silver salt solution using typical reducing agents, such as glucose, sodium borohydride, citrates, or ethylene glycol [83]. Apart from antibacterial activity, silver nanoparticles may also be used in several applications, such as the reduction of carbon dioxide [84] and printed electronics [85]. However, the toxicity of silver nanoparticles to bacteria suggests that the antimicrobial characteristics of silver nanoparticles may pose a threat to aquatic ecosystems. According to the United States Environmental Protection Agency, the permissible limit of silver nanoparticles in fresh water and salt water are 3.4—1.9 parts per billion (ppb), respectively. Therefore quantification of silver nanoparticles in waste water after washing of clothes is quite important from the point of view of environmental aspects [86].

Iron is one of the most abundant elements widely available on earth. Nanoparticles of iron may be used in batteries, adsorbents for environmental remediation, contrast agents for magnetic resonance imaging, magnetic fluids, and so on [87]. General preparation routes are based on chemical reduction [88] and thermal decomposition [89]. Even though the thermal decomposition route of synthesis yields small nanoparticles with high structural uniformity and purity, the chemical reduction route is more economical because it does not involve highly toxic precursors as are used in the thermal decomposition process [87].

Zinc nanoparticles have attracted research focusing on energy storage, electronic applications, and electric vehicles. They may be synthesized via laser ablation, in-liquid plasma, ball-milling, CVD, mechanical deformation, thermal vapor phase deposition, and so on [90].

3.2.3.2 Metal oxide nanoparticles

Titanium dioxide, or titania (TiO_2), nanoparticles are emerging as a promising material for advanced research because of its dielectric properties [91], gas sensitivity [92], photocatalytic characteristics [93], energy storage and conversion characteristics [94], easy synthesis process, high stability, and nontoxicity. Titania nanoparticles may be synthesized via sol-gel [95], vapor phase [96], microwave irradiation route [97], and so on. Even though titania nanoparticles have not yet been reported to have any remarkable carcinogenic effect, they have adverse effects on human cells and aquatic organisms. The permissible limit of titania in sewage is 0.1−3 mg/L, and the limit in effluent is 5−15 μg/L [98].

Core-shell hybrids, Janus-like nanoparticles, and two- and three-dimensional nanostructures of inorganic materials are used in biological applications [99]. Mesoporous (2−100 nm in diameter) silica (SiO_2) nanoparticles are promising drug delivery vehicles that are used to encapsulate various drugs for targeted delivery [100]. Silica nanoparticles are also used in lubricants [101] as well as in polymer composites and dielectric materials [102]. However, exposure to nanosilica has been proven to have adverse effects on human health, including different types of lung disease such as silicosis and pulmonary tuberculosis. Is it also adversely affects the human immune system [103].

Zinc oxide (ZnO) nanoparticles are among the most frequently produced inorganic nanoparticles. They are wide-band gap semiconductors and are used in printed electronics [104], antibacterial [105] and antifungal applications [106], catalysis [107], and so on. Indium tin oxide is widely used in photovoltaics, owing to its high transparency and electrical conductivity [108].

Among different other metal oxides with commercial applications are lithium oxide [109,110], tin oxide [111], copper oxide [112], iron oxide [113], aluminum oxide [114], magnesium oxide [115], zirconium oxide [116], and silver oxide [117]. They are noteworthy for their promising applications in optoelectronics, sensors, dielectrics, energy harvesting, energy conversion, cosmetics, pharmaceutical drug delivery, and thermal and electrical insulation [118].

3.3 Classification of nanowastes

The nanowastes that are generated as a result of the use and application of nanotechnology can be systematically classified on the basis of their toxicity levels and the stream of their generation into five broad categories, as shown in Table 3.2. This classification system was proposed with the thought that it is not perfect for waste streams with nano-sized materials. Also, it can provide assistance in nanowaste management via:

1. Separating the wastes that require distinct operational management practices throughout the various process of production;

Table 3.2 Classification of nanowastes based on nanomaterial toxicity and potency of exposure based on the status of nanomaterial in the nanotechnology employed.

Class	Nanotoxicity		Risk profile	Waste streams	Remarks
	Hazard	Exposure			
Class I	Nontoxic	Low to medium	None to very low	Transparent electrodes in OLED TVs as display backplanes, solar panels, nanocircuits, polishing agents	• If the bulk parent materials cause human and environmental hazard via accumulation over a threshold value, apprehensions may rise over management of waste. If not, the waste can be managed as safe/benign. • No special disposal requirements.
Class II	Harmful or toxic	Medium to high	Low to medium	Transparent electrodes in OLED TVs as display backplanes, solar panels, nanocircuits, polishing agents, surface coatings, paints	• To ascertain the optimum and most appropriate approach for nanowaste management while handling, carrying, and disposal, acute or chronic effects due to potential exposure have to be established.
Class III	Toxic to very toxic	Low to medium	Medium to high	Packaging and additives for food products, effluents and waste water, cosmetics, and personal care products, polishing agents, pesticides	• The protocols for the management of hazardous nanowaste streams are needed. • More study is required to ascertain whether the current management structure is capable of handling the nanowaste stream hazards.

(*Continued*)

Table 3.2 (Continued)

Class	Nanotoxicity		Risk profile	Waste streams	Remarks
	Hazard	Exposure			
Class IV	Toxic to very toxic	Medium to high	High	Paints and surface coatings, cosmetic and personal care products, pesticides	• Choice of suitable designated sites specialized for the disposal of hazardous nanowaste.
Class V	Very toxic to extremely toxic	Medium to high	High to very high	Pesticides, sunscreen lotions, fullerene-containing colloidal suspensions in food and drinking products	• Disposal in designated sites specialized for the disposal of hazardous nanowaste only. • Improper management may lead to nanopollution of various ecosystems and all components of the same, which may be cost and time intensive to remediate. • The most effective techniques to treat and manage are neutralization and immobilization.

Reproduced (adapted) with permission from S.A. Younis, E.M. El-Fawal, P. Serp, Nano-wastes and the environment: potential challenges and opportunities of nano-waste management paradigm for greener nanotechnologies, in: C.M. Hussain (Ed.), Handbook of Environmental Materials Management, Springer International Publishing, Cham, 2018, pp. 1–72. https://doi.org/10.1007/978-3-319-58538-3_53-1 [124].

2. Supervising the handling, transportation, and storage of wastes; and
3. Permitting required means of treating, recycling, and disposing of nanowastes to mitigate negative effects on human health and the environment on the basis of the extent of hazardous nature [119–121].

Nanoproducts pass through a variety of places; therefore the transfer coefficients throughout the service and subsequent end of life cycle vary for different nanomaterial-based products. This makes it all the more difficult to classify nanowastes into various categories. However, on the basis of the nanomaterial used, any particular nanomaterial-based product, in general, poses a wide range of risks. Thus the waste stream of products employing the same nanotechnology may be classified into five categories ranging from most benign (class I) to highly hazardous and toxic (class V). The risk profiles of the waste stream products and area-based examples are shown in Table 3.3.

A product that is beneficial to one can be hazardous to others. For instance, a farming activity promoter may support a germicidal and antimicrobial nanomaterial such as silver without understanding that the amount of research done on humans and the environment is not enough. In such a scenario, promoting silver nanomaterial–based germicides may not be the right thing to do, as it will not be approved on the basis of toxicology studies and the effect on living beings, especially humans who are likely to eat the farming produce. Also, the effects on the environment and the ecosystem may have to be understood before making the final decision on promotion and use of such products [122]. For toxic nanomaterials, this knowledge can aid in curtailing the unintentional adverse long-term effects of nanotechnology-enabled products during their disposal leading to different environmental systems [123].

When we look at the long- and short-term impacts of potential hazards based on use, the world of nanotechnology is certainly poised to induce new challenges and increase the cloudiness around nanowaste management, which in turn underlines the significance of suitable management of nanowaste. It is immensely important that promotion of safe, responsible, and sustainable development is pursued in a world where wonders such as nanotechnology will keep emerging and growing to make life easier, better, and more effective.

3.4 Sources of nanowastes

The majority of nanomaterials consumed across the globe are of inorganic or metallic origin and end up in the waste stream as wastes are generated from the electrical and electronics industries, energy and power, construction, healthcare, consumer goods, water filtration, packaging, transportation, aerospace, and so on. These wastes are also generated from the consumer end as the waste does not meet the intended disposal methodology [125]. The generation of such wastes via standard procedures such as combustion that are used on a day-to-day basis in industry and the households has been well

Table 3.3 Risk profiles at the end-of-life disposal phase for various nanomaterials employed in nanotechnology-oriented products.

Product generating waste stream	Nanomaterial	Hazard	Exposure potency	Disposal risk	Class
Personal care products	Silver	Medium	High	Medium	II—III
	Fullerenes	High		High	IV—V
	Fe_2O_3	Medium		Medium	II—III
	TiO_2	Low		Low	I
Food products/ beverages	TiO_2	Low	Medium	Low	I
	ZnO	Medium		Medium	II—III
	Fullerenes	High		High	IV—V
	Dendrimers	Medium		Medium	II—III
Sunscreens	ZnO	Medium	High	Medium	II—III
	TiO_2	Low		Low	I
	Fullerenes	High		High	IV—V
	Dendrimers	Medium		Medium	II—III
Automobile components	SWNCT	High	Medium	Medium	II—III
	MWNCT	High		Medium	II—III
	Nanoclays	Low		Low	I
	Fullerenes	High		Medium	II—III
Sports equipment	SiO_2	Low	Low	Low	I
	Silver	Medium		Low	I—II
	SWCNT	High		Low	I—III
	MWCNT	High		Low	I—III
Polishing agents	TiO_2	Low	High	Low	I
	ZnO	Medium		Medium	II—III
Pesticides	Fullerenes	High	High	High	IV—V
	Fe_2O_3	Medium	Medium	Medium	II—III
	TiO_2	Low	Low	Low	I
	SiO_2	Low	Low	Low	I
	CdSe	High	High	High	III—IV
Integrated chips	SWNCT	High	Low	Low	I—II
	CdSe	High			I—II
	Silicon nanowires	Low			I
	Fullerenes	High			I—II
Others	Silicon nanowires	Low	Low	Medium	I—II
	Dendrimers	Medium	Medium	Medium	II—III
	Gold	High	Medium	Medium	III—IV
	Silica	Low	Low	Medium	I—II
	Al_2O_3	Medium	Low	Low	I—II
	Yttrium iron oxide $(Y_3Fe_5O_{12})$	Low	Low	Medium	I—II

Adapted with permission from S.A. Younis, E.M. El-Fawal, P. Serp, Nano-wastes and the environment: potential challenges and opportunities of nano-waste management paradigm for greener nanotechnologies, in: C.M. Hussain (Ed.), Handbook of Environmental Materials Management, Springer International Publishing, Cham, 2018, pp. 1—72. https://doi.org/10.1007/978-3-319-58538-3_53-1 [124].

reported [126−128]. Some of the processes that involve combustion, such as automotive operations, industrial boilers, the cooking of food, have revealed via air effluent studies that the size distributions of aerosols with respect to mass distribution are lower than 200 nm and the number distribution is lower than 100 nm. In addition to such activities, there are other sources, such as dynamic sources, industrial effluents, research and manufacturing facilities, and atmospheric chemical reactions, including radioactive decay of vapor precursors, that have been reported to emit nanowaste [128,129].

Nanoparticles enter the ecosystem via natural or anthropogenic (accidental or intentional) activities. The anthropogenic materials and processes can pollute the environment when nanomaterials are released into the air, ground, or water and pollute them. Nanomaterials enter the environment in four ways viz. raw material manufacture, manufacture of nano-enabled products, consumption, and management process [122]. Once these nanoparticles are in the environment, their size makes them very easily transportable through various streams, such as suspensions in water, aerosols in the air, and sediments in the ground, thus making them more dangerous. As they are highly reactive, it is easy for them to react with other pollutants and form compounds that can be much harder to mitigate and thus more hazardous [130]. There can also be consequences related to health for living beings as they enter the food chain through crops in soil contaminated by nanoproducts and water used for irrigation-related activities [131,132].

3.4.1 Stationary sources

In the advanced technological world in which we live, the conventional systems involved in industrial combustion such as power generation, boilers, and incinerators have been complemented with emissions from advanced facilities such as refining industries and chemical manufacturing units. Furthermore, the use of advanced nanomaterials as catalysts in and for process improvements [133], such as SiO_2 encapsulated Ni [134], Pr_6O_{11} nanorod supported palladium [135], and NiCoP/OPC-300 [136], end up generating unintentional nanoemissions in direct or indirect form. The characteristics in emissions quality such as density, reactivity, particle size, and number of particles, differ with respect to the involved process.

Maguhn et al. studied the particle size distribution of ultrafine aerosol particles in flue and stack gases of municipal waste incinerators. They reported that the maximum particle size of flue gas was between 90 and 140 nm when the temperature of the incinerator unit was varied from 700°C to 300°C, respectively, inside the combustor. There was an increase in particle size via condensation, aggregation, nucleation, and physicochemical sorption of gases over the particulate matter that serves as a core [137]. More recently, Mertens et al. reported the results of on-line measurements of flue gas treatment system numbers of remaining nanoparticles in medium- to large-scale industrial setups. They

48   Nanomaterials Recycling

reported the formation of H_2SO_4 aerosol with particle sizes of $<$50 nm [138]. The composition of fuel employed for combustion also plays an important role in the resulting flue gas peak size of emission particles with various authors studying diesel engine performance with the use of alumina nanoparticles as additives with poultry litter–based biodiesel [139], cashew nut shell–based biodiesel [140], and so on.

Chang et al. reported on the peak particle size for a variety of fuels, that is, coal, oil, and gas. The sizes were in the range of 40–50, 70–100, and 15–25 nm for coal, oil, and gas, respectively. They also reported on the dependence of other parameters such as dilution ratio, aging time, and exhaust temperatures on the effluent nanoparticle size [141]. The nanowastes are generated by personal habits such as smoking of cigarettes and omnipresent indoor activities such as cooking [142–144]. Glytsos et al. studied the number and size distribution and mass concentration of the particulate matter resulting from a host of controlled indoor activities, including candle burning, hot plate heating, boiling of water, cooking, vacuuming, hair drying, hair spraying, smoking, and burning of incense sticks. Of these, candle burning was reported to have the highest particle emission rate with $>$85% of the total emitted particles being below 50 nm along with cooking (onion frying) with a geometric mean diameter of particle below 15 nm, incense stick burning, and smoking exhibiting a particle size of around 100 nm. A change in particle size was observed due to the coagulation effect and condensation [145].

Roy et al. studied burning of cigarettes, incense sticks, mosquito coils, and dhoop (a thick form of incense stick) as sources and reported their particle size to be in between 4.6 and 157.8 nm. The incense stick emissions exhibited values between 79 and 89 nm, cigarette and mosquito coils had values around 150–160 nm, and dhoop sticks had values around 240 nm. There was a high amount of lead in mosquito coils and trace amounts of lead in cigarettes along with the presence of cadmium and zinc in all reported sources [146]. A custom-built aspiration condenser ion mobility spectrometer and a sequential mobility particle sizer with a counter system were employed to characterize ultrafine aerosol particles (1.1–300 nm) in varying burning conditions of paraffin wax tea candles with particle sizes of 10–30 nm in normal burn mode and 100–300 nm in sooting burn mode. Also, a particle size of 2.5–9 nm was observed for soot precursor species, and 0.1–2 nm size was observed when a fan was placed behind the burning candle [147]. Other industrial processes viz. powder coating, printing, milling, plasma processing, smelting, welding, and refining, have been identified as processes with high chances of emission of ultrafine particles [148].

3.4.2 Dynamic sources

The largest application industry that can be included in this category is the transportation industry, in which diesel engines are in the top producers of nanoparticle-containing emissions. Beatrice et al. performed detailed size characterization from a modern Euro 5 automotive engine and reported that the peak observation of particle numbers was primarily due to nanoparticles with an average size of below 30 nm during the regeneration stage, possibly formed as a result of oxidative fragmentation of soot cake leading to the formation of elementary carbon. There is a further increase in the size by 2—4 times during the nucleation mode over normal operation [149]. La Rocca et al. investigated soot-in-oil taken from oil sump of a gasoline direct injection engine (modern light-duty EURO IV engine) operated between 1600—3700 rpm and 30—120-Nm torque. The soot aggregates were reported to be around 153 and 59 nm in terms of soot skeleton length and width, respectively, with the primary particles largely spherical in shape with some surface irregularities having an average diameter of 36 nm. They represented a core-shell structure like diesel soot and had an amorphous layer, as shown in Fig. 3.1 [150].

Several factors are responsible for the physical and chemical composition of nanoparticulate matter, such as specifications of fuel and oil used, operating conditions, and after-treatment systems of the exhaust. Also, soot and ash are coated with various additives, such as aldehydes and polycyclic aromatic hydrocarbons, which are toxic and hazardous, as they are cancerous in nature. An engine oil with higher viscosity can reduce particulate emissions as a result of its low volatile nature. The lubricant recipe should consider the durability factor of power trains and after-treatment systems [151]. Thiruvengadam et al. reported the effect of the practical engine load conditions on nanoparticle emissions from a heavy-duty diesel engine equipped with a diesel particulate filter and selective catalytic reduction. The emission contained nucleation mode particles with 6—15-nm size at higher exhaust temperatures in excess of 380°C [152].

The particle measurement program has defined a 23 nm cutoff, so the particles below 23 nm are considered to be a health hazard [153]. The measurement processes for particulate matter below the limit are not standardized, and volatile nucleation increases the particle size for sub-23-nm particles. The sensitivity of current techniques is not effective, owing to the action of oxygenated fuels producing lower particle sizes.

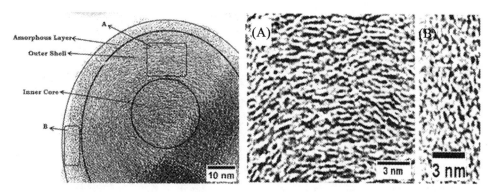

Figure 3.1 Core-shell structure of soot-in-oil primary particles exhibiting a 25-nm core and a 16-nm shell with an amorphous layer of 5 nm. (A, B) Images with enhanced contrast for visibility of fringes.

Maier et al. tested dimethyl carbonate and methyl formate against gasoline as a fuel in a single cylinder, direct injection spark ignition engine for primary and secondary gaseous emissions to successfully quantify sub-23-nm particles [154].

The presence of ultrafine particles of iron, nickel, copper, and other metallic compounds originates from lubricating oil, which moves these particles into the chamber where combustion takes place, thereby partially vaporizing them and forming particles <50 nm, or by abrasion, which causes wear and tear of moving engine parts such as piston rigs, valves, and cylinders [155]. These metal-based components of an engine as well as catalytic converters have also proved to be contributors of ultrafine metal particles in the engine exhaust stream. The emission of platinum nanomaterials has been reported to be present in three-way catalytic converters, with a scope of palladium and rhodium also present in the same [156]. There has been suggestive literature for reporting transition metals such as barium and cobalt as replacements for a noble metal such as platinum for NO_x catalytic converters [157].

The exhaust from a dynamic nanowaste-producing source, such as a diesel engine, comprises volatile and nonvolatile materials, such as sulfates, nitrates, unused hydrocarbons, lubricating oil, carbonaceous fraction, and ash, as shown in Fig. 3.2. The volatile fraction

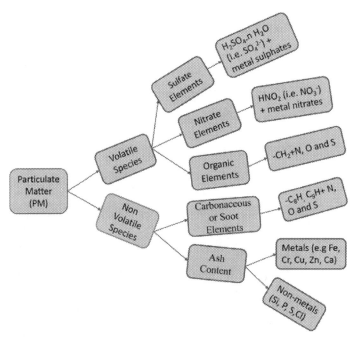

Figure 3.2 Classification of particulate matter emitted from a diesel engine. *Reproduced (adapted) with permission from M. Raza, L. Chen, F. Leach, S. Ding, A review of particulate number (PN) emissions from gasoline direct injection (GDI) engines and their control techniques, Energies. 11 (2018) 1417. https://doi.org/10.3390/en11061417 [158].*

consists of thousands of compounds, such as aromatics, esters, alcohols, esters, ketones, sulfate ions, sulfuric acid, metal nitrates, and nitric acid, and the nonvolatile fraction consists of carbonaceous material made in the engine before the opening of the exhaust valve. Later, this fraction oxidizes at high temperatures in the exhaust. The ash consists of metals and nonmetallic material, which form incombustible slag [158].

3.4.3 Miscellaneous sources

The formation of ultrafine particles in the atmosphere through nucleation is a known phenomenon. It is popularly known as Aitken mode because particles sized 25–50 nm were first reported by Aitken [159]. There are reports in the literature about the natural effect that may be supplemented by some anthropogenic methods, such as ammonium sulfate particles by alpha-pinene ozonolysis [160], the effects of continental clouds on Aitken and accumulation modes [161], and the 2015 atmospheric particle report of suburban Beijing [162]. The formation of nuclei mode particles occurs in the outflow of free troposphere clouds, which eventually come to the surface, contributing significantly to particulate formation and climate change across the planet by encouraging optical effects in the atmosphere. The life of a 10-nm particle is only about 15 minutes, but the particles have a tendency to be fine and thus easily transportable over very long distances even in that short time [163].

The synthesis of nanomaterials is done via a top-down or bottom-up approach [164]. The use of closed systems for both of these methodologies makes it safe in terms of operator exposure to nanomaterials. However, there can be a system failure or breakdown in the continuous process due to intermittent emergency shutdown that may lead to accidental release. Any spillage and the subsequent handling and transportation after production can lead to exposure, especially if the nanomaterial is in dry form. The filtration, exhaust, and ventilation systems become prone to trapped nanoparticles as well, and their handling has to be careful in terms of cleaning, change, recycling, and disposal. For some manufacturing processes, the cleaning and physical removal can be via a simple use of brushes, sponges, tissues, and vacuum or by chemical methods such as etching with the use of solvents, which leads to the generation of nanowaste streams via solid waste, air, and water effluents [165].

Many consumer goods in daily use contain nanomaterials for incorporating or enhancing functional effects. Some of the common areas of their use are pharmaceutical, cosmetics, health, fitness, home, and agriculture. Products many incorporate silver nanoparticles for antibacterial properties, TiO_2 and ZnO for UV absorbance and reflection, gold nanoparticles for antiaging properties, fullerene as antioxidants, SiO_2 and TiO_2 for durability improvement and hiding, and so on. These nanomaterials eventually find their way into various waste streams and from there to aquatic and terrestrial ecosystems through washing, bathing, swimming, and so on. Also, some products, such as cosmetics, have higher concentrations of nanomaterials than others, and it is estimated that 95% of these will end up into waste water treatment facilities [166,167].

3.5 Conclusion

The increased use of nanotechnology is leading to an increase in the generation of nanowaste. Because of the technological potential, there has been a huge influx of funds into nanotechnology research. However, the lack of clear enough guidelines for use, handling, disposal, and recycling by the local, national, and international administrative bodies has restricted the exploration of disruptive technologies like these in terms of commercial use. The issues involve the health hazards posed by nanomaterials and/or the technologies that incorporate them especially after the end of their life cycle, which may be due to limited research conducted on nanowaste disposal and recycling. The sources for the generation of nanowaste can be human-made, which may include stationary and dynamic sources and others that lead to nanowaste generation. These nanowastes may comprise of nanomaterials based on carbonaceous, polymeric, and inorganic materials. Their presence in the environment in the form of nanowaste can be potentially hazardous for nature and the health of living beings depending on the risk profile to biological systems that can be understood by nanomaterial toxicity and potency of exposure. The presence of nanomaterials in the waste stream can be ascertained by the employment of various characterization techniques based on surface, bulk, morphological, topological analysis, and so on, which can be employed either individually or in combination.

References

[1] K. Gajanan, S.N. Tijare, Applications of nanomaterials, Mater. Today Proc. 5 (2018) 1093–1096.

[2] N. Kumar, A. Dixit, Nanomaterials-enabled lightweight military platforms, Nanotechnology for Defence Applications, Springer, 2019, pp. 205–254.

[3] M.J. Pitkethly, Nanomaterials – the driving force, Mater. Today 7 (2004) 20–29. Available from: https://doi.org/10.1016/S1369-7021(04)00627-3.

[4] Global Industry Analysts, Global Nanotechnology Industry, 2020. https://www.reportlinker.com/p0326269/Global-Nanotechnology-Industry.html?utm_source = GNW (accessed 04.01.21).

[5] Research and Markets Ltd, Global Nanotechnology Market (by Component and Applications), Funding & Investment, Patent Analysis and 27 Companies Profile & Recent Developments – Forecast to 2024, 2018. https://www.researchandmarkets.com/reports/4520812/global-nanotechnology-market-by-component-and (accessed 04.01.21).

[6] A.J. Haider, R.H. Al–Anbari, G.R. Kadhim, C.T. Salame, Exploring potential environmental applications of TiO_2 nanoparticles, Energy Procedia 119 (2017) 332–345.

[7] T. Wu, X. Zhu, Z. Xing, S. Mou, C. Li, Y. Qiao, et al., Greatly improving electrochemical N2 reduction over TiO_2 nanoparticles by iron doping, Angew. Chem. Int. (Ed.) 58 (2019) 18449–18453.

[8] T. Wu, H. Zhao, X. Zhu, Z. Xing, Q. Liu, T. Liu, et al., Identifying the origin of Ti^{3+} activity toward enhanced electrocatalytic N_2 reduction over TiO_2 nanoparticles modulated by mixed-valent copper, Adv. Mater. 32 (2020) 2000299.

[9] W. Hu, W. Zhou, X. Lei, P. Zhou, M. Zhang, T. Chen, et al., Low-temperature in situ amino functionalization of TiO_2 nanoparticles sharpens electron management achieving over 21% efficient planar perovskite solar cells, Adv. Mater. 31 (2019) 1806095.

[10] R. Wang, L. Xie, S. Hameed, C. Wang, Y. Ying, Mechanisms and applications of carbon nanotubes in terahertz devices: a review, Carbon 132 (2018) 42−58.

[11] N. Bagotia, V. Choudhary, D.K. Sharma, A review on the mechanical, electrical and EMI shielding properties of carbon nanotubes and graphene reinforced polycarbonate nanocomposites, Polym. Adv. Technol. 29 (2018) 1547−1567.

[12] X. Wang, A. Dong, Y. Hu, J. Qian, S. Huang, A review of recent work on using metal−organic frameworks to grow carbon nanotubes, Chem. Commun. 56 (2020) 10809−10823.

[13] J. Xu, Z. Cao, Y. Zhang, Z. Yuan, Z. Lou, X. Xu, et al., A review of functionalized carbon nanotubes and graphene for heavy metal adsorption from water: preparation, application, and mechanism, Chemosphere 195 (2018) 351−364.

[14] J. Ahmed, Use of graphene/graphene oxide in food packaging materials: thermomechanical, structural and barrier properties, Reference Module in Food Science, Elsevier, 2019. Available from: https://doi.org/10.1016/B978-0-08-100596-5.22499-2.

[15] N.F. Atta, A. Galal, E.H. El-Ads, Graphene-a platform for sensor and biosensor applications, in: T. Rinken (Ed.), Biosensors − Micro and Nanoscale Applications, InTech, 2015. Available from: https://doi.org/10.5772/60676.

[16] S. Abdolhosseinzadeh, H. Asgharzadeh, H. Seop Kim, Fast and fully-scalable synthesis of reduced graphene oxide, Sci. Rep. 5 (2015) 10160. Available from: https://doi.org/10.1038/srep10160.

[17] Y.A. Arfat, J. Ahmed, M. Ejaz, M. Mullah, Polylactide/graphene oxide nanosheets/clove essential oil composite films for potential food packaging applications, Int. J. Biol. Macromol. 107 Part A (2018) 194−203. Available from: https://doi.org/10.1016/j.ijbiomac.2017.08.156.

[18] W. Han, Z. Wu, Y. Li, Y. Wang, Graphene family nanomaterials (GFNs)—promising materials for antimicrobial coating and film: a review, Chem. Eng. J. 358 (2019) 1022−1037.

[19] M. Tukmachi, M. Moudhaffer, Effect of nano silicon dioxide addition on some properties of heat vulcanized maxillofacial silicone elastomer, IOSR J. Pharm. Biol. Sci. 12 (2017) 37−43.

[20] H. Ezzat, S. El-Badawy, A. Gabr, S. Zaki, T. Breakah, Predicted performance of hot mix asphalt modified with nano-montmorillonite and nano-silicon dioxide based on Egyptian conditions, Int. J. Pavement Eng. 21 (2020) 642−652.

[21] Y. Tzeng, R. Chen, J.-L. He, Silicon-based anode of lithium ion battery made of nano silicon flakes partially encapsulated by silicon dioxide, Nanomaterials 10 (2020) 2467.

[22] S. Qu, M. Li, C. Zhang, Y. Sun, J. Duan, W. Wang, et al., Sulfonated poly (ether ether ketone) doped with ammonium ionic liquids and nano-silicon dioxide for polymer electrolyte membranes, Polymers 11 (2019) 7.

[23] M.B. Tahir, H. Kiran, T. Iqbal, The detoxification of heavy metals from aqueous environment using nano-photocatalysis approach: a review, Environ. Sci. Pollut. Res. 26 (2019) 10515−10528.

[24] A.H. Abdalsalam, E. Şakar, K.M. Kaky, M.H.A. Mhareb, B.C. Şakar, M.I. Sayyed, et al., Investigation of gamma ray attenuation features of bismuth oxide nano powder reinforced high-density polyethylene matrix composites, Radiat. Phys. Chem. 168 (2020) 108537.

[25] H. Mamur, M.R.A. Bhuiyan, F. Korkmaz, M. Nil, A review on bismuth telluride (Bi2Te3) nanostructure for thermoelectric applications, Renew. Sustain. Energy Rev. 82 (2018) 4159−4169.

[26] N. Akbarianrad, F. Mohammadian, M. Alhuyi Nazari, B. Rahbani Nobar, Applications of nanotechnology in endodontic: a review, Nanomed. Journal. 5 (2018) 121−126.

[27] M. Amiri, Z. Etemadifar, A. Daneshkazemi, M. Nateghi, Antimicrobial effect of copper oxide nanoparticles on some oral bacteria and candida species, J. Dental Biomater. 4 (2017) 347.

[28] V. Rajput, T. Minkina, B. Ahmed, S. Sushkova, R. Singh, M. Soldatov, et al., Interaction of copper-based nanoparticles to soil, terrestrial, and aquatic systems: critical review of the state of the science and future perspectives, Rev. Environ. Contam. Toxicol. 252 (2019) 51−96.

[29] V.D. Rajput, T. Minkina, S. Suskova, S. Mandzhieva, V. Tsitsuashvili, V. Chapligin, et al., Effects of copper nanoparticles (CuO NPs) on crop plants: a mini review, BioNanoScience 8 (2018) 36−42.

[30] H.K. Patel, S. Kumar, Experimental analysis on performance of diesel engine using mixture of diesel and bio-diesel as a working fuel with aluminum oxide nanoparticle additive, Therm. Sci. Eng. Prog. 4 (2017) 252−258.

[31] H.K. Hami, R.F. Abbas, E.M. Eltayef, N.I. Mahdi, Applications of aluminum oxide and nano aluminum oxide as adsorbents, Samarra J. Pure Appl. Sci. 2 (2020).

[32] S. Rajeshkumar, P. Naik, Synthesis and biomedical applications of cerium oxide nanoparticles—a review, Biotechnol. Rep. 17 (2018) 1—5.

[33] A. Dhall, W. Self, Cerium oxide nanoparticles: a brief review of their synthesis methods and biomedical applications, Antioxidants 7 (2018) 97.

[34] M. Nadeem, R. Khan, K. Afridi, A. Nadhman, S. Ullah, S. Faisal, et al., Green synthesis of cerium oxide nanoparticles (CeO_2 NPs) and their antimicrobial applications: a review, Int. J. Nanomed. 15 (2020) 5951.

[35] B. Khorshidi, S.A. Hosseini, G. Ma, M. McGregor, M. Sadrzadeh, Novel nanocomposite polyethersulfone-antimony tin oxide membrane with enhanced thermal, electrical and antifouling properties, Polymer 163 (2019) 48—56.

[36] S. Sreekumar, A. Joseph, C.S. Kumar, S. Thomas, Investigation on influence of antimony tin oxide/silver nanofluid on direct absorption parabolic solar collector, J. Clean. Prod. 249 (2020) 119378.

[37] Y. Hu, H. Zhong, Y. Wang, L. Lu, H. Yang, TiO_2/antimony-doped tin oxide: highly water-dispersed nano composites with excellent IR insulation and super-hydrophilic property, Sol. Energy Mater. Sol. Cell 174 (2018) 499—508.

[38] N. Elahi, M. Kamali, M.H. Baghersad, Recent biomedical applications of gold nanoparticles: a review, Talanta 184 (2018) 537—556.

[39] V. Amendola, R. Pilot, M. Frasconi, O.M. Maragò, M.A. Iatì, Surface plasmon resonance in gold nanoparticles: a review, J. Phys. Condens. Matter 29 (2017) 203002.

[40] H.D. Beyene, A.A. Werkneh, H.K. Bezabh, T.G. Ambaye, Synthesis paradigm and applications of silver nanoparticles (AgNPs), a review, Sustain. Mater. Technol. 13 (2017) 18—23.

[41] S.P. Deshmukh, S.M. Patil, S.B. Mullani, S.D. Delekar, Silver nanoparticles as an effective disinfectant: A review, Mater. Sci. Eng. C 97 (2019) 954—965.

[42] M. Bąkowski, B. Kiczorowska, W. Samolińska, R. Klebaniuk, A. Lipiec, Silver and zinc nanoparticles in animal nutrition—a review, Ann. Anim. Sci. 18 (2018) 879—898.

[43] V.L.R. Pullagurala, I.O. Adisa, S. Rawat, B. Kim, A.C. Barrios, I.A. Medina-Velo, et al., Finding the conditions for the beneficial use of ZnO nanoparticles towards plants—a review, Environ. Pollut. 241 (2018) 1175—1181.

[44] H. Agarwal, S.V. Kumar, S. Rajeshkumar, A review on green synthesis of zinc oxide nanoparticles—an eco-friendly approach, Resource-Efficient Technol. 3 (2017) 406—413.

[45] Q. Gao, X. Niu, L. Shao, L. Zhou, Z. Lin, A. Sun, et al., 3D printing of complex GelMA-based scaffolds with nanoclay, Biofabrication 11 (2019) 035006.

[46] F. Guo, S. Aryana, Y. Han, Y. Jiao, A review of the synthesis and applications of polymer—nanoclay composites, Appl. Sci. 8 (2018) 1696.

[47] N. Khatoon, M.Q. Chu, C.H. Zhou, Nanoclay-based drug delivery systems and their therapeutic potentials, J. Mater. Chem. B 8 (2020) 7335—7351.

[48] Z. Li, H. Cong, Z. Yan, A. Liu, B. Yu, The potential human health and environmental issues of nanomaterials, Handbook of Nanomaterials for Industrial Applications, Elsevier, 2018, pp. 1049—1054. https://doi.org/10.1016/B978-0-12-813351-4.00060-2.

[49] G. Yang, L. Li, W.B. Lee, M.C. Ng, Structure of graphene and its disorders: a review, Sci. Technol. Adv. Mater. 19 (2018) 613—648. Available from: https://doi.org/10.1080/14686996.2018.1494493.

[50] S. Rao, J. Upadhyay, K. Polychronopoulou, R. Umer, R. Das, Reduced graphene oxide: effect of reduction on electrical conductivity, J. Compos. Sci. 2 (2018) 25. Available from: https://doi.org/10.3390/jcs2020025.

[51] T. Mondal, A.K. Bhowmick, R. Krishnamoorti, Butyl lithium assisted direct grafting of polyoligomeric silsesquioxane onto graphene, RSC Adv. 4 (2014) 8649—8656. Available from: https://doi.org/10.1039/c3ra47373b.

[52] D. Yadav, F. Amini, A. Ehrmann, Recent advances in carbon nanofibers and their applications — a review, Eur. Polym. J. 138 (2020) 109963. Available from: https://doi.org/10.1016/j.eurpolymj.2020.109963.

[53] E. Hammel, X. Tang, M. Trampert, T. Schmitt, K. Mauthner, A. Eder, et al., Carbon nanofibers for composite applications, Carbon 42 (2004) 1153—1158. Available from: https://doi.org/10.1016/j.carbon.2003.12.043.

[54] J. Huang, Y. Liu, T. You, Carbon nanofiber based electrochemical biosensors: a review, Anal. Methods 2 (2010) 202–211. Available from: https://doi.org/10.1039/b9ay00312f.

[55] Z. Wang, S. Wu, J. Wang, A. Yu, G. Wei, Carbon nanofiber-based functional nanomaterials for sensor applications, Nanomaterials 9 (2019). Available from: https://doi.org/10.3390/nano9071045.

[56] E.J. Ra, E. Raymundo-Piñero, Y.H. Lee, F. Béguin, High power supercapacitors using polyacrylonitrile-based carbon nanofiber paper, Carbon 47 (2009) 2984–2992. Available from: https://doi.org/10.1016/j.carbon.2009.06.051.

[57] M. Singh, R.L. Vander Wal, Nanostructure quantification of carbon blacks, J. Carbnon Res. 5 (2019) 2. Available from: https://doi.org/10.3390/c5010002.

[58] C. Dwivedi, S. Manjare, S.K. Rajan, Recycling of waste tire by pyrolysis to recover carbon black: alternative & environment-friendly reinforcing filler for natural rubber compounds, Compos. Part B Eng. 200 (2020) 108346. Available from: https://doi.org/10.1016/j.compositesb.2020.108346.

[59] N. Doğan-Sağlamtimur, A. Bilgil, A. Güven, H. Ötgün, E.D. Yıldırım, B. Arıcan, Producing of qualified oil and carbon black from waste tyres and pet bottles in a newly designed pyrolysis reactor, J. Therm. Anal. Calorim. 135 (2019) 3339–3351. Available from: https://doi.org/10.1007/s10973-018-7576-1.

[60] K.-M. JaÈger, D.H. McQueen, Fractal agglomerates and electrical conductivity in carbon black polymer composites, Polymer 42 (2001) 9575–9581.

[61] V. Collin, I. Boudimbou, E. Peuvrel-Disdier, New insights in dispersion mechanisms of carbon black in a polymer matrix under shear by rheo-optics, J. Appl. Polym. Sci. 127 (2013) 2121–2131. Available from: https://doi.org/10.1002/app.37769.

[62] M. Bera, P. Gupta, P.K. Maji, Structural/load-bearing characteristics of polymer – carbon composites, in: M. Rahaman, K. Dipak, A.K. Aldalbahi (Eds.), Carbon-Containing Polymer Composites. Springer Series on Polymer and Composite Materials, Springer, Singapore, 2019, pp. 457–502. Available from: https://doi.org/10.1007/978-981-13-2688-2.

[63] P. Sae-Oui, K. Suchiva, C. Sirisinha, W. Intiya, P. Yodjun, U. Thepsuwan, Effects of blend ratio and sbr type on properties of carbon black-filled and silica-filled SBR/BR tire tread compounds, Adv. Mater. Sci. Eng. 2017 (2017) 8. Available from: https://doi.org/10.1155/2017/2476101. Article ID 2476101.

[64] S.B. KWAK, N.S. Choi, Thermo-oxidative degradation of a carbon black compounded EPDM rubber hose, Int. J. Automot. Technol. 12 (2011) 401–408. Available from: https://doi.org/10.1007/s12239-011-0047-3.

[65] J.-M. Degrange, M. Thomine, P. Kapsa, J.M. Pelletier, L. Chazeau, G. Vigier, et al., Influence of viscoelasticity on the tribological behaviour of carbon black filled nitrile rubber (NBR) for lip seal application, Wear 259 (2005) 684–692. Available from: https://doi.org/10.1016/j.wear.2005.02.110.

[66] A. Mostafa, A. Abouel-Kasem, M.R. Bayoumi, M.G. El-Sebaie, Effect of carbon black loading on the swelling and compression set behavior of SBR and NBR rubber compounds, Mater. Des. 30 (2009) 1561–1568. Available from: https://doi.org/10.1016/j.matdes.2008.07.043.

[67] M. Rahaman, T.K. Chaki, D. Khastgir, Development of high performance EMI shielding material from EVA, NBR, and their blends: effect of carbon black structure, J. Mater. Sci. 46 (2011) 3989–3999. Available from: https://doi.org/10.1007/s10853-011-5326-x.

[68] R. Ganesamoorthy, G. Sathiyan, P. Sakthivel, Review: fullerene based acceptors for efficient bulk heterojunction organic solar cell applications, Sol. Energy Mater. Sol. Cell 161 (2017) 102–148. Available from: https://doi.org/10.1016/j.solmat.2016.11.024.

[69] Y. Liu, D. Zhu, X. Zhu, G. Cai, J. Wu, M. Chen, et al., Enhancing the photodynamic therapy efficacy of black phosphorus nanosheets by covalently grafting fullerene C_{60}, Chem. Sci. 11 (2020) 11435–11442. Available from: https://doi.org/10.1039/d0sc03349a.

[70] E.A. Kyzyma, A.A. Tomchuk, L.A. Bulavin, V.I. Petrenko, L. Almasy, M.V. Korobov, et al., Structure and toxicity of aqueous fullerene C_{60} solutions, J. Synch. Investig. 9 (2015) 1–5. Available from: https://doi.org/10.1134/S1027451015010127.

[71] J.R. McCarthy, J.M. Perez, C. Brückner, R. Weissleder, Polymeric nanoparticle preparation that eradicates tumors, Nano Lett. 5 (2005) 2552–2556. Available from: https://doi.org/10.1021/nl0519229.

[72] T. Liu, L. Finn, M. Yu, H. Wang, T. Zhai, X. Lu, et al., Polyaniline and polypyrrole pseudocapacitor electrodes with excellent cycling stability, Nano Lett. 14 (2014) 2522–2527. Available from: https://doi.org/10.1021/nl500255v.

[73] Y. Qiao, S. Bao, C.M. Li, X. Cui, Z. Lu, J. Guo, Nanostructured polyaniline/titanium dioxide composite anode for microbial fuel cells, ACS Nano 2 (2008) 113−119. Available from: https://doi.org/10.1021/nn700102s.

[74] W. Zou, B. Quan, K. Wang, L. Xia, J. Yao, Z. Wei, Patterned growth of polyaniline nanowire arrays on a flexible substrate for high-performance gas sensing, Small 7 (2011) 3287−3291. Available from: https://doi.org/10.1002/smll.201100889.

[75] A. Zielinska, F. Carreiró, A.M. Oliveira, A. Neves, B. Pires, D.N. Venkatesh, et al., Polymeric nanoparticles: production, characterization, toxicology and ecotoxicology aleksandra, Molecules 25 (2020) 3731. Available from: https://doi.org/10.3390/molecules25163731.

[76] A. Gupta, D.F. Moyano, A. Parnsubsakul, A. Papadopoulos, L.S. Wang, R.F. Landis, et al., Ultrastable and biofunctionalizable gold nanoparticles, ACS Appl. Mater. Interfaces 88 (2016) 14096−14101. Available from: https://doi.org/10.1021/acsami.6b02548.

[77] A.A. Yaqoob, K. Umar, M.N.M. Ibrahim, Silver nanoparticles: various methods of synthesis, size affecting factors and their potential applications—a review, Appl. Nanosci. (Switz.) 10 (2020) 1369−1378. Available from: https://doi.org/10.1007/s13204-020-01318-w.

[78] E. Murgueitio, L. Cumbal, M. Abril, A. Izquierdo, A. Debut, O. Tinoco, Green synthesis of iron nanoparticles: application on the removal of petroleum oil from contaminated water and soils, J. Nanotechnol. 2018 (2018) 8. Available from: https://doi.org/10.1155/2018/4184769. Article ID 4184769.

[79] D. Pedone, M. Moglianetti, E. De Luca, G. Bardi, P.P. Pompa, Platinum nanoparticles in nano-biomedicine, Chem. Soc. Rev. 46 (2017) 4951−4975. Available from: https://doi.org/10.1039/c7cs00152e.

[80] C. Kan, C. Wang, J. Zhu, X. Zeng, X. Wang, H. Li, et al., Synthesis of high-yield gold nanoplates: fast growth assistant with binary surfactants, J. Nanomaterials 2010 (2010) 9. Available from: https://doi.org/10.1155/2010/969030. Article ID 969030.

[81] P. Zhao, N. Li, D. Astruc, State of the art in gold nanoparticle synthesis, Coord. Chem. Rev. 257 (2013) 638−665. Available from: https://doi.org/10.1016/j.ccr.2012.09.002.

[82] M. Grzelczak, J. Pérez-Juste, P. Mulvaney, L.M. Liz-Marzán, Shape control in gold nanoparticle synthesis, Chem. Soc. Rev. 37 (2008) 1783−1791. Available from: https://doi.org/10.1039/b711490g.

[83] S. Chernousova, M. Epple, Silver as antibacterial agent: ion, nanoparticle, and metal, Angew. Chem. − Int. (Ed.) 52 (2013) 1636−1653. Available from: https://doi.org/10.1002/anie.201205923.

[84] A. Salehi-Khojin, H.R.M. Jhong, B.A. Rosen, W. Zhu, S. Ma, P.J.A. Kenis, et al., Nanoparticle silver catalysts that show enhanced activity for carbon dioxide electrolysis, J. Phys. Chem. C 117 (2013) 1627−1632. Available from: https://doi.org/10.1021/jp310509z.

[85] S. Joo, D.F. Baldwin, Adhesion mechanisms of nanoparticle silver to substrate materials: identification, Nanotechnology 21 (2010) 055204. Available from: https://doi.org/10.1088/0957-4484/21/5/055204.

[86] T.M. Benn, P. Westerhoff, Nanoparticle silver released into water from commercially available sock fabrics, Environ. Sci. Technol. 42 (2008) 4133−4139. Available from: https://doi.org/10.1021/es7032718.

[87] K.C. Huang, S.H. Ehrman, Synthesis of iron nanoparticles via chemical reduction with palladium ion seeds, Langmuir 23 (2007) 1419−1426. Available from: https://doi.org/10.1021/la0618364.

[88] F. Li, C. Vipulanandan, K.K. Mohanty, Microemulsion and solution approaches to nanoparticle iron production for degradation of trichloroethylene, Colloids Surf. A: Physicochem. Eng. Asp. 223 (2003) 103−112. Available from: https://doi.org/10.1016/S0927-7757(03)00187-0.

[89] D.L. Huber, E.L. Venturini, J.E. Martin, P.P. Provencio, R.J. Patel, Synthesis of highly magnetic iron nanoparticles suitable for field structuring using a β-diketone surfactant, J. Mag. Magn. Mater. 278 (2004) 311−316. Available from: https://doi.org/10.1016/j.jmmm.2003.12.1317.

[90] N. Amaliyah, S. Mukasa, S. Nomura, H. Toyota, T. Kitamae, Plasma in-liquid method for reduction of zinc oxide in zinc nanoparticle synthesis, Mater. Res. Express 2 (2015) 25004. Available from: https://doi.org/10.1088/2053-1591/2/2/025004.

[91] D. Dastan, A. Banpurkar, Solution processable sol−gel derived titania gate dielectric for organic field effect transistors, J. Mater. Sci. Mater. Electron. 28 (2017) 3851−3859. Available from: https://doi.org/10.1007/s10854-016-5997-9.

[92] J. Trimboli, M. Mottern, H. Verweij, P.K. Dutta, Interaction of water with titania: implications for high-temperature gas sensing, J. Phys. Chem. B 110 (2006) 5647–5654.

[93] M. Adachi, Y. Murata, M. Harada, S. Yoshikawa, Formation of titania nanotubes with high photocatalytic activity, Chem. Lett. (2000) 942–943. Available from: https://doi.org/10.1246/cl.2000.942.

[94] W.Q. Wu, Y.F. Xu, J.F. Liao, L. Wang, D. Bin Kuang, Branched titania nanostructures for efficient energy conversion and storage: a review on design strategies, structural merits and multifunctionalities, Nano Energy 62 (2019) 791–809. Available from: https://doi.org/10.1016/j.nanoen.2019.05.071.

[95] M. Hema, A.Y. Arasi, P. Tamilselvi, R. Anbarasan, Titania nanoparticles synthesized by sol-gel technique, Chem. Sci. Trans. 2 (2013) 239–245. Available from: https://doi.org/10.7598/cst2013.344.

[96] M. Kamal Akhtar, S.E. Pratsinis, Dopants in vapor phase synthesis of titania powders, J. Am. Ceram. Soc. 75 (1992) 3408–3416. Available from: https://doi.org/10.1111/j.1151-2916.1992.tb04442.x.

[97] X. Wu, Q.-Z. Jiang, Z.-F. Ma, M. Fu, W.-F. Shangguan, Synthesis of titania nanotubes by microwave irradiation, Solid State Commun. 136 (2005) 513–517. Available from: https://doi.org/10.1016/j.ssc.2005.09.023.

[98] R. Zhang, Y. Bai, B. Zhang, L. Chen, B. Yan, The potential health risk of titania nanoparticles, J. Hazard. Materials 211–212 (2012) 404–413. Available from: https://doi.org/10.1016/j.jhazmat.2011.11.022.

[99] Y. Gao, Z. Tang, Design and application of inorganic nanoparticle superstructures: current status and future challenges, Small 7 (2011) 2133–2146. Available from: https://doi.org/10.1002/smll.201100474.

[100] F. Wang, C. Li, J. Cheng, Z. Yuan, Recent advances on inorganic nanoparticle-based cancer therapeutic agents, Int. J. Environ. Res. Public Health 13 (2016) 1182. Available from: https://doi.org/10.3390/ijerph13121182.

[101] B.A. Kheireddin, W. Lu, I.C. Chen, M. Akbulut, Inorganic nanoparticle-based ionic liquid lubricants, Wear 303 (2013) 185–190. Available from: https://doi.org/10.1016/j.wear.2013.03.004.

[102] S. Babanzadeh, S. Mehdipour-Ataei, A.R. Mahjoub, Effect of nanosilica on the dielectric properties and thermal stability of polyimide/SiO_2 nanohybrid, Des. Monomers Polym. 16 (2013) 417–424. Available from: https://doi.org/10.1080/15685551.2012.747159.

[103] L. Chen, J. Liu, Y. Zhang, G. Zhang, Y. Kang, A. Chen, et al., The toxicity of silica nanoparticles to the immune system, Nanomedicine 13 (2018) 1939–1962. Available from: https://doi.org/10.2217/nnm-2018-0076.

[104] K.S. Suganthi, K. Harish, N.M. Nair, P. Swaminathan, Formulation and optimization of a zinc oxide nanoparticle ink for printed electronics applications, Flex. Print. Electron. 3 (2018) 015001. Available from: https://doi.org/10.1088/2058-8585/aaa166.

[105] K.S. Siddiqi, A. ur Rahman, Tajuddin, A. Husen, Properties of zinc oxide nanoparticles and their activity against microbes, Nanoscale Res. Lett. 13 (2018) 141. Available from: https://doi.org/10.1186/s11671-018-2532-3.

[106] Q. Sun, J. Li, T. Le, Zinc oxide nanoparticle as a novel class of antifungal agents: current advances and future perspectives, J. Agric. Food Chem. 66 (2018) 11209–11220. Available from: https://doi.org/10.1021/acs.jafc.8b03210.

[107] T.B. Rawal, A. Ozcan, S.H. Liu, S.V. Pingali, O. Akbilgic, L. Tetard, et al., Interaction of zinc oxide nanoparticles with water: implications for catalytic activity, ACS Appl. Nano Mater. 2 (2019) 4257–4266. Available from: https://doi.org/10.1021/acsanm.9b00714.

[108] A.B. Chebotareva, G.G. Untila, T.N. Kost, S. Jorgensen, A.G. Ulyashin, ITO deposited by pyrosol for photovoltaic applications, Thin Solid Films 515 (2007) 8505–8510. Available from: https://doi.org/10.1016/j.tsf.2007.03.097.

[109] M. Tanaka, T. Kageyama, H. Sone, S. Yoshida, D. Okamoto, T. Watanabe, Synthesis of lithium metal oxide nanoparticles by induction thermal plasmas, Nanomaterials 6 (2016) 60. Available from: https://doi.org/10.3390/nano6040060.

[110] H. Sone, T. Kageyama, M. Tanaka, D. Okamoto, T. Watanabe, Induction thermal plasma synthesis of lithium oxide composite nanoparticles with a spinel structure, Jpn. J. Appl. Phys. 55 (2016) 07LE04. Available from: https://doi.org/10.7567/JJAP.55.07LE04.

[111] Y.S. Feng, S.M. Zhou, Y. Li, C.C. Li, L.D. Zhang, Synthesis and characterization of tin oxide nanoparticles dispersed in monolithic mesoporous silica, Solid State Sci. 5 (2003) 729–733. Available from: https://doi.org/10.1016/S1293-2558(03)00080-3.

[112] N. Verma, N. Kumar, Synthesis and biomedical applications of copper oxide nanoparticles: an expanding horizon, ACS Biomater. Sci. Eng. 5 (2019) 1170–1188. Available from: https://doi.org/10.1021/acsbiomaterials.8b01092.

[113] A.S. Teja, P.Y. Koh, Synthesis, properties, and applications of magnetic iron oxide nanoparticles, Prog. Cryst. Growth Charact. Mater. 55 (2009) 22–45. Available from: https://doi.org/10.1016/j.pcrysgrow.2008.08.003.

[114] S.R. Smith, R. Rafati, A. Sharifi Haddad, A. Cooper, H. Hamidi, Application of aluminium oxide nanoparticles to enhance rheological and filtration properties of water based muds at HPHT conditions, Colloids Surf. A: Physicochem. Eng. Asp. 537 (2018) 361–371. Available from: https://doi.org/10.1016/j.colsurfa.2017.10.050.

[115] A.J. Noori, F.A. Kareem, The effect of magnesium oxide nanoparticles on the antibacterial and antibiofilm properties of glass-ionomer cement, Heliyon 5 (2019) e02568. Available from: https://doi.org/10.1016/j.heliyon.2019.e02568.

[116] C. Liu, T.J. Hajagos, D. Chen, Y. Chen, D. Kishpaugh, Q. Pei, Efficient one-pot synthesis of colloidal zirconium oxide nanoparticles for high-refractive-index nanocomposites, ACS Appl. Mater. Interfaces 8 (2016) 4795–4802. Available from: https://doi.org/10.1021/acsami.6b00743.

[117] B.N. Rashmi, S.F. Harlapur, B. Avinash, C.R. Ravikumar, H.P. Nagaswarupa, M.R. Anil Kumar, et al., Facile green synthesis of silver oxide nanoparticles and their electrochemical, photocatalytic and biological studies, Inorg. Chem. Commun. 111 (2020) 107580. Available from: https://doi.org/10.1016/j.inoche.2019.107580.

[118] M.S. Chavali, M.P. Nikolova, Metal oxide nanoparticles and their applications in nanotechnology, SN Appl. Sci. 1 (2019) 607. Available from: https://doi.org/10.1007/s42452-019-0592-3.

[119] N. Musee, L. Lorenzen, C. Aldrich, New methodology for hazardous waste classification using fuzzy set theory: part I. Knowledge acquisition, J. Hazard. Mater. 154 (2008) 1040–1051.

[120] N. Musee, C. Aldrich, L. Lorenzen, New methodology for hazardous waste classification using fuzzy set theory: part II. Intelligent decision support system, J. Hazard. Mater. 157 (2008) 94–105.

[121] I. Resent, Risk of nanowastes, Inżynieria i OchrSrodowiska 19 (2016) 469–478.

[122] T. Tolaymat, A. El Badawy, R. Sequeira, A. Genaidy, A system-of-systems approach as a broad and integrated paradigm for sustainable engineered nanomaterials, Sci. Total Environ. 511 (2015) 595–607.

[123] N. Musee, Nanotechnology risk assessment from a waste management perspective: are the current tools adequate? Hum. Exp. Toxicol. 30 (2011) 820–835.

[124] S.A. Younis, E.M. El-Fawal, P. Serp, Nano-wastes and the environment: potential challenges and opportunities of nano-waste management paradigm for greener nanotechnologies, in: C.M. Hussain (Ed.), Handbook of Environmental Materials Management, Springer International Publishing, Cham, 2018, pp. 1–72. Available from: https://doi.org/10.1007/978-3-319-58538-3_53-1.

[125] E. Inshakova, O. Inshakov, World market for nanomaterials: structure and trends, MATEC Web Conf. 129 (2017) 02013. Available from: https://doi.org/10.1051/matecconf/201712902013.

[126] S. Kotsilkov, E. Ivanov, N.K. Vitanov, Release of graphene and carbon nanotubes from biodegradable poly (lactic acid) films during degradation and combustion: risk associated with the end-of-life of nanocomposite food packaging materials, Materials 11 (2018) 2346.

[127] F. Part, G. Zecha, T. Causon, E.-K. Sinner, M. Huber-Humer, Current limitations and challenges in nanowaste detection, characterisation and monitoring, Waste Manage. 43 (2015) 407–420. Available from: https://doi.org/10.1016/j.wasman.2015.05.035.

[128] J.C. Chow, J.G. Watson, N. Savage, C.J. Solomon, Y.-S. Cheng, P.H. McMurry, et al., Nanoparticles and the environment, J. Air Waste Manage. Assoc. 55 (2005) 1411–1417.

[129] J. Allan, S. Reed, J. Bartlett, et al., Comparison of methods used to treat nanowaste from research and manufacturing facilities, in: Australian Institute of Occupational Hygienists Annual Conference, Canberra, Australia, Citeseer, 2009, pp. 5–9.

[130] D. Soni, P.K. Naoghare, S. Saravanadevi, R.A. Pandey, Release, transport and toxicity of engineered nanoparticles, in: D. Whitacre (Ed.), Reviews of Environmental Contamination and

Toxicology (Continuation of Residue Reviews), vol. 234, Springer, Cham, 2015, pp. 1−47. Available from: https://doi.org/10.1007/978-3-319-10638-0_1.

[131] A. El-Henawy, I. El-Sheikh, A. Hassan, A. Madein, A. El-Sheikh, A. El-Yamany, et al., Response of cultivated broccoli and red cabbage crops to mineral, organic and nano-fertilizers, Environ. Biodivers. Soil. Security 2 (2018) 221−231.

[132] C.M. Rico, S. Majumdar, M. Duarte-Gardea, J.R. Peralta-Videa, J.L. Gardea-Torresdey, Interaction of nanoparticles with edible plants and their possible implications in the food chain, J. Agric. Food Chem. 59 (2011) 3485−3498.

[133] R.M. Mohamed, D.L. McKinney, W.M. Sigmund, Enhanced nanocatalysts, Mater. Sci. Eng. R Rep. 73 (2012) 1−13. Available from: https://doi.org/10.1016/j.mser.2011.09.001.

[134] H. Liu, X. Meng, T.D. Dao, L. Liu, P. Li, G. Zhao, et al., Light assisted CO_2 reduction with methane over SiO_2 encapsulated Ni nanocatalysts for boosted activity and stability, J. Mater. Chem. A 5 (2017) 10567−10573.

[135] N. Jiang, X. Zhou, Y.-F. Jiang, Z.-W. Zhao, L.-B. Ma, C.-C. Shen, et al., Oxygen deficient Pr 6 O 11 nanorod supported palladium nanoparticles: highly active nanocatalysts for styrene and 4-nitrophenol hydrogenation reactions, RSC Adv. 8 (2018) 17504−17510.

[136] X. Qu, R. Jiang, Q. Li, F. Zeng, X. Zheng, Z. Xu, et al., The hydrolysis of ammonia borane catalyzed by NiCoP/OPC-300 nanocatalysts: high selectivity and efficiency, and mechanism, Green. Chem. 21 (2019) 850−860.

[137] J. Maguhn, E. Karg, A. Kettrup, R. Zimmermann, On-line analysis of the size distribution of fine and ultrafine aerosol particles in flue and stack gas of a municipal waste incineration plant: effects of dynamic process control measures and emission reduction devices, Environ. Sci. Technol. 37 (2003) 4761−4770. Available from: https://doi.org/10.1021/es020227p.

[138] J. Mertens, H. Lepaumier, P. Rogiers, D. Desagher, L. Goossens, A. Duterque, et al., Fine and ultrafine particle number and size measurements from industrial combustion processes: primary emissions field data, Atmos. Pollut. Res. 11 (2020) 803−814. Available from: https://doi.org/10.1016/j.apr.2020.01.008.

[139] D.K. Ramesh, J.D. Kumar, S.H. Kumar, V. Namith, P.B. Jambagi, S. Sharath, Study on effects of alumina nanoparticles as additive with poultry litter biodiesel on performance, combustion and emission characteristic of diesel engine, Mater. Today Proc. 5 (2018) 1114−1120.

[140] S. Radhakrishnan, D.B. Munuswamy, Y. Devarajan, A. Mahalingam, Effect of nanoparticle on emission and performance characteristics of a diesel engine fueled with cashew nut shell biodiesel, Energy Sources A Recovery Util. Environ. Eff. 40 (2018) 2485−2493.

[141] M.-C.O. Chang, J.C. Chow, J.G. Watson, P.K. Hopke, S.-M. Yi, G.C. England, Measurement of ultrafine particle size distributions from coal-, oil-, and gas-fired stationary combustion sources, J. Air Waste Manage. Assoc. 54 (2004) 1494−1505. Available from: https://doi.org/10.1080/10473289.2004.10471010.

[142] J.S. Lighty, J.M. Veranth, A.F. Sarofim, Combustion aerosols: factors governing their size and composition and implications to human health, J. Air Waste Manage. Assoc. 50 (2000) 1565−1618.

[143] W. Li, P.K. Hopke, Initial size distributions and hygroscopicity of indoor combustion aerosol particles, Aerosol Sci. Technol. 19 (1993) 305−316.

[144] C.-S. Li, F.-T. Jenq, W.-H. Lin, Field characterization of submicron aerosols from indoor combustion sources, J. Aerosol. Sci. 23 (1992) 547−550.

[145] T. Glytsos, J. Ondráček, L. Džumbová, I. Kopanakis, M. Lazaridis, Characterization of particulate matter concentrations during controlled indoor activities, Atmos. Environ. 44 (2010) 1539−1549. Available from: https://doi.org/10.1016/j.atmosenv.2010.01.009.

[146] A.A. Roy, S.P. Baxla, T. Gupta, R. Bandyopadhyaya, S.N. Tripathi, Particles emitted from indoor combustion sources: size distribution measurement and chemical analysis, Inhal. Toxicol. 21 (2009) 837−848.

[147] M.D. Wright, A.P. Fews, P.A. Keitch, D.L. Henshaw, Small-ion and nano-aerosol production during candle burning: size distribution and concentration profile with time, Aerosol Sci. Technol. 41 (2007) 475−484. Available from: https://doi.org/10.1080/02786820701225812.

[148] J.H. Vincent, C.F. Clement, Ultrafine particles in workplace atmospheres, Ultrafine Particles in the Atmosphere, World Scientific, 2000, pp. 141−154.

[149] C. Beatrice, S.D. Iorio, C. Guido, P. Napolitano, Detailed characterization of particulate emissions of an automotive catalyzed DPF using actual regeneration strategies, Exp. Therm. Fluid Sci. 39 (2012) 45−53. Available from: https://doi.org/10.1016/j.expthermflusci.2012.01.005.

[150] A. La Rocca, F. Bonatesta, M.W. Fay, F. Campanella, Characterisation of soot in oil from a gasoline direct injection engine using transmission electron microscopy, Tribol. Int. 86 (2015) 77−84. Available from: https://doi.org/10.1016/j.triboint.2015.01.025.

[151] C.L. Myung, A. Ko, S. Park, Review on characterization of nano-particle emissions and PM morphology from internal combustion engines: part 1, Int. J. Automot. Technol. 15 (2014) 203−218.

[152] A. Thiruvengadam, M.C. Besch, D.K. Carder, A. Oshinuga, M. Gautam, Influence of real-world engine load conditions on nanoparticle emissions from a DPF and SCR equipped heavy-duty diesel engine, Environ. Sci. Technol. 46 (2012) 1907−1913. Available from: https://doi.org/10.1021/es203079n.

[153] United Nations Economic Commission for Europe's (UNECE), Group des, Rapporteurs de Pollution et Energie (GRPE), Report of the GRPE Particle Measurement Programme (PMP) Government1 Sponsored Work Programmes July, 2003. https://unece.org/fileadmin/DAM/trans/doc/2003/wp29grpe/TRANS-WP29-GRPE-specinf01e.pdf.

[154] T. Maier, M. Härtl, E. Jacob, G. Wachtmeister, Dimethyl carbonate (DMC) and methyl formate (MeFo): emission characteristics of novel, clean and potentially CO_2-neutral fuels including PMP and sub-23 nm nanoparticle-emission characteristics on a spark-ignition DI-engine, Fuel 256 (2019) 115925. Available from: https://doi.org/10.1016/j.fuel.2019.115925.

[155] A. Mayer, J. Czerwinski, M. Kasper, A. Ulrich, J.J. Mooney, Metal Oxide Particle Emissions from Diesel and Petrol Engines, SAE Technical Paper 2012-01-0841, 2012. DOI: https://doi.org/10.4271/2012-01-0841.

[156] S. Artelt, O. Creutzenberg, H. Kock, K. Levsen, D. Nachtigall, U. Heinrich, et al., Bioavailability of fine dispersed platinum as emitted from automotive catalytic converters: a model study, Sci. Total. Environ. 228 (1999) 219−242.

[157] R. Vijay, R.J. Hendershot, S.M. Rivera-Jiménez, W.B. Rogers, B.J. Feist, C.M. Snively, et al., Noble metal free NO_x storage catalysts using cobalt discovered via high-throughput experimentation, Catal. Commun. 6 (2005) 167−171.

[158] M. Raza, L. Chen, F. Leach, S. Ding, A review of particulate number (PN) emissions from gasoline direct injection (GDI) engines and their control techniques, Energies 11 (2018) 1417. Available from: https://doi.org/10.3390/en11061417.

[159] J. Aitken, On the formation of small clear spaces in dusty air, Proc. R. Soc. Edinburg 12 (1884) 440−448.

[160] J.M. Krasnomowitz, M.J. Apsokardu, C.M. Stangl, L. Tiszenkel, Q. Ouyang, S. Lee, et al., Growth of Aitken mode ammonium sulfate particles by α-pinene ozonolysis, Aerosol. Sci. Technol. 53 (2019) 406−418.

[161] S.R. Noble, J.G. Hudson, Effects of continental clouds on surface Aitken and accumulation modes, J. Geophys. Res. Atmos. 124 (2019) 5479−5502.

[162] P. Du, H. Gui, J. Zhang, J. Liu, T. Yu, J. Wang, et al., Number size distribution of atmospheric particles in a suburban Beijing in the summer and winter of 2015, Atmos. Environ. 186 (2018) 32−44.

[163] W. Junkermann, J.M. Hacker, Ultrafine particles in the lower troposphere: major sources, invisible plumes, and meteorological transport processes, Bull. Am. Meteorol. Soc. 99 (2018) 2587−2602.

[164] B.P. Isaacoff, K.A. Brown, Progress in Top-Down Control of Bottom-Up Assembly, ACS Publications, 2017.

[165] J. Saleem, U.B. Shahid, G. McKay, Environmental nanotechnology, in: C.M. Hussain (Ed.), Handbook of Environmental Materials Management, Springer International Publishing, Cham, 2018, pp. 1−32. Available from: https://doi.org/10.1007/978-3-319-58538-3_94-1.

[166] R.J. Aitken, M.Q. Chaudhry, A.B.A. Boxall, M. Hull, Manufacture and use of nanomaterials: current status in the UK and global trends, Occup. Med. 56 (2006) 300−306.

[167] N.C. Mueller, B. Nowack, Exposure modeling of engineered nanoparticles in the environment, Environ. Sci. Technol. 42 (2008) 4447−4453.

CHAPTER 4

General regulations for safe handling of manufactured nanomaterials

Maria Batool[1], Muhammad Nadeem Zafar[1] and Muhammad Faizan Nazar[2]
[1]Department of Chemistry, Faculty of Science, University of Gujrat, Gujrat, Pakistan
[2]Department of Chemistry, University of Education, Lahore, Pakistan

4.1 Introduction

Nanoscience is an emerging technology dealing with particles measuring $1-100$ nm in diameter at the atomic, molecular, and macromolecular levels. However, in the biomedical sciences, owing to the presence of particles in some cases (e.g., the vascular space around a tumor), the size range is expanded to >100 nm [1]. Novel systems and devices have been created using nanomaterials, owing to their ability to be modified at the atomic level [2]. Nanotechnology provides entirely different materials at nano-scale in terms of properties and activity in comparison to their respective bulk materials. At nano-scale, particles may be confined to small structures or distributed over a large surface area as compare to bulk materials; this range enables nanoparticles to offer diverse characteristics [3]. As nanotechnology is a revolutionary field of this era, it is being utilized in almost every sector of life, such as medical, consumer products, and industries. This automatically means that it will lead to unwanted problems. A major problem is the toxicity of some nanoparticles and their effects on workers' and consumers' health. Nanomaterials have been playing major roles in the eradication, detection, and prevention of different pollutants along with major developments in designing of safe products with less toxicity. However, there is need of through study of the impacts of nanomaterials on human health and the environment to get answers to important questions. Currently, huge amounts of research are being conducted for investigation of nanomaterials in almost every field of science, so workers' health is at more risk from nanomaterials than is the case for consumers, who are less exposed to nanomaterials in comparison. Most investigation regarding exposure has been related to ultrafine particles, nonengineered nanoparticles, and soot. There is much less data on exposure through engineered nanomaterials, owing to inadequate analytical methods for engineered nanoparticles [4−6]. So there is a high necessity to determine the exposures that cause serious health problems in the workplace where nanomaterials are

Nanomaterials Recycling
DOI: https://doi.org/10.1016/B978-0-323-90982-2.00004-4

© 2022 Elsevier Inc.
All rights reserved.

engineered or handled. The measurement and assessment of exposure to nanomaterials [7,8] were started to address the following questions:
1. What methods should be adopted for identification of nanomaterials in air?
2. How can nanomaterials be quantified in air?
3. What are possible ways to differentiate between natural, engineered, and accidentally produced nanomaterials?

The workplace was selected to get answers to these questions as in this case the type and amount of nanomaterials are clearly defined along with appropriate working conditions to investigate. Since 2004, greater focus has been shifted toward consumer exposure and its possible effects [9]. Consumers interact with nanomaterials during their production, processing, use, and end of lifecycle. Consumer exposure assessment is somewhat trickier than workplace exposure assessment, owing to less availability of important information. Exposure via the environment is the least investigated type of exposure with regard to measurement and assessment [10]. This is due to a lack of knowledge and to the difficulty of identification as nanomaterials get agglomerated in natural water or soil and thus cannot be easily identified. Thus measurement methods and assessment strategies must be selected carefully to account for its implications in practical setup.

4.1.1 Precautionary principles

In 1992 during a conference on the environment and development, precautionary principles were formulated to protect the environment. World leaders were directed to enforce these principles in the places where there is serious threat, using cost-effective measurement methods [11]. While this principle has primarily been used internationally with a focus on environmental health issues, other groups are adopting this philosophy to protect the health of workers. The American Public Health Association approved a resolution for enforcing precautionary principles for dangerous chemicals with known toxicity and limiting exposure to these chemicals. The European Commission issued a commission letter on the precautionary principle, which eventually led to European Union (EU) policy [12]. Different types of measures have been taken to ensure the safe production of nanomaterials, as shown in Fig. 4.1.

Figure 4.1 Different types of measures for safe production of nanomaterials.

4.2 Precautionary measures

4.2.1 Technical measures

Technical measures include the use of F-7 filters (EN 779—European Standard for Ventilation Filters) with upto 80%—90% average efficacy for 400-nm nanoparticles.

4.2.2 Organizational measures

Organizational measures are quite similar in almost all type of laboratories. There should be a nanotechnology safety officer in each laboratory. Entry of pregnant women should be restricted strictly at workplaces where there is nanomaterial exposure. Health and safety specialists should perform lab safety audits. Permanent laboratory members should be subjected to periodic medical inspections, especially in regard to the respiratory tract and cardiovascular system.

4.2.3 Personal measures

These measures ensure safe exposure of workers through the use of protective equipment according to different hazard levels. Workers who have been exposed to nanomaterials for >2 hours should use masks with powered air respirators, while workers who have been exposed for shorter periods should use a P3 (EN 143) or FFP3 (EN 149)/P-100 (USA NIOSH) filter/filtering mask. Protective gloves and glasses are compulsory for anyone who is exposed to nanomaterials while working.

External cleaning staff members with protective equipment may work in a chemical laboratory when nanomaterials are less hazardous. Lab employees themselves under the supervision of a lab superintendent or safety officer should clean where highly toxic nanomaterials are used.

4.3 Health hazards

The state of nanoparticles controls their tendency to affect the health of workers or consumers. Nanoparticles fused with solid matrix have been observed to be less hazardous to health if the material is not grounded or cut. Nanoparticles in form of slurry can affect health by forming a nanoaerosol that is aroused during mixing, stirring, or sonication. Loose powdered nanoparticles with high mobility can cause major health problems associated with the respiratory tract or circulatory system during their handling. The main routes through which nanoparticles can move into the human body are as follows (also depicted in Fig. 4.2):

1. Inhalation into lungs
2. Absorption through the skin
3. Ingestion through the gastrointestinal system

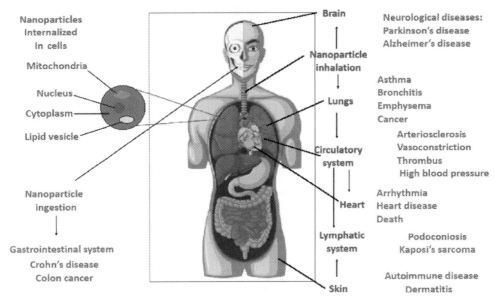

Figure 4.2 Distribution of nanoparticles in different body parts and associated diseases.

Exposure through the inhalation pathway is of especially high concern, owing to its effect on occupational health. Dermal exposure can be lessened if skin health is maintained, as the skin is a better barrier against nanoparticles as compared to the respiratory tract unless particles size is <5–10 nm. Dermal exposure can be harmful when there is a wound in the skin or strong mechanical stress is applied. Oral uptake of nanoparticles can be avoided by good personal hygiene, such as washing hands regularly after every work assignment and before eating food. Small particles diffuse more deeply into the system and are difficult to remove, which can lead to many health problems, such as liver cancer, Crohn's disease, and respiratory tract infection, which can cause bronchitis, asthma, lung cancer, and emphysema. Neural disease associated with exposure to nanomaterials are Parkinson's disease and Alzheimer's disease [13].

High aspect ratio and water solubility are deciding factors for level of toxicity and effects of nanomaterials on health. Less soluble nanomaterials present more risk as compared to more soluble nanomaterials. There are some nanomaterials that may be insoluble in water but soluble in biological media. Table 4.1 represents the classification of nanomaterials on the basis of their solubility and toxicity [14].

4.3.1 Exposure routes

4.3.1.1 Inhalation

Inhalation is one of the more lethal routes for exposure to nanoparticles when the particles are in the air, especially when they are in the breathable size range. This is a very

Table 4.1 Different types of nanoparticles with their hazard level.

Risk level	Properties of nanoparticles	Examples
High	Poorly soluble or insoluble	Carbon nanotubes
Medium-high	Poorly soluble or insoluble with specific toxicity	Ag nanoparticles, Au nanoparticles, ZnO nanoparticles
Medium-low	Poorly soluble with no specific toxicity	Carbon black, titanium dioxide
Low	Soluble	NaCl nanoparticles, lipid nanoparticles, flour nanoparticles, amorphous silica

dangerous pathway, as nanoparticles not only diffuse in the lungs but also can be transported to other organs, thus leading to bioaccumulation. This is possible because the capillaries and alveoli have thin walls (0.5 μm). Microparticles with size >2.5 μm reside mostly in the upper part of the respiratory tract and are discharged into mucous lining of the bronchi or coughed out. Microparticles with size <2.5 μm enter the alveolar region, and only alveolar macrophages can remove them. Nanoparticles with size <100 nm behave more like gas than like solid particles and can penetrate the pulmonary system. After entering the bloodstream, these nanoparticles accumulate in body organs and engulfed by cells, or damage the DNA thus leading to cancer.

4.3.1.2 Dermal exposure

Damaged skin is prone to nanoparticle penetration through skin pores and hair follicles. After penetration, nanoparticles pass into the bloodstream, followed by their accumulation in organs. Nanoparticles damage the tissues through enhanced oxidative stress and inflammation of cytokines.

4.3.1.3 Ingestion

Exposure to nanoparticles through ingestion includes accidental oral consumption followed by passage through the digestive tract. There is a higher possibility of their absorption in the bloodstream as compared to skin exposure, owing to the increased surface area, as only smaller nanoparticles are absorbed in this way, although their extent of absorption is unknown. Very little data have been gathered for nanoparticle exposure through ingestion, so their adverse effects on health are not understood completely. However, nanoparticles are found accumulated in the spleen, kidney, and liver.

4.4 Fire and explosion hazards

Nanomaterials are more likely to explode as compared to bulk materials when present in high concentration in the atmosphere in form of an aerosol. This happens especially in the presence of carbonaceous nanomaterials, metal-containing nanomaterials, igniting materials, or oxygen. Although there is less risk of explosion in research laboratories, lab workers should still avoid formation of dense aerosol of explosive nanoparticles. Even inert materials developed by decreasing their particle size into the nanometer range show a higher rate of combustion. For example, aluminum powder with particles sizes >80 nm is found to be inert as compared to aluminum particles with sizes smaller than 80 nm.

4.5 Environmental hazards

There has been very little research on the effects of nanomaterials on living organisms. This is a result of the limited analytical techniques for measurement, thus leading to a lack of evidence about nanomaterials' existence in the environment. So an important item on the agenda is to detect toxicity levels of nanomaterials in the environment. The hazard level will increase if someone is exposed to nanomaterial. Consumption of a living organism that has been exposed to nanomaterial can also cause harm. Even plants can be added into the food chain as the source of nanomaterial if the plants have been exposed to nanomaterial. Plants have been seen exposed to nanomaterials through several pathways, such as landfills, fertilizers, and wear from consumer products. Although most of the release of nanomaterials into the environment is unintentional, there is also intentional release into the environment.

4.6 Risk assessment and safety precautions for nanomaterial use

Exposure of nanomaterial is likely upon the release of nanomaterial. There is a high need to investigate engineered nanoparticles to check their toxicity. There should be investigation in relation to their potential exposure to reduce the risk. Release of nanomaterial occurs through daily workplace activities and through unintentional spills. In these cases the exposure level should be well defined. These workplace activities include material packing, unpacking, shipment, weighing, cleaning, storage, and waste management.

4.6.1 Risk evaluation

Risk evaluation is the systematic procedure of investigating the risk level of an activity or object. A safe approach during handling of engineered nanomaterial is highly required. This is accomplished through identification, assessment, and controlling of risk followed by evaluation of the efficacy of control measures, as highlighted in Fig. 4.3.

Risk evaluation can be based on the type of nanoparticles and the safety data sheet. The data sheet is likely to lack toxicological data for nanomaterials. However, until proven otherwise, materials in nanoform will be considered toxic if their bulk form has been considered toxic or carcinogenic.

4.6.2 Controlling exposure

Nanomaterial exposure can be controlled by several methods if the material has been proven hazardous, as illustrated in Fig. 4.4.

4.6.2.1 Elimination or substitution

Hazardous material can sometimes be changed by elimination and substitution reactions. However, the feasibility of this method is compromised by the toxicity of these materials; thus handling them will not be free from risk. In this case, risk factors will be decreased by less toxic solvents and by using a liquid suspension in the place of loose powder.

4.6.2.2 Engineering controls

Chemical fume hoods are used to protect workers from toxic fumes generated during lab work. Nanomaterials have been tested in various types of fume hoods, such as

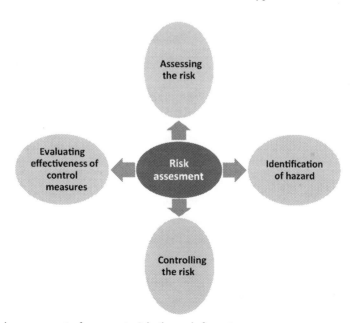

Figure 4.3 Risk assessment of nanomaterials through four steps.

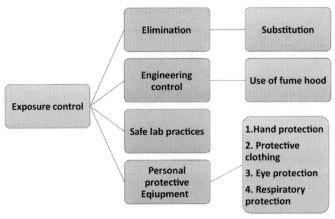

Figure 4.4 Exposure control of nanomaterials by different methods.

constant air volume and variable air volume, during their handling. Results suggest that even when fume hoods are used, airborne nanoparticles are detected in the laboratory environment. There are many parameters that control release of nanoparticles. These results led to the formulation of the following safety recommendations in relation to the use of fume hoods to get the maximum protected environment:

1. Escape of nanoparticles under a fume hood can be prevented only if it is operated with face velocities in the range of 80–120 ft./min.
2. While nanoparticles are being manipulated, sash heights should remain as low as possible.
3. A variable air volume hood has been proved to work better during the handling of dry samples as compared to conventional fume hoods, owing to the maintenance of hood face velocity in the desired range irrespective of sash height.
4. Air currents in the vicinity of the fume hood is also considered to be responsible for nanomaterial escape. Therefore there should be efforts to reduce or eliminate air currents.
5. The handling method in regard to the nanoparticles should be the least energetic one possible. The lowers amount of nanoparticles should be used inside fume hoods during operations.
6. The exhaust system in laboratories should be equipped with air filters such as high efficiency particulate air and ultralow penetration air filters. Examples of exhaust systems that are equipped with such filters include nano safety cabinets and biosafety cabinets.
7. An alternative to using a fume hood is the use of a glove box or an enclosed system.

4.6.2.3 Safe laboratory work practices

Safe laboratory practices contribute strongly in reducing exposure to nanomaterial. Some important safety practices are the following:

1. Students, staff members, and employees should be trained to deal with nanomaterials if they are at risk of exposure to nanomaterials.
2. Standard operating procedures should be developed, modified as needed, and applied while working with specific nanoparticles.
3. Unauthorized personnel should be forbidden to enter the working area.
4. Edibles should be forbidden in the working area of the lab.
5. Frequent hand washing should be encouraged.
6. Working of nanomaterials should be specified with warning signs or marking tape to avoid exposure to nanomaterials.

In addition, clean room mats or tacky mats can be used to reduce the spread of nanoparticles. Tacky mats help to trap impurities for locations that require stringent dust and dirt control. The tacky surface removes particles from shoes, therefore decreasing the spread of nanoparticles to nanoparticle-free areas. Antistatic devices can also be used in handling dry powders to avoid spreading airborne particles.

4.6.2.4 Personal protective equipment

Even though personal protective equipment (PPE) should never be considered the first line of defense against contaminants, nevertheless they must always be used to minimize users' exposure to nanoparticles. Basic PPE for use in working with nanoparticles should include the following.

4.6.2.4.1 Hand protection

Gloves should be considered mandatory during lab work, especially during handling of nanoparticles and nanoparticles in suspension. There are various types of gloves, such as latex, nitrile, and vinyl. Most gloves are used to avoid nanoparticles in powder form, while nitrile gloves have been proven best for handling nanoparticles in suspension. Gloves can be used more than once if they are in good condition. However, in case of excessive use, double gloves are suggested, as gloves material can wear out, thus enabling invasion of nanoparticles through the gloves. It is best to use disposable gloves because they can be replaced whenever there is sign of contamination or invasion. The same gloves should not be used for >2 hours.

4.6.2.4.2 Eye protection

Eyes are a very sensitive part of the body, so safety glasses are obligatory in most university laboratories. In the presence of a nano cabinet or glove box, safety glasses can be considered enough for minimal eye protection. If there is no engineering control, then safety goggles give increased eye protection as compared to safety glasses. Safety goggles are used in the presence of large amount of nanoparticles, a suspension of nanoparticles in solvents or aerosol, and during heavy use of nanoparticles.

4.6.2.4.3 Protective clothing

Nanoparticles can penetrate regular fabric, so regular lab coats are not enough to protect against nanoparticles exposure. So disposable material such as Tyvek is used in lab coats, sleeves, and shoe covers.

4.6.2.4.4 Respiratory protection

If nano safety cabinets are not available in the lab, then respiratory protection should be used, especially in the case of dry powdered nanoparticles. Different types of masks are available for respiratory protection against nanoparticles. However, there is a restriction in use of masks if there is less availability of oxygen. Moreover, regular masks cannot filter dust and nanoparticles, so they are not suitable to use in the lab if dealing with exposure to nanomaterials.

4.6.2.4.4.1 Filtering facepiece respirators There are many types of disposable respiratory masks, including filtering facepiece respirators (FFRs). According to National Institute for Occupational Safety and Health (NIOSH) certification, FFRs are used to protect against dust and aerosols.

FFRs do not act as filters but rather seize nanoparticles through different filtering mechanisms, such as impaction, interception, dispersion, and electrostatic attraction (in the presence of electric filter). It is an irreversible process; when particles are trapped inside FFRs, they cannot be removed. NIOSH, the Occupational Health and Safety Administration, and Institut de Recherche Robert-Sauvé en Santé et en Sécurité du Travail (Occupational Health and Safety Research Institute) have performed different tests on FFRs to show that N95 masks met the NIOSH filtering standards, with penetration <5% for particles <100 nm, and therefore provide suitable protection against nanoparticles. However, their effectiveness of filtration can get compromised in the presence of humidity and by breathing rate.

Disposable FFRs have the issue of providing compromised sealing around the nose and mouth. Worker carelessness, such as putting the mask on top of the head, can decrease the effectiveness of FFRs, and a loosely fitted mask can allow entrance to nanoparticles. So the use of FFRs for working with fine and high amount of nanoparticles is not recommended.

4.6.2.4.4.2 Half- or full-face respirators Half- or full-face respirators consist of a silicone mask that can meet most of the drawbacks of FFRs, as they can be held in place around the mouth and nose with the help of straps. Respirators are available with two replaceable filter cartridges. Although the filter with these elements becomes very heavy as compared to disposable FFRs, it comes with an enhanced level of protection for the user. Fit testing for any half- or full-face respirator is a must. In exposure to hazardous nanoparticles in the absence of proper engineering controls, the use of a half-face respirator is recommended because of its better seal. When a half-face respirator is equipped with P100 filter cartridges, it provides ultraprotection against nanoparticles with sizes down to 2.5 nm.

4.7 Storage, waste handling and spills

4.7.1 Storage

It is pertinent to store nano-sized substances in sealed containers with proper ventilation. All containers should be marked with identification in the lab. A lab cabinet should be reserved for nanoparticles only. If any nanoparticles are brought with a plastic bag covering, this bag should be placed into a sealable container. Nanoparticles must not come into contact with any acids, oxidizers, or other metals.

4.7.2 Waste handling

Nanoparticles are considered toxic, so they should be disposed of cautiously. Although there are no proper guidelines for waste eradication in relation to nanoparticles, disposal of waste with nanoparticles should not be done in the regular manner, and they should not be flushed into the drain. Some environmental health and safety chemical waste guidelines for the disposal of nanoparticles are as follows:

1. Any polluted objects, such as gloves and disposable lab coats, should be disposed of as solid chemical waste.
2. Solid nano-sized substances, including those in powder form, should be disposed of as chemical waste.
3. Nano-sized substances in solution or suspension should be disposed of as liquid chemical waste.

There should be proper labeling of waste containers to mention the presence of nanoparticles. Sealing of containers that contain nanoparticles should not be compromised, especially at the time of their pickup. EHS at hazardouswaste@concordia.ca provides more information for the handling of chemical waste.

4.7.3 Spills

If nanoparticles have the tendency to become airborne as a result of spillage, there will be a high chance of exposure through inhalation, especially if the particles are in powder form rather than the solution form. Spillage of nanoparticles on nonporous surfaces such as linoleum, stainless steel, or hardwood flooring can be washed by qualified lab personnel. However, spillage on a porous mat can pose difficulty in cleanup of the surface, thus leading to disposal of the mat along with the nanoparticles. If the spillage is small and the lab user is trained for this situation, then cleanup will happen through the following steps:

1. Solid nanoparticles should be wiped up through wet wiping.

2. Dry sweeping should be discouraged. If it happens through dry vacuuming, then the vacuum cleaner should be equipped with a HEPA filter.
3. For liquid suspensions, an adsorbent wipe should be used.

The procedure includes cleaning up the spill as early as possible. Spread of spilled nanomaterials should be prevented, and other people should be restricted from entering the area of spillage. During cleanup, the worker should wear disposable gloves, shoe covers, and goggles. Nanoparticles collected through cleanup should be labeled and sealed into a chemical waste container.

However, in the case of a large spillage or a spillage of highly hazardous nanomaterial, cleanup should be handled differently from cleanup of a small spillage of nanoparticles. In the case of emergency spillage, which is difficult for lab users to handle properly, a lab user should contact security. The procedure for handling a hazardous emergency spill includes advice and warning to coworkers. Contaminated and injured people should be rescued first, and then the area should be evacuated. All individuals should avoid touching any hazardous material at the spillage site. Restrict Entry to the site should be restricted, and security should be provided with all the information about the incident: location, name of spilled material, amount of spilled material, possible health problems, and required precautions. A material safety sheet or safety data sheet should be made available.

4.8 Regulations

Toxicological hazards have been studied in workplace and occupational areas. The study focus has been mostly on fibrous particles with sizes ≤ 100 nm. So far, these studies have shown a lack of good data on assessment of nanomaterials and their effects on human health. Health, safety, and environment organizations have critically studied the existing data on the physicochemical and toxicological hazards of nanomaterials in the workplace and the occupational exposure. Although the available data are limited, they suggest that a cautious approach is required for handling nanomaterials, especially for researchers and workers who has been exposed to nanoparticles for a longer time [15]. The Health and Safety at Work Act was passed in the United Kingdom in 1974 to regulate the hazards associated with industrial chemicals. This legislation was later standardized across the EU.

The European Regulation on Registration, Evaluation, Authorisation and Restriction of Chemicals (REACH) came into force in 2007, replacing earlier EU regulations with regard to chemicals [16].

The current working regulatory framework and the roles involved are summarized below [17−20].

4.8.1 Legislation for suppliers

Under the EU directed policy for chemicals in regard of supply under hazard information and packaging for supply regulations (CHIP) [21], the supplier should be responsible enough to deliver information about any physiochemical and toxicological hazards of chemicals to consumers. This delivery is done through EU standardized safety data sheet classification and labeling. In appreciation of the scarcity of reliable hazard information for many industrial chemicals, the 7th modification to the EU dangerous substances directive was applied in the United Kingdom as the notification of new substances (NONS) regulations, which involved homogeneous testing of hazardous properties of industrial chemicals that are new to the market [22]. There is also accountability for suppliers to report the risks and their management. Since the 1990s, EU has been trying to gain and advance the hazard information available for older industrial chemicals and to measure the risks to health and safety via the existing substances regulation (ESR) [23].

EU directive 98/8/EC direct policy for biocidal supply led to the formation of a regulated market through cooperative authorization and evaluation of risk. This directive was implemented and named as biocidal products regulation (BRP). These regulations are expected to replace the currently implemented UK scheme for nonagriculture pesticides through pesticide regulation. This new scheme will be enforced by examination of substances; until it goes into effect, the existing scheme will be used. Under the BRP scheme in the EU, active substances in industries will be evaluated in regard to their hazardous characteristics, classification, tagging and risk management. It was recomended that member states and a risk assessment team will perform proper evaluation of all the recommended procedures but this target cannot be met.

4.8.2 Legislation for recipients and users of chemicals

Consumers and suppliers of chemicals are well aware of their situation in relation to the hazardous effects of chemicals and their overall exposure. The supplier's knowledge is used for assessment and management of exposure level to hazardous chemicals. This will help in the reduction or elimination of health and safety risk. These procedures relate to carcinogens and the EU chemical agents directive for legislation formulation in the dangerous substances and explosive atmospheres regulations (DSEAR) and substances hazardous to health (COSHH) regulations [24].

4.8.3 Legislation for regulatory authority

Health, safety, and environment organizations control these regulations and are involved in negotiations, agreements, and disagreements about these regulations. Development of a regulatory framework is also performed by these organizations. They are involved in specification and classification of substances and aim to limit exposure to airborne particles in the workplace. In the United Kingdom, the Health and Safety Executive (HSE) enforces the regulations of hazardous material under control of major accident hazards (COMAH) regulations [25]. COMAH evaluates risk levels and the effects on the surrounding population. This deals with explosives in the environment and the resulting release of toxins in accidental situations.

Duties regulation is done by documented rules known as approved codes of practice and approved guidance. These documented rules must be followed by duty holders; the duty holders will be questioned in case of any deviation. Additional guidance is also provided to make implementation easier or to make any necessary changes. Effective implementation of regulations is possible if the characteristics of the substance under observation are known. However, many substances have been commercialized and used without complete knowledge of their characteristics. So NONS regulation requires the formulation of data sets to provide the characteristics of new substances before their commercialization into the market. So the most important job is to characterize a material as new or not so as to decide on the implementation of NONS.

In 1970 the concept of a new substance was defined to implement regulation of NONS successfully. Moreover, a data set of already supplied chemicals in the EU was prepared. This data set is named the European inventory of existing commercial chemical substances (EINECS) [26]. The data set was prepared through convention naming and specific criteria. Thus substances in this data set were considered to be existing substances, and any new entry after this was considered liable to notification before introduction into market except foods, foods additives, pesticides and pharmaceuticals. It is the suppliers' responsibility to determine whether their substance is on the EINECS list or not. When a substance is difficult to identify, there will be also difficulty in being spotting on the EINECS list.

In this case the competent authority (CA) in the relevant member state is consulted. In the United Kingdom, HSE and the environmental agency (EA) act as CA. Many problems have been raised in the enforcement of NONS in many member states with recorded decisions regarding throughout the process in the manual of decisions. This recorded data, which are without legal binding, help in guidance and legislation with regard to new substances.

Nanomaterials are fabricated by two types of methods: top-down and bottom-up. In the top-down method, grinding down from bulk material to nanomaterial forms nanomaterials. Top-down nanomaterials face fewer restrictions as compared to the bottom-up method in acquiring notification under NONS. To get notified under NONS is no easy thing; it requires a structured procedure.

There is a need to establish a relationship between particle size and shape. Regulation formulation and implementation are possible only if the physical properties of particles, especially size in different dimensions, are known. Commonly used granulometric tests for particle size determination are not enough. Most of time, electron microscopy—scanning electron microscopy or transmission electron microscopy—is used. Thus information about physical and chemical features, toxicology, and ecotoxicology is required for notification of nanomaterials through standard test methods. HSE has many times recommended timing and amount of testing that are different in comparison to standard NONS regulations. For example, it will be more useful to have early assessment of absorption of nanomaterials across the respiratory tract, skin, and gastrointestinal tract. These early assessments are not included in the standard NONS setup, but if they are useful, they should be recommended. NONS guidance recommends toxicological testing through the most relevant route of exposure. So NONS recommends exposure through inhalation route for toxicity testing. In contrast to NONS, many tests are done through the oral route (less common) for repeated exposure to toxicity.

4.8.3.1 Exposure assessment in NONS

Possible exposure to nanomaterials should be assessed properly to follow regulations effectively. Most assessment is qualitative through semiquantitative modeling using the estimation and assessment of substance exposure model. This model's ability to assess is regulated with a margin of further improvement when more data on the properties of the nanomaterial are made available. A CA can be used for risk assessment if necessary. A CA under NONS regulation can request more information to evaluate risk properly to create safety for the environment and human health.

4.8.4 Chemicals (hazard information and packaging for supply) regulations

The CHIP Regulations made supplier responsible for the identification and categorization hazards of materials through EU agreed rules. The CHIP regulations give data in relation to hazards and required precautionary steps for users, which is done by a safety data sheet and labeling of the products.

CHIP regulations are applied to the majority of substance and mixtures of substances. These regulations do not apply to cosmetics and medicines, which are assessed through their own legislation. Issues of materials for classification and their definition under different legislation are covered by CHIP. For example, material is determined either to be included in EINECS or not. Under CHIP, there is need of packaging and labeling for materials that, after their exposure to consumers, cause the release of toxic substances.

4.8.4.1 Safety data sheet requirements

Hazardous materials require safety data sheets with specific patterns under CHIP regulations. Safety data sheets come with a lot of concerns, as there is no guarantee of relevance of the data used for the data sheet. Interpretation of data sheets by users is also important issue. It has been observed that safety data sheets for carbon nanotubes included no explanation of the relevance of data regarding hazard information for respirable graphite.

4.8.4.2 Classification and labeling of an individual substance

Nanomaterial classification also depends upon regulations such as NONS and BPR used for examining these materials. Thus this classification is in agreement with EU level and is endorsed into Annex I of the dangerous substances directive. Annex I is a UK-approved supply list that includes classification and labeling for existing substances. This list was approved after going through n intense procedure of EU-agreed rules. Suppliers are restricted to using this classification if the substance is in the Annex I list. If a substance in the nano-size range has not gone through the EU system, the supplier is considered responsible for gathering all the data in relation to that substance and classifying it under the available guidance. This procedure is referred to as self-classification. Regulatory authorities have observed a serious lack of data on hazards of substance in the nano-size range. This absence of data on hazard levels is considered to be an absence of hazard in relation to these nanomaterials, eventually leading to no classification or labeling of these substances. In this case, read-across policy is used, in which suppliers of these nano-sized substances seek data in relation to their structure at micro-size and predict their toxicity. Even regulatory authorities in EU use this approach until appropriate data are made available. This approach should not be used extensively, as the properties of nano-sized material can be very different from those of micro-sized material. These issues demonstrate the need for proper guidance.

4.9 Workplace risk management

4.9.1 Control of substances hazardous to health regulations

The COSHH regulations restrict workers to avoid exposure to nanomaterials if possible or, if not possible, they control the potential level of risk due to exposure in the working area. The pattern of COSHH requires assessment of the hazard level, amount of exposure, and level of risk imposed on the worker. These steps led to a possible strategy for risk management and specification of measurement tools for adequate control. COSHH establishment happened in 2004 for the control of exposure to nanomaterials under the light of its regulatory principles. The following criteria are designed in the COSHH regulations to consider a material hazardous:

1. The material has been listed with the classification of toxic, very toxic, harmful, corrosive, or irritant.
2. The material is respirable or inhalable with a weighted average concentration greater $4-10$ mg/m^3, respectively, at time period of 8 hours.
3. An occupational exposure limit to the material has been approved.
4. There is the possibility of risk to human health in the presence of certain material.
5. The material can be hazardous if it is supplied or produced at the working area.

4.9.1.1 Assessment of hazards and exposure

Assessment of risk in relation to exposure to nanomaterials is made mandatory by COSHH regulation through already available knowledge of hazards and condition of exposure. In COSHH there are certain elements that are necessary for risk assessment: hazard level, exposure level, conditions and type of work, results after observation of exposure, and results of any health surveillance. Thus proper assessment requires a considerable amount of information. However, there is currently a major shortage of information to accomplish proper assessment. There is lack of information on toxicological hazards for nanoparticles, appropriate exposure measurement tools, methods for characterization, and an appropriate unit for dosage during investigation of hazard and exposure levels.

Without this information, carrying out a complete COSHH risk assessment is very problematic. Moreover, implementation of appropriate health surveillance is difficult if there is less available data on hazard. Assessment in COSHH is done to show safety representatives and enforcing authorities the validity of steps taken toward measurement to avoid or control exposure of workers to nanomaterials and critical consideration of all the important factors required for risk management. Employers mostly rely on safety data sheets to accomplish their assessments. Risk assessment cannot be carried out until there is availability of knowledge on toxicological hazards, a unit to be used for dosage of nanomaterials, and level of exposure.

In COSHH regulation, the definition of inhalable material is the definition for further appropriate treatment of dust particles. Thus it would be foolish to consider exposure to respirable dust at or below 4 mg/m^3 (at an average exposure of 8 hours) to be safe, as it is authentic knowledge to rely on.

4.9.1.2 Prevention or control of exposure

COSHH regulation wants surety from employers in regard to exposure of employees to nanomaterials. Exposure to hazardous materials should be either avoided or controlled adequately. COSHH regulations provide a list of measurement tools with priority order to use for suitable conditions. Carcinogens and mutagens are discussed separately for controlling exposures to these substances. COSHH regulation number 7, which was amended in 2004, explains prevention and control of exposure. According to this regulation, good practice to control exposure is required, exposure should not exceed the workplace limit, and exposure to mutagens and carcinogens should be as low as possible.

4.9.1.3 Prevention of exposure

Employers should be responsible for prevention of exposure by all means other than PPE. The main goal is complete eradication of hazardous materials, as the need for actual nanomaterials is very low and specific. So in certain procedures, nanomaterials can be used after appropriate substitution, which will lead to a substance that is less hazardous.

4.9.1.4 Control of exposure

Complete prevention of the use of nanomaterials is not practical, so employers are ordered to control exposure to nanomaterials by all means. This is achieved through the following steps:

1. By suitable and appropriate system, procedures, and controls.
2. By implementation and use of appropriate tools and materials.
3. By controlling exposure at the site.
4. In case of lack of engineering control, strict use of PPE.
5. By good practice of suitable steps for control of exposure to hazardous nanomaterials.
6. By strict implementation of the principle of not exceeding the workplace limit with additional consideration of carcinogens and mutagens.

Although employer should practice all the principles for control to exposure, it is not mandatory in some cases to follow all control principles. However, issues in regard to performance and effectiveness of control practices are difficult, including use of PPE when required for nanomaterials. Suggestions with regard to occupational exposure limit for newly manufactured nanomaterial are very necessary. The current occupational limit, which is functional in the United Kingdom, is workplace exposure limit. This exposure limit is based on inhaled particles through the information provided for larger particle size of the bulk materials. However, health agencies are less satisfied with the exposure limit, as it can deviate in cases of nano-sized particles. As we do not have sufficient data and listing two exposure limits for the same material would increase consumers' confusion, this problem will be revisited in future when enough data on nanomaterials become available.

4.9.1.5 Monitoring exposure

COSHH guidelines recommend observation of exposure through valid and appropriate methods. However, available methods for monitoring of exposure are not practical, as they require expensive equipment that is not portable. To overcome this problem a lot of research has been conducted with the goal of manufacture of a portable sampler for monitoring. Research is underway to try to develop a portable personal sampler. A decision is required for the appropriate unit for exposure metrics used for exposure assessments. There is also a requirement for specific exposure methods for fibrous nanoparticles, such as carbon nanotubes. Currently, light microscopy is used for carbon nanotubes, but it is not a suitable method for their measurement.

4.9.1.6 Health surveillance

Protection of workers is emphasized in COSHH regulation number 11. According to this regulation, workers' health should be observed critically whenever it is considered necessary. The main objective of this regulation is to prevent employees from major loss as soon as possible. Changes in health that may be caused by exposure to nanomaterials are monitored regularly. Health monitoring is also performed to determine the efficacy of measures to control exposure. Health surveillance is required to gather and use data for the evaluation of hazardous properties of the substance. At present, there is a shortage of data on risk to workers' health due to nanoparticle exposure. Such data are required to make valid assessments in terms of health surveillance implementation.

4.9.1.7 Instruction and training

Employees should be given a complete instruction set and training by employer. This is necessary to allow them to perform their work safely in a nanoparticle-exposed area. The instruction set includes information on risk to health and how employees should behave to minimize their exposure to nanomaterials.

4.9.1.8 Risk management

In COSHH regulations, employees are investigated for exposure to nanomaterials. People in nearby areas also checked to determine whether or not they have been affected. Employers and other shareholders are instructed to equip themselves to measure the risks posed by nanomaterials.

4.9.1.9 Issues under COSHH

Overall, the basic elements and procedures in the COSHH regulation are enough to tackle most of issues of nanomaterials in working areas. However, there is a shortage of knowledge in almost all areas of COSHH regulation, including assessment, measurement, control, and investigation of nanomaterials in general. This shortage creates problems for employers in carrying out assessment and for regulators in attempting to judge the assessment's reliability. Thus when further knowledge becomes available, more guidance in support of COSHH is needed to take account of all issues.

4.9.2 Dangerous substances and explosive atmospheres regulations

The DSEAR regulations are used to manage and assess the risk of explosion, so they deal with all substances that are capable of explosion. Identification of such substances is done along with their explosion limits. Explosions can create airborne dust, which can be dangerous if it is inhaled. Fire and explosion hazards have been studied extensively, and results have shown that it is impossible to predict on the basis of knowledge of larger-sized particles that certain materials can become either respirable or not on explosion [19]. In view of the already existing lack of knowledge in relation to nanomaterials, it is not surprising that there is a shortage of knowledge in regard to their flammability and explosivity.

4.9.3 Existing substances regulation

The ESR requires details about existing substances with their hazardous characteristics from employer. The details are submitted to the European Commission. ESR is required for assortment through risk assessment and prioritization, leading to deductions about implementation of risk management measures in addition, to currently applied measures. ESR is necessary to guard all sectors of the human population and the environment. The UK proficient agencies for the ESR are HSE and the EA. Till now, very few substances in form of nanoparticles have been recorded in the ESR, but there is hope for this to happen in the future. The issues that would rise are analogous to those covered above in relation to other regulations.

4.9.4 Biocidal products regulations

HSE is the UK skilled agency for the directive of biocidal products across the EU under Biocidal Products Directive 1998, endorsed in the United Kingdom as the BPRs (2002, and amendments). Till now, no nano-scale substances have been regulated under the BPR, and there are less chances of this happening in the future. Issues faced in regard to nanomaterials in the BPR are same as in other legislations.

4.9.5 Control of major accident hazards regulations

COMAH regulation involves announcement to HSE of locations where substances with hazardous properties are collected above specified amounts. These substances are those that relate to certain CHIP classification criteria. This notification procedure is based on accomplishment of certain results of settlement between the site worker and the regulator on a risk assessment. This requires identification of the essential measures to diminish the risks of a release so that it is as low as rationally practicable. The responsibility is on the site worker to recognize the suitable hazard classification of the material(s) stored to decide implementation of COMAH. Still, in this legislation there is an issue of identification of the hazards of nanomaterials. However, at this time when there is major growth happening in regard of nanotechnology industries, it seems doubtful that the amounts of materials being formed and kept, will be pertinent to this legislation.

4.10 Conclusion

Despite the potential dangers, exposure to artificial nanoparticles is increasing exponentially, especially among manufacturing workers. However, it is clear that nanotechnology is the most emerging and revolutionary technology of this era. Considering the huge demands in the current era, it would be absurd and impractical to completely ban the use of nanomaterials. Therefore taking strong preventive measures according to WHO guidelines and national legislation is the best strategy.

References

[1] S.T. Stern, S.E. McNeil, Nanotechnology safety concerns revisited, Toxicol. Sci. 101 (1) (2008) 4−21.

[2] J. Morris, J. Willis, D. De Martinis, B. Hansen, H. Laursen, J.R. Sintes, et al., Science policy considerations for responsible nanotechnology decisions, Nat. Nanotechnol. 6 (2) (2011) 73−77.

[3] C. Ostiguy, G. Lapointe, L. Ménard, Y. Cloutier, M. Trottier, M. Boutin, et al., Nanoparticles: Actual Knowledge About Occupational Health and Safety Risks and Prevention Measures, Institut de recherche Robert-Sauvé en santé et en sécurité du travail, Montreal, CN, 2006.

[4] W.J. Peijnenburg, M. Baalousha, J. Chen, Q. Chaudry, F. Von der kammer, T.A. Kuhlbusch, et al., A review of the properties and processes determining the fate of engineered nanomaterials in the aquatic environment, Crit. Rev. Environ. Sci. Technol. 45 (19) (2015) 2084−2134.

[5] M. Baalousha, G. Cornelis, T. Kuhlbusch, I. Lynch, C. Nickel, W. Peijnenburg, et al., Modeling nanomaterial fate and uptake in the environment: current knowledge and future trends, Environ. Sci. Nano. 3 (2) (2016) 323−345.

[6] G. Cornelis, K. Hund-Rinke, T. Kuhlbusch, N. Van den Brink, C. Nickel, Fate and bioavailability of engineered nanoparticles in soils: a review, Crit. Rev. Environ. Sci. Technol. 44 (24) (2014) 2720−2764.

[7] T. Kuhlbusch, S. Neumann, H. Fissan, Number size distribution, mass concentration, and particle composition of PM_1, $PM_{2.5}$, and PM_{10} in bag filling areas of carbon black production, J. Occup. Environ. Hyg. 1 (10) (2004) 660−671.

[8] A.D. Maynard, P.A. Baron, M. Foley, A.A. Shvedova, E.R. Kisin, V. Castranova, Exposure to carbon nanotube material: aerosol release during the handling of unrefined single-walled carbon nanotube material, J. Toxicol. Environ. Health Part A 67 (1) (2004) 87−107.

[9] A.P. Dowling, Development of nanotechnologies, Mater. Today 7 (12) (2004) 30−35.

[10] V.L. Colvin, The potential environmental impact of engineered nanomaterials, Nat. Biotechnol. 21 (10) (2003) 1166−1170.

[11] D. Mark, Nanomaterials—A Risk to Health at Work? Health and Safety Laboratory, Buxton, 2004.

[12] G. Amoabediny, A. Naderi, J. Malakootikhah, M. Koohi, S. Mortazavi, M. Naderi, et al., Guidelines for safe handling, use and disposal of nanoparticles, J. Phys. Conf. Ser. 170 (2009) 012037.

[13] K. Hegde, S.K. Brar, M. Verma, R.Y. Surampalli, Current understandings of toxicity, risks and regulations of engineered nanoparticles with respect to environmental microorganisms, Nanotechnol. Environ. Eng. 1 (1) (2016) 5.

[14] A. Groso, A. Petri-Fink, A. Magrez, M. Riediker, T. Meyer, Management of nanomaterials safety in research environment, Part Fibre Toxicol. 7 (1) (2010) 1−8.

[15] A. Azadeh, I.M. Fam, M. Khoshnoud, M. Nikafrouz, Design and implementation of a fuzzy expert system for performance assessment of an integrated health, safety, environment (HSE) and ergonomics system: the case of a gas refinery, Inf. Sci. 178 (22) (2008) 4280−4300.

[16] T. Petry, R. Knowles, R. Meads, An analysis of the proposed REACH regulation, Regul. Toxicol. Pharmcol. 44 (1) (2006) 24−32.

[17] N.E. Hazard, Force, R. A. T. ANNEX C: Written/Oral Evidence Collected, 2016. < https://www.osha.gov/sites/default/files/publications/OSHA3844.pdf > .

[18] V. Murashov, P. Schulte, C. Geraci, J. Howard, Regulatory approaches to worker protection in nanotechnology industry in the USA and European Union, Ind. Health 49 (3) (2011) 280−296.

[19] D. Pritchard, Literature Review: Explosion Hazards Associated With Nanopowders, Health and Safety Laboratory, 2004.

[20] D.M. Bowman, More than a decade on: mapping today's regulatory and policy landscapes following the publication of nanoscience and nanotechnologies: opportunities and uncertainties, NanoEthics 11 (2) (2017) 169−186.

[21] M.E. Vance, T. Kuiken, E.P. Vejerano, S.P. McGinnis, M.F. Hochella Jr, D. Rejeski, et al., Nanotechnology in the real world: Redeveloping the nanomaterial consumer products inventory, Beilstein J. Nanotechnol. 6 (1) (2015) 1769−1780.

[22] K.D. Grieger, A. Baun, R. Owen, Redefining risk research priorities for nanomaterials, J. Nanopart. Res. 12 (2) (2010) 383−392.

[23] K.G. Steinhäuser, P.G. Sayre, Reliability of methods and data for regulatory assessment of nanomaterial risks, NanoImpact 7 (2017) 66−74.

[24] P. Swuste, D. Zalk, Risk management and nanomaterials, Nanotechnol. Fundam. Appl. 1 (2013) 155−173.

[25] M. Riediker, C. Ostiguy, J. Triolet, P. Troisfontaine, D. Vernez, G. Bourdel, et al., Development of a control banding tool for nanomaterials, J. Nanomaterials 2012 (2012) 8. Available from: https://doi.org/10.1155/2012/879671. Article ID 879671.

[26] H. Fiedler, O. Hutzinger, J. Giesy, Utility of the QSAR modeling system for predicting the toxicity of substances on the European inventory of existing commercial chemicals, Toxicol. Environ. Chem. 28 (2−3) (1990) 167−188.

CHAPTER 5

Safety and global regulations for application of nanomaterials

Md Abdus Subhan[1] and Tahrima Subhan[2,3]

[1]Department of Chemistry, Shahjalal University of Science and Technology, Sylhet, Bangladesh
[2]Department of Social Work, Shahjalal University of Science and Technology, Sylhet, Bangladesh
[3]The Sylhet Khajanchibari International School and College, Sylhet, Bangladesh

5.1 Introduction

Nanotechnology deals with the science and application of objects smaller than 100 nm. Besides the numerous beneficial applications, including environment, health, and medicine, concerns exist about adverse health effects of unintended human exposure to nanomaterials. The 2010 Parma Declaration on Environment and Health of the 53 member states of the World Health Organization (WHO) regional office of Europe listed the health implications of nanotechnology and nanoparticles among the key environment and health challenges. The WHO regional office in Europe originated a critical assessment of the current state of knowledge and the key evidence on the possible health consequences of nanomaterials, with a view to identifying options for risk assessment and policy formulation, and organized an expert meeting to address the issue [1].

The current signs of risk with the use of nanomaterials is not conclusive. As complexity and uncertainty are great, risk assessment is challenging, and formulation of evidence-based policies and regulations is difficult. Advanced models and frameworks for risk assessment and risk governance are being developed and applied to organize the available evidence on biological and health effects of nanomaterials in ways that will inform policy [1].

Nanomaterials exist in nature, result from human activity, or can be manufactured for use in different applications. The power of nanotechnology stems from the fact that physical, chemical, electromagnetic, and other properties of nanomaterials change with the process and the source of production. Nanomaterials bring perceptible benefits to society. They are currently used in a wide range of products from electronic batteries to paints and cosmetics, as they enhance functions or properties of materials or products. Besides their use in everyday consumer goods, they also provide a key enabling technology for groundbreaking products and processes in many sectors,

Nanomaterials Recycling
DOI: https://doi.org/10.1016/B978-0-323-90982-2.00005-6

© 2022 Elsevier Inc.
All rights reserved.

including car manufacturing, construction, electronics, nanotechnology, biotechnology, and healthcare [2].

A central ethical and policy issue regarding minimizing and managing the risks of engineered nanomaterials (ENMs) within a legal framework is how to defend public health and the environment. Policymakers should (1) use existing laws to regulate ENMs and the best available evidence to inform appropriate levels of regulation and (2) support additional research on the risks of ENMs. The public health and environmental risks of ENMs could be minimized and managed without sacrificing their potential clinical, social, and economic benefits. Nanotechnologies are the driving force behind the new industrial revolution. With the mass production of these nanomaterials and their applications, there has been a dramatic increase in the number of workers dealing with ENM and becoming exposed to resulting hazardous effects. Moreover, there is an increased environmental problem due to leaks of these materials from industrial processes. Therefore it is essential to consider the importance of the safety of ENMs and their applications. It is important for these materials and products to be safe to enable successful advancement of nanotechnology for useful applications. The influence of the ENM on both humans and the environment and the risk and the risk assessment of ENMs should be addressed [3,4]. Fig. 5.1 represents the main areas of nanoparticle translocation and accumulation after administration in humans [5,6].

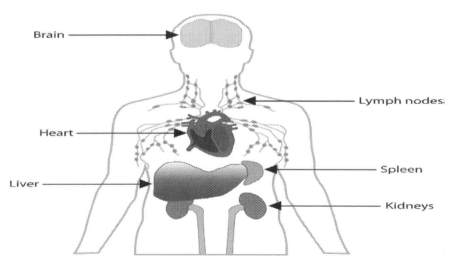

Figure 5.1 The main areas of nanoparticle translocation and accumulation after administration in humans. *Adapted from R. Foulkes, E. Man, J. Thind, S. Yeung, A. Joy, C. Hoskins, The regulation of nanomaterials and nanomedicines for clinical application: current and future perspectives, Biomater. Sci. 8 (2020) 4653, an open access article.*

Nanotechnology and nanotechnology-based products merit special government regulation [7]. This is mainly related to the need to assess new substances prior to their release into the market, community, and environment. Nanotechnology is incorporated in an increasing number of commercially available products, including socks, trousers, tennis racquets, and cleaning cloths for daily life and other uses [8]. There are numerous sources of nanoparticles in daily life (Fig. 5.2). The nanotechnology and corresponding companies have induced demands for greater public participation and regulatory provisions [2,9]. However, these demands have currently not led to inclusive regulation to direct research and the commercial application of nanotechnologies or to any extensive labeling for products that contain nanoparticles or are derived from nanoprocesses [2,9].

The application of nanomaterials in medicine has greatly increased over the past 10–15 years, with many nanoparticle-based systems being utilized within the clinic. After inadequate initial achievement in clinical trials, polymeric, metallic, and lipid-based nanoparticles have all found places in medicine. There is a great expectation encompassing the field of nanomedicine and its impact on the pharmaceutical industry and relevant medical industry. Despite frequent demands from the research community, manufacturers, policymakers, healthcare providers, and the general public, there is currently very little regulatory guidance in this area. This is reflected in the absence of an international definition of nanomaterial. Several bodies, such as the National Institutes of Health in the United States and the European Science Foundation and European Technology Platform, have different definitions, and the U.S. Food and Drug Administration (FDA) has no clear definition. This ambiguity may ultimately affect funding, research, and development of such products negatively, thus damaging public acceptance and perception of nanoproducts. Here, we aim to discuss the use of nanomaterials, including different industries, environment, medicine, and the clinical setting; why regulation of these materials is so important; and the challenges faced in regulating these materials generally as well as the current regulations used in different nations [5].

The regulatory bodies, including the U.S. Environmental Protection Agency (EPA) and the FDA in the United States and the Health and Consumer Protection Directorate of the European Commission have been working to determine the likely risks exerted by nanoparticles. However, neither engineered nanoparticles nor the products and materials that incorporate them are subject to any distinct regulation concerning production, handling, or labeling. In most cases, regulations for the bulk chemicals and toxic materials are being used as regulations for the corresponding nanomaterials. Table 5.1 shows the use of nanomaterials in different consumer products.

This chapter emphasizes the following points for sustainable development of nanotechnology: risk assessment and management, societal impact of nanotechnology, regulation of nanotechnology, biomedicine, and the environmental impact of nanotechnology.

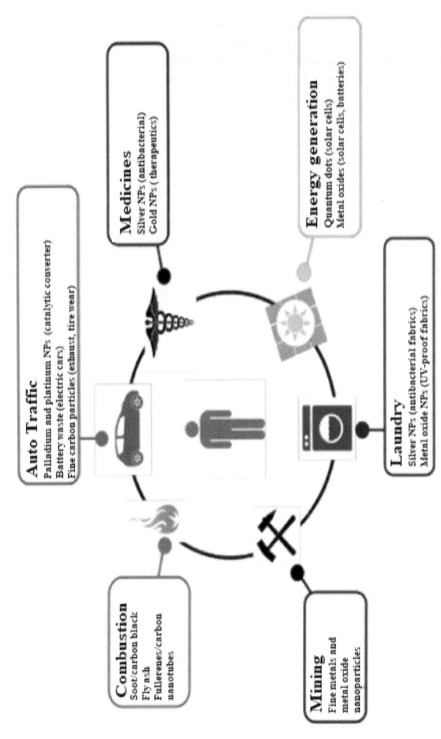

Figure 5.2 Sources of nanoparticles in daily life. Adapted from E.M. Osman, Environmental and health safety considerations of nanotechnology: nano safety, Biomed. J. Sci. Tech. Res. 19 (4) (2019) 14501–14515 open access article.

Table 5.1 Nanomaterials in some consumer products.

Product	Nanomaterials	Uses of products
Appliances	Silver, iron, and carbon	Refrigerators, air humidifiers, and washing machines
Agricultural products	Silver, gold, silica, titanium dioxide, zinc oxide, and carbon nanotubes	Plant germination and growth Plant protection products Plant pathogens and pesticides Water purification and pollutant remediation Water retention in soils
Air cleaners	Silver, titanium dioxide, and activated carbon	Solid- and spray-based air cleaners Ultraviolet (UV) light or ozone treatments
Coating and impregnation	Silver, titanium dioxide, silicon dioxide, and Teflon	Liquids and spray Cloth/textiles, paints, and shoe polish
Construction materials	Silver, titanium dioxide, silicon dioxide	Façade protection Plaster, cement, mortar, sealant, and soil stabilization products Coating for stone and tile protection Tiles, surface layer for roofs, surface protection and pavements
Cosmetics	Silver, gold, platinum, fullerenes, nanopeptides, silicon oxide, zinc oxide, titanium dioxide, calcium peroxide, copper, copper peptides, and carbon black	Sprays, liquids, and gels Sunscreen cream, face powder, lotions, lipstick, eye shadow, and mascara Mouthwash solution, toothpaste, and soap
Electronic devices and electronic products	Silver, gold, silicon dioxide, and zinc oxide	Computer keyboards and protection coating for metals Processor or cooling liquids Semiconductors and other solid matrices
Food and beverages	Silver, platinum, palladium, gold, and silica	Nanomaterial ingredients Food supplements Nanomaterial uses in food packaging
Fuel and lubrication oil additives	Gold, cerium oxide, and tungsten disulfide	Engine oil and fuel catalyst Diesel catalyst Lubricant for engine oil
Maintenance products (for cars and boats)	Silver, titanium dioxide, silicon dioxide, nanoboron, aluminum oxide, nanoceramics, and carbon nanotubes	Antimicrobial shield for cars Polish, sealant, and window coating for cars Motor protection and maintenance for cars Odor remover inside car Molding into automotive parts for repair
Medical devices	Silver, copper, silicate, and zirconia	Wound dressings Ostomy bags Dental fillings, coatings for implants
Textiles	Silver, titanium dioxide, bamboo charcoal, and Teflon	Textiles for antimicrobial effects UV protection

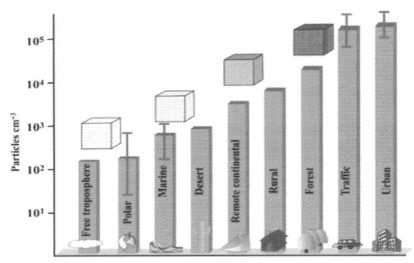

Figure 5.3 Concentrations (particles/cm^3) of nanoparticles in the atmosphere in various environments. Adapted from E.M. Osman, Environmental and health safety considerations of nanotechnology: nano safety, Biomed. J. Sci. Tech. Res. 19 (4) (2019) 14501−14515.

5.2 Risks management for environment and health safety

Investigations into the health effects of airborne particles revealed that among the toxic materials, usually smaller particles are highly toxic. For the same mass per volume, the dose in terms of particle numbers increases as particle size decreases.

The available data for current risk management approaches are not sufficient for the hazards related to nanoparticles. In particular, prevailing toxicological and ecotoxicological methods are not upto the task. Exposure evaluation or dose needs to be expressed as quantity of nanoparticles and/or surface area rather than mass. The instruments for regular identification and quantification of nanoparticles in air, water, or soil is insufficient; also, little is known about the human physiological responses to nanoparticles. Fig. 5.3 shows concentrations of nanoparticles in the atmosphere in various environments [2].

The regulatory organizations in the United States as well as in the European Union (EU) have agreed that nanoparticles pose a new risk for humans and the environment. It is essential to perform a substantial analysis of the risk. The difficulties for the regulators are whether a matrix can be established to detect nanoparticles and more complex nanoformulations, which are likely to have unusual toxicological properties, or whether it is more sensible for each particle or formulation to be tested distinctly. The following Scheme 5.1 shows a general health risk evaluation process for nanomaterials [2].

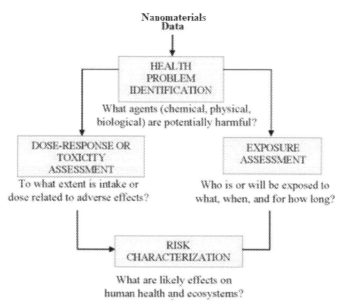

Scheme 5.1 The proposed health risk evaluation process for nanomaterials. *Adapted from E.M. Osman, Environmental and health safety considerations of nanotechnology: nano safety, Biomed. J. Sci. Tech. Res. 19 (4) (2019) 14501−14515.*

The International Council on Nanotechnology has a database and a virtual journal of scientific papers on environmental, health and safety research on nanoparticles. The database currently has over 2000 entries indexed by particle type, exposure pathway and other criteria. The Project on Emerging Nanotechnologies (PEN) lists >1600 manufacturer-identified nanotechnology-based consumer products introduced to the market that manufacturers have voluntarily identified. The FDA requires no labeling, so that number could be remarkably higher [10]. The use of nanotechnology in consumer products and industrial applications is increasing fast, with the products listed in the PEN inventory showing just the tip of the iceberg. A list of the products that have been voluntarily disclosed by their manufacturers is located at http://www.nanotechproject.tech/cpi/.

The Material Safety Data Sheet (MSDS) that must be issued for certain materials often does not distinguish between bulk and nano-scale size of the material. Even when it does, these MSDSs are advisory only.

5.3 Approaches of democratic governance to nanotechnology

Democratic government has an accountability to afford chances for the public to be engaged in the growth of new forms of science and technology, including nanotechnology [11]. Community involvement can be reached by different processes, such as referenda, consultation documents, debates, public meetings, workshops on nanotechnology, constructive dialogs, and advisory committees that include other stakeholders. More contemporary engagement processes that have been employed to include community members in decision making about nanotechnology include citizens' juries and consensus conferences.

It is necessary to increase awareness both in government and in society of the need to create a more constructive and trusting relationship between science and society. Governments need to reconsider their approach to science and science governance by encouraging the public, community members, and other stakeholders who are more scientifically aware to have confidence in policy issues, research, and decision making about nanotechnologies that affect them. It would revive the mechanism of policy making and risk management to be more aligned with public needs and aspirations for science and technology. This engagement of upstream in science and technology can help to overcome cultural barriers among scientists, members of the public, and decision makers as well as create space for scientists and decision makers to reflect on the wider social implications of their work, thus helping to put science into context [12].

The U.S. National Nanotechnology Initiative (NNI) is committed to fostering the development of a community of experts on ethical, legal, and societal issues (ELSI) related to nanotechnology and to building collaborations among ELSI communities, such as consumers, engineers, ethicists, manufacturers, nongovernmental organizations, regulators, and scientists. These stakeholder groups will consider potential benefits and risks of research breakthroughs and provide their perspectives on new research directions. Upstream engagement in this sense is meant to "create the best possible conditions for sound policy making and public judgments based on careful assessment of objective information" [12,13].

Limited nanotechnology labeling and regulation may exacerbate potential human and environmental health and safety issues associated with nanotechnology [14,15]. To secure a future for nanotechnologies, governments should emphasis funding, policy, and research on the potential applications of nanotechnology and nanomaterials that contribute to a wider social good and their possible environmental, health and safety, social, and ethical implications to create effective laws and regulations. Comprehensive laws and regulation of nanotechnology are key to confirming that the likely risks associated with the research and commercial application of nanotechnology do not shade its significant advantages [16]. Regulation may also be essential to meet community demands for accountable progress of nanotechnology and to confirm that public welfare is taken into account in determining the expansion of nanotechnology [17,18]

Public engagement is a mechanism through which public trust can be restored by increasing the transparency and accountability of scientific governance and policy development. The stance that the research, development, and use of nanotechnology should be subject to control by the public sector is sometimes referred to as "nanosocialism." The public engagement can be applied in nanotechnology through expected discussion with peers. Discussion may particularly catalyze upstream public engagement by inducing responsibility for people to pursue and process further information [13].

5.4 Nano inventiveness

Nanotechnology has been around for over 20 years. It is still considered new technology. As with many new technologies, the health and environmental effects of nanomaterials have not been thoroughly investigated. Consequently, the uncertainty surrounding the toxicity of nanomaterials merits an alert for people who work with them. For this reason, discussions on nano safety often deal with toxicology, ecotoxicology, exposure assessment, mechanism of interaction, risk assessment, and standardization. Thus the nanotechnology must follow the best regulations.

In 2007 an alliance of over 40 groups demanded that nanomaterials be classified as new materials and be regulated as such. However, chemicals containing nanoparticles that have formerly been subject to evaluation and regulation may be exempt from regulation, irrespective of the potential for various risks and effects. On the other hand, nanomaterials are often documented as "new" from the viewpoint of intellectual property rights, and as such are commercially secured via patenting laws. There is a substantial argument about who is accountable for the regulation of nanotechnology [19]. Some nonnanotechnology-specific regulatory agencies currently cover some products and processes. This enables some nanotechnology applications to occur without being covered by any regulations. Such gaps in regulation are anticipated to remain along with the growth and commercialization of progressively multifaceted second- and third-generation nanotechnologies.

So far, global regulations still fail to discriminate between materials in their nanoscale and bulk forms. Thus nanomaterials remain unregulated. There is no regulatory requirement for nanomaterials to undergo new health and safety testing or environmental effect evaluation before their use in commercial products if these materials have already been approved in bulk form. The health risks of nanomaterials are of specific apprehension for workers who may face professional exposure to nanomaterials at higher levels and on a more routine basis than the general public.

5.5 International law on nanomaterials

There is no international regulation of nanomaterial-based products and nanotechnology [20]. Currently, there is no internationally accepted definition or terminology for nanotechnology, no internationally approved protocols for toxicity testing of nanomaterials, and no standard protocols for assessing the environmental impacts of nanoparticles [21]. Moreover, nanoparticles do not fall within the space of prevailing international accords for regulation of toxic chemicals [22].

Since products that are formed using nanotechnologies will most probably enter international trade, it is necessary to harmonize nanotechnology standards across national borders. Notably, developing countries will be excluded from international standards negotiations. The Institute for Food and Agricultural Standards notes that "developing countries should have a say in international nanotechnology standards development, even if they lack capacity to enforce the standards" [23].

5.6 Arguments against regulation of nanomaterials

The vast use of nanotechnology in recent years has generated the sense that regulatory frameworks are necessary to cope with entirely new challenges. New substances or products are assessed for safety by many regulatory systems all over the world on a case-by-case basis before they are permitted on the market. These regulatory systems have been appraising the safety of nanomaterials for many years, and many substances containing nanoparticles have been in use for decades, including carbon black, titanium dioxide, zinc oxide, aluminum silicate, iron oxides, silicon dioxide, diatomaceous earth, talc, magnesium oxide, and copper sulfate.

The current endorsement settings most commonly use the best accessible science to evaluate safety and do not accept substances or products with an unacceptable risk-benefit profile. The anticipated applications of nanomaterials with the greatest impact are far in the future, and it is unclear how to regulate technologies whose viability is speculative at this point. The direct applications of nanomaterials raise encounters not much different from those of introducing any other new material and can be dealt with by minor amendment to existing regulatory schemes rather than sweeping regulation of entire scientific fields [24].

A strict protective approach to regulation could hinder growth in the field of nanotechnology. However, safety studies are essential for every nanoscience application. While the result of these studies can form the basis for government and international regulations, a more sensible approach might be the development of a risk matrix that recognizes probable nano offenders.

5.7 Response from governments all over the world

5.7.1 The United States

Development of nanomaterials has given the world the benefits of the fast-developing discipline of nanomaterials and nanotechnology. It is projected that over 1500 manufacturer-identified nanotechnology products are publicly to be had, with brand-new ones coming onto the market every week. The considerable range of products and programs offers nanotechnology its enormous boom prospects. Nanotechnology and nanomaterials are certain to play a vital role in future development.

Rather than adopting a new nano-specific regulatory framework, the FDA organizes an interest group each quarter with representatives of FDA centers that have accountability for assessment and regulation of various substances and products. This group ensures organization and communication [25]. There is a September 2009 FDA document for detecting sources of nanomaterials; how they move in the environment; the problems they might cause for people, animals, and plants; and how these problems could be circumvented or alleviated [26].

In the United States, nano regulations have been developed by a variety of organizations, including the EPA, the FDA, and the Consumer Product Safety Commission (CPSC) [25]. The EPA's policy for nanomaterials is controlled by the current regulations of the Toxic Substances Control Act (TSCA) and the Federal Insecticide, Fungicide, and Rodenticide Act (FIFRA). The TSCA requires that information be gathered about manufactured nanomaterials. The EPA has announced a significant new use rule of premanufacture notices for 13 chemicals, including carbon nanotubes and fullerenes. In addition, the manufacturers of new nanomaterials must notify of any previous production. The updated information can be found at http://www.epa.gov/oppt/nano/. In case of the manufacture and production of nanomaterials, the manufacturers must supply the EPA with information about the nanomaterials within 90 days. Moreover, FIFRA requires that pesticide products that contain nanomaterials be registered. This registration was determined on the basis of the EPA's assessment of the impacts of silver on human health and environmental safety. The FDA's regulation of nanotechnology products is disseminated through the FDA's website (https://www.fda.gov/home). The FDA's regulation approach is described as follows:

1. The FDA maintains both product-focused and regulatory policies for nanotechnology products based on a scientific approach.
2. The FDA respects variations of nanotechnology products in legal standards. Different nanomaterial product classes and application fields affect human safety differently.
3. The FDA conducts premarket reviews of nanotechnology products. Premarket reviews include various items relevant to nanotechnology products such as drugs, dietary supplements, food, and cosmetics. In addition, the FDA will continue postmarket monitoring. Therefore the FDA makes an effort to reduce the risks to human or animal health posed by nanotechnology products.
4. The FDA is responsible for ensuring that industry safety standards meet all of the applicable legal requirements. Additionally, the FDA will cooperate with its domestic and international counterparts in terms of regulatory policy.

5. The FDA will provide technical advice and guidance for industry (i.e., regulatory and statutory obligations).

The CPSC collaborates with the EPA to identify the safety risks of consumer products. The CPSC formally joined the NNI in 2011. It is working with several agencies to establish the following: (1) evaluation protocols for the airborne release of nanomaterials from consumer products and the effects of human exposure, (2) improved sports safety equipment using nanomaterials, (3) expanding consumer product testing to establish credible protocols to evaluate the risks of potential exposure to nanomaterials, and (4) research studies related to how consumer products that use nanomaterials affect humans.

5.7.1.1 California policy

Berkeley, California, is the first city policy regulating nanotechnology in the United States [18]. The National Research Council released a report calling for more regulation of nanotechnology on December 10, 2008 [27]. The California State Assembly Bill 289 (2006) authorized the Department of Toxic Substances Control (DTSC), which is part of the California EPA, and other agencies to request information on environmental and health impacts from chemical manufacturers and importers, including testing procedures. The task of the DTSC is to defend public health and the environment in California from toxic harm. updated information is available at http://www.dtsc.ca.gov/TechnologyDevelopment/Nanotechnology/index.cfm. DTSC associates with other institutions, including the University of California, Los Angeles, Santa Barbara, and Riverside; the University of Southern California: Stanford University: the Center for Environmental Implications of Nanotechnology; and the National Institute for Occupational Safety and Health on safe nanomaterial handling practices.

According to the Control of Substances Hazardous to Health regulations, one must identify which harmful substances are present in the workplace; understand how people could be exposed to them and be harmed; put measures in place to prevent harm and review whether one is doing enough; provide information, instruction, and training; and, where necessary, monitor health.

Materials or substances that are used or created at work that could be harmful include anything from paints and cleaners to blood and waste. Dusts, gases, fumes, liquids, gels, or powders that could come into contact with eyes or skin can all be harmful. Microorganisms can cause infections and allergic reactions, and some are toxic. Many substances can harm health, but if used properly, they rarely do. However, if one fails to control their use, one could be guilty of offenses under Section 33 of the Health and Safety at Work Act. DTSC in a close collaboration with EPA administers both the state and federal hazardous waste management programs.

5.7.2 The United Kingdom

In its seminal 2004 report *Nanoscience and Nanotechnologies: Opportunities and Uncertainties*, the Royal Society has suggested that nanomaterials be controlled as new chemicals, that research laboratories and factories treat nanomaterials "as if they were hazardous," that release of nanomaterials into the environment should be avoided as far as possible, and that products containing nanomaterials be subject to new safety testing necessities prior to their commercial release [28].

The 2004 report by the Royal Society and Royal Academy of Engineers noted that current UK regulations did not necessitate supplementary testing when existing substances were produced in nanoparticulate form [28]. The Royal Society suggested that such regulations be revised so that "chemicals produced in the form of nanoparticles and nanotubes be treated as new chemicals under these regulatory frameworks." They also suggested that existing regulation be modified on a precautionary basis because they expected that "the toxicity of chemicals in the form of free nanoparticles and nanotubes cannot be predicted from their toxicity in a larger form and in some cases, they will be more toxic than the same mass of the same chemical in larger form" [28].

The improved Regulation Commission's earlier 2003 report [1] had endorsed that the UK Government (1) through an informed debate, enable the public to consider the risks for themselves and help them to make their own decisions by providing suitable information; (2) be open about how it makes decisions and acknowledge when there are uncertainties; (3) communicate with and involve as far as possible the public in the decision-making process; (4) ensure that it develops two-way communication channels; and (5) take a strong lead in the handling of any risk issues, particularly information provision and policy implementation.

These recommendations were documented by the UK government perceiving that there was "no obvious focus for an informed public debate of the type suggested by the Task Force." The government's response was to accept the recommendations. The Royal Society's 2004 report [28] recognized two distinct governance issues: (1) the "role and behaviour of institutions" and their ability to "minimise unintended consequences" through adequate regulation and (2) the extent to which the public can trust and play a role in determining the trajectories that nanotechnologies may follow as they develop.

5.7.3 The European Union

The EU's definition of a nanomaterial was determined in 2011 by the Commission recommendation [1]. In EC regulation No. 1223/2009, article 2 (1) (k), nanomaterials are defined as insoluble or biopersistent and intentionally manufactured materials with one or more external dimensions or an internal structure, on the scale of $1-100$ nm.

On the basis of this definition, soluble or degradable/nonpersistent nanomaterials such as liposomes and emulsions are not considered within this classification.

The EU's regulation committee is Registration, Evaluation, Authorization and Restriction of Chemicals (REACH), which is employed to record substances that are manufactured or imported at a rate higher than one ton per year. These materials include nanomaterials as well as carcinogenic, mutagenic, and reproductive, and persistent, bioaccumulative and toxic substances. In agreement with Classification, Labelling, and Packaging (CLP), the European Chemicals Agency (ECHA) must classify nanomaterials with regard to how hazardous they are. Moreover, inventory that contains nanomaterials must be labeled to notify consumers [1]. In recent years the ECHA has made an effort to build a framework for the legislation of nanomaterials. The REACH Implementation Project on Nanomaterials (RIPoN) under REACH was launched in 2009 and provided the following three-step final report: substance identification of nanomaterials (RIPoN1), information requirements (RIPoN2), and chemical safety assessment (RIPoN3). In addition, the Group Assessing Already Registered Nanomaterials (GAARN) had a meeting in 2013. The GAARN focused on human exposure to nanomaterials and assessed the risks that nanomaterials pose to human health and the environment. In the GAARN meetings, International Uniform Chemical Information Database (IUCLID), software manuals for REACH registration presented the best practices, which should evaluate the safety assessment of nanomaterials. The ECHA has been updated to reflect version 5 of the IUCLID, which includes nanomaterials ("CASG Nano," composed of member states and stakeholder experts) of the Competent Authorities for REACH and Classification and Labelling (CARACAL). Alternatively, EU legislation covering electrical and electronic equipment falls under the Restriction of Hazardous Substances Directive (RoHS). RoHS restricted the use of hazardous substances such as heavy metals (e.g., lead, cadmium, mercury, and hexavalent chromium) and flame retardants (e.g., polybrominated biphenyls, polybrominated diphenyl ethers). Although RoHS was proposed to regulate nanosilver and carbon nanotubes in electronic equipment, the proposal was excluded from nanomaterial legislation [29]. The EU's cosmetic products were evaluated by the Scientific Committee on Consumer Products (SCCP). The SCCP was applied to animal testing in order to evaluate the safety of cosmetic products that contain nanomaterials [29]. The European Commission proposed that nanomaterials in cosmetics should be regulated. The Commission should be notified about a nanoproduct six months in advance of its distribution [30]. The cosmetic regulation (EU Regulation 1223/2009) has tightened the control of nanomaterials in the EU market since 2013. The nanomaterial regulation (Article 16) of cosmetic products in the EU is described by EU regulation 1223/2009, which can be found at https://knowledge.ulprospector.com/5760/pcc-nanomaterials-eu-regulation-1223-2009-update/ [31].

Nanomaterials are produced to have novel features relative to the same material at larger scale. Numerous products containing nanomaterials are now in use in the world markets, such as batteries, coatings, and antibacterial clothing. Analysis shows markets to raise significantly in the near future. A nano revolution is expected to be seen in many sectors, such as public health, employment and occupational safety and health, information technology, innovation, environment, energy, transport, security, and space.

Nanomaterials have the promise to enhance life and to contribute to industrial competitiveness in Europe and worldwide. However, the new materials may also pose risks to the environment and increase health and safety concerns. These risks and degree to which they can be managed by the current risk evaluation procedures in the EU, have been the focus of some views of the Scientific Committee on Emerging and Newly Identified Health Risks (SCENIHR). Even though nanomaterials are not dangerous per se, there still is scientific uncertainty about the safety of nanomaterials in many aspects; therefore the safety assessment of the substances must be done on a case-by-case basis.

Information on nanotechnologies in general can be found on the Europa website on nanotechnologies (https://ec.europa.eu/jrc/en/research-topic/nanotechnology) as well as on the website of the European Observatory for Nanomaterials (EUON) (https://euon.echa.europa.eu/). Information on how EU regulation commonly applies to nanomaterials can be found in the Commission Communication on the Second Regulatory Review on Nanomaterials and in the Commission Staff Working Document.

To define the term *nanomaterials*, the European Commission has provided a suggestion based exclusively on the size of the constituent particles of a material without regard to hazard or risk. This definition covers natural, incidental, or manufactured materials and supports the execution of regulatory provisions for this group of materials. Nevertheless, in some legislative areas the driver for legal obligations for nanomaterials is that they may have different properties compared to larger particles. For a harmonious and efficient application of regulatory provisions specific to nanomaterials, the Commission adopted the recommendation on the definition of a nanomaterial.

The EU formed SCENIHR to study the implications of nanotechnology, and it has published a list of risks associated with nanoparticles. Manufacturers and importers of carbon products, including carbon nanotubes, will have to submit full health and safety data within a year or so in order to conform registration in REACH.

A number of European member states have called for the formation of either national or European nanomaterials registries. France, Belgium, Sweden, and Denmark have developed national registries of nanomaterials. Moreover, the European Commission requested the ECHA to create EUON, which aims at collecting publicly available information on the safety and markets of nanomaterials and nanotechnology.

In the EU, nanomaterials are protected by the same rigorous regulatory framework that ensures the safe use of all chemicals and mixtures: the REACH and CLP regulations. Thus the hazardous properties of nanoforms of substances will have to be evaluated, and their safe use needs to be guaranteed. There are also precise provisions for nanomaterials in legislation specific to sectors such as food, biocides, and cosmetics.

To be legally manufactured or imported into the EU, all substances within the scope of REACH have to be registered. Depending on the volume of material placed on the market, manufacturers and/or importers, as part of their registration, must submit information on both human, health, and environmental impacts and harmful nanoforms, including an estimate of exposure throughout the life cycle. The same requirements apply to nanomaterials. When substances have harmful properties, the CLP needs them to be communicated to the ECHA and labeled and packaged so that the substances can be used safely.

Companies should be transparent in their REACH registration to clearly designate how the safety of nanoforms has been addressed, including what actions are desirable to effectively control the likely risk. ECHA guidance documents afford further support to companies on how to detect and report properties of their nanoforms.

In addition, to REACH and CLP, there is also sector-specific legislation in the EU for definite groups of products. They cover, for example, biocides, plant protection products, cosmetics, pharmaceuticals, toys, food, and electronic goods.

Legislation for environmental, worker, and consumer defense are applied in the EU through directives. If nanomaterials pose a risk to the environment, workers, or consumers, the generic rules set in the legislation apply in the same way to nanomaterials as to other forms of a substance. Examples of directives are the Water Framework Directive, the Directive on the Protection of Workers from the Risks Related to Exposure to Carcinogens and Mutagens at Work, and the Directive on the Safety of Toys.

5.7.3.1 Nanomaterials in REACH and CLP

REACH is the overarching legislation appropriate to the manufacture, placing on the market, and use of substances on their own, in preparations, or in products. Nanomaterials are covered by the definition of a substance in REACH, even though there is no obvious reference to nanomaterials. The general responsibilities in REACH, such as registration of substances manufactured at 1 ton or more and providing information in the supply chain, apply as for any other substance. Information on the implementation of REACH for nanomaterials, including guidance and the application of the REACH assessment procedures, can be found at https://ec.europa.eu/environment/chemicals/reach/reach_en.htm.

Nanomaterials that satisfy the standards for classification as hazardous under Regulation 1272/2008 on CLP of substances and mixtures must be classified and labeled. This applies to nanomaterials as substances in their own right or nanomaterials as special forms of the substance. Many of the associated provisions, including safety data sheets and classification and labeling, are applied now independently of the tonnage in which the substances are manufactured or imported. Substances, including nanomaterials, that meet the classification criteria as hazardous should have been communicated to the ECHA. Any update to the classification must also be communicated without unjustified delay. It is possible that the classification is functional only to precise form(s), including nanoforms, of the substance. On the basis of the information received under REACH registrations and CLP notifications, the ECHA publishes a classification and labeling inventory.

The pertinency of the provisions of the UNECE Globally Harmonized System of Classification and Labelling of Chemicals for nanomaterials was evaluated in 2015–16 by a subcommittee of specialists. As suitable, the outcomes will be reflected under CLP.

5.7.3.2 CARACAL and CARACAL subgroup on nanomaterials

In 2008–17 the European Commission prepared advice on how to manage nanomaterials in agreement with REACH and the CLP regulation in close cooperation with the competent authorities for REACH and the CARACAL Subgroup on Nanomaterials (CASG Nano, composed of member states and stakeholder experts). The first paper, "Nanomaterials in REACH," provides an impression of how the provisions of REACH apply to nanomaterials. The second paper, "Classification, Labelling and Packaging of Nanomaterials in REACH and CLP," emphasizes the classification of nanomaterials in harmony with REACH and predominantly the CLP regulation. CASG Nano also suggested RIPoN reports as an initial advice and basis for ECHA in its additional work on appropriate guidance documents.

CARACAL, chiefly through CASG Nano, has been following the growth of major creativities of the Commission on nanomaterials, such as the revision of REACH and the evaluation of effects of EU registry of nanomaterials, and assisted as the professional group and sounding board in the ongoing review of the Commission recommendation on the definition of nanomaterial.

The last mandate of CASG Nano, which expired in January 2018, has not been extended. CARACAL concluded that following the Commission's proposal on modification of REACH Annexes in 2017, the principal discussion moved to the REACH committee, while all the execution features are now discussed in the ECHA nanomaterial professional group. Nanomaterials, however, continue to be an important agenda point in every meeting of CARACAL.

5.7.4 Canadian policy on nanotechnology

Health Canada has recognized a definition of nanomaterials as "considers any manufactured product, material, substance, ingredient, device, system or structure to be nanomaterial if it is at or within the nanoscale (1–100 nm) in at least one spatial dimension, or is smaller or larger than the nanoscale in all spatial dimensions and exhibits one or more nanoscale phenomena." Concerning the endorsement of nanotechnology products, Canada trusts prevailing regulatory contexts. Health Canada directs manufacturers to coordinate with the accountable regulatory consultant during the early development process to determine and evaluate the product's risks and properties [32]. Health Portfolio Nanotechnology Working Group was developed in Canada for the assembly and debate of issues associated with nanotechnology, which comprises legislatures from regulatory bodies such as Health Canada and the Canadian Institutes of Health Research. An overall direction on nanotechnology-based health products and food has also been released by Health Canada [33].

5.7.5 Japanese nano policy

The Council for Science, Technology and Innovation (CSTI) has a principal role in developing a basic science and technology plan in Japan every five years [1]. The growth of nanomaterials was one of the four major plans. These plans projected three advancement approaches: (1) develop scientific technologies to resolve social problems with innovative materials by "True Nano," (2) develop scientific technologies to make innovations by "True Nano," and (3) develop technology-based construction for accelerating innovation by "True Nano." However, there is currently no legal control related to imposing specific nanomaterial safety regulations. The Japanese government used the Ministry of Economy, Trade and Industry to collect information about the nano industry and evaluated the harmful impacts of nanomaterials with the ministry of environment in order to improve nano safety [34]. They are currently examining a new system to deal with nanomaterials for safety precautions. The Japanese government has conducted several research projects analyzing the harmful risks of nanomaterials over the last few years.

On January 23, 2020, Japan held the 48th meeting of the CSTI. At the meeting, discussions were held on the comprehensive package to strengthen research capacity and support young researchers; goal setting for the Moonshot Research and Development Program, and other issues; and institutional reform policies to promote science, technology and innovation (STI). Japan intend to promote moonshot research in pursuit of six ambitious targets toward the resolution of issues that humanity is facing, such as a rapidly aging society, climate change, and quantum technology, which provides the foundation of next-generation encryption and other technologies.

5.7.6 South Korean policy on nanotechnology

Worldwide measures are now focused on safe applications of nanomaterials in terms of their effect on human health and the environment. Because of the safety concerns about numerous nanomaterials, international trade is heightened by regulating the trade of prevailing and new nanomaterials in the market. The government projected the act on the

Registration and Evaluation of Chemicals in the Republic of Korea (Korea REACH) to introduce the EU's REACH program in 2015 [35]. Furthermore, the Republic of Korea passed the Nanotechnology Development Promotion Act (NDPA, 2008, No. 8852) for advancing nano research and industrial applications of nanotechnology [36—38]. On the other hand, some major departments in the Republic of Korea have been able to manage nanomaterials. Though laws related exclusively to nano safety have yet to be decided, government-wide efforts in several departments are currently being made to execute nano safety management plans. The native policy about nanomaterials in the Republic of Korea is being advanced according to the first nano safety management plan (2012—16). The Korean government has focused on four fields: (1) construction of nano measurements, analysis techniques, and databases; (2) construction of nano safety evaluation techniques; (3) institutionalization of safety management and laying the foundation for introduction; and (4) professional worker training and building partnerships [38,39]. The Korean government continued with the second nano safety management plan (2017—21). According to the result of the first plan, the second plan was to draw conclusions related to the purpose, vision, and promoted strategy.

The second plan proposed more advanced and detailed fields, relative to the first plan, such as safety assessments and safety management. This will be accomplished by the construction of various nanomaterial and nanoproduct databases. The second plan is aimed at promoting the legal institutionalization of the life cycle assessment of nanomaterials and nanoproducts as well as preparing implementation methods for safety management [39]. In addition, to its own policy, the Korean government is trying to work with other international organizations. Korea has joined the OECD Working Party on Manufactured Nanomaterials (WPMN), which plays a dynamic role in the safety testing of nanomaterials. It also participates in the joint research of OECD WPMN SG 7 with the Korea Research Institute of Standards and Science (KRISS) and the EU's Joint Research Centre and in the collaborative research of OECD WPMN SG 8 with KRISS and the National Metrology Institute of South Africa. Korea also take part in the ISO/TC 229 for ISO standardization of nanotechnologies, the NANOREG project for the regulatory testing of manufactured nanomaterials, and cooperative research projects with KRISS and the Swiss Federal Laboratories for Materials Testing and Research of Nanotoxicology [40]. The Korean government has been constructing an organized approach to develop a nano safety plan. However, many companies in the Republic of Korea have continued to increase the production of nanomaterials and have sold a diversity of nanoproducts on the market in the absence of a nanomaterials policy [1].

5.7.7 Application of nanotechnology in Thailand

The results of a 2014—15 survey of nanotechnology status of the private sector by the Office of Science Policy in Thailand have been explored. The objective of this project is to investigate the status of nanotechnology in Thailand and to gather ideas for analysis. It also serves as a policy guide for promoting and supporting research development and innovation in nanotechnology. The surveyed groups were divided into three groups: government agencies and research institutes, educational institutes, and the private sector. Surveys were conducted in the form of questionnaires. The research and development cooperation between private agencies and other agencies is largely non-targeted. There is a lack of direction and clear policy on what to develop together and in what direction. At the same time, there has been an increase in collaborative projects with the government and universities. The major factor that the private sector sees as a conducive to the implementation of nanotechnology research with other organizations is that the private sector intends to develop products and services that are consistent with consumers' needs. The key problem to limiting private sector research collaboration with government and universities is that they have to take the financial risks if the project is not successful.

The Nanosafety and Ethics Strategic Plan (2017–21) is the second edition of the first Strategic Plan (2012–16) by the Office of the National STI Board together with the National Nanotechnology Center, National Science and Technology Development Agency, and Ministry of Science and Technology. It is designed to ensure continuous monitoring and management of safety and ethics in national nanotechnology development.

Recently, many other countries, including Australia, China, and India, have come forward with nanotechnology-based policies for safe and sustainable use either locally or globally. Updates can be found at https://statnano.com/policydocuments.

5.7.8 Response from advocacy groups

In January 2008 an alliance of over 40 civil society groups recommended a statement of principles calling for protective measure associated with nanotechnology [29]. The group has recommended action based on eight principles: (1) a precautionary foundation, (2) mandatory nano-specific regulations, (3) health and safety of the public and workers, (4) environmental protection, (5) transparency, (6) public participation, (7) inclusion of broader impacts, and (8) manufacturer liability.

Some nongovernmental organizations, including Friends of Earth, are calling for the creation of a separate nanotechnology-specific regulatory context for the regulation of nanotechnology. In Australia, Friends of the Earth suggested the formation of a Nanotechnology Regulatory Coordination Agency, directed by a Foresight and Technology Assessment Board. It is also claimed that a unified regulatory method would simplify the regulatory environment, thereby supporting industry innovation [17]. A national nanotechnology regulator could coordinate prevailing regulations associated with nanotechnology. Regulatory processes could vary from hard law at one extreme through licensing and codes of practice to soft self-regulation and negotiation in order to influence behavior [17]. The development of national nanotechnology regulatory frameworks may also assist in creating global regulatory contexts [17].

5.7.9 Some technical aspects of nanomaterials

For the regulation of nanotechnology a definition of the size at the nano-scale is necessary. The size-defining feature of nanotechnology is the issue of noteworthy argument and fluctuates to include particles and materials in the scale of at least 100−300 nm. Australia recommends defining nanoparticles as being upto 300 nm in size. They argue that "particles upto a few 100 nm in size share many of the novel biological behaviors of nanoparticles, including novel toxicity risks" and that "nanomaterials upto approximately 300 nm in size can be taken up by individual cells." The UK Soil Association claims to include nanomaterials with mean particle sizes of 200 nm or smaller. The U.S. NNI defines nanoparticles as being roughly 1−100 nm. Thus the usefulness of regulating nanotechnologies on the basis of their size or weight is inadequate because the toxicity of nanoparticles is associated more with surface area than with weight, and developing regulations should take such factors into account.

5.7.10 The regulation of nanomaterials for clinical application

Nanomedicines are currently entering drug regulatory processes, and within a few decades they could constitute a leading group of state-of-the-art pharmaceuticals. Nanomedicines may create unique or heightened policy challenges for government systems of cost-effectiveness as well as safety regulation [15,41].

The use of nanomaterials in biomedicine has greatly increased over the past two decades. Currently, many different nanoparticle systems are being utilized in clinics. However, for the use of nanomaterials in clinics, regulation of these materials is vital. The encounters commonly faced in regulating these materials and the current regulations used in different nations are the crucial points to address for the proper utilization of nanomaterials worldwide in future nano-based technologies.

There is great enthusiasm surrounding the emergent field of nanomedicine. Numerous nanomedicines function by direct interaction with genetic materials or by interaction with biomolecules that are essential for normal genome function and cell division [42]. Many of these nanomedicines can cause genotoxicity and mutagenicity [43]. Such toxicity is facilitated by the inflammatory response of neutrophils and macrophages by the creation of reactive oxygen and nitrogen species, which cause oxidative and nitrosative stress [44]. The buildup of such free radicals can cause extensive damage to the body [45]. There are numerous ways in which this damage can occur, including prompting oxidative DNA damage leading to strand breakage, protein denaturation and lipid peroxidation causing cancer, causing damage to mitochondrial membranes leading to cell death and necrosis, and transcription of genes responsible for carcinogenesis and fibrosis [46]. When delivered intravenously, these particles accumulate in the liver and are translocated to areas such as the central nervous, cardiovascular, and renal systems (Fig. 5.1) [47]. For particles that cannot be traced after internalization, there are too many unknown effects that may pose threats to health and safety. The specific interactions of many nanomedicines with biological systems are not yet fully understood yet. Therefore it is difficult to understand, identify, or draw conclusions about the physico-chemical and toxicological properties of nanomedicines.

One size certainly does not fit all in this process, as the nanomaterials are highly dependent upon synthesis procedure, nanoparticle type, surface properties, administration route, and nanoparticle morphology, which can be diverse. The regulatory agencies need to be more attentive. However, an overly cautious approach might lead to great apathy within the field. Guidance is critical, as without it, manufacturers, healthcare providers, the public, and policymakers are without clarity and legal certainty.

Some initiatives within local communities have been put together and funded, such as the REFINE project, which pursues defining the criteria for regulatory needs for nanomedicines and nanomaterials for clinical use [48]. However, many perceive that no firm and reliable lines have been drawn in order for regulation to be consistent worldwide. In their white paper published in 2019, the REFINE project outlines their objectives, including "Development and validation of new analytical or experimental methods" [49]. This is crucial, as those nano-based interventions reach the clinic and subsequently fail, owing to lack of reliable or proper preclinical testing models [50].

When the impact of nanomedicines is being anticipated, their possible environmental effects after use, upon disposal, and during production need to be considered [51]. Conventional pharmaceuticals are ultimately found in the environment, so it is expected that nanomedicines will behave similarly. Therefore they could negatively affect the environment [52]. The FDA quotes the lack of data to determine the safety to humans and the environment, and it is struggling to articulate a criterion to ensure safe and effective development of nanoproducts, whether they are drugs, devices, or biological agents. The FDA released its first draft guidance in June 2011. However, a final guidance document has not yet been generated for nanoparticles in medicine [53]. Despite the great need for a formal regulatory document, the FDA continues to ignore already assembled data on toxicity profiles. The agency is taking a precautionary approach to the regulation of nanomedicines, treating them as identical counterparts to their bulk equivalents. This is negatively affecting the development of nanomedicine and constraining future use of these nanomaterials. The ambiguity may affect future funding, research, and development while limiting public acceptance. This may also lead to a delay in the commercialization of nanoproducts [54].

In the assessment of medical products in the United States and the EU, there are inclusion and exclusion criteria based on likely environmental effects. In the EU, all marketing approval applications are required to undergo an environmental risk assessment and a prescreening stage involving an estimation of the predicted environmental concentration in surface water, the acceptable limit being 0.01 parts per billion (ppb). In the United Sates the FDA uses an environmental assessment for new drug applications unless they are exempt from it. However, if the expected concentration in the environment exceeds 1 ppb, an exemption cannot be made. Generally, these criteria also apply to nanomedicines [5].

5.8 Conclusion and future perspectives

The uses of nanomaterials are continuously rising as consumer products in worldwide sales. However, most of the global market has no distinct regulations regarding consumer products of nanomaterials. Considering nano safety of humans and the environment, international organizations and the developed countries (the United States, the UK, the EU, Canada, Japan, South Korea, etc.) are trying to design guidelines and standards for toxic assessment and the regulation of plans or research projects for nanotechnology. They are enhancing the efforts on international activities including international organization work, collaborative research, and domestic plans for nanotechnology and nano safety. Nevertheless, they should focus more on nano safety policies and regulation plans in connection with consumer products of nanomaterials, including nanomedicines. The ambiguity in nano regulations may affect future funding, research, and development while limiting public acceptance. This may also lead to a delay in the commercialization of products derived from nanomaterials. This will highly negatively affect the development of nanotechnology and constrain future use of these technology-based outcomes worldwide.

The problems that are commonly faced in regulating the nanomaterials and the existing regulations used in different countries are the vital points to address for the appropriate application of nanomaterials in future nano-based technologies.

An overly strict approach to regulation could hinder growth in the field of nanotechnology. Nevertheless, safety studies are essential for every nanoscience application. The results of these studies can form the basis for government and international regulations. A more sensible approach might be the development of a risk matrix that recognizes probable nano offenders.

References

[1] H.G. Park, M.K. Yeo, Nanomaterial regulatory policy for human health and environment, Mol. Cell Toxicol. 12 (2016) 223–236.
[2] E.M. Osman, Environmental and health safety considerations of nanotechnology: nano safety, Biomed. J. Sci. Tech. Res. 19 (4) (2019) 14501–14515.
[3] D.B. Resnik, Policy forum: how should engineered nanomaterials be regulated for public and environmental health? AMA J. Ethics 21 (4) (2019) E363–E369.
[4] A. Valavanidis, T. Vlachogianni, Engineered nanomaterials for pharmaceutical and biomedical products new trends, benefits and opportunities, Pharm. Bioprocess 4 (1) (2016) 013–024.

[5] R. Foulkes, E. Man, J. Thind, S. Yeung, A. Joy, C. Hoskins, The regulation of nanomaterials and nanomedicines for clinical application: current and future perspectives, Biomater. Sci. 8 (2020) 4653.

[6] C. Raab, M. Simkó, A. Gazsó, U. Fiedeler, M. Nentwich, How nanoparticles enter the human body and their effects there, nano trust dossiers, No. 002en. <http://hw.oeaw.ac.at/nanotrust-dossier>, 2011.

[7] O. Renn, M.C. Roco, Nanotechnology and the need for risk governance, J. Nanopart. Res. 8 (2006) 153−191. Available from: https://doi.org/10.1007/s11051-006-9092-7.

[8] P. Dario, Tiny things with a huge impact, the international regulation of nanomaterials, Mich. J. Environ. Adm. Law 7 (2) (2018) 456−458.

[9] G. Marchant, D. Sylvester, Transnational models for regulation of nanotechnology, J. Law Med. Ethics 34 (4) (2006) 714−725.

[10] K. Dixon, FDA says no new labeling for nanotech products, Reuters News Service. Planet Ark., 2007 (https://www.reuters.com/article/us-fda-nanotechnology-idUSN2514226320070725, retrieved 19.10.07).

[11] G. Rowe, T. Horlick-Jones, J. Walls, N. Pidgeon, Difficulties in evaluating public engagement initiatives: reflections on an evaluation of the UK GM nation? Public Underst. Sci. 14 (4) (2005) 331−352.

[12] K. Gavelin, R. Wilson, R. Doubleday, Democratic technologies? The final report of the Nanotechnology Engagement Group (NEG), Involve 2007, 212 High Holborn London WC1V 7BF.

[13] X. Michael, B. Amy, A. Ashley, B. Dominique, S. Dietram, Stimulating upstream engagement: an experimental study of nanotechnology information seeking, Soc. Sci. Q. 92 (5) (2011) 1192−1214.

[14] D.M. Bowman, G.A. Hodge, A small matter of regulation: an international review of nanotechnology regulation, Columbia Sci. Technol. Law Rev 8 (2007) 1−32.

[15] T.A. Faunce, Toxicological and public good considerations for the regulation of nanomaterial-containing medical products, Expert. Opin. Drug Saf. 7 (2) (2008) 103−106.

[16] D.M. Bowman, M. Fitzharris, Too small for concern? Public health and nanotechnology, Aust. N. Z. J. Public Health 31 (4) (2007) 382−384.

[17] D.M. Bowman, G.A. Hodge, Nanotechnology: mapping the wild regulatory frontier, Futures 38 (9) (2006) 1060−1073.

[18] D. Lerer, Big Things in Small Packages: Evaluating the City of Berkeley's Nanotechnology Ordinance Effectiveness as a Model of Targeted Transparency, Pace Envtl. L. Rev. 30 (2) (2013) 523−557. Available from: https://digitalcommons.pace.edu/pelr/vol30/iss2/6.

[19] C.A. Auplat, The challenges of nanotechnology policy making part 2. Discussing voluntary frameworks and options, Glob. Policy 4 (1) (2013) 101−107.

[20] Food and Drug Administration, FDA and Nanotechnology Products: Frequently Asked Questions, Food and Drug Administration, 2007.

[21] International Standardisation for Nanotechnologies, Institute for Food and Agricultural Standards. Archived from the original on 06-07-08.

[22] P. Dario, Tiny things with a huge impact, the international regulation of nanomaterials, Mich. J. Environ. Adm. Law 7 (2) (2018) 459−463.

[23] An Issues Landscape for Nanotechnology Standards, Report of a Workshop, Institute for Food and Agricultural Standards, Michigan State University, East Lansing, 2007.

[24] K. Rollins, Nanobiotechnology regulation: A proposal for self-regulation with limited oversight, 6 Nanotechnology L. & BUS, 2009, 221, 224.

[25] Task Force Report, FDA Nanotechnology Task Force, 2007.

[26] Small concerns: nanotech regulations and risk management, SPIE Newsroom, 2009.

[27] United States National Research Council, Review of Federal Strategy for Nanotechnology-Related Environmental, Health, and Safety Research, National Academies Press, 2008.

[28] Royal Society and Royal Academy of Engineering, Nanoscience and Nanotechnologies: Opportunities and Uncertainties, 2004.

[29] Principles for the Oversight of Nanotechnologies and Nano-materials, International Center for Technology Assessment, 2008.

[30] Broab International Coalition Issues Urgent Call for Strong Oversight of Nanotechnology (Press release), International Center for Technology Assessment, 31-07-07.

Safety and global regulations for application of nanomaterials   107

[31] P. John, Nanotechnology: no free lunch, Platter 1 (1) (2010) 9–17.

[32] Health Canada, Nanotechnology-based Health Products and Food, 2011. Available from: https://www.canada.ca/en/healthcanada/services/drugs-health-products/nanotechnology-basedhealth-products-food.html.

[33] Health Canada, Current Issues with Nanomedicines in Canada, 2010. Available from: https://www.ema.europa.eu/en/documents/presentation/presentation-nanomedicines-currentinitiatives-canada-duc-vu-health-canada_en.pdf.

[34] COMPASS, International survey of nanomaterials in leading countries in 2013. Compliance in Advance and Supporting System, COMPASS Report 310-13-009, 2013, pp. 1–9.

[35] CIRS, Available from: https://www.cirs-reach.com/news-and-articles/revised-korea-reach-the-act-on-the-registration-and-evaluation-of-chemicals.html.

[36] V. Milanović, A. Bučalina, Position of the countries in nanotechnology and global competitiveness, Management (1820-0222) 68 (2013) 69–79.

[37] M.D. Karim, A.B. Munir, Nanotechnology in Asia: a preliminary assessment of the existing legal framework, KLRI J. Law Legis 4 (2014) 75–131.

[38] NNPC, Nanotechnology Development Act Plan, National Nanotechnology Policy Center, NNPC Report. Available from: <http://www.nnpc.re.kr/attachfile/download?af_seqno = 392>, 2015, pp. 1–44.

[39] Ministry of Environment, A comprehensive plan study of nano-safety management by the 2nd governmental departments ('17-'21), Ministry of Environment of Republic of Korea, Chemical Safety Division, ME Report, 2015, pp. 1–10.

[40] MSIP, Nanotechnology Industry Strategies, Ministry of Science, ICT and Future Planning, Office of R&D Policy, MSIP Report, 2015, pp. 1–33.

[41] T.A. Faunce, Policy challenges of nanomedicine for Australia's PBS, Aust. Health Rev. 33 (2) (2009) 258–267.

[42] X.-Q. Zhang, X. Xu, N. Bertrand, E. Pridgen, A. Swami, O.C. Farokhzad, Interactions of nanomaterials and biological systems: implications to personalized nanomedicine, Adv. Drug Deliv. Rev. 64 (2012) 1363–1384.

[43] N. Singh, B. Manshian, G.J. Jenkins, S.M. Griffiths, P.M. Williams, T.G. Maffeis, et al., NanoGenotoxicology: the DNA damaging potential of engineered nanomaterials, Biomaterials 30 (2009) 3891–3914.

[44] B. Smolkova, M. Dusinska, A. Gabelova, Nanomedicine and epigenome. Possible health risks, Food Chem. Toxicol. 109 (2017) 780–796.

[45] V. Lobo, A. Patil, A. Phatak, N. Chandra, Free radicals, antioxidants and functional foods: impact on human health, Pharmacogn. Rev. 4 (2010) 118–126.

[46] A. Manke, L. Wang, Y. Rojanasakul, Mechanisms of nanoparticle-induced oxidative stress and toxicity, BioMed Res. Int. (2013) 942916. Available from: https://doi.org/10.1155/2013/942916.

[47] A. Kermanizadeh, D. Balharry, H. Wallin, S. Loft, P. Møller, Nanomaterial translocation – the biokinetics, tissue accumulation, toxicity and fate of materials in secondary organs – a review, Crit. Rev. Toxicol. 45 (2015) 837–872.

[48] <http://refine-nanomed.eu/>.

[49] B. Halamoda-Kenzaoui, H. Box, M. van Elk, S. Gaitan, R.E. Geertsma, E. Gainza Lafuente, et al., The REFINE White Paper, Publications office of the European Union, Luxemburg, 2019.

[50] D. Landesman-Milo, D. Peer, Transforming nanomedicines from lab scale production to novel clinical modality, Bioconjugate Chem. 27 (2016) 855–862.

[51] I. Mahapatra, J.R.A. Clark, P.J. Dobson, R. Owen, I. Lynch, J.R. Lead, Expert perspectives on potential environmental risks from nanomedicines and adequacy of the current guideline on environmental risk assessment, Environ. Sci. Nano 5 (2018) 1873–1889.

[52] A. Baun, S.F. Hansen, Environmental challenges for nanomedicine, Nanomedicine 3 (2008) 605–608.

[53] Food and Drug Administration, Drug Products, Including logical Products, that Contain Nanomaterials Guidance for Industry, 2017. Available from: https://www.fda.gov/regulatory-information/search-fda-guidance-documents/drug-products-including-biological-products-contain-nanomaterials-guidance-industry.

[54] R. Bawa, Regulating nanomedicine – can the FDA handle it? Curr. Drug Deliv. 8 (2011) 227–234.

CHAPTER 6

Nanowaste disposal and recycling

Sakshi Gupta[1] and Manish Kumar Bharti[2]
[1]Department of Civil Engineering, Amity School of Engineering & Technology, Amity University Haryana, Gurugram, India
[2]Department of Aerospace Engineering, Amity School of Engineering & Technology, Amity University Haryana, Gurugram, India

6.1 Introduction

Particulates that have their external dimensions in the nanometer (10^{-9} m) scale are called nanoparticles [1]. The prefix *nano-* stems from the Greek word *nános*, which means "dwarf." The domains that deal with the science, engineering, and technology of nanoparticles of approximate size between 1 and 100 nm are collectively termed nanotechnology. The International Organization for Standardization (ISO) defines nanotechnology as applying knowledge to manipulate and control nanometer-size matter [2]. The field of nanotechnology has brought about a new age of miniaturization with immense potential. To get an idea of how small a nanometer is, 1 nm is one-billionth of a meter, and 1 inch consists of 25,400,000 nm. A nanometer represents a dimension that is around 10,000 times smaller than a human hair [3].

Nanoparticles occur naturally through various cosmological, meteorological, geological, physical, chemical, and biological processes [4]. Nanoparticles fall onto the surface of the earth in the form of interplanetary dust in enormous quantities per year [5]. Nanoparticles are known to be formed in volcanic ash, as carbon soot from large-scale forest fires, or as the unintentional by-products of combustion processes [6,7]. Such naturally occurring nanoparticles are physically and chemically heterogeneous and are generally termed ultrafine particles. Nanoparticles can also be intentionally manufactured to exploit their peculiar physical, structural, morphological, chemical, electrical, electrochemical, mechanical, optical, or magnetic properties compared to their micro-scale counterparts [8]. Such deliberately manufactured and manipulated nanoparticles are aptly termed engineered nanoparticles (ENPs) and are specifically designed to deliver particular purposes or functions [9]. Nanoparticles are often characterized by their informal names depending on their variety of shapes, such as nanospheres, nanorods, nanofibers, nanorings, nanobelts, and nanosheets. According to ISO, nanomaterials can be categorized on the basis of their dimensions, as depicted in Fig. 6.1 [10,11].

Nanoparticles have been in use since the prehistoric age, and the applications of nanotechnology can be traced back almost 2000 years as an ingredient in the recipe of

Nanomaterials Recycling
DOI: https://doi.org/10.1016/B978-0-323-90982-2.00006-8

© 2022 Elsevier Inc.
All rights reserved.

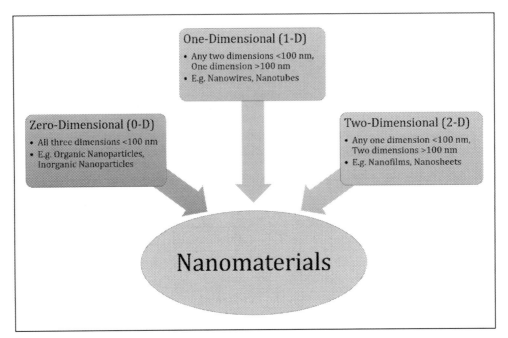

Figure 6.1 Categories of nanoparticles based on dimensions.

hair dye used by Romans [12]. Use of nanoparticles in the 4th—9th centuries BCE for surface glittering of pots is mentioned in historical literature. Artisans of the 10th century used gold and silver nanoparticles as quantum dots embedded in stained glass to obtain reflected red and yellow light [12,13]. Nevertheless, nanotechnology is considered to be a relatively modern branch of science and technology. Nanotechnology is deemed to have been conceptualized at Caltech on December 29, 1959, during the American Physical Society meeting. Renowned American physicist Dr. Richard P. Feynman, during his lecture titled "There Is Plenty of Room at the Bottom," stimulated his audience by describing the prospect of manipulation of individual atoms and molecules [14]. He suggested and discussed how scientists could obtain specific properties from materials manipulated and controlled at the molecular and atomic levels. These molecular manipulations could facilitate productive alterations in the physical, structural, morphological, chemical, electrical, electrochemical, mechanical, optical, or magnetic properties of any given material. Historians suggest that the talk was not that impactful at that time, but it is indeed considered the first milestone in the development course of nanotechnology [15].

Professor Norio Taniguchi, a Japanese scientist from Tokyo University of Science, is credited with coining the term "nano-technology" during a conference in 1974. He introduced the term while discussing the notion of precision manufacturing with dimensional tolerances in the nanometer range [16]. The year 1981 is considered the next major milestone of nanotechnology, owing to the invention of the scanning tunneling microscope by Gerd Binning and Heinrich Rohrer of the IBM Zurich research lab, for which both scientists were awarded the Nobel Prize in Physics in 1986. The invention of the scanning tunneling microscope was a pivotal advancement in science and technology, as it empowered the technical feasibility to image surfaces at the atomic level [17]. In parallel, a research scholar named K. Erich Drexler published his first paper on nanotechnology in 1981. His article highlighted the concept of molecular nanotechnology [18]. The next breakthrough in nanotechnology came with the discovery of fullerenes by Harry Kroto, Richard Smalley, and Robert Curl in 1985. They were awarded the Nobel Prize in Chemistry in 1996 for their discovery [19]. Unknowingly, K. Eric Drexler used a term relative to nanotechnology in his book *Engines of Creation* in 1986. His book had a significant impact and attracted the attention of researchers to the field of nanotechnology at a much larger magnitude [20]. By the early 2000s, nanotechnology had experienced a boost as nanomaterials started finding their passive applications in commercial sectors. It is reported that silver nanoparticles, titanium dioxide, zinc oxide, cerium oxide, etc. had begun to be used in versatile applications such as clothing disinfectants, food packaging, fuel catalyst, etc. by the early 2000s [21]. Since then, nanotechnology has emerged as one of the most rapidly developing science domains and is finding application in almost all fields of science, engineering, and technology. Owing to its versatility and multidisciplinary applicability, nanotechnology is often termed an enabling technology, as it materialized through the convergence of multiple spheres of science and engineering [22]. Given the cluster of advantages that nanotechnology had to offer, it aptly switched from passive use to active applications.

The exponential increase in the application areas of nanotechnology results from numerous peculiar properties of ENPs and their advantages. The smaller particles size, alterable aspect ratio, high chemical stability, high reactivity, high carrier capacity, high porosity, large quantum effects, and magnetic, electrical, and optical sensitivity along with distinctive physicochemical properties, chemical and geometric tractability, and surface functionalization, are a few of the commendable characteristics exhibited by ENPs, making them preferred candidates for a broad range of applications [23]. Today, this wide spectrum of applications of nanotechnology includes electronics [24], sensors [25,26], storage devices [27], medical imaging [28], targeted drug delivery [29], nanomedicine [30], cosmetics [31], aviation and space [32,33], food science [17], catalysts [34], water treatment [35,36], batteries [18], construction materials [37], and many more. According to the reports, the production of ENPs increased from 1000 tons in 2011 to around 58,000 tons by 2020 [38]. Such a massive upsurge in production reflects the increasing demand for ENPs in these industrial segments. According to reports published by Nanotechnology Products Database (NPD), almost 9000 types of ENPs had been manufactured by the end of the year 2020. A total of approximately 2500 companies from 63 countries have been actively participating in producing and utilizing ENPs [39]. Reports reveal that approximately 800–600 types of ENPs are used to manufacture sensors and cosmetics products, respectively [26]. Silver nanoparticles are the most preferred ENPs, owing to their superior performance for medical, healthcare, disinfectant, antimicrobial, food, and industrial purposes [20].

However, too much of anything becomes problematic. This uncontrolled upsurge in the production and use of ENPs in all fields and domains has given rise to a new waste group called nanowaste. As the name implies, it consists of the waste that materializes from the production and application of ENPs in one way or another. According to the British Standards Institution's Guide PD 6699-2, nanowaste can be categorized into four classes, as depicted in Fig. 6.2 [40]. Because of their smaller particle size, ENPs are prone to become airborne and penetrate the skin of living beings. These characteristics make ENPs more hazardous and reactive than their micro-scale or milli-scale counterparts [41]. Heavy metals such as cadmium or chromium used in the synthesis of various kinds of ENPs have proved to be less benign to the humans, animals, environment and aquatic ecosystems than was thought earlier [35].

6.2 Classifications of nanowaste

The classifications of nanowaste vary from researcher to researcher, but the primary classifications remain the same. Nanowaste can be classified on the basis of the nature of the ENPs in terms of exposure, hazard, and risk profile. Five classes of nanowaste have been identified, as shown in Fig. 6.3. Class I includes memory chips, solar panels, polishing agents, and so on. The nanowaste in this class has very low or negligible toxic effects on humans and the environment, primarily owing to nontoxic constituent

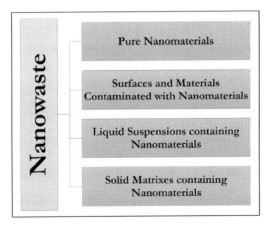

Figure 6.2 Forms of nanowaste.

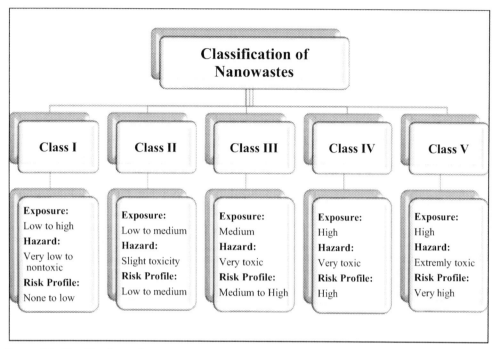

Figure 6.3 Classifications of nanowaste.

nanomaterials. Exposure potency for such nanowaste does not influence the waste stream's hazardousness even if the nanomaterials are embedded on the surface or inside the bulk of the material. Class II ENPs are generally found in paints, display backplanes, coatings, and so on. The ENPs in this class are likely to have slightly harmful or toxic effects on humans and the environment, owing to their characteristic toxicity ranging from low to high. The overall hazard potential of this class of nanowaste is found to be strongly linked to the exposure potency. If the exposure potency is low or unlikely, such wastes may be handled as nontoxic even when they contain considerably toxic materials. Class III comprises personal care products, waste water, food additives, and pesticides. The toxicity of class III nanowaste can be regarded as toxic to very toxic. This class of nanowaste has low to medium potential exposure during disposal. Class IV ENPs, which are very toxic, comprise pesticides and personal care products. The toxicity of this type of nanowaste ranges from toxic to very toxic. The exposure potential for class IV nanowaste is ranked from medium to high because the ENPs in the product are predicted to be freely bound on the products. The waste streams of class IV ENPs are viewed as extremely hazardous for humans and the ecosystem. Such waste streams need dedicated handling and appropriate treatment by either immobilizing or neutralizing the ENPs. Class V, a class of exceptionally harmful ENPs, includes sunscreen lotions, pesticides, food, and beverages. These nanowastes are tremendously hazardous because of their high degree of exposure and toxicity. The toxicity of class V nanowaste ranges from very toxic to extremely toxic. They need dedicated handling and effective and efficient treatment processes and must be disposed of only in well-designed and selected disposal sites. Continuous monitoring of the sites is advised to ensure that leachates from the disposal site are handled adequately. Immobilization and neutralization methods are among the most appropriate technologies for handling such waste. Not only is the waste stream highly toxic, but when released into the atmosphere, it is likely to have a very high contamination potential because it is in liquid form, potentially facilitating simple encounters with environmental species [42]. Class I is less to moderately dangerous, while class V is very hazardous. Such classifications of nanowaste are advantageous; however, there is no globally accepted nanowaste risk assessment framework, and there is a lack of policies for disposal and recycling of ENPs belonging to different classes [43].

6.3 Disposal and recycling of nanowaste

Although nanotechnology is essential to the development and advancement of many new products, there is a deficiency of information about the nanowaste produced during the production and application of these new ENP-based materials [42]. A new issue has come up in the waste management of ENPs in terms of their disposal and recycling. When nanowaste generation came into the picture in the past decade, it was assumed that the prevailing technologies would be adequate for the effective and efficient elimination of the ENPs from waste. However, current research studies have shown that the existing waste treatment and management systems are not adequate for removing ENPs, and much exploration is required in this field for disposal or recycling of such materials. The treatment of ENPs in water and sludge and their behavior in waste water treatment plants have been extensively studied in the past decade, and this research is still going on. However, the existing data and information are not adequate to regulate the general designs essential for the treatment, disposal, or recycling of nanowaste [40,42–44]. This clearly shows a dearth of internationally accepted classifications of nanowaste that can govern the level of risks involved. However, there are classifications that are being widely accepted these days, as was mentioned in Section 6.2.

Although the amount of ENPs in solid waste is proliferating, there is a shortage of knowledge about the effect and behavior of these pollutants in waste processing and disposal techniques, such as biological treatment, landfilling, and incineration [45]. However, several techniques, methodologies, or technologies have been introduced in recent studies that help in the disposal and recycling of nanowaste. A few of them are discussed in the subsequent sections.

6.3.1 Disposal of nanowaste

Nanowaste disposal requires widespread research and evaluation, and stringent and explicit norms and procedures must be adopted globally. Disposal of ENPs and ENPs-based products should be accomplished with specific care to ensure that ENPs that can threaten human health and the environment are not released. Whether hazardous, toxic, or chemically reactive, ENPs must be neutralized, and wherever possible, they must be recycled. Nanowaste can be the consequence or by-product of industrial or commercial developments. Owing to the extensive range of existing ENPs, a single disposal technique will not suffice for all classes. Hence it is imperative to comprehend the properties of specific nanowaste before developing effective disposal practices. The existing guidelines for the disposal of ENPs are vague, and not many techniques have been identified. In many cases, the organizations are left to determine whether a particular disposal method is required unless they fall under the jurisdiction of regulations for macro-sized chemicals [40,46]. Therefore organizations use the same techniques for disposal of ENPs as they do for any other chemicals. Usually, this means that they send the chemicals off to a specific company or group to deal with the disposal [47]. This technique can prove hazardous in some instances, as nanowaste cannot be treated similarly.

Although the primary source of ENPs is industrial products, waste treatment centers and research laboratories often also contribute. These ENPs eventually reach one of four sites, irrespective of the disposal outlet: a recycling facility, an incineration plant, a landfill, or a waste water treatment facility. Each of the sites has the potential, with some more preventable than others, to release ENPs into the environment. There are plastics with ENPs embedded in them in recycling facilities that can be released into the environment as dust. In an incineration facility, ENPs may be emitted as ash or dust at the moment of combustion. In an incinerator there are numerous paths that ENPs can follow; for example, the airborne particulates may flow through a filter intended to eliminate pollutants from the air. The disposal by incineration of any ENP-containing waste and the subsequent accumulation of particulate matter along with harmful by-products of combustion is unavoidable. The effect of ENPs on the toxicity of the particulate matter is also unknown. Therefore more research is being carried out in this area [48]. Many carbon-based ENPs, such as carbon nanotubes, can be destroyed in the incineration process, though only at adequately high temperatures. Carbon nanotubes are eradicated only by incineration at temperatures above $850°C$, so only modern waste incinerators functioning accurately can destroy them [47,49]. ENPs left in the bottom ash are generally of concern, as they are often recycled and can lead to health risks from working with them or environmental hazards when used in road construction [50].

Landfills have the most substantial likelihood for releasing ENPs, whether by air, water, or soil pollution. Waste water is converted into sludge, which is then burned or converted into agricultural fertilizer. Significantly less is known about the environmental consequences of fertilizer ENPs, while 100% of waste water sludge is typically incinerated. When recycling materials contain ENPs, there are three key concerns: the health of facility workers; the environmental impacts from residue that ends up in incinerators, landfills, or sewage treatment plants; and the contamination of products made from recycled materials. There is a possibility of the ENPs leaching from the products and affecting the surroundings [40]. According to a report by the Royal Commission on Environmental Pollution, during the manufacturing processes of fullerenes, only about 10% of the materials are usable, and the remaining 90% are disposed of in landfills. This introduces the ENPs into landfills.

The present procedures for testing and detection are insufficient or were not designed to deal with waste water streams comprising pollutants with nano-scale dimensions [42,51]. The available data and knowledge are inadequate to summarize the general necessities for documentation of nanowaste streams meriting treatment before release into the environment. This can be accredited to the absence of universally accepted classifications of nanowaste, which are vital in articulating their degree of hazardousness, or due to inadequate appreciation of the large of volumes of nanowaste that are being produced. It is unavoidable that some nanowaste will end up in waste water streams, even with recoverability. In recent years, researchers have suggested various ways to extract these ENPs from the waste water before they are released into the atmosphere and become challenging to filter out these ENPs through traditional filtration methods. Table 6.1 represents the types of nanowaste disposal or treatment processes and their probable leakage routes.

6.3.2 Recycling of nanowaste

During the past few years, the field of nanotechnology has grown significantly. From kilograms to thousands of tons of ENPs have been produced. Since ENPs are highly

Table 6.1 Types of nanowaste disposal or treatment processes and their probable leakage routes.

Type of nanowaste disposal and treatment processes	Probable leakage routes
Landfilling	• Release of gas into the atmosphere • Landfill surface discharge to the surrounding area • Various type of leaching
Waste water treatment	• Emissions to surface water • Waste water sludge to incinerators, landfills, and agriculture applications
Incineration	• Emission of the flue gas into the environment • Fly ash and bottom ash dumped at landfills
Recycling	• Entrenched in secondary materials

reactive and may have considerable toxicity, the environment will face many problems in the coming years. Owing to the small size of the ENPs, dealing with nanowaste is difficult. Conventional waste treatment processes do not decompose nanomaterials. Therefore several methods have recently been proposed to eliminate, reduce, or recycle ENPs from waste. Recovery and reuse of these expensive materials could further speed up nanotechnology applications to various fields by alleviating worries about nanowaste while also decreasing production costs. Recycling of waste is one of the strategies of waste minimization or waste prevention. Many individual countries and international bodies have set recycling goals and are taking measures to achieve them.

Recycling and reutilization of nanowastes are gaining universal acceptance for encouraging sustainable development. In the present situation, environmental conservation is an impetus for sustainability. The ENPs are recycled, reused, and recovered by filtration, centrifugation, ultracentrifugation, host-guest chemistry, solvent evaporation, pyrometallurgical and hydrometallurgical methods, electrochemical dissolution—electrodeposition, and magnetic seed recycling methods. The selection of recycling procedure for ENPs depends upon the nature of the nanowaste, the skill set of the operator, and the operator's knowledge of the process involved. In recycling, the conversion of nanowaste into useful products or derivatives is the simplest method of nanowaste management.

There are various methods and techniques for recycling of nanowaste. In terms of ease of recycling nanowaste containing ENPs and their possible effects on recycling processes, knowledge remains limited. Nowadays, more and more nanowaste is being generated that contains both toxic and valuable materials such as gold and silver. The successful recycling of such heavy metals from waste is indeed environmentally friendly. It is helpful in both avoiding heavy metal contamination and promoting sustainable development. Fig. 6.4 depicts the types of techniques used for the recycling of nanowaste.

Vacuum metallurgical separation processes are useful for separating and isolating heavy metal ions from electronic waste and nanowaste [52]. Another method of waste repurposing is to heat discarded compact discs (CDs) along with sand in a furnace to produce ENPs of silicon carbide, a nanomaterial with excellent thermal, chemical, and mechanical properties. The use of CDs is noteworthy because electronic waste such as CDs is expected to accumulate at three times the rate of regular household trash with growing urbanization [53]. Waste metallic film resistors, which are a significant part of electronics are of current interest. Since more than 1500 tons of waste from metallic film resistors, are produced every year, several countries like China are working on recycling the electronic waste. Waste metallic film resistors produce a ceramic matrix of about 33% by weight and 65% nickel by weight. If it is not adequately handled, the atmosphere and human health can be exposed to and harmed by nickel. Generally, magnetic separations and milling methods are used to recycle certain

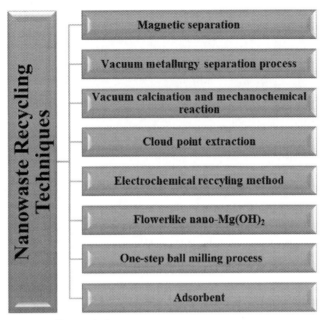

Figure 6.4 Nanowaste recycling techniques.

material forms. The magnetic separation technique is a primary method for separating nickel-based ENPs from there materials [54]. Ultracentrifugation and solvent evaporation methods document the recycling of usable ENPs and colloids. This technique is preferred for the separation, isolation, and recycling of ENPs such as Au, Ag, Ni, and Pd [55].

In recent years, the worldwide use of electrical and electronic equipment has increased dramatically, resulting in enormous amounts of electronic waste being produced from this electrical and electronic equipment. Different recycling techniques that allow the recovery of highly valuable ENPs have emerged as a new research area in this context [56]. Because of the need for useful ENPs and environmental protection concerns, the recovery of ENPs from electronic waste has been of great interest. Researchers have reported an aerosol-assisted method for the extraction of silicon ENPs from sludge wastes. Indeed, to directly recover silicon ENPs with high efficiency and high purity, the stated ultrasonic spray-drying method has been effectively applied. This low-cost and ecofriendly method is being used to manufacture recycled silicon ENPs that can be used directly for producing lithium-ion battery anodes. The transformation of waste into new products with dramatically improved values is a relatively low-cost solution [57].

Researchers have reported the separation of cadmium and zinc ENPs using a special solvent [58]. The solvent is a stable oil microemulsion in the water that breaks down into two layers when heated. All of the ENPs end up in one of the layers in the solution, allowing easy separation. The most widely used biological methods for disposing of solid organic waste from manufacturing and municipal operations are composting and anaerobic digestion. Household organic waste, garden and park waste, and sludge from waste water treatment plants are significant urban organic waste sources. Inorganic elements are conserved through both composting and anaerobic digestion and appear in the final outputs. Inorganic elements are not killed during biological waste treatment procedures but are likely to undergo transformation processes, such as complexation and aggregation with other compounds. Silver and zinc oxide, for instance, is converted into various more stable forms during biological treatment. In certain situations the presence of ENPs has had a beneficial impact on particular soil properties, such as the use of iron oxide to increase the supply of nutrients in saline soda [59,60].

In another instance, studies have been carried out to test incineration methods. In a study, 10 kg of cerium oxide ENPs with a diameter of 80 nm were kept in an incinerator for the process, and the incinerator was equipped with filters, and separation of fly ash was facilitated with the help of electrostatic filters and a scrubber of wet nature. The process was carried out at a rate of 8 tons of waste per hour [48]. A spraying technique was used for direct combustion of ENPs in the incineration chamber, and this experiment was based on the simulation of future cases in which the ENPs are expected to be released into the environment in high amounts [61]. In both of these studies, there was no effect of incineration on ENPs, and they were not removed or disintegrated. Other cases for nanowaste management include the recycling process of metal recovery from batteries, which requires quite a high temperature. However, the ENPs containing lithium batteries enabled with nanotechnology heating type of recovery process become inadequate. The smelting process that is generally applied in nano-enabled cases requires a very high temperature, which requires higher energy consumption, thereby increasing the overall load in terms of energy [62].

Some studies have reported promising methods of nanowaste treatment. An electroplating waste sludge method employing selective crystallization and formation of acid-insoluble nanowires has been used to recover SnO_2 NPs. Another example is the utilization of uranium-rich nanocrystals prepared by absorption-induced crystallization for uranyl enrichment. Additionally, thermoreversible fluid stage change and the cloud point extraction process guarantee fruitful division and recovery of necessary, high-worth, and asset-restricted materials from nanowaste [63].

6.4 Conclusion

Nanotechnology has launched a new era of miniaturization. Because of their unique properties, nanomaterials and ENPs are being utilized in various applications and consumer products, such as personal care products, electronics, textiles, pharmaceuticals, energy, water treatment, construction materials, and environmental applications. However, this has triggered the advent of a new waste group known as nanowaste that could create difficulties in existing waste management practices. Nanowaste includes waste groups containing ENPs; nano-scale synthetic by-products produced during production, storage, or distribution; end-of-life nanotechnological materials or products; and materials contaminated with ENPs, that is, as pipes, protective clothing, and so on. A scarcity of internationally accepted nanowaste classifications exists that could govern the level of risks of nanowaste. The lack of standard guidelines and frameworks related to the use and safe disposal, treatment, and recycling of ENPs is apparent. Toxicity and other hazards associated with ENPs and nanowaste are not yet entirely understood. Unfortunately, not much attention has been given to the assessment of toxicity and health and environmental hazards linked to the use of ENPs. Although the literature suggests some disposal and recycling methods, some dedicated technologies are urgently needed. Nanowaste disposal techniques should be judiciously chosen while keeping in mind the nature of the ENPs, posttreatment necessities, and disposal conditions. Currently, four methods are available: incineration, recycling, landfilling, and organic treatment. Landfilling is the ultimate disposal method; the other methods are usually followed by landfilling or recycling. Further analysis for the opposite characterization of ENPs must be carried out, taking full account of the latest advances in this domain. Moreover, improvement of the current disposal techniques to minimize hazards and disposal costs is essential along with the development of new techniques. It is expected, and there is a need for international and national organizations, R&D organizations, governments, and environmental agencies to extensively investigate comprehensive and effective ways and formulate stringent policies and guidelines for safe disposal and recycling of nanowaste.

References

[1] B.I. Kharisov, H.V.R. Dias, O.V. Kharissova, Mini-review: ferrite nanoparticles in the catalysis, Arab. J. Chem. 12 (2019) 1234–1246. Available from: https://doi.org/10.1016/j.arabjc.2014.10.049.

[2] The National Standards Authority of Ireland (NSAI), Nanotechnologies – Vocabulary – Part 2, Nano-Objects (ISO/TS 80004–2: 2015) (2017).

[3] P. Oberbek, P. Kozikowski, K. Czarnecka, P. Sobiech, S. Jakubiak, T. Jankowski, Inhalation exposure to various nanoparticles in work environment—contextual information and results of measurements, J. Nanopart. Res. 21 (2019). Available from: https://doi.org/10.1007/s11051-019-4651-x.

[4] S.K. Simakov, Nano- and micron-sized diamond genesis in nature: an overview, Geosci. Front. 9 (2018) 1849–1858. Available from: https://doi.org/10.1016/j.gsf.2017.10.006.

[5] J.M.C. Plane, Cosmic dust in the earth's atmosphere, Chem. Soc. Rev. 41 (2012) 6507–6518. Available from: https://doi.org/10.1039/c2cs35132c.

[6] M.S. Ermolin, P.S. Fedotov, N.A. Malik, V.K. Karandashev, Nanoparticles of volcanic ash as a carrier for toxic elements on the global scale, Chemosphere 200 (2018) 16–22. Available from: https://doi.org/10.1016/j.chemosphere.2018.02.089.

[7] G. Sigmund, C. Jiang, T. Hofmann, W. Chen, Environmental transformation of natural and engineered carbon nanoparticles and implications for the fate of organic contaminants, Environ. Sci. Nano 5 (2018) 2500–2518. Available from: https://doi.org/10.1039/C8EN00676H.

[8] P. Thakur, D. Chahar, S. Taneja, N. Bhalla, A. Thakur, A review on MnZn ferrites: synthesis, characterization and applications, Ceram. Int. 46 (2020) 15740–15763. Available from: https://doi.org/10.1016/j.ceramint.2020.03.287.

[9] D. Guo, G. Xie, J. Luo, Mechanical properties of nanoparticles: basics and applications, J. Phys. D Appl. Phys. 47 (2014). Available from: https://doi.org/10.1088/0022-3727/47/1/013001.

[10] S. Kralj, D. Makovec, Magnetic assembly of superparamagnetic iron oxide nanoparticle clusters into nanochains and nanobundles, ACS Nano 9 (2015) 9700–9707. Available from: https://doi.org/10.1021/acsnano.5b02328.

[11] X. Li, J. Wang, One-dimensional and two-dimensional synergized nanostructures for high-performing energy storage and conversion, InfoMat 2 (2020) 3–32. Available from: https://doi.org/10.1002/inf2.12040.

[12] G. Reiss, A. Hutten, Magnetic nanoparticles, Handbook of Nanophysics: Nanoparticles and Quantum Dots, CRC Press, 2010.

[13] N. Venkatesh, Metallic nanoparticle: a review, Biomed. J. Sci. Tech. Res. 4 (2018) 3765–3775. Available from: https://doi.org/10.26717/bjstr.2018.04.0001011.

[14] D. Schaming, H. Remita, Nanotechnology: from the ancient time to nowadays, Found. Chem. 17 (2015) 187–205. Available from: https://doi.org/10.1007/s10698-015-9235-y.

[15] S. Bayda, M. Adeel, T. Tuccinardi, M. Cordani, F. Rizzolio, The history of nanoscience and nanotechnology: from chemical-physical applications to nanomedicine, Molecules 25 (2020) 1–15. Available from: https://doi.org/10.3390/molecules25010112.

[16] M.F. Mady, M.A. Kelland, Review of nanotechnology impacts on oilfield scale management, ACS Appl. Nano Mater. 3 (2020) 7343–7364. Available from: https://doi.org/10.1021/acsanm.0c01391.

[17] S. Sharma, S. Jaiswal, B. Duffy, A.K. Jaiswal, Nanostructured materials for food applications: spectroscopy, microscopy and physical properties, Bioengineering 6 (2019) 1–17. Available from: https://doi.org/10.3390/bioengineering6010026.

[18] I. Kumar, C. Dhanasekaran, Nanomaterial-based energy storage and supply system in aircraft, Mater. Today Proc. 18 (2019) 4341–4350. Available from: https://doi.org/10.1016/j.matpr.2019.07.394.

[19] P. Innocenzi, L. Stagi, Carbon-based antiviral nanomaterials: graphene, C-dots, and fullerenes. A perspective, Chem. Sci. 11 (2020) 6606–6622. Available from: https://doi.org/10.1039/d0sc02658a.

[20] Z. Ferdous, A. Nemmar, Health impact of silver nanoparticles: a review of the biodistribution and toxicity following various routes of exposure, Int. J. Mol. Sci. 21 (2020) 2375. Available from: https://doi.org/10.3390/ijms21072375.

[21] J. Jeevanandam, A. Barhoum, Y.S. Chan, A. Dufresne, M.K. Danquah, Review on nanoparticles and nanostructured materials: history, sources, toxicity and regulations, Beilstein J. Nanotechnol. 9 (2018) 1050–1074. Available from: https://doi.org/10.3762/bjnano.9.98.

[22] S. Kargozar, M. Mozafari, Nanotechnology and nanomedicine: start small, think big, Mater. Today Proc. 5 (2018) 15492–15500. Available from: https://doi.org/10.1016/j.matpr.2018.04.155.

[23] I. Khan, K. Saeed, I. Khan, Nanoparticles: properties, applications and toxicities, Arab. J. Chem. 12 (2019) 908–931. Available from: https://doi.org/10.1016/j.arabjc.2017.05.011.

[24] Z. He, Z. Zhang, S. Bi, Nanoparticles for organic electronics applications, Mater. Res. Express. 7 (2020). Available from: https://doi.org/10.1088/2053-1591/ab636f.

[25] V. Vogel, Nanosensors and particles: a technology frontier with pitfalls, J. Nanobiotechnol 17 (2019) 7–9. Available from: https://doi.org/10.1186/s12951-019-0542-7.

[26] N.M. Noah, Design and synthesis of nanostructured materials for sensor applications, J. Nanomater. 2020 (2020) 8855321. Available from: https://doi.org/10.1155/2020/8855321.

[27] H. Tabassum, A. Mahmood, B. Zhu, Z. Liang, R. Zhong, S. Guo, et al., Recent advances in confining metal-based nanoparticles into carbon nanotubes for electrochemical energy conversion and storage devices, Energy Environ. Sci. 12 (2019) 2924–2956. Available from: https://doi.org/10.1039/c9ee00315k.

[28] S.D. Anderson, V.V. Gwenin, C.D. Gwenin, Magnetic functionalized nanoparticles for biomedical, drug delivery and imaging applications, Nanoscale Res. Lett. 14 (2019) 188. Available from: https://doi.org/10.1186/s11671-019-3019-6.

[29] D. Lombardo, M.A. Kiselev, M.T. Caccamo, Smart nanoparticles for drug delivery application: development of versatile nanocarrier platforms in biotechnology and nanomedicine, J. Nanomater. 2019 (2019) 3702518. Available from: https://doi.org/10.1155/2019/3702518.

[30] A. El-Sayed, M. Kamel, Advances in nanomedical applications: diagnostic, therapeutic, immunization, and vaccine production, Environ. Sci. Pollut. Res. 27 (2020) 19200−19213. Available from: https://doi.org/10.1007/s11356-019-06459-2.

[31] G. Fytianos, A. Rahdar, G.Z. Kyzas, Nanomaterials in cosmetics: recent updates, Nanomaterials 10 (2020) 1−16. Available from: https://doi.org/10.3390/nano10050979.

[32] J.A. Samareh, E.J. Siochi, Systems analysis of carbon nanotubes: opportunities and challenges for space applications, Nanotechnology 28 (2017). Available from: https://doi.org/10.1088/1361-6528/aa7c5a.

[33] D.A. Vartak, B. Satyanarayana, B.S. Munjal, K.B. Vyas, P. Bhatt, A.K. Lal, Potential applications of advanced nano-composite materials for space payload, Aust. J. Mech. Eng. 00 (2020) 1−9. Available from: https://doi.org/10.1080/14484846.2020.1733176.

[34] S.M. Shakil Hussain, M.S. Kamal, M.K. Hossain, Recent developments in nanostructured palladium and other metal catalysts for organic transformation, J. Nanomater. 2019 (2019) 1562130. Available from: https://doi.org/10.1155/2019/1562130.

[35] M.K. Bharti, S. Gupta, S. Chalia, I. Garg, P. Thakur, A. Thakur, Potential of magnetic nanoferrites in removal of heavy metals from contaminated water: mini review, J. Supercond. Nov. Magn. 33 (2020) 3651−3665. Available from: https://doi.org/10.1007/s10948-020-05657-1.

[36] P. Punia, M.K. Bharti, S. Chalia, R. Dhar, B. Ravelo, P. Thakur, et al., Recent advances in synthesis, characterization, and applications of nanoparticles for contaminated water treatment − a review, Ceram. Int. 47 (2021) 1526−1550. Available from: https://doi.org/10.1016/j.ceramint.2020.09.050.

[37] A. Mohajerani, L. Burnett, J.V. Smith, H. Kurmus, J. Milas, A. Arulrajah, et al., Nanoparticles in construction materials and other applications, and implications of nanoparticle use, Materials (Basel) 12 (2019) 1−25. Available from: https://doi.org/10.3390/ma12193052.

[38] S.H. Joo, D. Zhao, Environmental dynamics of metal oxide nanoparticles in heterogeneous systems: a review, J. Hazard. Mater. 322 (2017) 29−47. Available from: https://doi.org/10.1016/j.jhazmat.2016.02.068.

[39] I.D.L. Cavalcanti, M. Cajubá de Britto Lira Nogueira, Pharmaceutical nanotechnology: which products are been designed against COVID-19? J. Nanopart. Res. 22 (2020) 276. Available from: https://doi.org/10.1007/s11051-020-05010-6.

[40] H. Amoabediny, A. Naderi, J. Malakootikhah, M. Koohi, A. Mortazavi, M. Naderi, et al., Guidelines for safe handling, use and disposal of nanoparticles, J. Phys. Conf. Ser. 170 (2009) 012037. Available from: https://doi.org/10.1088/1742-6596/170/1/012037.

[41] V. De Matteis, Exposure to inorganic nanoparticles: routes of entry, immune response, biodistribution and in vitro/in vivo toxicity evaluation, Toxics 5 (2017) 29. Available from: https://doi.org/10.3390/toxics5040029.

[42] N. Musee, Nanowastes and the environment: potential new waste management paradigm, Environ. Int. 37 (2011) 112−128. Available from: https://doi.org/10.1016/j.envint.2010.08.005.

[43] A. Boldrin, S.F. Hansen, A. Baun, N.I.B. Hartmann, T.F. Astrup, Environmental exposure assessment framework for nanoparticles in solid waste, J. Nanopart. Res. 16 (2014) 2394. Available from: https://doi.org/10.1007/s11051-014-2394-2.

[44] G.G. Leppard, I.G. Droppoa, M.M. West, Compartmentalization of metals within the diverse colloidal matrices comprising activated sludge microbial flocs, J. Environ. Qual. 32 (6) (2003) 2100−2108. Available from: https://doi.org/10.2134/jeq2003.2100.

[45] C.R. Ratwani, Nanowaste: tiny waste that matters a lot, Int. J. Curr. Res. 10 (2018) 70262−70268.

[46] State Secretariat for Economic Affairs (SECO), Safety Data Sheet (SDS): Guidelines for Synthetic Nanomaterials, Switzerland, 2016.

[47] A.L. Holder, E.P. Vejerano, X. Zhou, L.C. Marr, Nanomaterial disposal by incineration, Environ. Sci. Process. Impacts 15 (2013) 1652−1664. Available from: https://doi.org/10.1039/c3em00224a.

[48] E.P. Vejerano, Y. Ma, A.L. Holder, A. Pruden, S. Elankumaran, L.C. Marr, Toxicity of particulate matter from incineration of nanowaste, Environ. Sci. Nano 2 (2015) 143−154. Available from: https://doi.org/10.1039/c4en00182f.

[49] A.R. Köhler, C. Som, A. Helland, F. Gottschalk, Studying the potential release of carbon nanotubes throughout the application life cycle, J. Clean. Prod. 16 (2008) 927−937. Available from: https://doi.org/10.1016/j.jclepro.2007.04.007.

[50] L. Andersen, F.M. Christensen, J.M. Nielsen, Nanomaterials in Waste − Issues and New Knowledge, 2014. Available from: http://www2.mst.dk/Udgiv/publications/2014/10/978-87-93283-10-7.pdf.

[51] H.P. Jarvie, H. Al-Obaidi, S.M. King, M.J. Bowes, M.J. Lawrence, A.F. Drake, et al., Fate of silica nanoparticles in simulated primary wastewater treatment, Environ. Sci. Technol. 43 (2009) 8622−8628. Available from: https://doi.org/10.1021/es901399q.

[52] L. Zhan, Z. Xu, State-of-the-art of recycling E-wastes by vacuum metallurgy separation, Environ. Sci. Technol. 48 (2014) 14092−14102. Available from: https://doi.org/10.1021/es5030383.

[53] R. Rajarao, R. Ferreira, S.H.F. Sadi, R. Khanna, V. Sahajwalla, Synthesis of silicon carbide nanoparticles by using electronic waste as a carbon source, Mater. Lett. 120 (2014) 65−68. Available from: https://doi.org/10.1016/j.matlet.2014.01.018.

[54] J. Ruan, J. Huang, L. Dong, Z. Huang, Environmentally friendly technology of recovering nickel resources and producing nano-Al_2O_3 from waste metal film resistors, ACS Sustain. Chem. Eng. 5 (2017) 8234−8240. Available from: https://doi.org/10.1021/acssuschemeng.7b01900.

[55] O. Myakonkaya, Z. Hu, M.F. Nazar, J. Eastoe, Recycling functional colloids and nanoparticles, Chem. A Eur. J. 16 (2010) 11784−11790. Available from: https://doi.org/10.1002/chem.201000942.

[56] T. Dutta, K.H. Kim, A. Deep, J.E. Szulejko, K. Vellingiri, S. Kumar, et al., Recovery of nanomaterials from battery and electronic wastes: a new paradigm of environmental waste management, Renew. Sustain. Energy Rev. 82 (2018) 3694−3704. Available from: https://doi.org/10.1016/j.rser.2017.10.094.

[57] H.D. Jang, H. Kim, H. Chang, J. Kim, K.M. Roh, J.H. Choi, et al., Aerosol-assisted extraction of silicon nanoparticles from wafer slicing waste for lithium ion batteries, Sci. Rep. 5 (2015) 1−5. Available from: https://doi.org/10.1038/srep09431.

[58] M.J. Hollamby, J. Eastoe, A. Chemelli, O. Glatter, S. Rogers, R.K. Heenan, et al., Separation and purification of nanoparticles in a single step, Langmuir 26 (2010) 6989−6994. Available from: https://doi.org/10.1021/la904225k.

[59] C.O. Dimkpa, P.S. Bindraban, Fortification of micronutrients for efficient agronomic production: a review, Agron. Sustain. Dev. 36 (2016) 1−26. Available from: https://doi.org/10.1007/s13593-015-0346-6.

[60] S. Bolisetty, M. Peydayesh, R. Mezzenga, Sustainable technologies for water purification from heavy metals: review and analysis, Chem. Soc. Rev. 48 (2019) 463−487. Available from: https://doi.org/10.1039/c8cs00493e.

[61] D.M. Mitrano, K. Mehrabi, Y.A.R. Dasilva, B. Nowack, Mobility of metallic (nano)particles in leachates from landfills containing waste incineration residues, Environ. Sci. Nano 4 (2017) 480−492. Available from: https://doi.org/10.1039/c6en00565a.

[62] S. Olapiriyakul, R.J. Caudill, Thermodynamic analysis to assess the environmental impact of end-of-life recovery processing for nanotechnology products, Environ. Sci. Technol. 43 (2009) 8140−8146. Available from: https://doi.org/10.1021/es9006614.

[63] Z. Zhuang, X. Xu, Y. Wang, Y. Wang, F. Huang, Z. Lin, Treatment of nanowaste via fast crystal growth: With recycling of nano-SnO_2 from electroplating sludge as a study case, J. Hazard. Mater. 211−212 (2012) 414−419. Available from: https://doi.org/10.1016/j.jhazmat.2011.09.036.

CHAPTER 7

Management of nanomaterial wastes

Nakshatra B. Singh[1], Martin F. Desimone[2], Ratiram Gomaji Chaudhary[3] and W.B. Gurnule[4]

[1]Department of Chemistry and Biochemistry, SBSR, Sharda University, Greater Noida, India
[2]Universidad de Buenos Aires, Consejo Nacional de Investigaciones, Científicas y Técnicas (CONICET), Instituto de Química y Metabolismo del Fármaco (IQUIMEFA), Facultad de Farmacia y Bioquímica, Buenos Aires, Argentina
[3]P.G. Department of Chemistry, S.K. Porwal College, Kamptee, India
[4]Department of Chemistry, Kamla Nehru Mahavidyalaya, Nagpur, India

7.1 Introduction

Nanomaterials (NMs) have size less than 100 nm in one, two, or three dimensions. Devices using NMs lead to nanotechnology. NMs show enhanced and different properties than the conventional bulk materials. Nanotechnology is a developing area of science and technology, and much research is being carried out in the fields of energy, manufacturing, healthcare, waste treatment, and so on [1–3]. A variety of diverse materials, such as nanofibers, nanoclays, graphene, and carbon nanotubes (CNTs), are being produced that are lighter and stronger as well as having more prominent chemical reactivity at the nano scale. Because of the special properties of NMs, they have applications in fabrication of various materials and products such as thermoelectric materials, sensors, dye-sensitized solar cells, photocatalysts, cosmetics, packing materials, sprays, paints, cleaning agents, plastics, coatings, sunscreens, films, and nutraceuticals. Applications of nanotechnology in our daily life have become important. The contributions of nanotechnology show the starting of a new era, leading to a point of significant discontinuity between the past and the present [4,5]. However, when they enter the end-of-life phase after use, these materials and products become NW. Different NMs used for different applications are converted to different types of NW (Fig. 7.1). Safe disposal of NW is a big concern.

The effects of NW on environment and health have not yet been studied in detail. The literature is scattered. In this chapter, attempts are made to summarize the available literature on NW and its management.

7.2 Types of nanomaterials and nanowaste

NMs can be classified on the basis of their dimension, morphology, composition, and uniformity and agglomeration (Fig. 7.2) [2]. NW can be classified into five categories (Table 7.1) [1].

Different nanotechnologies using nanoproducts, generation of NW, and recycling of NW can be seen in Fig. 7.3 [2].

Nanomaterials Recycling
DOI: https://doi.org/10.1016/B978-0-323-90982-2.00007-X

© 2022 Elsevier Inc.
All rights reserved.

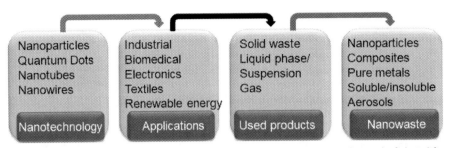

Figure 7.1 Nanomaterials used in nanotechnology become nanowaste at the end of their life.

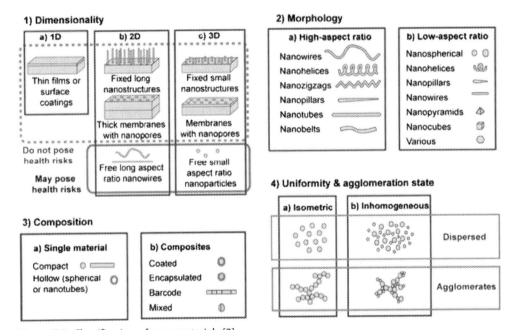

Figure 7.2 Classification of nanomaterials [2].

Table 7.1 Classification of nanowaste [1].

Class	Example	Nature		
		Exposure	Hazard	Risk profile
Class 1	Memory chips, solar panels, and polishing agents	Low to high	Very low to nontoxic	None to low
Class 1I	Polishing agents, paints, display backplanes, and coatings	Low to medium	Harmful effect	Low to medium
Class 1II	Personal care products, wastewater, food additives, pesticides	Medium	Very toxic	Medium to high
Class 1V	Pesticides, personal care products.	High	Very toxic	High
Class V	Sunscreen lotions, pesticides, and food and beverages	High	Extremely toxic	Very high

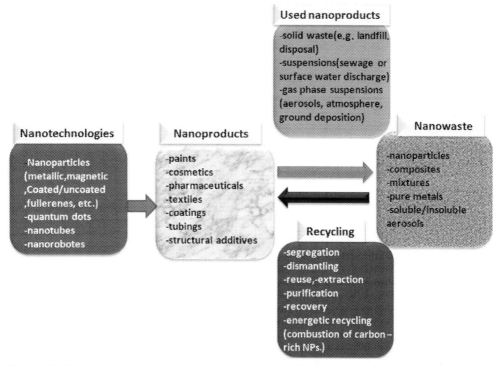

Figure 7.3 Pathways from nanotechnology to nanowaste [2].

7.3 Synthesis of nanomaterials

There are top-down and bottom-up approaches for the synthesis of NMs (Fig. 7.4). Considering the economics and toxicity of NMs, more emphasis is being given to green route synthesis (Fig. 7.5). In the green method of NMs synthesis, much emphasis is being given to plant extracts and biogenic methods. In recent years, wastes generated from different sources are also being used for the manufacture of NMs. Different type of wastes containing NMs and their effects on living systems are show in Fig. 7.6 [6].

7.4 Toxicity of nanomaterials and their release to the environment

Engineered NMs are being produced in large quantities and are being used in consumer products [7]. The driving forces for these and future developments are the possibility to design NMs with uniform size and structure that are amenable for specific grafting to fulfill the requirements for different applications [8,9]. Accordingly, the release of NMs into the environment can occur during different stages of the product life cycle, including production, transport, storage, use, and disposal. These NMs may possess new properties, including toxic behavior that is not yet completely known. Bachetta et al. [10] reported the effects of Ag NPs of 50 nm on *Piaractus mesopotamicus*, which is a fish,

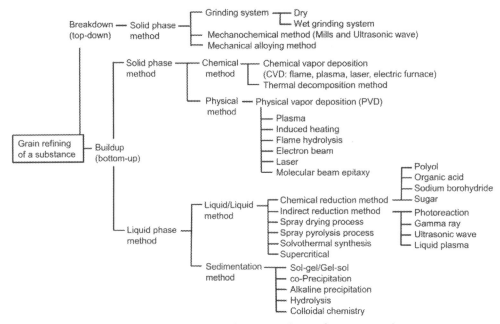

Figure 7.4 Top-down and bottom-up approaches to synthesis of nanomaterials.

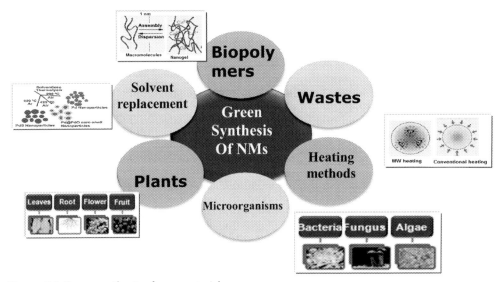

Figure 7.5 Green synthesis of nanomaterials.

widely distributed in South America and known as *pacú*. Silver accumulation was observed in the liver, brain, and gills. Moreover, affected activity of various antioxidant enzymes, including glutathione S-transferase, glutathione reductase, glutathione peroxidase, and catalase was observed in the brain, gills, and liver tissues of *P. mesopotamicus*

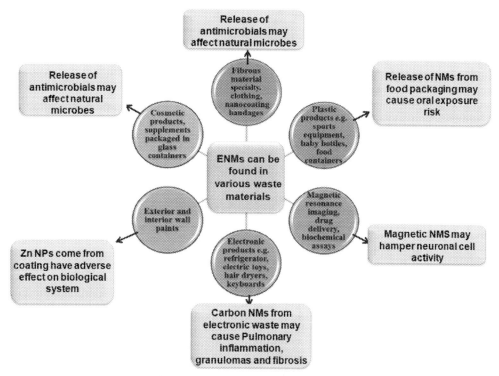

Figure 7.6 Nanomaterials in different wastes affecting living systems [6].

after exposure to silver NMs. The increased levels exhibited a dose-dependent response and clear relationship with silver concentration. Finally, blood erythrocytes of fish exposed to NMs also showed DNA damage. In conclusion, oxidative stress works as the main toxicological mechanism.

Ale et al. [11] analyzed the effect of sublethal concentrations of silver NMs on branchial multiple biomarkers of the neotropical fish *Prochilodus lineatus*. The common name of this fish is *sábalo*. Significant accumulation of silver in the gills was observed after 5 days, and this value increased five times after 15 days. Significant decreases in superoxide dismutase, catalase, glutathione S-transferase, glutathione reductase, and glutathione peroxidase, which are antioxidant enzymes, were observed from day 5. Moreover, pathological changes (i.e., aneurisms, hemorrhage, the presence of fusions, and hypertrophy) were observed in the gills [11].

Similar accumulation and toxicity of silver NMs in the marine mussel *Mytilus galloprovincialis* have been reported [12]. This marine bivalve is distributed mainly in the south of Europe and North America. It is a shelled mussel 14 cm in length that has the ability to filter great amounts of water. This behavior is particularly interesting to evaluate the retention of NMs and its consequent accumulation. Indeed, a concentration-dependent increase of Ag levels in mussel soft tissue after exposure to silver NMs of 30 nm was observed. The amount of silver detected was approximately 5 µg/g dry weight in whole soft tissues of mussels exposed to 10 µg/L of silver NMs. In addition, the hemocytes of mussels showed lysosome membrane destabilization, the frequency of micronucleus increased, the activity of glutathione S-transferase was higher, and there was more malondialdehyde accumulation.

In parallel, the nematode *Caenorhabditis elegans* was employed to evaluate the potential environmental hazard of nanosilver-coated dressings. *C. elegans* is a nematode 0.1 cm in length that lives in temperate soil environments. L1 larval worms exposed to nanosilver-coated dressing showed reductions in growth, fertility, and reproduction with a high reactive oxygen species (ROS) concentration. The authors suggested that the high ROS concentration observed in *C. elegans* exposed to nanosilver is the main toxicological mechanism generated by these particles. Furthermore, exposure to silver NMs reduced fertility and reproduction and also the size of adults.

In a different approach, Santo-Orihuela et al. [13] studied the effect of silica NPs on the viability of *Spodoptera frugiperda* cells (Sf9 cell line). *S. frugiperda* is one of the most abundant and widely distributed lepidopteran species in the world. The authors evaluated NMs of various sizes (14, 131, 240, 380, 448, and 1430 nm) and with different surface modifications (silanol and amine) with the aim of extending knowledge of their possible environmental toxicity. The nanotoxicological effects were assessed by the estimation of the lethal dose (LD_{50}). It was observed that the LD_{50} decreases with reduction in particle size. Alternatively, modification of the NMs with amine moieties significantly increased the LD_{50}, further confirming the lower toxic effect of these NMs. In conclusion, the study confirmed that the NPs possess toxicological effect that is highly influenced by their size, charge, and concentration [13].

Recently, Mitarotonda et al. presented an interesting review on the responses produced by cells of the immune system exposed to different NMs [14]. The focus of their work was on the physicochemical properties of the NMs that affect their interaction with the cells of the immune system. These interactions would contribute to understanding the possible toxicological effects of NMs. Indeed, there is evidence of size-dependent effect of silica microparticles and NPs on basal and specialized monocyte functions [15]. NMs increased the production of IL-8, IL-12, and nitrite.

There are a number of factors, such as size, mass, number, surface area, functionality, aspect ratio, and surface coating, that affect the toxicity of NMs. NMs may enter the body through different routes. CNTs were reported to be more toxic than silica, carbon black, Ag, TiO_2, Fe_2O_3, Al_2O_3, ZrO_2, and Si_3N_4. Single-walled CNTs and multiwalled CNTs (MWCNTs) have different toxicities [16]. CNTs are widely used in energy, biomedicine, electronics, photoelectricity, analysis, and catalysis, and they are toxic in nature. Other carbon NMs are also toxic. The mechanism of toxic effects of carbon NMs on cells is mainly the mechanism of oxidative stress toxicity and is shown in Fig. 7.7 [17].

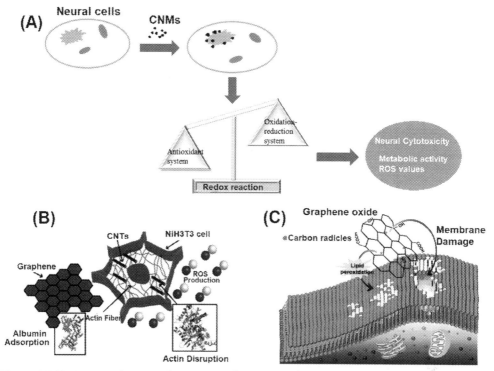

Figure 7.7 Toxicity mechanism of CNMs to cells [17]. (A) The equilibrium state of the ROS in the cells is broken by the addition of CNMs, resulting in the production of large amounts of oxidants and affecting cellular metabolic activity. (B) Toxic effects of CNMs on NIH3T3 cells: agonistic protein destruction and ROS production. (C) Toxic effects of GO on macrophages: lipid peroxidation and membrane damage.

7.5 Generation of nanowaste

NW is generated from various nano-based products of manufacturing industries, such as alloys and metals, paper, cardboard, plastics, textiles, leather, pesticides, electronics, batteries, paints and coatings, solar cells, tires, medical devices, and construction materials. The sources of NW generation are illustrated in Fig. 7.8. NMs are released as demolition wastes from the above-mentioned products and float easily in the air, eventually entering animals, human beings, aquatic life, and plant cells, causing lethal effects. As the nanotechnology is rapidly growing in various sectors, it is high time to monitor the NW materials properly to save the environment.

7.6 Impact of nanowaste on the environment

Large amounts of NM waste are found in the environment and are increasing exponentially in the commercial sector, causing risk to the environment and human health.

Figure 7.8 Sources of nanowaste generation.

They are also having chronic biological effects on aquatic organisms as a result of the accumulation of NW in their systems [18]. Environmental protection is today's burning issue for the entire globe. Scientists are constantly engaged in finding better and more cost-effective techniques with high efficiency to treat NW [19].

Nanotechnology continues to have a broad and fundamental impact on nearly all sectors of the global economy, including cosmetics, electronics, biomedicine, energy, defense, and agriculture [20]. Society has been facing the handling and management of NW for the last two decades [1]. NMs are important in three major sectors: chemicals and advanced materials, pharmaceuticals, and information and communication technology. Human health and the environment are affected to a large extent when they come into contact with NMs. A lot of NMs are being used in commercial products, such as sunscreen lotions, sports equipment, toothpaste, paints, electronics, energy, textiles, cosmetics, food and agriculture, construction, catalysis, and designer drugs for targeted drug delivery and disease diagnostics. Following are the main causes of NMs toxicity:

1. The chemical toxicity of materials from which they have been made. For example, Cd^{2+}, Hg^{2+}, Pb^{2+}, Al^{3+}, and Bi^{3+} are released from some NM waste.
2. Their small size. NPs may stick to cellular membranes and enter the cells.
3. Their shape. For example, CNTs can easily cut cell membranes.

Management of nanomaterial wastes   133

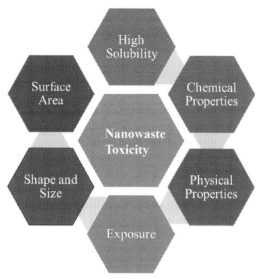

Figure 7.9 Causes of nanotoxicity.

4. The presence of NM-based coating on the surface of products.
5. The presence of another impurity in waste or disposal system that may lead to aggressive effects on human health and the ecosystem.

Al_2O_3 NPs show a substantial increase of biomass accumulation [21]. NW affects soil quality. Ag NPs affect soil microbial communities, and the amounts of *Acidobacteria*, *Actinobacteria*, *Cyanobacteria*, and *Nitrospirae* are considerably reduced with rising concentrations of Ag NPs. Nano-sized zero-valent iron decreases microbial biomass. These NPs show bactericidal effects on Gram-negative bacteria. MWCNTs may drop the microbial activity and biomass in soils. MWCNTs are found in numerous manufacturing goods and therefore may be released to soils and cause harmful effects on soil health [22]. Marine NW pollution affects several marine species [23].

7.7 Impact of nanowaste on health

Nanotechnology has lot of applications in biomedical sciences but the NMs present promote various types of human diseases. NMs may enter into the environment through automobile, mining, energy generation, biological agent, medicine, and combustion. The waste shows toxicity to human health (Fig. 7.9).

The acute toxicity of NMs is significantly associated with the physical dimensions, surface chemistry, and shape of the NMs. NMs enter through the nasal, oral, intraocular, intratracheal, tail vein, and other routes. The NMs used in sunscreens

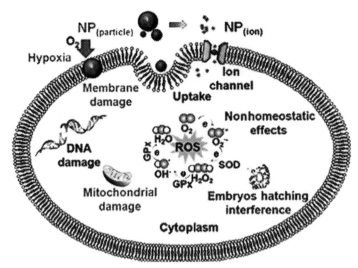

Figure 7.10 The mechanism of cell nanotoxicity [24].

and other cosmetics and consumer products generally penetrate through skin and have a toxic effect.

To evaluate the hazardous effects on the human body, biological systems, and the ecosystem, an understanding of the physicochemical properties of NMs and their performance is needed. Owing to agglomeration, the size distribution of NMs in a biosystem is continuously changing and affecting human health. NW from the burning of discarded textiles, batteries, and other nanotechnological products produces CNT, which is quite stable and does not degrade below 850°C. It has ability to enter the food chain, making it hazardous.

NMs contained in clothing penetrate into the body through human skin and become dangerous (Fig. 7.10) [24].

7.8 Biological treatment of nanowastes

The biogenic technique is a good option for the treatment of NW and involves microbes, algae, fungi protozoa, and plants to eliminate NW pollutants from an environment (Fig. 7.11). NW obtained from textile effluents can be removed using microbial cells and enzymes. Similarly, *Actinomycetes* demonstrated a superior capability to remove toxic metal ions by microbes and fungi [25]. Number of microorganisms, such as *Chlorella*, *Anabaena*, and marine algae, can be used to eliminate toxic NW. Moreover, biofilm degrades NW obtained from textile industry that contain copper, chromium, cadmium, mercury, and lead ions. Nitrifying bacteria also effectively remove copper and silver NPs [26,27]. Copper NP wastes are removed more effectively by using biomass.

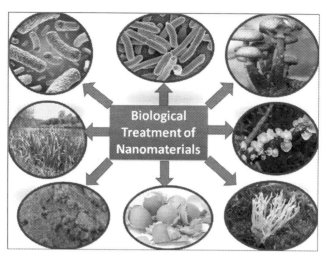

Figure 7.11 Biological treatments of nanomaterials by microbes, fungi, algae, and eggshells.

7.9 Recycling of nanowastes

During the last few years, the field of NMs and nanotechnology has grown considerably. From kilograms to thousands tons of NMs have been produced. Since NMs are highly reactive and toxic in nature, the environment will face lot of problems in the coming years. Owing to small size, dealing with NW is not easy. Conventional waste treatment processes do not decompose NMs. Because of this, a number of methods have recently been proposed to remove or recycle NMs from waste.

Recycling and reutilization of NW are gaining worldwide popularity for promoting sustainable development. In the current scenario the conservation of the environment is an urgent issue. The NPs are recycled, reused, and recovered by filtration, centrifugation, ultracentrifugation, solvent evaporation, host–guest chemistry, pyrometallurgical and hydrometallurgical methods, electrochemical dissolution-electrodeposition, and magnetic seed recycling methods. The selection of protocol for recycling of NPs depends upon the nature of the NW and the operator's skill and knowledge. In recycling, the conversion of NW into useful products or derivatives is the simplest method of NW management [28].

Fig. 7.12 illustrates the types of techniques that are used for recycling NW. Nowadays, more and more wastes are generated that contain both toxic and valuable materials such as silver, gold, and heavy metals. The successful recycling of heavy metals from e-waste is environmentally friendly. It is beneficial for both preventing heavy metal contamination and sustainable development. The vacuum metallurgical separation processes are useful for separation and isolation of heavy metal ions from e-waste and NW [29].

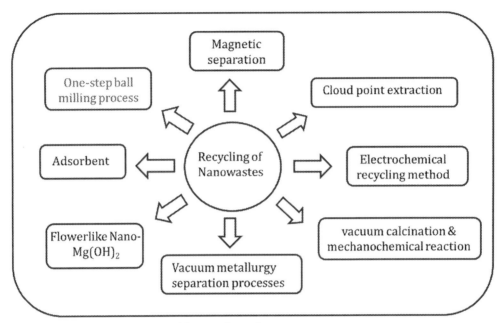

Figure 7.12 General methods used for recycling of nanowaste.

Waste metallic film resistors, a significant part of electronics, are of current interest around the globe. A number of countries, such as China, are working on the recycling of electronic waste, since 1500 tons of waste from metallic film resistors are generated every year. To recycle such types of material, magnetic separations and milling generally are used. Waste metallic film resistors contain about 33 wt.% ceramics matrix and 65 wt.% nickel. If it is not properly treated, nickel will be exposed to and damage the environment and human health. The magnetic separation technique is an efficient protocol for separating nickel NPs from materials [30].

Recycling of functional colloids and NPs has been reported by ultracentrifugation and solvent evaporation techniques. This technique is used for the separation, isolation, and recycling of NPs such as Au, Ag, Ni, and Pd [31].

The worldwide use of electrical and electronic equipment has risen considerably in the past years. Consequently, huge amounts of discarded electrical and electronic equipment (often referred to as e-waste) are being generated. In this context, different recycling methods that allow producing highly valuable NMs have emerged as a new scientific field [32]. Indeed, different procedures have been developed for the recovery of different types of NMs. Fig. 7.13 gives category-wise classification of the available recyclable resources and their recoverable targets (e.g., metals, oxides, NMs). Fig. 7.14 gives general steps involved in the recovery of high-purity NPs from spent alkaline batteries, through either hydrometallurgical or pyrometallurgical pathways [32].

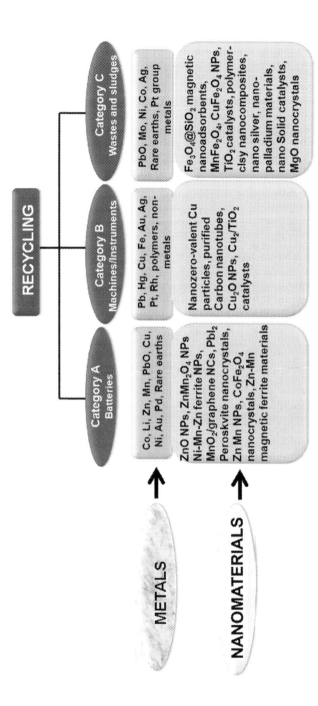

Figure 7.13 Recovery of metals and nanomaterials from recyclable wastes.

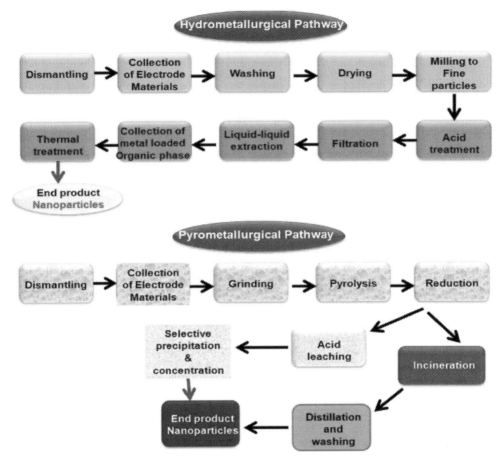

Figure 7.14 Recycling steps for the synthesis of nanomaterials.

Pins separated from microprocessors and treated by the hydrometallurgical method were used to obtain an Au(III)-enriched solution, from which gold NPs were obtained [33]. The method employs a combination of sodium citrate and ascorbic acid as reducing agents and polyvinylpyrrolidone as a polymeric stabilizer. These gold NPs were successfully applied in the production of optical sensing devices [33].

Recently, ZnO NPs were successfully recovered from spent Zn-carbon batteries via a facile thermal synthesis route. The developed method involves a thermal treatment (900°C) in an argon atmosphere. In this way, ZnO NPs were synthesized from a spent Zn-carbon battery. The resulting material was composed mainly of spherical NPs with diameters lower than 50 nm and a BET surface area of 9.2629 m^2/g [34]. In addition, fluorescent NM-carbon quantum dots were obtained by employing discharged batteries via a simple one-step recycling method [35]. These quantum dots could find applications in SERS-based sensing devices.

The recovery of NPs from waste has been of great interest, owing to the demand for valuable NMs and environmental protection issues. Jang et al. reported an aerosol-assisted method to extract silicon NPs from sludge waste. Indeed, the described ultrasonic spray-drying method was efficiently applied to directly recover silicon NPs with high efficiency and high purity. Furthermore, the proposed low-cost and ecofriendly method was employed to produce recycled Si NPs that can be directly used for making anodes of lithium –ion batteries [36]. It is a relatively low-cost approach to convert a waste into new materials with significantly enhanced values.

The textile industry uses a large number of nanotextiles for different purposes (Fig. 7.15). Textiles are used for ultraviolet radiation protection, antibacterial property, self-cleaning, flame retardancy, dirt repellency, water repellency, and so on. Nanofibers and nanocomposites are used in high-performance textiles. NMs used in the textile industry are (1) inorganic, such as metal oxide, metal, and nanoclays; (2) composite NMs; (3) carbon-based NMs, such as CNTs, carbon nanofibers, and graphene; (4) core-shell NPs; (5) polymeric NMs; and (6) hybrid NMs [37].

Textiles may be considered to be our second skin in some respects. Most textile materials are in direct and prolonged contact with our skin. NMs in textiles have the ability to modify the chemical and physical properties of textile materials. These NMs may enter the human body and may be dangerous for health [37]. However, there are contradictory reports about ill effects of cloths containing NMs.

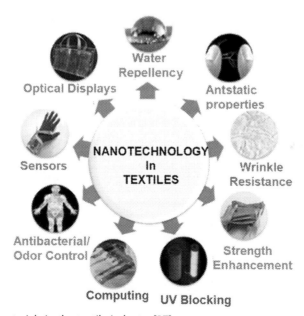

Figure 7.15 Nanomaterials in the textile industry [37].

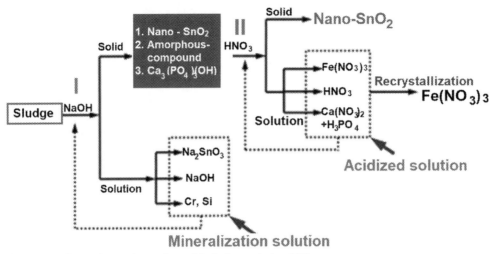

Figure 7.16 Extraction and recycling of SnO₂ from sludge [38].

Industrial sludge containing nanostructured materials are hazardous. Methods have been developed to extract and recycle electroplating sludge containing SnO_2 nanowires (Fig. 7.16) [38]

Fly ash (FA) produced from thermal power plants is a threat to the environment. Numbers of NMs are extracted from FA. In addition to the NMs in FA, nanocrystalline aluminosilicate is one of the promising nanocomposites the can be synthesized from FA (Fig. 7.17) [39].

7.10 Challenges in nanowaste management

Society is now entering the nanotechnology age, and NMs are being used in a number of technologies such as electronics, tissue engineering, biomedicine, healthcare, food technology, textile, and cosmetics. Different products after their use or life span become waste and generate NW. In general, NW containing NMs is not ecofriendly to humans or other natural biological systems. Generation of a huge amount of NW pollutes water bodies, soil, the air, and so on. Cosmetics containing NMs after use enter water bodies, sewage systems, and the environment, making them harmful. The negative impacts of NMs are not yet fully understood, and no definite procedure and policy have been established for treatment of NW. Experts agree that very little is known about suitable methods of cleaning and managing NMs from waste [1]. Therefore systematic and comprehensive studies need to be carried out so that methodologies and policies for NW treatment, management, and utilization can be formulated. Research organizations, national and international organizations, environmental agencies, and governments are expected to contribute in this area so that policies can be made for recycling and reuse of NW in the future.

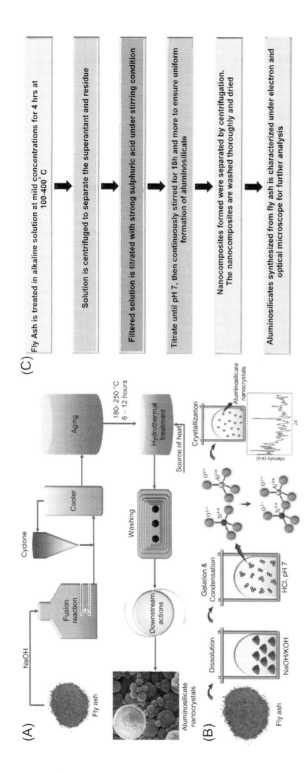

Figure 7.17 Nanostructured aluminosilicates from fly ash [39]. (a) Process flow diagram of hydrothermal method of nanostructured aluminosilicate extraction, (b) Schematic illustration of aluminosilicate nanocomposite synthesis at laboratory scale through the simplified alkaline treatment method, followed by gelation and condensation, (c) Experimental flowchart of synthesizing aluminosilicate nanocomposite using fly ash.

7.11 Conclusions

Synthesis and types of NMs and their pathways for conversion to NW have been described. Moreover, the generation of NW and its impact on the environment and human health have been discussed. Recycling of NW by different methods has been highlighted. There are many more challenges with regard to recycling of NW, which require further investigation.

References

[1] N. Musee, Nanowastes and the environment: potential new waste management paradigm, Environ. Int. 37 (2011) 112–128. Available from: https://doi.org/10.1016/j.envint.2010.08.005.

[2] G. Bystrzejewska-Piotrowska, J. Golimowski, P.L. Urban, Nanoparticles: their potential toxicity, waste and environmental management, Waste Manag. 29 (2009) 2587–2595. Available from: https://doi.org/10.1016/j.wasman.2009.04.001.

[3] E. Kabir, V. Kumar, K.-H. Kimc, A.C.K. Yipd, J.R. Sohn, Environmental impacts of nanomaterials, J. Environ. Manag. 225 (2018) 261–271. Available from: https://doi.org/10.1016/j.jenvman.2018.07.087.

[4] A. El Moussaouy, Environmental nanotechnology and education for sustainability: recent progress and perspective, in: C. Hussain (Ed.), Handbook of Environmental Materials Management, Springer, Cham, 2018, pp. 1–27. Available from: https://doi.org/10.1007/978-3-319-58538-3_96-1.

[5] R. Bakiu, Nanowaste classification, management, and legislative framework, in: C. Hussain (Ed.), Handbook of Environmental Materials Management, Springer, Cham, 2018, pp. 1–30. Available from: https://doi.org/10.1007/978-3-319-58538-3_151-1.

[6] P. Samaddar, S. Ok Yong, K.-H. Kim, E. Kwon Eilhann, C.W. Tsang Daniel, Synthesis of nanomaterials from various wastes and their new age applications, J. Clean. Prod. 197 (2018) 1190–1209. Available from: https://doi.org/10.1016/j.jclepro.2018.06.262.

[7] R. Kessler, Engineered nanoparticles in consumer products: understanding a new ingredient, Environ. Health Perspect. 119 (3) (2011) A120–A125. Available from: https://doi.org/10.1289/ehp.119-a120.

[8] A.M. Mebert, G.S. Alvarez, R. Peroni, C. Illoul, C. Hélary, T. Coradin, et al., Collagen-silica nanocomposites as dermal dressings preventing infection in vivo, Mater. Sci. Eng. C 93 (2018) 170–177. Available from: https://doi.org/10.1016/j.msec.2018.07.078.

[9] A.M. Wagner, J.M. Knipe, G. Orive, N.A. Peppas, Quantum dots in biomedical applications, Acta Biomater. 94 (2019) 44–63. Available from: https://doi.org/10.1016/j.actbio.2019.05.022.

[10] C. Bacchetta, A. Ale, M.F. Simoniello, S. Gervasio, C. Davico, A.S. Rossi, et al., Genotoxicity and oxidative stress in fish after a short-term exposure to silver NMs, Ecol. Indic. 76 (2017) 230–239. Available from: https://doi.org/10.1016/j.ecolind.2017.01.018.

[11] A. Ale, C. Bacchetta, A.S. Rossi, J. Galdopórpora, M.F. Desimone, F.R. de la Torre, et al., Nanosilver toxicity in gills of a neotropical fish: metal accumulation, oxidative stress, histopathology and other physiological effects, Ecotoxicol. Environ. Saf. 148 (2018) 976–984. Available from: https://doi.org/10.1016/j.ecoenv.2017.11.072.

[12] A. Ale, G. Liberatori, M.L. Vannuccini, E. Bergami, S. Ancora, G. Mariotti, et al., Exposure to a nanosilver-enabled consumer product results in similar accumulation and toxicity of silver nanoparticles in the marine mussel *Mytilus galloprovincialis*, Aquat. Toxicol. 211 (2019) 46–56. Available from: https://doi.org/10.1016/j.aquatox.2019.03.018.

[13] P.L. Santo-Orihuela, M.L. Foglia, A.M. Targovnik, M.V. Miranda, M.F. Desimone, Nanotoxicological effects of SiO_2 nanoparticles on spodoptera frugiperda Sf9 cells, Curr. Pharm. Biotechnol. 17 (2016). Available from: https://doi.org/10.2174/1389201017705160303165604.

[14] R. Mitarotonda, E. Giorgi, M.F. Desimone, M.C. De Marzi, NMs and immune cells, Curr. Pharm. Des. 25 (2019) 3960–3982. Available from: https://doi.org/10.2174/1381612825666190926161209.

[15] M.C. De Marzi, M. Saraceno, R. Mitarotonda, M. Todone, M. Fernandez, E.L. Malchiodi, et al., Evidence of size-dependent effect of silica micro-and nano-particles on basal and specialized monocyte functions, Ther. Deliv. 8 (2017) 1035−1049. Available from: https://doi.org/10.4155/tde-2017-0053.

[16] A. Saleh Tawfik, NMs: Classification, properties, and environmental toxicities, Environ. Technol. Innov. 20 (2020) 101067.

[17] Z. Peng, X. Liu, W. Zhang, Z. Zeng, Z. Liu, C. Zhang, et al., Advances in the application, toxicity and degradation of carbon nanomaterials in environment: a review, Environ. Int. 134 (2020) 105298. Available from: https://doi.org/10.1016/j.envint.2019.105298.

[18] S. Sharma, S. Chatterjee, Microplastic pollution, a threat to marine ecosystem and human health: a short review, Environ. Sci. Pollut. Res. 24 (2017) 21530−21547. Available from: https://doi.org/10.1007/s11356-017-9910-8.

[19] A. Quigg, W.-C. Chin, C.-S. Chen, S. Zhang, Y. Jiang, A.-J. Miao, et al., Direct and indirect toxic effects of engineered nanoparticles on algae: role of natural organic matter, ACS Sustain. Chem. Eng. 1 (2013) 686−702. Available from: https://doi.org/10.1021/sc400103x.

[20] L. Kaounides, H. Yu, T. Harper, Nanotechnology innovation and applications in textiles industry: current markets and future growth trends, Mater. Technol. 22 (2007) 209−237. Available from: https://doi.org/10.1179/175355507X250014.

[21] G. Juhel, E. Batisse, Q. Hugues, D. Daly, F.N.A.M. van Pelt, J. O'Halloran, et al., Alumina nano-particles enhance growth of Lemna minor, Aquat. Toxicol. 105 (2011) 328−336. Available from: https://doi.org/10.1016/j.aquatox.2011.06.019.

[22] H. Chung, Y. Son, T.K. Yoon, S. Kim, W. Kim, The effect of multi-walled carbon nanotubes on soil microbial activity, Ecotoxicol. Environ. Saf. 74 (2011) 569−575. Available from: https://doi.org/10.1016/j.ecoenv.2011.01.004.

[23] A. Ganesh Kumar, K. Anjana, M. Hinduja, K. Sujitha, G. Dharani, Review on plastic wastes in marine environment - biodegradation and biotechnological solutions, Mar. Pollut. Bull. 150 (2020) 110733. Available from: https://doi.org/10.1016/j.marpolbul.2019.110733.

[24] N. Mahmoud, S.S. Mohammad, Chapter 7 Risks of nanotechnology to human life, Interface Sci. Technol. 28 (2019) 323−336. Available from: https://doi.org/10.1016/B978-0-12-813586-0.00007-9.

[25] S.S. Ahluwalia, D. Goyal, Microbial and plant derived biomass for removal of heavy metals from wastewater, Bioresour. Technol. 98 (2007) 2243−2257. Available from: https://doi.org/10.1016/j.biortech.2005.12.006.

[26] V.C. Reyes, S.O. Opot, S. Mahendra, Planktonic and biofilm-grown nitrogen cycling bacteria exhibit different susceptibilities to copper nanomaterials, Environ. Toxicol. Chem. 34 (2015) 887−897. Available from: https://doi.org/10.1002/etc.2867.

[27] C.L. Alito, C.K. Gunsch, Assessing the effects of silver nanomaterials on biological nutrient removal in benchscale activated sludge sequencing batch reactors, Environ. Sci. Technol. 48 (2014) 970−976. Available from: https://doi.org/10.1021/es403640j.

[28] A. Patwa, A. Thiéry, F. Lombard, M.K.S. Lilley, C. Boisset, J.-F. Bramard, et al., Accumulation of nanomaterials in "jellyfish" mucus: a bio-inspired route to decontamination of nano-waste, Sci. Rep. 5 (2015) 11387. Available from: https://doi.org/10.1038/srep11387.

[29] L. Zhan, Z. Xu, State-of-the-art of recycling e-wastes by vacuum metallurgy separation, Environ. Sci. Technol. 48 (2014) 14092−14102. Available from: https://doi.org/10.1021/es5030383.

[30] J. Ruan, J. Huang, L. Dong, Z. Huang, Environmentally friendly technology of recovering nickel resources and producing nano-Al_2O_3 from waste metal film resistors, ACS Sustain. Chem. Eng. 5 (2017) 8234−8240. Available from: https://doi.org/10.1021/acssuschemeng.7b01900.

[31] O. Myakonkaya, Z. Hu, M.F. Nazar, J. Eastoe, Recycling functional colloids and nanoparticles, Chem. Eur. J. 16 (2010) 11784−11790. Available from: https://doi.org/10.1002/chem.201000942.

[32] T. Dutta, K.-H. Kim, A. Deep, J.E. Szulejko, K. Vellingiri, S. Kumar, et al., Recovery of nanoma-terials from battery and electronic wastes: a new paradigm of environmental waste management, Renew. Sustain. Energy Rev. 82 (2018) 3694−3704. Available from: https://doi.org/10.1016/j.rser.2017.10.094.

[33] V. Oestreicher, C.S. García, R. Pontiggia, M.B. Rossi, P.C. Angelomé, G.J.A.A. Soler-Illia, e-waste upcycling for the synthesis of plasmonic responsive gold nanoparticles, Waste Manag. 117 (2020) 9−17. Available from: https://doi.org/10.1016/j.wasman.2020.07.037.

[34] R. Farzana, R. Rajarao, P. Behera, K. Hassan, V. Sahajwalla, Zinc oxide nanoparticles from waste Zn-C battery via thermal route: characterization and properties, Nanomaterials 8 (2018) 717. Available from: https://doi.org/10.3390/nano8090717.

[35] P. Devi, K.N. Hipp, A. Thakur, R.Y. Lai, Waste to wealth translation of e-waste to plasmonic nanostructures for surface-enhanced Raman scattering, Appl. Nanosci. 10 (2020) 1615–1623. Available from: https://doi.org/10.1007/s13204-020-01273-6.

[36] H.D. Jang, H. Kim, H. Chang, J. Kim, K.M. Roh, J.-H. Choi, et al., Aerosol-assisted extraction of silicon nanoparticles from wafer slicing waste for lithium ion batteries, Sci. Rep. 5 (2015) 9431. Available from: https://doi.org/10.1038/srep09431.

[37] H. Saleem, J. Zaidi Syed, Sustainable use of nanomaterials in textiles and their environmental impact, Materials 13 (2020) 5134. Available from: https://doi.org/10.3390/ma13225134.

[38] Z. Zhuang, X. Xua, Y. Wang, Y. Wang, F. Huang, Z. Lin, Treatment of nanowaste via fast crystal growth: with recycling of nano-SnO_2 from electroplating sludge as a study case, J. Hazard. Mater. 211–212 (2012) 414–419. Available from: https://doi.org/10.1016/j.jhazmat.2011.09.036.

[39] S. Ramanathan, S.C.B. Gopinath, M.K.Md Arshad, P. Poopalan, Nanostructured aluminosilicate from fly ash: potential approach in waste utilization for industrial and medical applications, J. Clean. Prod. 253 (2020) 119923. Available from: https://doi.org/10.1016/j.jclepro.2019.119923.

SECTION II

Methods for recycling of nanomaterials

CHAPTER 8

General techniques for recovery of nanomaterials from wastes

Kuray Dericiler[1], Ilayda Berktas[1], Semih Dogan[1], Yusuf Ziya Menceloglu[1,2] and Burcu Saner Okan[1]

[1]Sabanci University Integrated Manufacturing Technologies Research and Application Center & Composite Technologies Center of Excellence, Teknopark Istanbul, Istanbul, Turkey
[2]Department of Materials Science and Nano Engineering, Faculty of Engineering and Natural Sciences, Sabanci University, Istanbul, Turkey

Abbreviations

4-NP	4-Nitrophenol
Ag NPs	Silver nanoparticles
Cd NP	Cadmium nanoparticle
CdS	Cadmium sulfide
CPE	Cloud point extraction
CR	Congo red
DCE	Dichloroethylene
DMCR	Differential magnetic catch and release
EVA	Ethyl vinyl acetate
Fe_2O_3	Magnetite
GO	Graphene oxide
Hal	Halloysite
HGMS	High-gradient magnetic separation
IL	Ionic liquid
M-GO	Magnetic graphene oxide
MB	Methylene blue
MO	Methyl orange
Mel	Melamine
MF	Magnetic flocculant
MWCNT	Multiwalled carbon nanotube
NM	Nanomaterial
NP	Nanoparticle
PA6	Polyamide 6
Pd NP	Palladium nanoparticle
PP	Polypropylene
PVA	Polyvinyl alcohol
rGO	Reduced graphene oxide
RhB	Rhodamine B
SAP	Supercritical antisolvent precipitation
SWCNT	Single-walled carbon nanotube

Nanomaterials Recycling
DOI: https://doi.org/10.1016/B978-0-323-90982-2.00008-1

© 2022 Elsevier Inc.
All rights reserved.

TCE	Trichloroethylene
TEA	Triethyleneamine
Zn NP	Zinc nanoparticle
ZnS	Zinc sulfide

8.1 Introduction

Every product reaches a point in its usage life at which it is no longer desired by the consumer. At this point, the product reaches the end of its life cycle and faces several options, ranging from disposal as waste to recycling to be reused elsewhere [1]. This applies to nanomaterials (NMs) and products that contain NMs as well. General products have been recycled at the end of their life cycle starting as early as the 11th century [2]. Since the discovery of NMs around the 1980s [3], a very limited number of reports have been published from the environmental and toxicology point of view [4,5]. As the use of NMs in industry increases, the amount of generated nanowastes also increases [6]. However, unlike traditional recycling procedures, here the end user has almost no input into the process. Therefore it is important to define and categorize the NMs that are used in consumer-grade products. Even though they are defined to be smaller than 100 nm [7], NMs can reach several sizes larger yet remain surface active. These NMs are classified on the basis of morphology, composition, dimensionality, and dispersion/agglomeration state, as the impact of these properties varies depending on the physical states of NMs, such as fixed structure, free-standing layers, aspect ratio, and composition [8]. Fixed nanostructures pose significantly less of a health risk when properly handled, compared to free nanoparticles (NPs). Fig. 8.1 categorizes the NMs according to their dimensionality, morphology, composition, and uniformity and agglomeration states.

This distinction of whether the NM is fixed or a free NP is important for issues of health as well. Because of Brownian motion at the atomic scale, NMs can move freely, be released into the environment, and enter living bodies. However, not all NMs are toxic [9,10]; some NMs have health benefits [11,12], and some can be rendered nontoxic [13]. It is nevertheless important not to release NMs to the environment, and it is best to recycle them. Nanotechnology and its implementation have been going on since the discovery and application of NMs in the 1980s. However, conducting a risk analysis and developing a feasible recycling roadmap for the even earlier generation of NMs have proven difficult [14]. The development of nanotechnology and the implementation of NMs are visualized.

Ending the life cycle of materials by landfilling or mixing through waste streams is not a preferable fate for NMs, as they have great added value and value per kilogram. Thus it is both economically and environmentally beneficial for these materials to be

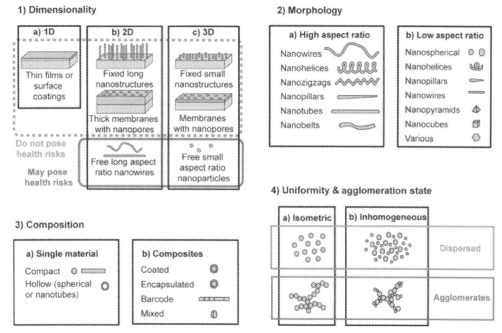

Figure 8.1 Categorization of nanostructures by their dimensions, morphology, composition, and uniformity and agglomeration states [8]. *Reproduced with the permission of AIP Publishing LLC.*

recycled and reutilized, which also reduces the consumption of already limited natural resources [15].

The purpose of this chapter is the evaluation of the efforts that have been made through centuries to recycle NMs from waste products such as carbon, ceramic, metal, and nanocomposite products. Moreover, understanding the kinetics of the recycling techniques is important for the effective design and recovery of the NMs. With this regard, the present study also reviews the methods and parameters affecting the recycling of NMs.

8.2 Types of nanomaterial wastes

In general, the recycling of NMs relies on the proper collection of used NMs, whether they are free-standing or fixed in structures, the separation of compounds, and the recovery and reutilization of these NMs in the same or different industries or products. Another output of the recycling aspect, depending on the process, is to obtain energy in the form of heat and to reduce environmental effects [16]. Most recyclable NMs can be classified on the basis of their structure and composition; such categories include carbon-based, ceramic-based, metal-based, and nanocomposite products.

8.2.1 Carbon-based nanomaterials

Carbon-based NMs offer great potential in terms of mechanical, electrical, and thermal properties [17] and have been used in a variety of industries, such as the marine and automotive industries [18–21]. Carbon nanotubes, especially because of their high surface area and surface activity levels, have been utilized in water filtration mediums [22]. One study from Cai et al. [23] described the utilization of commercial multi-walled carbon nanotubes (MWCNTs) as a solid phase extraction adsorbent to analytically determine bisphenol A, 4-n-nonylphenol, and 4-tert-octylphenol amounts. These organic pollutants in water sources such as rivers, seashores, taps, and sewage discharges were successfully adsorbed by the MWCNTs, which were then used as a solid phase packing material. Furthermore, the desorption and recycling (reutilization) of these MWCNTs were observed to be effortless such that the adsorbed molecules were easily desorbed by using methanol, and the MWCNTs were ready to be used again. The MWCNTs had no carryover in the next analysis cycle and were found to be capable of being used repeatedly.

Similarly, for single-walled carbon nanotubes (SWCNTs), Liu et al. [24] developed a nonionic surfactant-based approach with cloud point extraction (CPE). The authors had a broad focus on the recycling of various NMs, such as metal-, ceramic-, and carbon-based NMs, using commercial Triton X-114 as the surfactant. As is shown in Fig. 8.2, at a lower critical solution temperature, by centrifugation, SWCNTs (as well as other NMs) moved to a surfactant-rich phase from an aqueous sodium dodecylbenzene sulfonate dispersion to be separated and recovered. It was found that even after 10 times of recycling, the materials could be sustained at the surfactant-rich phase for over two months without aggregation. The authors also indicated that this method could find an important role such as recovery and storage of NMs such as catalysts that are used in the waste treatment industry.

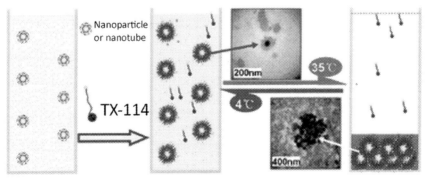

Figure 8.3 Triton X-114-based cloud point extracton method for recycling of nanomaterials [24]. *Reproduced with the permission of Royal Society of Chemistry.*

Graphene and its derivatives are extremely important in the nanotechnology world, owing to their outstanding properties [25]. Green synthesis of these materials combined with the recycling aspect is highly likely to yield significant benefits in the long term. When combined with metal NPs and their catalytic properties, graphene derivatives become a viable support material for these catalysts [26]. In one very recent study, Hemmati et al. [27] described a novel synthesis method for graphene oxide (GO) and its in situ reduction with silver nanoparticles (Ag NPs) to obtain reduced graphene oxide (rGO)/Ag catalyst. GO was synthesized by following the modified Hummers' method [28] and then dispersed in *Menthapulegium* flower extract. After reduction of GO in the flower extract, the Ag precursor was added, and the nano-composite catalyst formed. This nanocatalyst was found to efficiently reduce methyl orange (MO) and rhodamine B (RhB) environmental pollutants as well as withstanding the recycling process of centrifugation, washing, and drying for 10 successive runs while maintaining its catalytic activity. Carbonaceous NMs, because of their extraordinary properties, such as electrical and thermal conductivity, mechanical behavior, and dimensionality, are great subjects for further study on the recyclability aspect. As was discussed earlier, solution-based recycling processes seem to be the most promising methods for recycling carbon-based NMs.

8.2.2 Ceramic-based nanomaterials

Ceramic-based NMs, mainly in metal oxide form, have already found a wide variety of use cases in the industry [29−31]. One of the main use cases is catalyst support, in which their chemical and thermal resistivity is required where the material should not lose its abilities by chemically or thermally deforming during the process [32]. In a narrower sense, Pd-catalyzed organic synthesis reactions have great potential for the fabrication of therapeutic drugs, chemicals, and intermediary products. Coating magnetic γ-Fe_2O_3 with Pd and N-heterocyclic carbenes yields magnetic iron oxide−Pd systems [33]. Stevens et al. [34] reported that this homogenous catalyst and support complex acts as a prominent catalyst and soluble solvent for Suzuki, Heck, and Sonogashira reactions. This organometallic complex showed high thermal stability and was recovered, and thus recycled effortlessly, by using an external magnet while maintaining catalytic activity around 90% after five rounds of recycling. This study indicates that recycling such complexes may well become important in industrial-scale applications, since the use of several million tons of solvent can thus be avoided. The recovery process could also be performed repeatedly without a significant loss of catalyst, according to the authors.

The recovery and reuse of catalysts is a significant concern for the organic chemistry industry. Magnetic NMs are widely used as heterogeneous reactive catalysts. For instance, Ó Dálaigh et al. [35] found that 4-N,N-dimethylaminopyridine supported on a magnetic nanocatalyst demonstrates outstanding activity and rigidity, especially when magnetite (Fe_3O_4) NPs coated with a layer of SiO_2 do not aggregate and maintain a high surface area for catalysis. The authors also indicated that these nanocatalysts show over 90% activity even when recycled and reused over 30 times with a simple recovery method, such as an external magnetic field.

Another opportunity to recover ceramic-based NMs is electroplating sludge, in which the waste contains heavy metal by-products from the plating industry [36]. Tin is widely used in the industry as solders, plating, glasses, and ceramic glazes, leading to large amounts of Sn being discarded as industrial waste [37]. For example, Zhuang et al. [38] reported that through acid treatment of electroplating sludge, 98.5% of sludge can be dissolved, while the remaining solid part was found to contain a residue of crystalline SnO_2 and SiO_2 phases. Further base treatment and hydrothermal methods showed that nano-sized SnO_2 nanowires with around 90% purity can be obtained. Recycling of SnO_2 and the formation of SnO_2 nanowires are illustrated in Fig. 8.3.

Similarly, the chrome-plating industry generates toxic Cr^{VI} sludge that contains $Mg(OH)_2$ NPs [39]. Liu et al. [40] indicated that a typical Cr^{VI}-containing sludge consists of 30% nano-$Mg(OH)_2$ and 16% nano-$CaCO_3$ and that the nanophases can be aggregated, grown, and precipitated in a solution, which leads to the recovery of harmless NPs from toxic waste with the use of mineralizers and hydrothermal processes.

High-value technological applications of ceramic-based NPs, such as Cd NPs and Zn NPs, include similarly high-value colloidal suspensions [41,42]. It is therefore important to separate and recover these valuable NMs. Myakonkaya et al. [43] demonstrated a reversible liquid phase approach to recover and reuse inorganic CdS and ZnS NPs. CdS and ZnS NPs were synthesized by using a liquid phase microemulsion method and suspended in surfactant stabilized microemulsion systems after sonication. At a critical temperature and water-to-surfactant molar ratio, it was found that the systems could be separated and recovered (recycled) for further use. The authors noted the potential applications of this technique in purification technologies for recycling these valuable NMs, which had been previously ignored because of the irreversible aggregation of these NMs. To conclude, ceramic NMs have been found in various waste sites and are used in high-technology products. The recovery of these NMs is important to preserve the already limited natural resources that are needed to fabricate them. A significant amount of research has been put toward solvent-based recovery systems, which has proven ideal for the recycling of these NMs.

8.2.3 Metal-based nanomaterials

Metal-based NMs currently stand out as the material group to receive the most research attention for recycling at the nano scale. The storage of nanowaste containing metal NPs is not sustainable; the separation of NPs composed of metals may offer a

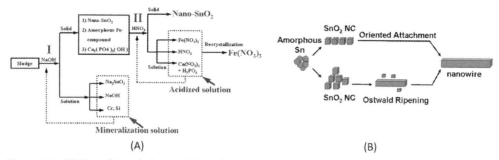

Figure 8.3 (A) Recycling of electroplating sludge to obtain solid nano-SnO$_2$. (B) Formation of SnO$_2$ nanowires from amorphous Sn. *Modified from Z. Zhuang, X. Xu, Y. Wang, Y. Wang, F. Huang, Z. Lin, Treatment of nanowaste via fast crystal growth: with recycling of nano-SnO$_2$ from electroplating sludge as a study case, J. Hazard. Mater. 211–212 (2012) 414–419. https://doi.org/10.1016/j.jhazmat.2011.09.036 (Elsevier).*

viable path toward more sustainable solutions. Batteries that contain heavy metals such as Ni and Cd should be recycled and prevented from releasing them into the environment. Bystrzejewska-Piotrowska et al. suggested using traditional procedures applied to batteries to obtain a viable scheme to recycle metal NMs [16,44]. However, batteries containing free-standing NMs or thin films of iron or titanium may be subjected to superheating and can cause problems during recycling [45,46]. This can result in NP emission to the air and contaminating feedstock [47].

Zero-valent metals such as Fe and Zn play a significant role in organic synthesis applications [48]. Degradation and dechlorination of chlorine-containing hazardous organic materials have been important factors in organic processes. Degradation of trichloroethylene (TCE) and dichloroethylene (DCE) using cellulose acetate membrane-immobilized iron and nickel NPs was reported by Meyer et al. [49]. The authors reported that although around 10 wt.% of the metals was lost during the membrane fabrication, with one-tenth of the metal loading used in the literature, the membrane degraded 75% of the TCE in 4.25 hours. In addition, the degradation rate of DCE was an order of magnitude greater than that of TCE. The authors also indicated that after a proper washing with deionized ultrafiltrated water, the membrane maintained performance nearly identical to that of fresh membranes, which shows a great reusability potential and recyclability of the NMs.

As was mentioned, metal NPs, especially transition metal NPs, have fascinated researchers, owing to their unique size-induced physicochemical properties. Ag NPs have drawn significant interest for their wide variety of applications, including catalysis [50] and biosensors [51]. In one of their earlier studies, Murugan and Jebaranjitham [52] reported the possible reduction of 4-nitrophenol (4-NP) by incorporating Ag NPs in a cross-linked poly(styrene)-co-poly(N-vinylimidazole) matrix. The catalyst complex was synthesized by using a solvothermal method AgNO$_3$ as the metal precursor and polymer beads. The obtained Ag NP—containing complex displayed very high recycling efficiency through repeated reduction cycles as the Ag NPs are regained and reused by a simple filtration method. Moreover, Rostami-Vartooni et al. [53] used *Bunium persicum* seed extract together with Ag NPs and seashells to obtain a nanocomposite catalyst. The nanocomposite catalyst was obtained by first dispersing the seashells in the extract and mixing the dispersion with the $AgNO_3$. The obtained catalyst exhibited significant catalytic reduction properties for MO, Congo red (CR), methylene blue (MB), and 4-PH without the loss of activity after five reusing and recycling steps, which included separation and washing with ethanol. In a more recent study, Veisi et al. [54] explored the green synthesis of Ag NPs without stabilizers or surfactants and found that by using *Thymbra spicata* extract, an Ag NP/*Thymbra* complex catalyst can be obtained. After obtaining the leaf extract, $AgNO_3$ aqueous solution could simply be added to the extract to synthesize Ag NP/*Thymbra* complex. The authors indicated that the green complex demonstrated excellent catalyzing properties at low catalyst loadings for the reduction of 4-NP, MB, and RhB as well as great recyclability with no appreciable catalytic activity loss even after eight reruns.

Similarly, liquid phase dispersions of palladium NPs (Pd NPs) have been a highly active research topic [55—57]. Recovery of these precious metals, especially at the nano scale, is an important subject [58]. Nazar et al. [59] reported the synthesis of Pd NPs from $Na_2PdCl_6 \cdot 0.4H_2O$ in the aqueous phase and stabilization of the particles using mercaptoethanesulfonate. It is found that with a mixture of certain commercial surfactants such as Triton X-100 and X-114, the NPs can be separated and recycled for further use. Fig. 8.4 displays the recycling of Pd NPs using phase separation and centrifugation.

In another study, utilizing the natural materials and benefiting from the plant extracts to promote green chemistry, Sadjadi et al. [60] used the combination of natural biocompatible halloysite clay and *Heracleum persicum* extract to fabricate

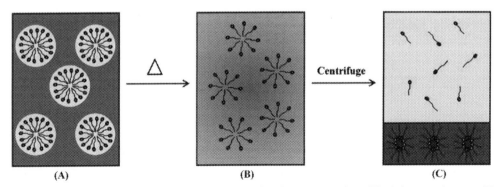

Figure 8.4 (A) Mercaptoethanesulfonate-stabilized Pd nanoparticles. (B) Solvent mixture. (C) Recycling by phase separation and centrifugation [59]. *Reproduced with the permission of Elsevier.*

magnetically recyclable Fe_3O_4/Hal-Mel-TEA(IL)-Pd catalyst. The halloysite nanotubes were first magnetized by using $FeCl_2$ and $FeCl_3$ to obtain a Fe_3O_4-halloysite complex, and the surface functionalization was carried out by using melamine, 1,4-dibromobutane, and triethanolamine to provide an ionic liquid (IL) presence on the halloysite surface. Pd NPs were reduced by using *H. persicum* extract and decorated on the former. This catalyst complex was used for the reductive degradation of MO and RhB, and it was discovered that it could be recycled and reused for eight reactions, losing only around 3% of its catalytic application.

Metal-containing nanowastes, especially high-value Pd and Ag NMs, became a focus for researchers. Recovery of these metals using various techniques is quite attractive, owing to the wide application areas of the NMs. When combined with green synthesis methods, recycling has become a prominent subject for metal-based NMs. Researchers indicate that magnetic separation and recovery as well as solution-based processes are the most used techniques for recycling metal NPs.

8.2.4 Nanomaterial-reinforced composite materials

Nanocomposites are the product of incorporating NMs in bulk materials to obtain various improved properties. According to their sizes, nanoadditives show catalytic activity, change magnetic behavior and refractive index, or improve mechanical properties by restricting dislocation movement in the matrix or, for the case of polymers, polymer chain movement [61]. The existence of natural nanocomposites was reported as early as 1996 [62]. The attraction of nanocomposites has been mostly due to a wide variety of nanoadditives with different aspect ratios, mechanical, electrical, thermal, and magnetic properties brought to the bulk material at very low loadings (0.5—5 wt. %) due to incredibly high surface ratio of nanoadditives [63,64]. The increasingly high popularity of nanocomposites brings a concern as well. As the nanocomposites reach the end of their life cycle, they can no longer be categorized and discarded or recycled as general products. Researchers suggest that they should be regarded as nanowaste and treated accordingly [65]. Fig. 8.5 displays a simplified view of the implementation of nanotechnology and the life cycle of NMs.

Fiber-type NMs with high aspect ratios are widely used as additives to polymer matrices [66,67] and can be recycled in standard recycling plants to obtain reusable products [68]. However, the toxic effect of nano-scale fibers (nanofibers, nanotubes, and nanorods) poses a great threat to human and environmental health [69—71]. It is therefore imperative to capture these materials before they are released into the environment and more beneficial if recycling and reusing are involved in this process. However, Khanna et al. [72] reported that the separation of materials and recycling are problematic because of the presence of multiphase components such as matrices, reinforcements, and additives. Additionally, the recycled material often displays poor

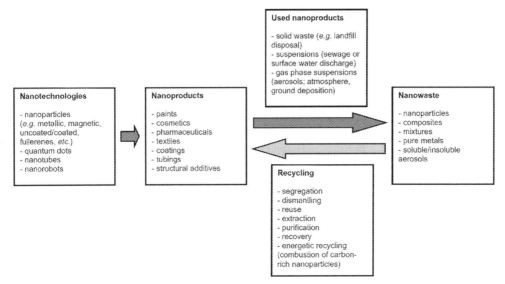

Figure 8.5 Life cycle of nanoproducts from development to end of life [16]. *Reproduced by the permission of Elsevier.*

quality in terms of properties compared to the virgin material, which also hinders the tendency of recycling these materials [73]. Busquets-Fité et al. [74] reported recovery and recycling of six largely produced NMs: SiO$_2$, TiO$_2$, ZnO, two varieties of organo-modified montmorillonite, and MWCNT from the most commercially used polymers polypropylene (PP) and polyamide 6 (PA6). Commercial NMs were used to fabricate 3 wt.% nanocomposites followed by acid dissolution and calcination to recover the NPs from PA6 and PP, respectively. Results showed that nearly 95% of the NMs had been recovered after the calcination of PP nanocomposites, compared to 40%–60% recovery from the dissolution of PA6 in formic acid. The authors indicated that the NMs obtained from the PA6 could be reused to prepare new nanocomposite, while the ones recovered from PP did not retain the functional groups; therefore a new compatibilization step is required to produce new PP nanocomposites. Lloyd et al. [75] analyzed the life cycle of automotive industry nanocomposites and concluded that montmorillonite-filled polymer nanocomposites can be recycled without significant changes to the material because the aspect ratio between the clay and the matrix will not be reduced as opposed to long fiber reinforcement, and therefore it can be used and performs similarly to the virgin material.

The recovery of NMs from composite wastes has become a difficult subject because of the complex nature of interactions between fillers and matrices. Considering the continuous increase in the fabrication of nanocomposites, the concern also grows. From the end user point of view, it is difficult to determine how to recycle the nanocomposites, as the relevant information is not available. It is up to industrial leaders to inform people and create guidelines for proper recycling of their products.

8.3 Types of techniques used for the recovery of nanomaterials from wastes

Various NMs and NPs that are commonly used in industry or commercial products are considered to be highly toxic when they enter ecosystems without recycling. A study by public health experts reported that nano-sized silver particles in wastewater are a major problem that could affect the conservation of biodiversity. It is also worth noting that the number of physical and chemical processes applied during the recycling of NMs is less than normal production processes, so a significant energy saving is achieved in the production of materials by recycling. The amount of energy saved by recycling NMs varies depending on the type and composition of the waste.

Separating NPs from the waste streams to recycle, recover, and reuse can be an area of interest for researchers. For instance, Živković et al. [76] reported the use of conventional separation techniques such as washing, precipitation, centrifugation, and solvent evaporation to recycle the NPs. Several other innovative methods have also been reported. At the same time, there are different innovative methods for the recovery and recycling of NMs. In that matter, Myakonkaya et al. [77] described several recycling methods, such as magnetic field interaction, pH change, application of molecular antisolvents, and colloidal solvents, that provide effective and efficient methodologies for recycling NPs, significantly reducing costs such as time and energy consumption. Furthermore, Wang et al. [78] directed their attention to enabling more efficient and economical recycling or reuse of nanowastes with several approaches, such as magnetic recycling of nanocatalysts, recovery of NMs using antisolvents, and absorbent NMs on recyclable carriers. Therefore it is important to emphasize the necessity of recycling valuable wastes from the synthesis process to the disposal process with a sustainable perspective.

Despite all these developments, there is not enough industrial practice in recycling and reuse of NMs and NPs. Considering the chemical pollution in the world and the use of nanoproducts (NMs, NPs, etc.), simple recycling and recovery studies, as well as the applicability of these techniques, have become necessary.

8.3.1 Magnetic separation technique

Magnetic-assisted separation is an important trend and is often used to separate and recycle NMs. Magnetic separation is an efficient and often cost-effective approach in many key areas of interest: catalyst recycling, environmental applications, large-scale water purification, sewage treatment, and varied applications in biomedicine activities [79]. In this simple, low-cost, and time-saving method, separation is induced by using

Figure 8.6 Magnetic separation of CoFe$_2$O$_4$ nanoparticles in hexane [80]. *Reproduced by the permission of Wiley.*

an external magnetic field using a block of magnet. For instance, Grass et al. [80] reported the magnetic separation of cobalt NPs on an industrial scale. The cobalt core has a high magnetic moment, which leads to efficient separation of the particles. Fig. 8.6 displays the magnetic separation process using cobalt NPs.

Nowadays, water purification using external magnetic fields has attracted significant interest [81–83]. Magnetic nanocrystals and NPs can also be easily removed and collected from solutions by using a regular strong magnet and different magnetic fields. Magnetic separations at very low magnetic field gradients (<100 tesla per meter) can be applied to various problems, such as water purification and the simultaneous separation of complex mixtures [84]. For instance, Xu et al. [85] investigated the application of magnetic NPs as demulsifying agents. In this study, a system composed of multiple demulsification processes was designed with the use of an external magnetic field to recycle magnetic NPs. Magnetic demulsification was found to be an environmentally friendly method to separate the oil from the emulsified wastewater. In another work, Liu et al. described a process to functionalize GO with iron carbonyl to obtain magnetic graphene oxide (M-GO) to demulsify oil/water emulsions. Within a few minutes, M-GO demulsified the emulsion and could be recycled six to seven times without losing

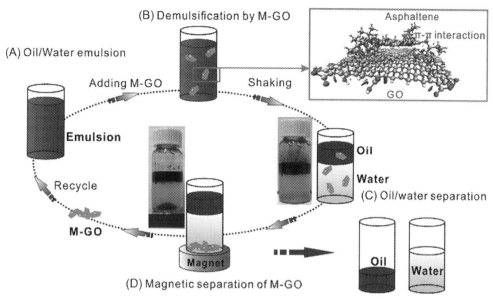

Figure 8.7 Schematic illustration of demulsification and recycling tests using magnetic graphene oxide as a demulsifier [86]. *Reproduced by the permission of Elsevier.*

its demulsification capability [86]. Fig. 8.7 presents a schematic illustration of demulsification and recycling tests using M-GO as a demulsifier in a stepwise procedure.

Latham et al. reported that capillary magnetic field flow fractionation for the purification of magnetic NPs could be a potential technique [87,88]. This method demonstrates the separation of a NP solution into size-monodisperse fractions. Another route for the magnetic separation technique is differential magnetic catch and release (DMCR). DMCR carries out a variable magnetic flux orthogonal to the flow direction in an open tubular capillary to trap and controllably release magnetic NPs. For instance, Beveridge et al. [89] described a design of a magnetic separation system combining an electromagnet with a capillary system. Magnetic separation was performed by utilizing the size-dependent magnetic moments of superparamagnetic NPs, which enables the separation and isolation of particles smaller than 10 nm in diameter. Moreover, Moeser et al. [90] showed that a high-gradient magnetic separation (HGMS) technique can be utilized to separate magnetic materials from nonmagnetic liquid mediums, as illustrated in Fig. 8.8. An HGMS system generally takes place in a column placed inside an electromagnet and magnetic field. The target solute is loaded into the system followed by applying a magnetic field across the column and creating magnetic field gradients in the fluid around the wires to hold particles in the column. This results in separation NPs to be reused elsewhere. HGMS has been shown to be an effective technique for separating and filtering fine and weakly magnetic particles.

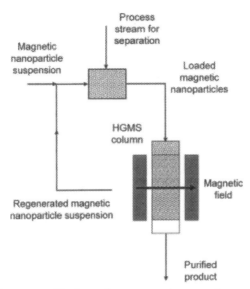

Figure 8.8 Schematic illustration of high-gradient magnetic separation process [90]. *Reproduced by the permission of Wiley.*

Magnetic flocculation is a new approach to NP recovery that combines the advantages of agglomeration with magnetic separation to enable a simple collection of nonmagnetic NPs [91]. Magnetic flocculants (MFs) can be created by coating flocculant polymers on magnetic NMs. Leshuk et al. [91] reported a method to fabricate reusable MFs by coating flocculant polymers on magnetic NPs (Fe_3O_4@SiO_2). When added to colloidal NP dispersions, MFs aggregate with the suspended NPs to form magnetically responsive flocs, which upon separation can be reversibly deflocculated for NP release and reuse in a closed-loop process, as shown in Fig. 8.9. In this work, high separation efficiency was attained in a variety of NP suspensions, including Au, Ag, Pd, Pt, and TiO_2, stabilized by different coatings and surface charges.

As these NMs can cause toxicity for all living creatures when released into the environment, the magnetic separation technique is one of the most scalable methods to recycle NMs generated by industrial wastes.

8.3.2 Antisolvent technique by using CO_2

Utilization of supercritical CO_2 as an antisolvent is carried out by precipitating the solute from an expanding solution while mixing with supercritical carbon dioxide, which allows controlling particle size in both critical and supercritical regions followed by nucleation and crystal growth. Researchers have used CO_2 as an antisolvent for the recovery of the NPs from the reverse micelles. It is well known that compressed CO_2

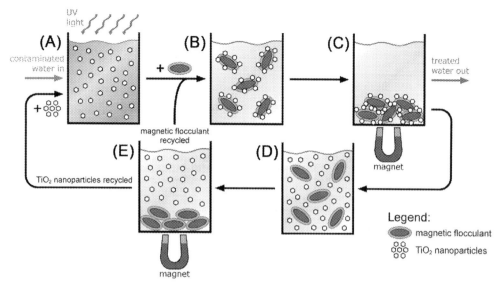

Figure 8.9 TiO$_2$ nanoparticle recycling steps with magnetic flocculants [91]. *Reproduced by the permission of Royal Society of Chemistry.*

can dissolve in many organic solvents. For instance, Liu et al. [92] investigated the possibility of recovering TiO$_2$ NPs from reverse micelles using supercritical carbon dioxide as an antisolvent. The solubility of CO$_2$ in the solvent is defined as a function of pressure; therefore the properties of the liquid solvent can be changed by pressure. The separation of the gas and the liquid solvent can be obtained simply by depression. The authors also proposed that this technique enables the NPs to be controlled efficiently under pressurized conditions and is not toxic. Furthermore, Zhang et al. [93] reported the recovery of the Ag NPs from the reverse micelles by dissolving antisolvent CO$_2$ in the micellar solution at suitable pressures. Similarly, Zhang et al. [94] described a process to separate ZnS NPs from reverse micelles by compressed CO$_2$. The general purpose of using the supercritical technique is to improve mass transfer by forming fine droplets with a high surface area.

Besides the manufacturing process, recovery of dry NPs is still a substantial challenge. The mechanisms of supercritical antisolvent precipitation (SAP), sedimentation, and single-unit recovery processes of NPs are known but have not yet been evaluated precisely. In one of the studies, Torino et al. [95] investigated the SAP technique for collecting nanopowders directly on a sintered metallic filter placed on the bottom of the precipitation vessel. The authors indicate that after the collection, NPs with diameters in the range of 50–150 nm could be easily separated into single nano units by final processing with ultrafiltration, ultracentrifugation, and ultrasound-based techniques. Overall, the antisolvent technique can find potential application areas from which NM separation using conventional methods proves difficult. The main advantage of the SAP technique is preventing the thermal degradation of materials due to near room temperature process conditions. However, pressure-based damages may still occur during recycling, depending on the properties of the NM.

8.3.3 Aqueous dispersion techniques

Thermodynamically stable dispersion of two immiscible liquids in the presence of an emulsifier or surfactant is called a microemulsion. The microemulsion method offers an alternative technique for recycling and reuse of NPs while providing ultralow interfacial tension, a large interface area, and the capacity to dissolve both water and oil components. Recycling methods based on aqueous dispersion are limited in the literature, and most of the research is based on adding an excess of water to a water/oil microemulsion, which induces a phase separation such that two distinct phases are formed, one mostly consisting of oil and the other dominated by water. For instance, Helfrich et al. [96] reported the separation of Au and Ag nanospheres by using aqueous two-phase systems that consist of polyethylene glycol/dextran and water or aqueous buffer interface. Surfactant, cosurfactant, and salt impurity concentrations increased in the lower aqueous-rich phases, while the surfactant-coated NPs remain dispersed in the upper oil-rich phases. This technique represents a simpler and more universally applicable route to separate the unwanted parent microemulsion components and thus purify NPs. The hydrophobic surfactant-coated NPs are more soluble in the upper oil-rich phase, whereas the free residual surfactant molecules partition strongly into the aqueous-rich phase. The NPs can then be separated from excess surfactant and hydrophilic reactants. The benefits of using water to achieve the separation are apparent: As the greenest possible solvent, it represents a clean method for particle recovery. There is great interest in the reversible transferring of NMs between two phases for recycling NMs. For instance, Liu et al. [24] reported a thermoreversible recovery, concentration, separation, and dispersion of various NMs in the aqueous phase with CPE by using a low-cost and commercially available nonionic surfactant, as illustrated in Fig. 8.10. In this study, Ag NPs were selectively concentrated from environmental water samples without disturbing their sizes and shapes by CPE. This proposed method provided a novel approach for the characterization and quantification of Ag NPs in environmental water samples, which can further contribute to understanding the toxicity of Ag NPs in aqueous systems [97]. Furthermore, Gao et al. [98] described a method to prepare ZnO nanorod structures with wet chemical methods from various heterogeneous substrates and provided recycling and reuse of these nanostructures by a hydrothermal route.

Separation by using an aqueous dispersion technique allows the reaction solutions to be recycled and reused a large number of times by simply supplementing with water. Therefore this method becomes an environmentally friendly alternative that can be deemed appropriate to be transferred to industrial applications.

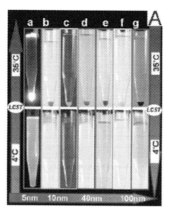

Figure 8.10 Image of separated and dispersed solutions of typical nanomaterials with dimensions ranging from 5 to 100 nm [24]. *Reproduced by the permission of Royal Society of Chemistry.*

8.3.4 Colloidal solvent technique

Recovery and purification of valuable NPs can also be achieved by a colloidal solvent technique in which solid substances are in a state of suspension in a fluid. This approach has various advantages compared to the current methods, such as low cost and the possibility of having a positive impact on the environment. For instance, Hollamby et al. [99] reported a novel method for the recovery of Ag NPs using the colloidal solvent technique from two coexisting phases: an octane-rich upper phase and an aqueous lower phase. The authors also stated that one of the most important features of the colloidal solvent technique is making a mild purification method accessible by supporting the NPs as a structured solvent, which is otherwise impossible with a normal molecular solvent. To conclude, this technique compares favorably with existing methods, as it does not use additional organic solvents and surfactants. Consequently, the important advance here is that the phase behavior of the promoter "colloidal solvent" is concluded in that the valuable NPs are recovered and purified by simply adding excess water.

8.3.5 Centrifugation/solvent evaporation technique

Solid NPs are usually separated from the solution by filtration or centrifugation. Although these methods are widely used, they have certain limitations due to time- and energy-consuming operations, blockage of filtration units, and the necessity of heavy-duty instruments. For large-scale industrial applications, Shahi et al. [100] reported that the conventional NP separation techniques are not economically viable, which hinders the widespread use of functional NPs for commercial applications. In this regard, Wang et al. [101] showed that high-quality iron oxide NPs could be prepared in recycled ILs that are separated from the initial synthesis products after washing with hexane followed by centrifugation and evaporation processes.

8.3.6 Other approaches for the recovery of nanomaterials

As the utilization of NMs in industry grows, the need for recycling of NMs also increases. The progress of new strategies requires effective recycling techniques. For this reason, alternative methods for large-scale industrial applications regarding the separation and recycling of NPs are being developed. Recently, Liu et al. [102] reported a system based on pH—controlled reversible covalent bonding between boronic acid—modified NPs and flexible polyvinyl alcohol (PVA) polymer chains. In this work, PVA was used as a particle agglomeration reagent to achieve controlled NP separation. The aggregation of NPs can respond to pH variation, allowing the NP assembly to be controlled independently of specific molecular binding and release. In another work, Oliveira et al. [103] described that the use of magnetically recoverable supports for the immobilization of gold NP catalysts allows easy, clean, fast, and efficient catalyst separation at the end of the reaction cycle. The commercial recovery of gold is normally dependent on aqueous solutions; however, a suitable dry process would greatly simplify the process of separation and would reduce the need for wastewater treatment. On this subject, Kakumazaki et al. [104] reported a dry process to volatilize target elements by heating with CO gas or solid carbon in a reducing atmosphere. However, the proposed technique is not economically viable for elements with a high boiling point.

The recovery of NPs and separation of remaining unreacted reagents from surfactant is still an applicable technique when done on reverse micelle systems. In a study regarding this technique, Abécassis et al. [105] used an inverse microemulsion based on catanionic surfactants to synthesize and recover gold NPs using a temperature-induced phase separation resulting from the catanionic properties. The structure of the microemulsion is preserved during the reaction, and it can be used to separate NPs from reaction by-products, thus allowing a simple and effective separation method for microemulsions that show liquid-liquid phase transition in the triple-phase prism. These systems provide an interesting and potentially useful recovery method for NPs. Similarly, metal NPs such as Pb, Hg, Cu, Fe, Au, Ag, and Pd can be recovered from electronic waste such as computer circuit boards, smartphones, and laptop computers [106]. For instance, Mdlovu et al. [107] showed that Cu NPs with microemulsion processes were prepared from acidic $CuCl_2$ waste etchers produced during the production process for printed circuit boards. Myakonkaya et al. in their study managed to separate CdS and ZnS NPs by a liquid-liquid phase transition method. They have shown that this easy method can potentially be applied to NPs dispersed in a supportive colloidal fluid in general [43]. Also, Deep et al. [108] obtained the recovery of pure ZnO NPs from Zn-Mn dry alkaline batteries. Recovery of other valuable metals from spent Zn-Mn batteries was also reported using the phase separation method [109].

Recovering NMs from polymer-based nanocomposites is a challenge because of the intricate interactions between filler and matrix. Busquets-Fité et al. [74] investigated the recovery of NMs from PP, PA6, and ethyl vinyl acetate (EVA)

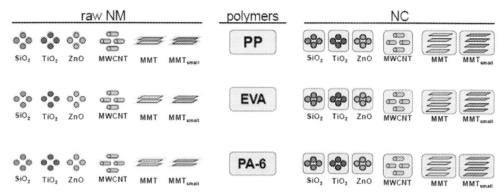

Figure 8.11 Representation of raw nanomaterials, polymers, and polymer nanocomposites from left to right [74]. *Reproduced by the permission of IOP publishing.*

nanocomposites. Eighteen different polymer-filler compositions were fabricated; however, EVA nanocomposites did not yield any recovery of NPs. PP and PA6 were subjected to calcination and acid digestion, respectively, to obtain NMs. The representation of nanocomposites is given in Fig. 8.11.

On the other hand, Liu et al. [110] investigated industrial sludge that contains NMs and found out that this so-called nanosludge incorporates heavy metals and displays NM properties such as high surface area and activity. This work suggests that through several chemical methods, recovery of heavy metal NPs from "nanosludge" is possible.

Numerous NMs are being produced, and the number will continue to grow. Developing novel techniques to recover and recycle these NMs is an important subject from both environmental and economic viewpoints. Since every NM and its properties are different, there is no single best method for recycling NMs. As researchers delve further into the NP kinetics in the waste media, optimal techniques are likely to appear. The most common recyclable NMs and NM recycling techniques with their advantages and disadvantages are presented in Table 8.1.

8.4 Conclusions and outlook

Recovering NMs from waste sources is an important topic, since the generation of NMs has been increased rapidly. The approach of recycling wastes for obtaining reusable NMs has beneficial not only to the environment but also to industries from an economic point of view [4,111,112]. Therefore in a longer-term perspective of the implementation of NMs, life cycle analysis and the effective strategy for reuse of nanotechnology products should be adopted for sustainability. Remarkable progress has been made in establishing new methods for recovering the NMs from several types of waste sources. The aim of this chapter was to discuss the comprehensive scientific progress of recycling NMs from waste sources and future perspectives. For this

Table 8.1 Summary of recyclable nanomaterials and recycling methods.

NM type	NM	Production method	Recycling technique	Advantages	Disadvantages	Reference
Carbonaceous NM	MWCNT	Commercial	Washing (desorption)	Effortless	—	[23]
	SWCNT	Commercial	CPE and centrifugation	Near room temperature	Appropriate solvent required	[24]
	rGO/Ag NP	Modified Hummers' and in situ reduction	Centrifugation and washing	10 + runs of recycling while maintaining activity	—	[27]
Ceramic-based NM	γ-Fe_2O_3/Pd complex	Hydrothermal method	Magnetic separation	Scalable	10% Loss of catalytic activity after five rounds	[34]
	Fe_3O_4 coated with SiO_2	Hydrothermal method	Magnetic separation	Effortless, high recyclability around 30 + runs	Additional coating to prevent agglomeration	[35]
	SnO_2	Generated as waste by electroplating	Acid and base treatment	High-purity nanowires	Harsh treatment	[38]
	$Mg(OH)_2$ and $CaCO_3$	Generated as waste by chrome plating	Mineralizer and hydrothermal treatment	Harmless NP recovery	NM changes phase and cannot be used further	[40]
	CdS and ZnS	Liquid phase microemulsion	Phase separation at critical conditions	Reversible approach	Surfactant required	[43]
Metal-based NM	Fe and Ni	Liquid phase preparation	Washing	High performance at low loading	Loss of NMs during membrane preparation	[49]
	Ag	Solvothermal method	Filtration	Very high recycling efficiency	—	[52]

		Solvothermal method	Separation and washing	No loss activity for five runs	Low recyclability	[53]
		Hydrothermal method	Separation	High catalyzing properties at low loading	Low recyclability	[54]
	Pd	Hydrothermal method	Phase separation and centrifugation	No change in the NM properties	Mixture of solvents and stabilizers required, low recovery	[59]
	Fe_3O_4/Hal-Mel-TEA (IL)-Pd complex	Hydrothermal method	Magnetic separation	High retention of catalytic properties	Complex production route	[60]
Nanocomposite products	NM-reinforced PA6	Commercial	Acid dissolution, centrifugation, and washing	NMs can be further used in the same nanocomposite	Low amounts of NM recovery	[74]
	NM-reinforced PP	Commercial	Calcination	High NM recovery	NMs should be compatibilized again	[74]

CPE, Cloud point extraction; MWCNT, multiwalled carbon nanotube; NM, nanomaterial; NP, nanoparticle; rGO, reduced graphene oxide; SWCNT, single-walled carbon nanotube.

purpose, various recovery processes of carbon-, ceramic-, metal-based NMs, and nanocomposite products were reviewed together with related issues, such as economical and processability perspectives. The chapter started with the identification of wastes that can be used as resources for the recycling of NMs, followed by discussing various methods of recovering techniques. Recycling of waste containing NMs characteristically contains different mechanical, physical, and chemical processes, such as washing, crushing, taking to pieces, pulping, melting, extraction, drying, thinning, and adjusting.

Recycling and reutilization of these materials may increase the potential of high-performance nanocomposite materials in numerous application areas, including catalysis, sports industries, photovoltaic, energy devices, environment, and biomedicine [113]. Similar to any other techniques developed at an early stage, to enable the next generation of progress in NM recycling, more innovative fabrication methods should be provided for the nanotechnology life cycle. Moreover, a variety of preparation methods can be developed with different pathways, depending upon the perceived application of the material. There is still room for extensive research for a fundamental understanding of the waste materials containing NMs, and more future commercial utilization on NM recycling can be advanced by enhanced technologies. A further in-depth and explicit investigation should be carried out to determine the real potential of the recycling of NMs from waste components and the performance of products that include recycled NM.

References

[1] E. Asmatulu, J. Twomey, M. Overcash, Life cycle and nano-products: end-of-life assessment, J. Nanopart. Res. 14 (3) (2012) 720. Available from: https://doi.org/10.1007/s11051-012-0720-0.

[2] C.J. Cleveland, C. Morris, Handbook of Energy Volume II: Chronologies, Top Ten Lists, and Word Clouds, Elsevier, 2014.

[3] Keiper, A., The nanotechnology revolution, The New Atlantis, 2003.

[4] D.G. Rickerby, M. Morrison, Nanotechnology and the environment: a European perspective, Sci. Technol. Adv. Mater. 8 (1–2) (2007) 19–24. Available from: https://doi.org/10.1016/j.stam.2006.10.002.

[5] N. Musee, Nanowastes and the environment: potential new waste management paradigm, Environ. Int. 37 (1) (2011) 112–128. Available from: https://doi.org/10.1016/j.envint.2010.08.005.

[6] Novel Materials in the Environment: The Case of Nanotechnology Twenty-Seventh Report, Royal Commission on Environmental Pollution, London, 2008.

[7] Terminology for Nanomaterials, British Standards Institute, 2007.

[8] C. Buzea, I.I. Pacheco, K. Robbie, Nanomaterials and nanoparticles: sources and toxicity, Biointerphases 2 (4) (2007) MR17–MR71. Available from: https://doi.org/10.1116/1.2815690.

[9] C.M. Goodman, C.D. McCusker, T. Yilmaz, V.M. Rotello, Toxicity of gold nanoparticles functionalized with cationic and anionic side chains, Bioconjugate Chem. 15 (4) (2004) 897–900. Available from: https://doi.org/10.1021/bc049951i.

[10] E.E. Connor, J. Mwamuka, A. Gole, C.J. Murphy, M.D. Wyatt, Gold nanoparticles are taken up by human cells but do not cause acute cytotoxicity, Small 1 (3) (2005) 325–327. Available from: https://doi.org/10.1002/smll.200400093.

[11] S. Bosi, T. Da Ros, G. Spalluto, M. Prato, Fullerene derivatives: an attractive tool for biological applications, Eur. J. Med. Chem. 38 (11−12) (2003) 913−923. Available from: https://doi.org/10.1016/j.ejmech.2003.09.005.

[12] D. Schubert, R. Dargusch, J. Raitano, S.W. Chan, Cerium and yttrium oxide nanoparticles are neuroprotective, Biochem. Biophys. Res. Commun. 342 (1) (2006) 86−91. Available from: https://doi.org/10.1016/j.bbrc.2006.01.129.

[13] A.M. Derfus, W.C.W. Chan, S.N. Bhatia, Probing the cytotoxicity of semiconductor quantum dots, Nano Lett. 4 (1) (2004) 11−18. Available from: https://doi.org/10.1021/nl0347334.

[14] A.D. Maynard, Nanotechnology: A Research Strategy for Adressing Risk, Woodrow Wilson International Center for Scholars, 2006.

[15] A. Gungor, S.M. Gupta, Issues in environmentally conscious manufacturing and product recovery: a survey, Comput. Ind. Eng. 36 (4) (1999) 811−853. Available from: https://doi.org/10.1016/S0360-8352(99)00167-9.

[16] G. Bystrzejewska-Piotrowska, J. Golimowski, P.L. Urban, Nanoparticles: their potential toxicity, waste and environmental management, Waste Manag. 29 (9) (2009) 2587−2595. Available from: https://doi.org/10.1016/j.wasman.2009.04.001.

[17] R. Asmatulu, S. Davluri, W. Khan, Fabrications of CNT based nanocomposite fibers from the recycled plastics, ASME Int. Mech. Eng. Cong. Expos. Proc. 12 (PART B) (2010) 859−864. Available from: https://doi.org/10.1115/IMECE2009-12338.

[18] X. Peng, Z. Luan, J. Ding, Z. Di, Y. Li, B. Tian, Ceria nanoparticles supported on carbon nanotubes for the removal of arsenate from water, Mater. Lett. 59 (4) (2005) 399−403. Available from: https://doi.org/10.1016/j.matlet.2004.05.090.

[19] Y.-H. Li, S. Wang, J. Wei, X. Zhang, C. Xu, Z. Luan, et al., Lead adsorption on carbon nanotubes, Chem. Phys. Lett. 357 (3−4) (2002) 263−266. Available from: https://doi.org/10.1016/S0009-2614(02)00502-X.

[20] K.W. Ng, W.H. Lam, S. Pichiah, A review on potential applications of carbon nanotubes in marine current turbines, Renew. Sustain. Energy Rev. 28 (2013) 331−339. Available from: https://doi.org/10.1016/j.rser.2013.08.018.

[21] I. Berktas, M. Hezarkhani, L. Haghighi Poudeh, B. Saner Okan, Recent developments in the synthesis of graphene and graphene-like structures from waste sources by recycling and upcycling technologies: a review, Graphene Technol. 5 (3−4) (2020) 59−73. Available from: https://doi.org/10.1007/s41127-020-00033-1.

[22] Y.H. Li, S. Wang, Z. Luan, J. Ding, C. Xu, D. Wu, Adsorption of cadmium(II) from aqueous solution by surface oxidized carbon nanotubes, Carbon 41 (5) (2003) 1057−1062. Available from: https://doi.org/10.1016/S0008-6223(02)00440-2.

[23] Y. Cai, G. Jiang, J. Liu, Q. Zhou, Multiwalled carbon nanotubes as a solid-phase extraction adsorbent for the determination of bisphenol A, 4-n-nonylphenol, and 4-tert-octylphenol, Anal. Chem. 75 (10) (2003) 2517−2521. Available from: https://doi.org/10.1021/ac0263566.

[24] J.F. Liu, R. Liu, Y.G. Yin, G. Bin Jiang, Triton X-114 based cloud point extraction: a thermoreversible approach for separation/concentration and dispersion of nanomaterials in the aqueous phase, Chem. Commun. 12 (2009) 1514−1516. Available from: https://doi.org/10.1039/b821124h.

[25] S.K. Tiwari, S. Sahoo, N. Wang, A. Huczko, Graphene research and their outputs: status and prospect, J. Sci. Adv. Mater. Devices 5 (1) (2020) 10−29. Available from: https://doi.org/10.1016/j.jsamd.2020.01.006.

[26] M. Hashemi Salehi, M. Yousefi, M. Hekmati, E. Balali, Application of palladium nanoparticle-decorated *Artemisia abrotanum* extract-modified graphene oxide for highly active catalytic reduction of methylene blue, methyl orange and rhodamine B, Appl. Organomet. Chem. 33 (10) (2019) e5123. Available from: https://doi.org/10.1002/aoc.5123.

[27] S. Hemmati, M.M. Heravi, B. Karmakar, H. Veisi, Green fabrication of reduced graphene oxide decorated with Ag nanoparticles (rGO/Ag NPs) nanocomposite: a reusable catalyst for the degradation of environmental pollutants in aqueous medium, J. Mol. Liq. 319 (2020) 114302. Available from: https://doi.org/10.1016/j.molliq.2020.114302.

[28] W.S. Hummers, R.E. Offeman, Preparation of graphitic oxide, J. Am. Chem. Soc. 80 (6) (1958) 1339. Available from: https://doi.org/10.1021/ja01539a017.

[29] L. Zhan, O. Li, Z. Wang, B. Xie, Recycling zinc and preparing high-value-added nanozinc oxide from waste zinc-manganese batteries by high-temperature evaporation-separation and oxygen control oxidation, ACS Sustain. Chem. Eng. 6 (9) (2018) 12104−12109. Available from: https://doi.org/10.1021/acssuschemeng.8b02430.

[30] K. Schmid, M. Riediker, Use of nanoparticles in swiss industry: a targeted survey, Environ. Sci. Technol. 42 (7) (2008) 2253−2260. Available from: https://doi.org/10.1021/es071818o.

[31] T. Lü, D. Qi, D. Zhang, K. Fu, Y. Li, H. Zhao, Fabrication of recyclable multi-responsive magnetic nanoparticles for emulsified oil-water separation, J. Clean. Prod. 255 (2020). Available from: https://doi.org/10.1016/j.jclepro.2020.120293.

[32] T. Das, G. Kalita, P.J. Bora, D. Prajapati, G. Baishya, B.K. Saikia, Humi-Fe_3O_4 nanocomposites from low-quality coal with amazing catalytic performance in reduction of nitrophenols, J. Environ. Chem. Eng. 5 (2) (2017) 1855−1865. Available from: https://doi.org/10.1016/j.jece.2017.03.021.

[33] W.A. Herrmann, N-heterocyclic carbenes: a new concept in organometallic catalysis, Angew. Chem. Int. Ed. 41 (8) (2002) 1290−1309. doi:10.1002/1521-3773(20020415)41:8 < 1290::AID-ANIE1290 > 3.0.CO;2-Y.

[34] P.D. Stevens, G. Li, J. Fan, M. Yen, Y. Gao, Recycling of homogeneous Pd catalysts using superparamagnetic nanoparticles as novel soluble supports for Suzuki, Heck, and Sonogashira cross-coupling reactions, Chem. Commun. 35 (2005) 4435−4437. Available from: https://doi.org/10.1039/b505424a.

[35] C. Ó Dálaigh, S.A. Corr, Y. Gun'ko, S.J. Connon, A magnetic-nanoparticle-supported 4-N,N-dialkylaminopyridine catalyst: excellent reactivity combined with facile catalyst recovery and recyclability, Angew. Chem. Int. Ed. 46 (23) (2007) 4329−4332. Available from: https://doi.org/10.1002/anie.200605216.

[36] N.P. Cheremisinoff, Handbook of Solid Waste Management and Waste Minimization Technologies, Butterworth-Heinemann, 2003.

[37] C. Kerr, Sustainable technologies for the regeneration of acidic tin stripping solutions used in PCB fabrication, Trans. Inst. Met. Finish. 82 (1−2) (2004). Available from: https://doi.org/10.1080/00202967.2004.11871544.

[38] Z. Zhuang, X. Xu, Y. Wang, Y. Wang, F. Huang, Z. Lin, Treatment of nanowaste via fast crystal growth: with recycling of nano-SnO_2 from electroplating sludge as a study case, J. Hazard. Mater. 211−212 (2012) 414−419. Available from: https://doi.org/10.1016/j.jhazmat.2011.09.036.

[39] J.A. Kent, Salt, Chlor-alkali, and Related Heavy Chemicals, Riegel's Handbook of Industrial Chemistry, vol. 112, Springer United States, Boston, MA, 2003, pp. 429−462. Available from: http://doi.org/10.1007/0-387-23816-6_12.

[40] W. Liu, F. Huang, Y. Liao, J. Zhang, G. Ren, Z. Zhuang, et al., Treatment of CrVI-containing $Mg(OH)_2$ nanowaste, Angew. Chem. Int. Ed. 47 (30) (2008) 5619−5622. Available from: https://doi.org/10.1002/anie.200800172.

[41] N. Bühler, K. Meier, J.F. Reber, Photochemical hydrogen production with cadmium sulfide suspensions, J. Phys. Chem. 88 (15) (1984) 3261−3268. Available from: https://doi.org/10.1021/j150659a025.

[42] K. Sooklal, B.S. Cullum, S.M. Angel, C.J. Murphy, Photophysical properties of ZnS nanoclusters with spatially localized Mn^{2+}, J. Phys. Chem. 100 (11) (1996) 4551−4555. Available from: https://doi.org/10.1021/jp952377a.

[43] O. Myakonkaya, C. Guibert, J. Eastoe, I. Grillo, Recovery of nanoparticles made easy, Langmuir 26 (6) (2010) 3794−3797. Available from: https://doi.org/10.1021/la100111b.

[44] Z. Rogulski, K. Klimek, A. Czerwiński, Założenia procesu utylizacji i recyklingu baterii cynkowo-węglowych i cynkowo-manganowych, Przem. Chem. 85 (8−9) (2006) 1208−1211.

[45] L. Zhang, L.H. Zhang, M.L. Sui, J. Tan, K. Lu, Superheating and melting kinetics of confined thin films, Acta Mater. 54 (13) (2006) 3553−3560. Available from: https://doi.org/10.1016/j.actamat.2006.03.045.

[46] W.H. Qi, M.P. Wang, Size- and shape-dependent superheating of nanoparticles embedded in a matrix, Mater. Lett. 59 (18) (2005) 2262−2266. Available from: https://doi.org/10.1016/j.matlet.2004.06.079.

[47] S. Olapiriyakul, R.J. Caudill, Thermodynamic analysis to assess the environmental impact of end-of-life recovery processing for nanotechnology products, Environ. Sci. Technol. 43 (21) (2009) 8140–8146. Available from: https://doi.org/10.1021/es9006614.

[48] L. Li, J. Hu, X. Shi, M. Fan, J. Luo, X. Wei, Nanoscale zero-valent metals: a review of synthesis, characterization, and applications to environmental remediation, Environ. Sci. Pollut. Res. 23 (18) (2016) 17880–17900. Available from: https://doi.org/10.1007/s11356-016-6626-0.

[49] D.E. Meyer, K. Wood, L.G. Bachas, D. Bhattacharyya, Degradation of chlorinated organics by membrane-immobilized nanosized metals, Environ. Prog. 23 (3) (2004) 232–242. Available from: https://doi.org/10.1002/ep.10031.

[50] N. Pradhan, A. Pal, T. Pal, Catalytic reduction of aromatic nitro compounds by coinage metal nanoparticles, Langmuir 17 (5) (2001) 1800–1802. Available from: https://doi.org/10.1021/la000862d.

[51] F. Frederix, J.M. Friedt, K.H. Choi, W. Laureyn, A. Campitelli, D. Mondelaers, et al., Biosensing based on light absorption of nanoscaled gold and silver particles, Anal. Chem. 75 (24) (2003) 6894–6900. Available from: https://doi.org/10.1021/ac0346609.

[52] E. Murugan, J.N. Jebaranjitham, Synthesis and characterization of silver nanoparticles supported on surface-modified poly(N-vinylimidazale) as catalysts for the reduction of 4-nitrophenol, J. Mol. Catal. A Chem. 365 (2012) 128–135. Available from: https://doi.org/10.1016/j.molcata.2012.08.021.

[53] A. Rostami-Vartooni, M. Nasrollahzadeh, M. Alizadeh, Green synthesis of seashell supported silver nanoparticles using *Bunium persicum* seeds extract: application of the particles for catalytic reduction of organic dyes, J. Colloid Interface Sci. 470 (2016) 268–275. Available from: https://doi.org/10.1016/j.jcis.2016.02.060.

[54] H. Veisi, S. Azizi, P. Mohammadi, Green synthesis of the silver nanoparticles mediated by *Thymbra spicata* extract and its application as a heterogeneous and recyclable nanocatalyst for catalytic reduction of a variety of dyes in water, J. Clean. Prod. 170 (2018) 1536–1543. Available from: https://doi.org/10.1016/j.jclepro.2017.09.265.

[55] Y. Shen, X. Bo, Z. Tian, Y. Wang, M. Xie, F. Gao, et al., Fabrication of highly dispersed/active ultrafine Pd nanoparticle supported catalysts: a facile solvent-free: in situ dispersion/reduction method, Green Chem. 19 (11) (2017) 2646–2652. Available from: https://doi.org/10.1039/c7gc00262a.

[56] F. Subhan, S. Aslam, Z. Yan, M. Yaseen, Unusual Pd nanoparticle dispersion in microenvironment for p-nitrophenol and methylene blue catalytic reduction, J. Colloid Interface Sci. 578 (2020) 37–46. Available from: https://doi.org/10.1016/j.jcis.2020.05.093.

[57] V.M. Shinde, E. Skupien, M. Makkee, Synthesis of highly dispersed Pd nanoparticles supported on multi-walled carbon nanotubes and their excellent catalytic performance for oxidation of benzyl alcohol, Catal. Sci. Technol. 5 (8) (2015) 4144–4153. Available from: https://doi.org/10.1039/c5cy00563a.

[58] J.A. Dahl, B.L.S. Maddux, J.E. Hutchison, Toward greener nanosynthesis, Chem. Rev. 107 (6) (2007) 2228–2269. Available from: https://doi.org/10.1021/cr050943k.

[59] M.F. Nazar, S.S. Shah, J. Eastoe, A.M. Khan, A. Shah, Separation and recycling of nanoparticles using cloud point extraction with non-ionic surfactant mixtures, J. Colloid Interface Sci. 363 (2) (2011) 490–496. Available from: https://doi.org/10.1016/j.jcis.2011.07.070.

[60] S. Sadjadi, P. Mohammadi, M. Heravi, Bio-assisted synthesized Pd nanoparticles supported on ionic liquid decorated magnetic halloysite: an efficient catalyst for degradation of dyes, Sci. Rep. 10 (1) (2020) 2–11. Available from: https://doi.org/10.1038/s41598-020-63558-8.

[61] O. Kamigaito, What can be improved by nanometer composites? J. Jpn. Soc. Powder Powder Metall. 38 (3) (1991) 315–321. Available from: https://doi.org/10.2497/jjspm.38.315.

[62] M. José-Yacamán, L. Rendón, J. Arenas, M.C. Serra Puche, Maya blue paint: an ancient nanostructured material, Science 273 (5272) (1996) 223–225. Available from: https://doi.org/10.1126/science.273.5272.223.

[63] Z. Tian, H. Hu, Y. Sun, A molecular dynamics study of effective thermal conductivity in nanocomposites, Int. J. Heat Mass Transf. 61 (1) (2013) 577–582. Available from: https://doi.org/10.1016/j.ijheatmasstransfer.2013.02.023.

[64] E. Manias, Nanocomposites: stiffer by design, Nat. Mater. 6 (1) (2007) 9–11. Available from: https://doi.org/10.1038/nmat1812.

[65] M.C. Powell, M.P.A. Griffin, S. Tai, Bottom-up risk regulation? How nanotechnology risk knowledge gaps challenge federal and state environmental agencies, Environ. Manage. 42 (3) (2008) 426−443. Available from: https://doi.org/10.1007/s00267-008-9129-z.

[66] S. Erden, K. Ho, Fiber reinforced composites, in: M. Özgür Seydibeyoğlu, A.K. Mohanty, M. Misra (Eds.), Woodhead Publishing Series in Composites Science and Engineering, Fiber Technology for Fiber-Reinforced Composites, Woodhead Publishing, 2017, pp. 51−79. Available from: http://doi.org/10.1016/B978-0-08-101871-2.00003-5.

[67] P. Ziehl, R. Anay, J.J. Myers, Durability of fiber-reinforced plastics for infrastructure applications, Durability of Composite Systems, Elsevier, 2020, pp. 271−288. Available from: http://doi.org/10.1016/b978-0-12-818260-4.00006-5.

[68] G. Marsh, Facing up to the recycling challenge, Reinf. Plast. 45 (6) (2001) 22−26. Available from: https://doi.org/10.1016/S0034-3617(01)80204-6.

[69] G. Jia, H. Wang, L. Yan, X. Wang, R. Pei, T. Yan, et al., Cytotoxicity of carbon nanomaterials: single-wall nanotube, multi-wall nanotube, and fullerene, Environ. Sci. Technol. 39 (5) (2005) 1378−1383. Available from: https://doi.org/10.1021/es048729l.

[70] C. Kirchner, T. Liedl, S. Kudera, T. Pellegrino, A.M. Javier, H.E. Gaub, et al., Cytotoxicity of colloidal CdSe and CdSe/ZnS nanoparticles, Nano Lett. 5 (2) (2005) 331−338. Available from: https://doi.org/10.1021/nl047996m.

[71] C.A. Poland, R. Duffin, I. Kinloch, A. Maynard, W.A.H. Wallace, A. Seaton, et al., Carbon nanotubes introduced into the abdominal cavity of mice show asbestos-like pathogenicity in a pilot study, Nat. Nanotechnol. 3 (7) (2008) 423−428. Available from: https://doi.org/10.1038/nnano.2008.111.

[72] V. Khanna, B.R. Bakshi, L.J. Lee, Assessing life cycle environmental implications of polymer nanocomposites, in: IEEE International Symposium on Electronics and the Environment, 2008. doi: 10.1109/ISEE.2008.4562903.

[73] S.J. Pickering, T.A. Turner, F. Meng, C.N. Morris, J.P. Heil, K.H. Wong, et al., Developments in the fluidised bed process for fibre recovery from thermoset composites, in: CAMX 2015 − Composites and Advanced Materials Expo, December 2015, pp. 2384−2394.

[74] M. Busquets-Fité, E. Fernandez, G. Janer, G. Vilar, S. Vázquez-Campos, R. Zanasca, et al., Exploring release and recovery of nanomaterials from commercial polymeric nanocomposites, J. Phys. Conf. Ser. 429 (1) (2013) 012048. Available from: https://doi.org/10.1088/1742-6596/429/1/012048.

[75] S.M. Lloyd, L.B. Lave, H.S. Matthews, Life cycle benefits of using nanotechnology to stabilize platinum-group metal particles in automotive catalysts, Environ. Sci. Technol. 39 (5) (2005) 1384−1392. Available from: https://doi.org/10.1021/es049325w.

[76] D. Živković, L. Balanović, A. Mitovski, N. Talijan, N. Štrbac, M. Sokić, et al., Nanomaterials environmental risks and recycling: actual issues, Reciklaza Odrzivi Razvoj 7 (1) (2015) 1−8. Available from: https://doi.org/10.5937/ror1401001z.

[77] O. Myakonkaya, Z. Hu, M.F. Nazar, J. Eastoe, Recycling functional colloids and nanoparticles, Chem. A Eur. J. 16 (39) (2010) 11784−11790. Available from: https://doi.org/10.1002/chem.201000942.

[78] J.-L. Wang, J.-W. Liu, S.-H. Yu, Recycling valuable elements from the chemical synthesis process of nanomaterials: a sustainable view, ACS Mater. Lett. 1 (5) (2019) 541−548.

[79] M. Iranmanesh, J. Hulliger, Magnetic separation: its application in mining, waste purification, medicine, biochemistry and chemistry, Chem. Soc. Rev. 46 (19) (2017) 5925−5934. Available from: https://doi.org/10.1039/c7cs00230k.

[80] R.N. Grass, E.K. Athanassiou, W.J. Stark, Covalently functionalized cobalt nanoparticles as a platform for magnetic separations in organic synthesis, Angew. Chem. Int. Ed. 46 (26) (2007) 4909−4912. Available from: https://doi.org/10.1002/anie.200700613.

[81] R.D. Ambashta, M. Sillanpää, Water purification using magnetic assistance: a review, J. Hazard. Mater. 180 (1−3) (2010) 38−49. Available from: https://doi.org/10.1016/j.jhazmat.2010.04.105.

[82] N.S. Zaidi, J. Sohaili, K. Muda, M. Sillanpää, Magnetic field application and its potential in water and wastewater treatment systems, Sep. Purif. Rev. 43 (3) (2014) 206−240. Available from: https://doi.org/10.1080/15422119.2013.794148.

[83] A.A. Adewunmi, M.S. Kamal, T.I. Solling, Application of magnetic nanoparticles in demulsification: a review on synthesis, performance, recyclability, and challenges, J. Pet. Sci. Eng. 196 (2021) 107680. Available from: https://doi.org/10.1016/j.petrol.2020.107680.

[84] C.T. Yavuz, J.T. Mayo, W.W. Yu, A. Prakash, J.C. Falkner, S. Yean, et al., Low-field magnetic separation of monodisperse Fe_3O_4 nanocrystals, Science 314 (5801) (2006) 964−967. Available from: https://doi.org/10.1126/science.1131475.

[85] H. Xu, W. Jia, S. Ren, J. Wang, Novel and recyclable demulsifier of expanded perlite grafted by magnetic nanoparticles for oil separation from emulsified oil wastewaters, Chem. Eng. J. 337 (2017) (2018) 10−18. Available from: https://doi.org/10.1016/j.cej.2017.12.084.

[86] J. Liu, H. Wang, X. Li, W. Jia, Y. Zhao, S. Ren, Recyclable magnetic graphene oxide for rapid and efficient demulsification of crude oil-in-water emulsion, Fuel 189 (2017) 79−87. Available from: https://doi.org/10.1016/j.fuel.2016.10.066.

[87] A.H. Latham, R.S. Freitas, P. Schiffer, M.E. Williams, Capillary magnetic field flow fractionation and analysis of magnetic nanoparticles, Anal. Chem. 77 (15) (2005) 5055−5062. Available from: https://doi.org/10.1021/ac050611f.

[88] A.H. Latham, M.E. Williams, Controlling transport and chemical functionality of magnetic nanoparticles, Acc. Chem. Res. 41 (3) (2008) 411−420. Available from: https://doi.org/10.1021/ar700183b.

[89] J.S. Beveridge, J.R. Stephens, M.E. Williams, Differential magnetic catch and release: experimental parameters for controlled separation of magnetic nanoparticles, Analyst 136 (12) (2011) 2564−2571. Available from: https://doi.org/10.1039/c1an15168a.

[90] G.D. Moeser, K.A. Roach, W.H. Green, T.A. Hatton, P.E. Laibinis, High-gradient magnetic separation of coated magnetic nanoparticles, AIChE J. 50 (11) (2004) 2835−2848. Available from: https://doi.org/10.1002/aic.10270.

[91] T. Leshuk, A.B. Holmes, D. Ranatunga, P.Z. Chen, Y. Jiang, F. Gu, Magnetic flocculation for nanoparticle separation and catalyst recycling, Environ. Sci. Nano. 5 (2) (2018) 509−519. Available from: https://doi.org/10.1039/c7en00827a.

[92] D. Liu, J. Zhang, B. Han, J. Chen, Z. Li, D. Shen, et al., Recovery of TiO_2 nanoparticles synthesized in reverse micelles by antisolvent CO_2, Colloids Surf. A Physicochem. Eng. Asp. 227 (1−3) (2003) 45−48. Available from: https://doi.org/10.1016/S0927-7757(03)00385-6.

[93] J. Zhang, B. Han, J. Liu, X. Zhang, J. He, Z. Liu, et al., Recovery of silver nanoparticles synthesized in $AOT/C_{12}E_4$ mixed reverse micelles by antisolvent CO_2, Chem. A Eur. J. 8 (17) (2002) 3879−3883. doi:10.1002/1521-3765(20020902)8:17 < 3879::AID-CHEM3879 > 3.0.CO;2-W.

[94] J. Zhang, B. Han, J. Liu, X. Zhang, G. Yang, H. Zhao, Size tailoring of ZnS nanoparticles synthesized in reverse micelles and recovered by compressed CO_2, J. Supercrit. Fluids 30 (1) (2004) 89−95. Available from: https://doi.org/10.1016/S0896-8446(03)00113-X.

[95] E. Torino, I. De Marco, E. Reverchon, Organic nanoparticles recovery in supercritical antisolvent precipitation, J. Supercrit. Fluids 55 (1) (2010) 300−306. Available from: https://doi.org/10.1016/j.supflu.2010.06.001.

[96] M.R. Helfrich, M. El-Kouedi, M.R. Etherton, C.D. Keating, Partitioning and assembly of metal particles and their bioconjugates in aqueous two-phase systems, Langmuir 21 (18) (2005) 8478−8486. Available from: https://doi.org/10.1021/la051220z.

[97] J.F. Liu, J.B. Chao, R. Liu, Z.Q. Tan, Y.G. Yin, Y. Wu, et al., Cloud point extraction as an advantageous preconcentration approach for analysis of trace silver nanoparticles in environmental waters, Anal. Chem. 81 (15) (2009) 6496−6502. Available from: https://doi.org/10.1021/ac900918e.

[98] P. Gao, Y. Chen, Y. Wang, Q. Zhang, X. Li, M. Hu, A simple recycling and reuse hydrothermal route to ZnO nanorod arrays, nanoribbon bundles, nanosheets, nanocubes and nanoparticles, Chem. Commun. 2 (19) (2009) 2762−2764. Available from: https://doi.org/10.1039/b900391f.

[99] M.J. Hollamby, J. Eastoe, A. Chemelli, O. Glatter, S. Rogers, R.K. Heenan, et al., Separation and purification of nanoparticles in a single step, Langmuir 26 (10) (2010) 6989−6994. Available from: https://doi.org/10.1021/la904225k.

[100] S.K. Shahi, N. Kaur, A. Kaur, V. Singh, Green synthesis of photoactive nanocrystalline anatase TiO_2 in recyclable and recoverable acidic ionic liquid [Bmim] HSO_4, J. Mater. Sci. 50 (6) (2015) 2443−2450. Available from: https://doi.org/10.1007/s10853-014-8799-6.

[101] Y. Wang, S. Maksimuk, R. Shen, H. Yang, Synthesis of iron oxide nanoparticles using a freshly-made or recycled imidazolium-based ionic liquid, Green Chem. 9 (10) (2007) 1051−1056. Available from: https://doi.org/10.1039/b618933d.

[102] C. Liu, H. Gong, W. Liu, B. Lu, L. Ye, Separation and recycling of functional nanoparticles using reversible boronate ester and boroxine bonds, Ind. Eng. Chem. Res. 58 (11) (2019) 4695−4703. Available from: https://doi.org/10.1021/acs.iecr.9b00253.

[103] R.L. Oliveira, P.K. Kiyohara, L.M. Rossi, High performance magnetic separation of gold nanoparticles for catalytic oxidation of alcohols, Green Chem. 12 (1) (2010) 144−149. Available from: https://doi.org/10.1039/b916825g.

[104] J. Kakumazaki, T. Kato, K. Sugawara, Recovery of gold from incinerated sewage sludge ash by chlorination, ACS Sustain. Chem. Eng. 2 (10) (2014) 2297−2300. Available from: https://doi.org/10.1021/sc5002484.

[105] B. Abécassis, F. Testard, T. Zemb, Gold nanoparticle synthesis in worm-like catanionic micelles: microstructure conservation and temperature induced recovery, Soft Matter 5 (5) (2009) 974−978. Available from: https://doi.org/10.1039/b816427d.

[106] F.R. Xiu, F.S. Zhang, Size-controlled preparation of Cu_2O nanoparticles from waste printed circuit boards by supercritical water combined with electrokinetic process, J. Hazard. Mater. 233−234 (2012) 200−206. Available from: https://doi.org/10.1016/j.jhazmat.2012.07.019.

[107] N.V. Mdlovu, C.L. Chiang, K.S. Lin, R.C. Jeng, Recycling copper nanoparticles from printed circuit board waste etchants via a microemulsion process, J. Clean. Prod. 185 (2018) 781−796. Available from: https://doi.org/10.1016/j.jclepro.2018.03.087.

[108] A. Deep, K. Kumar, P. Kumar, P. Kumar, A.L. Sharma, B. Gupta, et al., Recovery of pure ZnO nanoparticles from spent Zn-MnO_2 alkaline batteries, Environ. Sci. Technol. 45 (24) (2011) 10551−10556. Available from: https://doi.org/10.1021/es201744t.

[109] Y.A. El-Nadi, J.A. Daoud, H.F. Aly, Leaching and separation of zinc from the black paste of spent MnO_2-Zn dry cell batteries, J. Hazard. Mater. 143 (1−2) (2007) 328−334. Available from: https://doi.org/10.1016/j.jhazmat.2006.09.027.

[110] W. Liu, C. Weng, J. Zheng, X. Peng, J. Zhang, Z. Lin, Emerging investigator series: treatment and recycling of heavy metals from nanosludge, Environ. Sci. Nano 6 (6) (2019) 1657−1673. Available from: https://doi.org/10.1039/c9en00120d.

[111] C. Som, M. Berges, Q. Chaudhry, M. Dusinska, T.F. Fernandes, S.I. Olsen, et al., The importance of life cycle concepts for the development of safe nanoproducts, Toxicology 269 (2−3) (2010) 160−169. Available from: https://doi.org/10.1016/j.tox.2009.12.012.

[112] K.D. Grieger, A. Laurent, M. Miseljic, F. Christensen, A. Baun, S.I. Olsen, Analysis of current research addressing complementary use of life-cycle assessment and risk assessment for engineered nanomaterials: have lessons been learned from previous experience with chemicals? J. Nanopart. Res. 14 (7) (2012). Available from: https://doi.org/10.1007/s11051-012-0958-6.

[113] A. Bratovcic, Different applications of nanomaterials and their impact on the environment, Int. J. Mater. Sci. Eng. 5 (1) (2019) 1−7. Available from: https://doi.org/10.14445/23948884/ijmse-v5i1p101.

CHAPTER 9

Procedures for recycling of nanomaterials: a sustainable approach

Ajit Behera[1], Deepak Sahini[2] and Dinesh Pardhi[3]

[1]Department of Metallurgical & Materials EngineeringNational Institute of Technology, Rourkela, India
[2]Department of Production Engineering, Birla Institute of Technology, Ranchi, India
[3]Faculty of Health Sciences, School of Pharmacy, University of Eastern Finland, Kuopio, Finland

9.1 Introduction

Nanomaterials are utilized in a wide variety of commercial products from electronic components to healthcare applications. Nanotechnology will grow into a US$4.5 trillion industry over the next $10-15$ years if current demand is maintained. Nanotechnology involves nanomaterials, nanoparticles (NPs), and nanocompounds in the range of $1-100$ nm (10^{-9} m) in size [1,2]. Nanomaterials are used because of their remarkably strong properties, enabled by their nanoscale structure. Nanoscale materials are commonly used in electronic equipment, household appliances (containing nanosilver, nanotitania), packaging, construction materials, healthcare (nanosized silica dental fillings), cosmetics (UV protection), paints (antibacterial and anticorrosion coating), textiles and synthetics (water repellent, antibacterial surface), sporting equipment, and medical devices. Liberation of NPs from these materials induces various risks for environmental health and safety depending on their involvement in chemical and/or mechanical interaction during working. The percentage of risk depends mainly on the size of the nanomaterial (relative surface area) and the chemical composition (especially toxic material content). The size of NPs allows them to interact strongly with biological structures, so they cause potential risk to human health as well as the environment [3]. A huge percentage of release of nanoscale products is from nanowaste. Also, their nano size presents problems during their selective separation, recovery, and recycling. The term *nanowaste* indicates waste that includes materials with nanoscale range. Nanowaste minimization is a novel challenge for researchers. To understand these risks better, the current knowledge about four specific waste treatment processes—recycling, incineration, landfilling, and wastewater treatment—is required [4,5]. Fig. 9.1 shows various form of nanowaste generated from commercial activities, and how the NPs interact gradually with human life is shown in Fig. 9.2. Table 9.1 categorizes various types of potential nanowaste that pose environmental risk. Fig. 9.2 shows how the nanowaste interferes in the human life cycle.

Nanomaterials Recycling
DOI: https://doi.org/10.1016/B978-0-323-90982-2.00009-3

© 2022 Elsevier Inc.
All rights reserved.

176 Nanomaterials Recycling

Figure 9.1 Major nanowaste producer sectors.

There is a proverb "Small snake has more poison" that applies the extremely small nanomaterials, specifically to their capability to interact with biological cells. Their high surface area-to-volume ratio makes NPs highly reactive. Generally, NPs interacts with living cells in four ways: endocytosis, semiendocytosis, adhesion, and penetration. Endocytosis occurs when the NP wraps around the cell membrane and enters entirely in the cell. Semiendocytosis occurs when the NP is wrapped up but remains suspended in the cell membrane. Adhesion occurs when NPs stick to the cell membrane. Penetration occurs when the NPs are small enough and slip through the cell membrane. The most dangerous risk comes when the NPs are inhaled, as they can easily become airborne [6]. It is much easier for NPs to enter the bloodstream and cells from the lungs than through the dermis, and they have been shown to accumulate in other organs. Long fibrous NPs, such as CNTs behave similarly to asbestos and can have similar hazards. For example, research has indicated that the general symptom is inflammation in lungs. There is no standard process to find or remove NPs from the living body. The interaction with living systems is also affected by the dimensions of the NPs. For instance, NPs no bigger than a few nanometers may reach well inside biomolecules, which is not possible for larger NPs. Owing to penetration behavior,

Procedures for recycling of nanomaterials: a sustainable approach 177

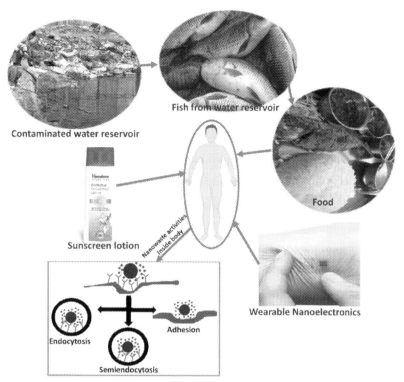

Figure 9.2 Direct and indirect entry of nanowastes and nanocontaminants to the human body from the environment.

Table 9.1 Categorization of potential exposed materials to the environment.

Nanomaterials	Type
Carbon nanotubes (CNTs) and their derivatives and fullerenes	Carbonaceous compound
Au, Ag, Fe, TiO$_2$ (titania), CeO (ceria), ZnO, Al$_2$O$_3$, FeO	Metals and metal oxides
Quantum dots, ZnSe, ZnTe, ZnS, CdS, CdTe, CdSe, GaAs, AlGaAs, PbSe, PbS, InP	Semiconductor device
Polymer nanoparticles (nanosphere, nanocapsules), polymer nanofibers	Polymer
Al$_2$O$_3$, SiO$_2$	Ceramic nanomaterials
Al$_2$O$_3$/SiO$_2$, Al$_2$O$_3$/CNT, SiO$_2$/CNT, polymer/CNT	Nanocomposite

inhaled NPs can reach the blood and may reach other target sites, such as the liver, heart, or blood cells. The fear is that because nanomaterials are smaller than cells, they might enter and create destruction or bioaccumulate in smaller creatures, then work their way up the food chain in ever-increasing concentrations until they cause problems for humans [7]. Buckyballs are strongly hydrophobic and strictly insoluble in water. Buckyballs dropped into the open environment cannot be transported by water but simply stick to the soil and other organic materials. Small concentrations of nano-C60 (20 parts per billion) kill half of the human liver and skin cells, damages brain cells in fish, and halts the growth of bacteria. Other human diseases that may be caused by nanomaterials are Crohn's disease, colon cancer and other cancers, autoimmune diseases, dermatitis, urticaria, vasculitis, neurological diseases, Parkinson's disease, Alzheimer's disease, asthma, bronchitis, emphysema, arteriosclerosis, vasoconstriction, thrombus, high blood pressure, arrhythmia, heart disease, kidney disease, liver disease, podoconiosis, and Kaposi's sarcoma [8−10]. Specific environmental conditions play an important role as well as the function of the nanomaterial.

We know very little about the behavior of nanomaterials or about environmental and health risks when these products enter various waste streams at the end of their life cycles. Nanomaterials could harm bacteria, the engine behind practically every ecosystem and food chain. Some nanomaterials may be pyrophoric or readily combustible, creating a risk of explosions and fires. NPs can be used for drug delivery purposes, either as the drug itself or as the drug carrier. The product can be administered orally, applied onto the skin, or injected [11]. The objective of drug delivery with NPs is to get more of the drug to the target cells, to reduce the harmful effects of the free drug on other organs, or both. NPs have the potential to cross the blood−brain barrier, which makes them extremely useful as a way to deliver drugs directly to the brain. On the other hand, this is also a major drawback because NPs that are used to carry drugs may be toxic to the brain. Metabolism and excretion of NPs depend upon the route of absorption and the particle surface properties. Some carbonaceous NPs have been metabolized in aquatic systems, and it is therefore assumed that those with branched side chains or hydrophilic groups are targets for normal metabolic processes driven by oxidative enzymes. Research using animal models has shown that certain polymer-based NPs are excreted via urine. Radiolabeled NPs administered to laboratory animals were found to be secreted in bile. Therefore it can be assumed, depending upon the properties of the NPs, that the feces of contaminated animals will contain NPs or NP metabolites. The metabolism and potential risks associated with NP use requires that all potentially contaminated carcasses, bedding, and other materials be disposed of in accordance with current regulations [12].

Tian et al. [13] showed that exogenous NPs, containing Zn and Al exert toxic effects on the germination and growth of roots in the seedlings of six agriculturally relevant plant species. Ryegrass biomass was significantly reduced, root tips shrank, and root epidermal and cortical cells were highly vacuolated or collapsed as an effect of adherence of ZnO NPs onto the root surface. Individual ZnO NPs have been observed to enter apoplast and protoplast of the root endodermis and stele. Baun et al. [14] studied the toxicity of C60, CNT, and TiO_2 to an aquatic invertebrate, *Daphnia magna*. The first forum dedicated to toxicity and the assessment of risks associated with the utilization of NPs took place in Stockholm in 2007. Since then, many academic groups have been involved in research concerning the interaction of NPs, such as CNTs, with cells. The group of Professor Shunho Jin from University of California, San Diego, United States, showed that in the presence of magnetic NPs, as long as they are smaller than 10 nm, neuronal cells stop responding to chemical signals, and instead of transducing signals, they enter a latent state. Several investigators have found that at high doses of different NPs, cytotoxic effects emerge in a dose- and time-dependent manner. As for the potential toxicity of NPs to the reproductive system, experimental data suggest that NPs cross the blood-testis barrier and are deposited in the testes and that there is a likelihood of adverse effects on sperm cells. Genaidy et al. [15] pointed out that the recycling of batteries with nanomaterials may generate nano-particulate dust once the battery casing is no longer intact. Recycling of textiles that contain nanomaterials may include various kinds of mechanical, thermal, and chemical treatments that may be associated with NP emissions.

9.1.1 Nanomaterials posing risks to humans and environment

Various nanomaterials responsible for potential risk on human life and for environment are discussed in this section [16–18].

Nanolead waste risk: Lead (Pb) is known as a heavy metal having a silvery bluish appearance that becomes gray by the action of air. Different type of origin of Pb pollution are listed in Table 9.2. Lead is becoming an environmental and health concern around the globe, owing to its widespread use. Lead is a carcinogenic element as declared by the US Environmental Protection Agency (EPA). *Lead poisoning* is a term used for its toxicity, and it may be acute or chronic. Lead poisoning can cause allergies, mental retardation, paralysis, birth defects such as autism, brain damage, dyslexia, and kidney damage and may also result in death.

Nanoarsenic waste risk: Arsenic (As) is a metalloid element found in mineral form generally compounded with sulfur, some other metals, salts of iron, calcium, sodium, and copper and also in pure elemental form. Water may be contaminated by the As-based materials listed in Table 9.2. As in the form of arsenate and arsenite is lethal to the environment and living creatures. As disturbs the protoplasm of cells by interacting with the sulfhydryl group of the cells, causing respiration malfunctioning and affecting mitosis and cell enzymes.

Nanocadmium waste risk: Cadmium (Cd) is a bluish-white soft metal having chemical behavior similar to that of mercury and zinc. Cd NPs are produced from the sources

Table 9.2 Nanomaterials risks and benefits.

Nanoparticles	Area of applications
Pb	Battery industries, fertilizers and pesticides, metal plating and finishing operations, exhaust, additives in gasoline, pigment in automobiles, smelting of ores
As	Pesticides, inappropriate disposal of arsenic-based reagents or chemicals
Cd	Smelting of ores, electroplating, batteries, plasticizers, pigments, nuclear industry
Cr (VI)	Particulates present in ground water and surface water
Ti	Tissue engineering, aviation industries, high-temperature equipment
Au	Catalysis industry, chemical synthesis, jewelry
Ag	Antimicrobial agent in domestic appliance (refrigerators, vacuum cleaners, air conditioning), antibacterial packing, antibacterial paints, textiles, plastics, varnish, cosmetics and personal care products, disinfectant sprays, deodorants, laundry soaps, wound dressings, toothpaste, baby products (milk bottles, teethers), medical instruments, hardware (computer, mobile phones), food storage containers, cooking utensils, food additives and supplements
Pt, Pd	Automotive exhaust converters, catalysts
Fe	Purification of ground water from PCBs, organochlorine pesticides, chlorinated organic solvents, degradation of PAH-based contaminants
Cd	Cd^{2+} is released from nanoparticles of cadmium selenide
ZnO	Skin care products, bottle coatings, gas purification, contaminant sensors, paints, cements, cosmetics, catalysts, UV protection, batteries, catalysis industry, photochemical devices, renewable energy storage
TiO_2	Paints, cements, sunscreen, catalysts, UV protection, batteries, cosmetics, personal care products, food processing, food packaging and labeling
SiO_2	Fireproof glass, bone reconstruction, UV protection, varnish, ceramics, electronics, pharmaceutical products, dentistry, polishing, biological tagging or labeling of genes, rubber tires industry, paints and coatings
Fe_2O_3	Concrete additive
Fe_3O_4	Biochemical assays, biomanipulation, removal of contamination
Al_2O_3	Batteries, fire protection, metal and biosorbent, cell separation and probing, nanopesticides/herbicides
CeO_2	Combustion catalyst in diesel fuels, solar cells, oxygen pumps, coatings, electronics, glass and ceramics, ophthalmic lenses
CaO	Catalysis industry, nanopesticides, herbicides
TiN	Development of nanocomposites, reinforced polymers

(*Continued*)

Table 9.2 (Continued)

Nanoparticles	Area of applications
Fe/Ni, Fe/Co, Fe/Pd, SiO_2/TiO_2	Remediation removal of toxic element and compounds
$CaCo_3$, MgO	Nanofertilizers
Fullerene	Cosmetics, personal care products
Carbon nanotubes	Electronics, additives to tires, lubricants, used in absorption of contaminants, sports equipment, electronics and electrical industry, drug delivery, implantable sensors
Graphene oxide films	Nanofertilizers, protective coating, microelectromechanical systems
Ferrofluidic particles	Magnetic resonance imaging, drug delivery, biochemical assays, microfluid coolants
Dendrimers	Drug delivery, tumor treatment, manufacture of macrocapsules, nanolatex, colored glass, chemical sensors, modified electrodes, coating, structural products
Quantum dots	Medical imaging, targeted therapeutics, solar cells, photovoltaic cells, security links, telecommunications

shown in Table 9.2. Generally, Cd is present at low levels in the environment; however, industrial wastes have greatly increased these levels. Cd-induced toxicity is responsible for degradation of the kidneys, respiratory system, and skeleton, and Cd is carcinogenic to humans.

Nanochromium waste risk: Chromium (Cr) is found in the air, water, and soil from different metallurgical and chemical industries. Cr is found in different forms such as divalent, tetravalent, pentavalent, and hexavalent states. Trivalent and hexavalent forms are the most stable form of Cr. Whereas, Cr(III) is an essential nutritional supplement for humans and animals. Cr(VI) form is highly toxic and carcinogenic in nature. The World Health Organization (WHO) has set the safe concentration limit of 50 µg/L for Cr(VI).

Nanozinc waste risk: Zinc (Zn) is an essential trace metal for humans. However, excessive intake of Zn can suppress iron (Fe) absorption. Zn ions are categorized as highly toxic to plants, vertebrate fishes, and invertebrates.

Nanosilver waste risk: Generally, silver (Ag)–NPs are used to kill bacteria. Ag-NPs are also is incorporated for their antibacterial, antiviral, and antifungal properties in a growing range of modern domestic or household items. They have the highest degree of commercialization of NPs in consumer products, including cosmetics, household appliances and cleaners, clothing (such as socks and underpants), personal care products (such as menstrual pads), refrigerator trays, filters, air conditioners and tubing, cutlery, and children's toys in the market. In 2010 Ag-NPs in wastewater were listed by a team of public health experts as one of 15 nascent issues that could deleteriously affect the conservation of biological diversity. Nanosilver-coated urethral and central line catheters and other implantable medical devices, such as infusion ports, orthopedic protruding fixateurs, endovascular stents, urological stents, endotracheal tubes, contact lens coatings, endoscopes, electrodes, peritoneal dialysis devices, subcutaneous cuffs, and surgical and dental instruments, are also used to prevent the growth of slime-containing biofilms that promote bacterial infection and sepsis.

NPs of TiO_2 and SiO_2 waste risk: NPs of titanium dioxide (TiO_2) are used in sunscreens, paints, and electronic circuits. TiO_2 can penetrate diseased or damaged skin in significant amounts and enter the bloodstream, where it can affect the central nervous system, resulting in permanent damage through intranasal instillation and neuroinflammation in the brain. TiO_2 dust is categorized as a Group 2B carcinogen by the International Agency for Research on Cancer. The presence of TiO_2 waste in the environment can kill beneficial soil microbes and bacteria, completely changing the ecosystem balance. Unlike crystalline TiO_2, amorphous TiO_2 and SiO_2 particles are known to be hazardous when inhaled and have led to water pollution affecting fish in the tested areas. Research has even found that certain cardiovascular diseases and cytotoxicity are related to these particles, which can also cause central nervous disorders. The soil also gets polluted by TiO_2 as it releases fumes during degradation, and it is responsible for Cd accumulation in fish. To reduce the hazardous effects, the nanocomposites must be recycled and should be used to the fullest extent of their life cycle.

Nanoplatinum waste risk: It has found that nanodispersed platinum group elements can be transferred to animal tissues. Nowadays, the majority of vehicles are equipped with catalysts containing noble metals, such as Pt. A few years ago, it was announced that Pt NPs, with dimensions in the range 0.8–10 nm, are released from car catalysts during their lifetime. The newest catalysts also contain NPs of metals such as Pt and Pd that have size several times smaller than those in conventional catalysts. Recently, it has been found that the addition of Al or Al_2O_3 NPs to diesel fuel improves its properties.

Carbon-based nanomaterials waste risk: CNTs, fullerenes, graphene, and carbon nano-dots are carbon-based nanomaterials that have applications in electronics, medicine, drug delivery, implantable sensors, and other fields. However, reports suggest the potentially hazardous, toxic, or even carcinogenic nature of these materials, making them unsuitable for use in biological applications. Indeed, the use of CNT for light-weight strengthening of building materials such as pavers could create asbestos-like problems when those materials are cut with electric saws and the particles are inhaled. CNT is extremely toxic to humans, producing more damage to the lungs than carbon black or silica. Varieties of CNT aggregates and some carbon blacks have been shown to be as cytotoxic as asbestos. Diesel combustion from vehicles represents an important source of emission of NPs. Cars still emit carbon-based aerosol NPs as a result of incomplete combustion, as well as lead compounds. CNTs produce mesotheliomas, damaging the lung epithelium in a way similar to asbestos. They are toxic, but the hazardous effects are largely dependent on size and production method. The nanomaterials generally cause proliferation inhibition and cell death. Generally, CNTs are less toxic than carbon nanofibers (CNFs) or nanospheres. Much of their underlying toxicity comes from oxidative stress, inflammatory responses, malignant transformation, DNA damage and mutation, and interstitial fibrosis.

Fig. 9.3A shows the recent TiO_2-nanowaste generation in consumer electronics, paints, batteries, metals, and other materials with nanomaterials recovery of 54%, 96%, 97%, 98%, and 100%, respectively. Fig. 9.3B is showing Ag-nanowaste generation in consumer electronics, healthcare, textiles, and metal coatings with their recovery of 76%, 84%, 91%, 97%, and 99.95%, respectively. According to the "Risk Assessment Characterization" the environmental risk depends on three factors: chemical present (hazard), amount of contact with the chemical (exposure), and the effect of the chemical on the individual (vulnerability). Risks are usually split into two categories: known risks and predictable risks. The EPA has set out a definition for known risks with a clear cause and effect of their impact. Predictable risks pose an unknown, and it is unclear how much of the hazard, exposure, and vulnerability is present. Engineered NPs fall under this predictable risk category, making them hard to define.

9.2 Classification of nanowaste

Classifications of the source of generation for the nanowaste are given in Fig. 9.4. Nanowaste can be generated from industrial products as well as natural activities. There are three major phases of industrial nanomaterials to generate waste: during production, during the working period, and at the time of disposal

1. *Nanowaste during production*: Waste generates during fabrication of products using nanoscale raw materials.
2. *Nanowaste during application*: Detachment of nanoscale particulates/layers by any mechanical action like friction or abrasion.

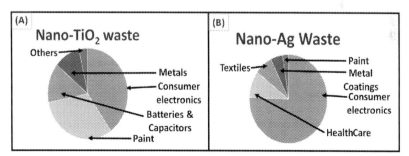

Figure 9.3 (A) TiO_2-nanowaste generation in consumer electronics, paints, batteries, metals and others. (B) Ag-nanowaste generation in consumer electronics, healthcare, textiles, metal coatings.

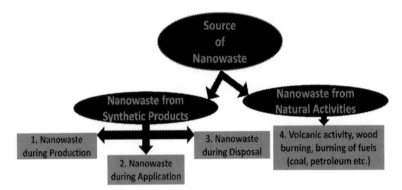

Figure 9.4 Classification of nanowaste generation.

3. *Nanowaste during disposal:* Waste generated at the end of life of nanomaterials or nanotechnology products. The useful lifespan of NPs or NP-containing products will last in landfills, wastewater treatment plants or waste incineration plants.
4. *Nanowaste from natural source:* By-products in volcanic activity, wood burning, burning of fuels such as coal, petroleum etc in nanoform can be consider as nanowaste.

Musee presented a classification of nanowastes according to their toxic effects on humans and other ecosystems [19]. On the basis of the intensity of their toxicity, nanowaste can be categorized as class I, class II, class III, class IV, or class V. Nanomaterials in the class I nanowaste category present hazards but very low or no toxic effects and are considered to have a very low risk profile. Examples of class I nanowastes include display backpanes of television screens, solar panels, memory chips, polishing agents. This type of nanomaterial does not have special disposal requirements. Hazard during waste management may be caused by toxicity of the bulk parent materials to humans and the environment in case of exceeding the concentration limit. Nanomaterials in the class II nanowaste category have toxic effects and are hazardous. They have a low to medium risk profile. This group includes used products such as display backplanes, solar panels, memory chips, polishing agents, paints, and coatings. Nanowaste of class II due to toxicity can cause acute or chronic effects, so appropriate and optimum waste management is recommended during handling, transportation, or disposal processes. Nanomaterials in the class III nanowaste category have toxic to very toxic effects and are hazardous. They have a medium to high risk profile. Examples of used products in class III are food packing, food additives, wastewater containing personal care products, polishing agents, and pesticides. For this class of nanowaste, proper protocols for managing of hazardous waste streams in the entire waste management chain are recommended. Furthermore, there is a need to determine whether the current waste management infrastructure is adequate to safely dispose of the hazardous waste. Nanomaterials in the class IV nanowaste category have toxic to very toxic effects and are hazardous. Their risk profile is high. Examples of nanowaste classified in this group are paints and coatings, personal care products, and pesticides. Owing to their toxic properties, nanowaste in this class should be disposed of in specialized hazardous wastes designated sites. The mismanagement of class IV nanowaste can result in a significant threat to humans and the environment. Nanomaterials in the class V nanowaste category have very toxic to extremely toxic effects and are hazardous. Their risk profile is high to very high. This group includes used products such as pesticides, sunscreen lotions, and food and beverages containing fullerenes in colloidal suspensions. Such nanowaste should be disposed of only in specialized designated hazardous waste stream sites. Inadequate waste management can lead to significant pollution of various ecosystems. For this group of nanowaste, immobilization and neutralization are recommended as the most effective treatment techniques.

Most of the procedures to fabricate nanomaterials are directly or indirectly related to chemical synthesis. The chemical synthesis of nanomaterials from various sources can be categorized into three parts, as shown in Fig. 9.5. The low productivity of chemical processing is generally caused by the extraction of by-products (feature A), and it also describes the high-value reaction medium in chemical reactions (feature B) [20]. Feature C is the scarified nanotemplate that existed in the form of ions in solution after template reactions.

Recently, porous or hollowed nanostructures have shown wide potential applications in fields of catalysis, sensing, biomedicine, and so on because of their highly open structure and low coordination sites on the surface. Among various strategies the nanotemplate reaction and chemical etching have been demonstrated to be the most straightforward and commonly used methods for the synthesis of porous or hollowed nanostructures. For example, Ag nanostructures with various morphologies are widely used as templates to fabricate Au, Pt, and Pd hollow and porous nanostructures for further applications in the fields of catalysis or biomedicine. To improve the performance of the catalyst, Niu et al. [21] fabricated porous Pt-rich nanoframes (Pt-Ni) by interior erosion of synthesized bimetallic Ni-rich nanocrystals (Pt-Ni), and Park et al.

Figure 9.5 Schematic illustration of waste generated during the chemical synthesis of nanomaterials.

[22] fabricated an Ir-based multimetallic nanoframe structure by etching Ni and Cu components from the IrNiCu bilayered nanoframes. For both nanotemplate and etching reactions, the more active elements were selectively etched as sacrificial agents to help complete the desired nanostructures during the synthesis process. Normally, the value of the sacrificial agents is much lower in comparison to that of the desired porous or hollowed nanomaterials. Thus this method has become one of the most popular routes for designing various porous or hollow noble metal catalysts to enhance their catalytic reaction. Tellurium (Te) nanowires with around kilogram production have been demonstrated to be excellent chemical and physical nanotemplates for the processing of different one-dimensional nanostructures and hollow CNFs. As a chemical template, Te nanowires were usually oxidized to get the pure noble metal nanostructures with much superior properties and transferred into tellurite in solution after template reactions. Even used as a physical template, Te nanowires could also be etched to complete the hollow nanostructures to enhance the performance of the nanomaterials. As a consequence, these synthesis strategies have shown great advantages in the fabrication of high-quality nanostructures to make the best of their active sites. However, to some extent, the process is still not faultless, because the sacrificial elements that are generated during the synthesis process inevitably increase the cost, especially when the sacrificial elements are toxic [23].

9.3 Typical safety guidelines for handling nanoparticles

The practices for safely working with NPs are essentially the same as would be used in working with any chemical of unknown toxicity. Following are the guidelines to avoid nanowaste contamination [24–26]:

1. Write a safety operating procedure to outline the safety processes to be followed in the handling of NPs in lab protocols. Use good general laboratory safety practices as stated in the chemical hygiene plan (CHP). For example, wear double gloves that are specific for the chemical or material, safety glasses or goggles, and appropriate protective clothing.

2. All personnel participating in research involving nanoscale materials need to be trained concerning the potential hazards as well as proper techniques for handling NPs. As with all safety training, documentation in the CHP needs to be maintained to indicate who has been trained in the safe use of the nanoscale material.

3. No eating and drinking are allowed in lab areas where chemicals or NPs are used.

4. When purchasing commercially available nanoscale materials, be sure to obtain the material safety data sheet (MSDS) and to review the information in the MSDS with all personnel who will be working with the material. (Note that given the lack of toxicological data on the effects of NPs, the information on an MSDS may be more applicable to the properties of the bulk material.)

5. In some cases, the making of NPs involves the use of chemicals that are known to be hazardous or toxic. Be sure to consider the hazards of the precursor materials when evaluating the process hazard or final product. Users of any chemicals should make themselves familiar with the known chemical hazards by reading the MSDS or other hazard literature.

6. To minimize airborne release of engineered NPs to the environment, NPs are to be handled in solutions or attached to substrates so that dry material is not released. When this is not possible, nanoscale materials should be handled with engineering controls such as a HEPA-filtered local capture hood, biosafety cabinet, or glove box. If none of these are available, work should be performed inside a laboratory fume hood. HEPA-filtered local capture systems should be located as close to the source of NPs as possible, and the installation must be properly engineered to maintain adequate capture ventilation.

7. Use fume hoods to expel any NPs from tube furnaces or chemical reaction vessels. Do not exhaust aerosols containing engineered NPs inside buildings.

8. Never work outside of a ventilated area with nanomaterials that could become airborne. If a research protocol requires work to be done outside of a ventilated area, contact the EH&S Laboratory Safety Program at prior to proceeding; a respirator may be required.

9. Lab equipment and exhaust systems used with nanoscale materials should be wet-wiped and HEPA vacuumed prior to repair, disposal, or reuse. Construction and maintenance crews should contact the EH&S Laboratory Safety Program for assistance.

10. Spills of engineered NPs are to be cleaned up right away. The person or persons cleaning up should wear appropriate personal protective equipment (PPE), including double gloves to prevent contamination by the spilled material. Depending on the size of the spill and the material involved, the spill area can be either vacuumed with a HEPA-filtered vacuum and/or the area can be wet-wiped with towels or a combination of the two. For spills that might result in airborne NPs, proper respiratory protection should be worn. Do not brush or sweep spilled or dried NPs. Place a Tacky-Mat at the exit to reduce the likelihood of spreading NPs on footwear. For assistance with cleanup of large chemical spills or those of toxic or unknown content, contact EH&S Hazardous Materials Management.

11. Many engineered NPs are not visible to the naked eye, and surface contamination may not be obvious. Work surfaces should be wet-wiped regularly. Daily is recommended. Alternatively, disposable bench paper can be used. Wet wipes or bench paper must be placed in a plastic bag and secured before being removed from the work area. This bag will then need to be placed into a second bag for disposal as hazardous waste.

12. All waste-engineered NPs should be treated as hazardous materials unless they are specifically known to be nonhazardous. Dispose of and transport waste NPs in solution according to hazardous waste procedures for the solvent.

13. If animals are subjected to an aerosol containing NPs, they should be housed in environmentally controlled cages. If the animals are dosed with NPs via ingestion or injection, they can be housed in conventional housing. Staff members that care for the animals in either condition should wear appropriate PPE to prevent exposure to airborne materials or materials on surfaces from urine, feces, and so on. Workers who are disposing of contaminated bedding should wear the appropriate PPE, including but not limited to protective eyewear, disposable gloves, dust mask, closed front disposable gown, hair cover, and shoe covers.

9.4 Disposal of nanoparticle waste

Although the major source of NPs in general waste treatment centers is from commercial products, research laboratories sometimes contribute as well. Whatever the outlet of disposal, the NPs ultimately arrive at one of four places: a recycling facility, an incineration plant, a landfill, or a wastewater treatment facility. Each place has the potential for the release of NPs, with some more preventable than others. In recycling facilities there are plastics with NPs in them that can be released as dust, but the dust is usually controlled so as to not expose any humans or the environment. NPs can be liberated in the time of combustion in an incineration plant as ash or dust. Most plants have filters to collect pollutants, but many filters are not rated for NPs or can be damaged by them. Landfills have the most potential for NP release by emission into air, water, or soil. Wastewater is turned into sludge that is then burned or sometimes made into fertilizer for agriculture. Little is known about the environmental impact of NPs in fertilizer, although 100% of wastewater sludge is generally incinerated. There are multiple pathways that NPs can follow in an incinerator. Airborne particulates in the facility flow through a filter designed to remove pollutants from the air. When materials containing NPs are being recycled, there are three main concerns: the health of facility workers; the environmental impacts from residue that ends up in incineration, landfills, or sewage treatments; and the contamination of products made from recycled materials. There is a possibility of the NPs leaching from the products and affecting their surroundings [27]. Table 9.3 lists the disposal types and their risk by various possible leakage routes.

9.5 Various processes for nanowaste recycling

General practices such as landfill and incineration are not the final means to stop the nanocontamination. Disposing of solid nanowaste in landfills can cause a potential threat to the environment if the waste is not managed properly, contaminating both the soil

Table 9.3 Possible leakage routes in waste treatment.

Type of nanowaste disposal or treatment processes	Possible leakage routes
Landfilling	• Gas release to the atmosphere • Landfill surface release to the surrounding • Various type of leaching
Incineration	• Flue gas emissions to the environment • Fly ash and bottom ash dumped in landfills
Recycling	• Embedded in secondary materials
Wastewater treatment	• Emissions to surface water • Wastewater sludge to incinerators • Wastewater sludge to landfills • Wastewater sludge to agricultural applications

system and ground water. Hence recycling is a better choice than landfilling. Recycling is defined as the process of collecting and processing materials that would otherwise be thrown away as trash but turning them into new products. The potential risks of nanomaterial exposure depend on the specific recycling procedures. During the waste recycling process, three types of exposure should be considered [28]: (1) exposures to nanoobjects present in liquid media (water, solvent) following the cleaning of recycling equipment or products before mechanical recycling; (2) exposures to nanoobjects produced in the combustion gases or in ambient air with thermal processes (pyrolysis, heating, and soldering); and (3) exposures to fine or ultrafine dusts containing loose nanoobjects emitted during transportation, sorting, and shredding. Very little information is available on the transfer of nanomaterials into new products during the recycling process. Different techniques have been adopted for the processing of nanomaterials from wastes after specific pretreatment either physically or chemically or combinations of both. Various kind of processes including leaching solvent extraction, electrolysis, ion exchange, membrane separation, and microbiological methods, are discussed in this section. Depending on the nature of the waste, thermal or chemicophysical processing of the waste are possible solutions. For example, metal NPs (Pb, Hg, Cu, Fe, Au, Ag, Pd, Pt, Rh, etc.) and polymers can be recovered from electronic wastes such as computer circuit boards, laptops, cellphones, automobiles, and supercapacitors. However, other forms of e-waste are also good sources of a variety of valuable materials. Again, these recovered nanomaterials are used as a source in the required applications.

9.5.1 Physical processes

Physical isolation and purification processes, centrifugation and sedimentation, filtration, crystallization, and evaporation are the main tools that are used to separate one substance from compounds in solution. After the solution has been separated from the compounds, it can be subjected to further applications instead of being dumped out. The most common physical pretreatment methods are grinding and milling, from which the required concentration is separated out. For example, a very high percentage of worn tires are recycled, and the granulated material is used as filler on artificial grass on football fields and for rubber tiles and carpeting used on sports arenas and on playgrounds for children. Potentially, the presence of nanomaterials in the tires may lead to leaching of nanomaterials from the recycled products [29].

9.5.2 Chemical processes

Chemical recycling methods are more favorable for recycling the waste ions by transforming them into insoluble sediments from the reaction solution. The chemical processing of nanomaterials has been shown to be among the most powerful tools to fabricate nanomaterials, enabling us to control their shape, morphology, size, and composition to engineer their physicochemical properties. In practice, these strategies are always combined in recycling the valuable by-products from chemical synthesis. Since Te nanowires are widely used as a sacrificial nanotemplate for various one-dimensional nanostructures, recycling the Te nanowire template can be done after their template synthesis [30]. By adjusting the pH value of the produced solution with NaOH solution, the traditional waste element of Te, in the form of sedimental tellurous acid, can be obtained and, when mixed with the synthesized products, can be transferred to soluble Na_2TeO_3. Following centrifugation, 65%−81% of the Te was successfully collected in different template reaction systems and was then successfully used to produce high-quality Te nanowires for further applications [31]. In the case of silver recovery, after various template reactions the shape-controlled Ag nanostructures were mostly transferred to AgCl, which was then dissolved to form the transparent Tollens' reagent and centrifuged from the valuable metal nanostructures with the assistance of NH_4OH. The Ag particles were then obtained and collected by reducing the Tollens' reagent with N_2H_4. More important, the recycled Ag particles with a recycling rate of 70%−84% in different reactions can be oxidized to high-quality $AgNO_3$ for further synthesis of Ag nanowires by the polyol method [32].

In contrast, Lee et al. [33] reported a general recyclable redox metallothermic route to obtain composites of porous carbon and electrochemically active metal particles for rechargeable battery anode materials. In the general process, the Zn-MOF was used as a template to react with GeO_2 powder at temperatures of $600°C−900°C$ to get the Ge/hierarchically porous carbon, during which the GeO_2 was reduced by the Zn. The oxidized ZnO was automatically reduced to metallic Zn via carbothermic reduction during the conversion process. Thus the conversion experiments can be finished completely, and even the molar ratios of Zn-MOF to GeO_2 reached a 667% excess of GeO_2

$(Zn^{2+}/Ge^{4+} = 0.3)$ from stoichiometrically required amounts $(Zn^{2+}/Ge^{4+} = 2)$. Moreover, this recyclable strategy is suitable for many other metal oxides, including In_2O_3, Bi_2O_3, and SnO. By precipitation and chemical transfer reactions, including ion precipitation, acid–base method, and oxidation–reduction reaction, the solute and ions will be further separated from the reaction reagents. With multiple solutes dissolved in the solvent, the valuable elements of solutes can be isolated by distillation, fractionation, extraction, and so on. Developing sustainable methods to obtain nanomaterials, which was not limited to the chemical synthesis, would also notably reduce the processing cost. Recently, separation and purification of high-purity semiconducting single–walled CNTs have been widely studied by polymer wrapping [34].

Chemical pretreatment has been applied to separate any contaminants present in the waste sample by heating or treating with reagents such as strong acids (e.g., H_2SO_4, HNO_3, and HCl) to prepare Fe NPs from steel pickling waste materials released from a steel plant. The most popular synthesis method for metallic NPs is reduction with sodium borohydride. Generally, the sodium borohydride solution is freshly prepared and rapidly added to the solution of waste materials. NPs produced via reaction are washed repetitively with ultrapure water and absolute alcohol for the removal of excess $NaBH_4$ [35]. Silver, copper, and bimetallic Ag/Cu NPs were synthesized from a leachate solution. This solution was derived from a solution of leached metalized acrylonitrile butadiene styrene from plastic wastes combined with nitric acid and ascorbic acid in the presence of chitosan at $60^\circ C$. Researchers have also reutilized NPs from blood plasma. For example, functionalized NPs were successfully recycled to capture toxins from spiked blood plasma samples using a glycine buffer to free up the NPs [36]. Chertok et al. [37] reported that high-quality iron oxide NPs could be prepared in the recycled 1-butyl-3-methylimidazolium bis(trifluoromethylsulfonyl) imide ionic liquid that separated from the initially processed products and washing agent of hexane by a centrifugation and evaporation process. Similarly, it was reported that the ionic liquid of 1-ethyl-3-methylimidazolium acetate could be recycled for the second synthesis of nanosheet-constructed Pd electrocatalyst or Ag-NPs via a centrifugation process. By using techniques of centrifugation, filtration, and evaporation, Shahi et al. [38] successfully recovered ionic liquid of 1-buty-3-methylimidazolium hydrogen sulfate for the green synthesis of photoactive nanocrystalline anatase TiO_2 upto three cycles. The ionic liquids as reaction reagents could be almost fully recovered for the second synthesis, giving the sustainable synthesis of nanomaterials with much lower cost. Besides the high-value solvents, the excess solutes can also be recycled for further applications to reduce the cost of nanomaterials and promote their scalable production. By applying crystallization technology, Du et al. developed a green and environmentally friendly manufacturing method to produce biodegradable cellulose nanomaterials with unique mechanical and optical properties. Compared with the traditional strategies that use the concentrated mineral acid hydrolysis process or oxidation methods, Du et al. used concentrated solid dicarboxylic acid hydrolysis to produce cellulose nanocrystals. After the hydrolysis reaction the dicarboxylic acids were easily recovered, since they had low solubility at much lower temperatures, which greatly decreased the cost of the reaction. It is well known that the lithium-containing chemical reagent always represents the highest cost in the preparation of materials for lithium-ion batteries. Therefore it is of great importance to recycle the lithium source in view of the applied chemistry and environment [39]. In this case, Yang et al. [40] reported a green and scalable strategy to synthesize $LiFePO_4$ in which the raw materials of LiOH solute could be recycled. Typically, the obtained product of $LiFePO_4$ nanocrystals was separated by filtration, and then the LiOH solute was recovered from the filtrate by separating the impurities of $BaSO_4$ precipitate with the addition of $Ba(OH)_2 \cdot 8H_2O$. In this study, 90% of the LiOH solute was recovered and reused for the secondary synthesis of $LiFePO_4$ nanocrystals after adjusting the concentration of recycled LiOH solute to the value of the original run.

Graphene can be recovered from various waste plastics by using a variety of techniques. Gong et al. [41] were able to create high yields of graphene flakes. Their technique used waste polypropylene (PP) catalyzed by organically modified montmorillonite. A uniform mixture of PP (~ 89 wt.%), talcum (~ 11 wt.%), and modified montmorillonite was placed in a crucible and heated to $700°C$ for 15 minutes to obtain carbonized char. After cooling, the carbonized char was immersed in hydrofluoric acid and HNO_3. The hydrofluoric acid dissolved the impurities, and HNO_3 oxidized the amorphous carbon. After a repeated process of centrifuging and isolating from solution, graphene flakes were obtained.

Pati et al. [42] developed a laboratory-scale method to recover gold nanowaste. These types of methods reduce costs of nanotechnology, making it more convenient. In this process, α-cyclodextrin (α-CD) is used to facilitate host-guest inclusion complex formation involving second-sphere coordination of $[AuBr_4]^-$ and $[K(H_2O)_6]^+$, which are used for Au recovery, and the recovered metal is then used to produce new NPs. The selective Au recovery by α-CD shows an important advantage over traditional methods for selective Au recovery in that it does not involve the use of toxic cyanide or mercury. The procedure of selective Au recovery by α-CD is explained as follows: A nanowaste composed of citrate-reduced Au NPs was prepared. The nanowaste suspension was precipitated by adding KCl and was then dissolved by using a 3:1 volume:volume mixture of HBr and HNO_3. HBr was employed to ensure that gold was present as the square planar complex [AuBr4]. The pH of the resultant clear red solution was adjusted to approximately 5 by using KOH. If all the gold that was originally in the Au NP suspension was precipitated and redissolved in the HBr-HNO3 mixture, a calculated mass of α-CD was added, sufficient to achieve a 2:1 molar ratio of gold:α-CD. Almost immediately, the clear red solution became turbid. After 30 minutes the reaction mixture was filtered through a 0.22-μm polytetrafluoroethylene (PTFE) filter. The retentate was resuspended in deionized water by sonification, which yielded a clear brown solution. An aliquot of 50 mM of $Na_2S_2O_5$ was then added to precipitate and recover the Au. To recycle the freshly recovered Au for synthesizing new Au NPs, the gold was dissolved in aqua regia, and the resulting yellow solution was boiled to remove HNO_3, adding HCl intermittently. Boiling was stopped after ebullition of brown gas concluded. The final solution was analyzed and used to synthesize new citrate-reduced Au NPs. This technique has high efficiency, and it is possible to use it for recovery of Au not only from nanowaste but also from other waste streams that contain Au, such as gold alloy scraps and electronic waste. The chemical process for Au recovery is shown in Fig. 9.6.

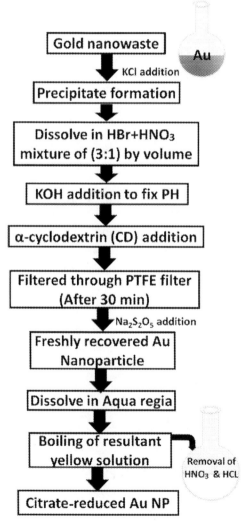

Figure 9.6 Method of Au recovery.

9.5.3 Thermal processes

Wang et al. has investigated a general strategy to recover Ag, Ag_2S, and Ag_2Se nanocrystals via the thermal decomposition of $AgNO_3$ in octadecylamine (ODA) to construct popular mesoporous structures. Because of the high boiling point (349°C) and chemical stability of ODA, there is almost no loss of ODA, as it was heated to 180°C during the reaction. As a result, the residual ODA after the sedimentation of Ag–NPs was collected for the next synthesis of Ag–NPs with almost the same quality. The recycling and reuse of ODA allow the process to become very economical and can be easily scaled up without obvious increase in cost. Lead NPs were processed from solders of waste printed circuit boards (WPCB) via evaporation under vacuum with forced flow inert gas condensation. Copper-tin (CuSn) NPs were fabricated from the WPCBs of spent computers by using selective thermal transformation while also separating toxic Pb and Sb [43]. Silver was completely recovered from incinerated organic solar cells. Various materials with and without nanomaterials are incinerated at high temperature in thermal treatment plants. For incineration the general agreement is that the nanoinorganics, TiO_2, ZnO, and Ag will end up primarily in the slag and to some degree in the filter ash. In this process, all the existing nanomaterials are fused away as a result of their higher surface energy (lower melting point) and agglomerate to form a macroaggregate [44]. In another study, ZnO NPs were synthesized from spent Zn–C batteries via a thermal technique at 900°C under an Ar atmosphere using a horizontal quartz tube furnace [45].

CNTs can be recovered from waste plastics in diverse systems, such as autoclaves, quartz tubes, crucibles, and muffle furnaces. Each method aims to detach the solid-state polymers into their carbon precursors via pyrolysis. Polyolefin is a polymeric material that is created from olefin monomer (C_nH_{2n}) and is present in the large majority of plastic waste. Two waste polyolefins, PP and polyethylene (PE), can be used for carbon materials processing. Other plastics integrated with additional elements such as polyvinyl alcohol (PVA) and polyethylene terephthalate (PET), can also be used for CNTs synthesis [46]. Zhou et al. reported another combustion method for polyethylene decomposition. This method generates light hydrocarbon in situ during an exothermic process. This method utilized a stainless steel wire mesh as a catalyst and a substrate. A ceramic filter was placed before the stainless steel wire mesh to eliminate the soot that could deactivate the mesh that served as a catalyst. The yield of CNTs was higher than 10 wt.%. In another synthesis process, mixed PP, maleated PP, and Ni catalysis powders were used in an autoclave. The resulting mixture was then heated by using an electric stove followed by cooling at ambient temperature. Carbon solid spheres were the only product obtained in the absence of Ni from the reaction system, which indicated that Ni powder catalyzed the decomposition of PP. MA-PP contributed to two actions in the growth process of CNTs, first improving the dispersion of Ni in PP and then forming a homogenous system between carbon atoms and Ni catalysts. Ni particles were separated to form carbon-surrounded Ni particles, and a high surface packing density of Ni particles enabled the CNT to grow along a consistent direction [47]. This method of production of 160-nm CNTs had a yield of 80%. Bajad et al. attained an approximately 45.8% yield in the production of multi-walled CNTs (MWCNTs). These MWCNTs were derived from PP waste catalyzed by Ni/Mo/MgO by using a combustion technique. The powdered catalysts and PP were placed in a covered silicon crucible and heated to 800°C in a muffle furnace. The Ni/Mo ratio was found to affect the yield and size of the CNTs. Increasing the Mo content resulted in large-diameter CNTs, while lower Mo content gave a higher yield with short-radius CNTs [48]. Khachatur et al. developed an energy-saving combustion technique to prepare graphene sheets using waste PTFE and silicon carbide (SiC). The process mechanism was similar to epitaxial growth on SiC, in which Si was removed by C_2F_4 through an exothermal reaction [49]. Deng et al. developed a green synthesis technique in which raw wastes (e.g., polystyrene) were transformed into high-quality single-layered graphene. In this method, 10 mg of start materials were placed on a slightly bent piece of Cu foil held by a quartz boat in a CVD quartz tube. After low-pressure annealing at a temperature of approximately 1050°C in an inert atmosphere (Ar and H_2), graphene growth was observed on the back of the Cu foil. Combustion of materials with nano-CeO_2 in municipal waste incinerators (with energy recovery) may lead to the presence of nano-CeO_2 in flue gases, but such NPs can be largely recovered from these gases by advanced flue gas treatment [50]. The solvent thermal method involves chemical reactions in solvents contained in a closed autoclave reactor and heated to a critical temperature. When water is used as a solvent, the process is known as hydrothermal. As an example, Fe_2O_3 NPs were synthesized via a solvent thermal method using ethanol as a solvent and heating the solution of the starting materials mixture ($FeCl_3$ and NaOH) at 150°C for 2 hours to produce NPs [51].

9.5.4 Electrodeposition deposition and electrokinetic process

Another technique that uses voltage to promote chemical reactions in aqueous solutions for the synthesis of NPs is electrodeposition. This technique has been applied to produce nanowires, nanoporous materials, and nanocylinders. In addition to these techniques, green synthesis processes based on the use of environmentally friendly and biocompatible materials, such as plants and microorganisms, have recently arisen [52]. Cu_2O/TiO_2 photocatalysts were recovered by using electrokinetic processes. Cu_2O NPs of different sizes were also created from WPCBs using supercritical water oxidation technique and electrokinetic processes [53].

9.5.5 Sludge treatment process

According to various studies, approximately 90%—95% of the assessed nanomaterials in wastewater are retained in sewage treatment plants. Sewage sludge is partly dried and incinerated in thermal treatment plants. The application of sewage sludge on agricultural soils is possible only under certain conditions. Because of the sometimes high concentrations of hazardous substances, ending the use of sewage sludge on agricultural soils is currently under discussion. It involves several critical aspects regarding the content of nanomaterials in sewage sludge that is used on agricultural soils, including nanosilver (and potentially other, similar nanomaterials) that are firmly bound in a sludge-soil mixture and can hence accumulate in soils. Nanosilver and iron oxide particles inhibit the activity of microorganisms, partly already in very low concentrations, and can therefore disturb natural processes in soils. Research results indicate that the use of sewage sludge on agricultural soils may lead to negative impacts of nanomaterials on soils.

Liu et al. developed a strategy to recycle nano-SnO_2 from tin plate electroplating sludge. Sn is an important material in solder, tin plating, ceramic glazes, and glass, and studies have shown that a large amount of Sn is disposed as waste in industry. It is estimated that the world supply of Sn might run out in 20 years or less at current global consumption rates. Because of the presence of Sn in waste and the importance and limited quantity of Sn, the recycling of this material from sludge is an urgent and nonnegligible task. This method was designed as a strategy for recycling SnO_2 from tin plate electroplating sludge that mainly contains amorphous Sn and Fe compounds. It presents some disadvantages; for example, it has a lower recovery efficiency of Sn, it presents difficulties in the disposal of residual hazardous solid waste, and it is hampered for its application in a real system. The procedure of this method is divided into analysis of the original electroplating sludge, mineralization processing of the sludge, and acid treatment of sludge after the mineralization. With this method, nano-SnO_2 powders with purity close to 90% can be obtained. Only approximately 12% Sn was dissolved in 0.1 M NaOH solution at 230°C for 48 hours. To reach the highest possible amount of dissolution and recycling of Sn, a longer process time and higher energy cost may be necessary. Also, it is possible to use microwave at an industrial scale to save the energy consumption during the realization of the recycling of nano-SnO_2 [54]. The method is summarized in Fig. 9.7.

It has been recognized recently that some traditional industrial sludges are actually nanowastes with adsorbed heavy metals or organic pollutants. The $Mg(OH)_2$ nanowastes generated by the sodium and potassium chlorate industries, for example, are typically adsorbed with the carcinogenic heavy metal Cr(VI). Transforming $Mg(OH)_2$ NPs into bulk materials releases the adsorbed Cr(VI) into solution. This Cr(VI)-containing solution can subsequently be recycled in the chlorate process, and the detoxified solids can potentially be reutilized as additives in other applications, such as ceramics, paint, flame-retardant engineering plastics, or lubricants. Typical Cr(VI)-

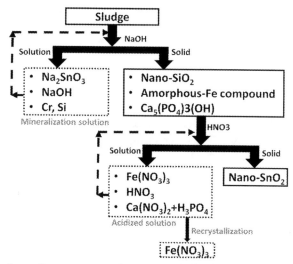

Figure 9.7 Method of recycling nano-SnO₂ from electroplating sludge.

containing nanowastes consist of about 50% water, 2048 mg kg^{-1} of Cr(VI), 30% 20-nm Mg(OH)$_2$, and 16% 100-nm CaCO$_3$. The pH of the waste is around 9. The viscous solid is quite difficult to separate from chromate solutions by any centrifugation, washing, or filtration methods. However, the addition of a suitable mineralizer to the nanowastes during hydrothermal coarsening might accelerate the crystal growth speed of NPs or transform them into compounds with reduced surface adsorption properties. The most appropriate mineralizer was found to consist of 0.5 m Na$_2$CO$_3$ and 1.5 m NaHCO$_3$, as it provides the highest removal efficiency of Cr(VI), and the mineralizer remaining in the supernatant solution can be transformed into essential reagents for industrial recycling [55].

9.5.6 Microemulsion process

Copper NPs were prepared from acidic CuCl$_2$ waste etchants generated in the manufacturing process for printed circuit boards (PCBs) by microemulsion processes. Likewise, aqueous solutions of CuSO$_4$ from PCBs were chemically reduced to create organically stabilized Cu NPs. WPCBs have also been recycled to obtain metals such as Cu, Pb, Fe, Au, and Hg. The metals that are recovered from WPCB can be used to create NPs. Cu from WPCB has acted as a starting material in the synthesis of Cu NPs. WPCBs can also be used to create NPs directly [56].

9.5.7 Microbiological process

A large share of household wastes consists of biological materials which are partly disposed of in composting and fermentation plants. Via the use of compost and fermentation products in agriculture and horticulture nanomaterials potentially contained in the bio wastes could be released to the environment. In addition, microorganisms responsible for composting and fermentation could be disturbed by nanomaterials in the treatment plants. According to an assessment on the potential input of nanomaterials into biological waste treatment plants it was identified, that only few nanomaterials are authorized for use in food and feed stuff. Nanomaterials could enter the waste stream through packaging materials and household appliances. Another important property of NPs which influences the treatment of wastewaters in biological plant is some NPs have the potential to alter the functionality of the microorganisms used in treating the waste streams especially in biological treatment plants as most of them have antibacterial properties. Therefore conventional chemical and biological contaminants treated at plants may pass untreated including the nanowastes after microbial functionalities have been compromised due to the NPs. Use of physical barriers and filters can also prove helpful in removal of nanowastes from wastewater but more research needs to be done to prove its efficacy [57].

9.5.8 Coagulation technique

Coagulation methods can be applied for purification of wastewater from various natural colloids and make it free from nanowastes. For instance, polyaluminum chloride can be used to remove NPs from industrial wastewaters by coagulation technique. Apart from coagulation, adhesion can be used for extraction of nanowastes from the wastewater in waste treatment plant. Moreover, this technique has its own limitations as it cannot effectively remove the wastes completely. Although a large proportion of the nanowaste can be removed from the waste treatment plant through the adhesion process, a significant percentage still escapes the waste treatment plant clearing systems [58].

9.5.9 Nanoporous materials and membrane separation

Nanoporous materials and membranes can able to separate the nanomaterials from contaminated waters. Walden and Zhang found that the presence of SiO_2 NPs in wastewater streams can be remediated using the small-angle neutron scattering technique in a waste treatment plant. Several factors controlled the colloidal stability of SiO_2 nanowastes, the composition of the aqueous matrix, influence of the large particles, and the effects of surface functionalization of the NPs. The findings suggested that both functionalized (using a thin coating of nonionic surfactant Tween 20) and unfunctionalized SiO_2 nanowastes were found to be stable in nano pure water. Therefore it was evident that flocculation can be an effective technique for removing SiO_2 from the wastewater [59].

9.5.10 Glucose reduction process

CNFs as promising candidates for catalysts, flexible electronics, thermal insulation, adsorbing materials, and so on have been synthesized by reducing the glucose. We have realized the large-scale production of CNFs, while the amount of the added glucose during the synthesis is much larger than the production of CNFs as the reacted solution containing a large amount of micromolecule organics in brown color. The by-products of micromolecule organics generated during the synthesis of CNFs are treated as waste [60].

9.5.11 Layer-by-layer assembling

Yuan et al. developed a new layer-by-layer assembling mechanism. In this method, waste PP was catalyzed via activated carbon and Ni_2O_3 in a quartz tube reactor. Activated carbon was effective in breaking PP into light hydrocarbons. Activated carbon also provided highly efficient catalytic conversion with Ni by encouraging dehydrogenation and aromatization with the formation of different aromatic groups. CNTs growth was accomplished based on benzene rings. The carboxylic moieties of activated carbon also enabled synergistic catalysis with Ni_2O_3. The highest carbon yield ($\sim$50 wt.%) was reached at 820°C with a proportion of raw materials of $PP:10Ni_2O_3:8AC$ (wt.%). Considering the high energy demand of this method, the recovery of Ni_2O_3 could be an alternative for more sustainable scale-up manufacturing [61].

9.6 Various nanowaste recycling products

9.6.1 Nanomaterials in concrete production

The amount of recycled material that is used in construction is important because of the growing need for the industry to join the circular economy. On the other hand, how building construction is being managed in a more environmentally friendly way can be seen, for example, in a construction site that involves the conversion of a large industrial building and surrounding land into a residential area with offices and shops. In this project, the concretes and concrete products that are used meet the strictest standards, and it is currently possible to produce concrete made from 100% recycled aggregates. Another approach to improve concrete is through the use of nanomaterials that, depending on the process and materials that are used, can add properties such as strength or crack prevention [62]. As Lev Lyapeikov, the company's product development manager, explains, "Nanostructured materials include solutions based on aqueous suspensions designed to treat concrete mixtures by increasing the strength of the concrete structure. These patented processes include solutions with NPs which allow the use of construction site waste as a raw material for new building structures." The

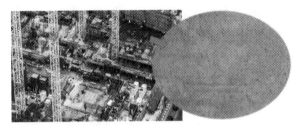

Figure 9.8 Nanomaterials in concrete.

company has already outlined plans for commercialization of the nanomaterials with a view to expanding its current raw material supplying business in the construction industry [63] (Fig. 9.8).

Construction industries include CNTs, which can boost mechanical and thermal properties in ceramics, be used in sensor production for real-time monitoring of the structural condition of materials and buildings (known as nanoelectromechanical systems or NEMS), and improve strength and crack prevention in cement and concrete. In comparison with plain cement mortar, the addition of 0.1% concentration of MWCNTs increased the compression strength by 56%, whereas concentrations of 0.1% of CNT increased it by 31% [64]. Silicon dioxide is perhaps the most commonly used NP additive in concrete and cement, owing to its nano size and pozzolanic reaction. This gives added strength and durability to the concrete by stimulating the hydration reaction and filling in micropores that are present in a cement paste structure. It is a reaction that decreases the concrete's porosity. TiO_2 NPs can provide faster hydration and self-cleaning properties in concrete, while Fe_2O_3 NPs increase compression strength and abrasion resistance [65].

9.6.2 Nanomaterials applied in suspensions

Potential applications of suspended engineered inorganic and carbon nanomaterials include adsorption, use as sensor, heterogeneous catalysis in organic chemistry, and the photocatalytic degradation of poorly biodegradable compounds. Suspended inorganic and carbon particles may be part of more complex nanomaterials generated by, for example, modification, functionalization, capping, or coating or in combination with stabilizing substances. Options for nanomaterial recovery from suspensions have been widely studied. Such options include (ultra)centrifugation, phase separation, solvent evaporation, antisolvent addition, the use of colloidal solvents, pH-induced aggregation, size-exclusion chromatography, gel electrophoresis, flocculation/coagulation, temperature manipulation, and advanced filtration. A study on the removal of commercial metal oxide NPs from water with alum (a standard drinking water treatment process) found removal of less than 80% [66]. Liu et al. [67] reported approximately 99% removal of silica NPs (with diameters of around 30 and 75 nm) from suspension in water by treatment with $AlCl_3$, leading to aggregation and sedimentation, but they noted that this removal usually takes a long time (about two weeks), which makes it unsuitable for many applications. Nazar et al. [68] reported a maximum recovery efficiency of 52% for phase separation of Au and Pd NPs in mixtures of water and nonionic surfactants (cloud point extraction). When nanomaterials are (or are made) magnetic, the use of magnetic fields is a recovery option, whether or not combined with other separation technologies.

Reduction of catalytic activity is the formation of NP assemblies with reduced surface area available for functionality. For instance, Phan and Jones [69] found a substantial reduction in activity of diaminosilane-functionalized cobalt spinel ferrite magnetic NPs in a Knoevenagel condensation and suggested that this might be linked to NP clustering. Graphene nanosheets with attached metal NPs easily aggregate in the presence of ions. Such clustering or assembly formation may be counteracted by the use of stabilizers, spacers, or shells or by firmly immobilizing NPs on large-sized supporting materials. Often, substances that are leached from nanomaterials used as catalysts contribute to catalytic activity. NPs with this characteristic will not be further considered here. Rather, the focus will be on heterogeneous catalysis, for which functionality resides in the nanomaterial [70]. Leaching may in this case contribute to loss of functionality. For instance, Beletskaya and Tyurin [71] found that metal leaching negatively affected the performance over time of several nanocomposite metal catalysts for organic synthesis. Navalon et al. [72] found that catalysis of the Fenton reaction by iron and manganese NPs was characterized by a large extent of leaching and limited reusability. The use of Fe NPs to allow for magnetic separation may give rise to dissolution of Fe, which might negatively affect catalytic or adsorptive activity. Such dissolution is linked to Fe oxidation; thus prevention thereof may reduce leaching. An additional problem of leaching in nanocatalysis is the contamination of the reaction product. Also, leaching might give rise to the emission of hazardous substances [73].

9.6.3 Low-cost sensors for energy storage applications

Researchers are developing low-cost sensors to monitor pollutants using recycled NPs. Jiao et al. [74] developed an optical sensor that uses recycled carbon NPs from waste pomelo peels. Pomelo (*Citrus maximus*) has a flavor similar to that of grapefruit but with a much thicker peel and is a popular fruit from Brazil to Southeast Asia. Abdelbasir et al. [75] developed an electrochemical sensor for environmental protection applications. The sensor was fabricated by anchoring copper NPs to laser-scribed graphene electrodes. The nanocuprous oxide was synthesized from electrical waste. Yin et al. have reported that various carbon-based functional materials can be fabricated from waste for further water remediation and energy storage applications. Using the disposable PE glove waste as raw material, Yin et al. [76] successfully fabricated a PE-coated mesh for efficient oil-water separation via a simple dissolution and immersion method.

9.7 Recycling of nanocomposites

There is a wide variety of potential applications for nanocomposites from fire-retardant materials to sensors, from antibacterial wound dressings to batteries, and from tennis rackets to solar cells. Some of the composites of NPs and large-sized materials have been used for a considerable time [77]. This applies, for example, to tires with combinations of rubber and nanoparticulate carbon black and/or silica and to catalytic converters in motor vehicles, which utilize composites of Pt group NPs and ceramics. Dry cutting of nanocomposites consisting of CNT and alumina gave rise to a (short-term) workplace CNT number concentration of 1.6 cm^{-3}. This number concentration is in excess of current recommended exposure levels [78]. Raynor et al. [79] assessed shredding of a nanoclay-PP nanocomposite and found that the release of NP was lower than in the case of neat PP. Unfortunately, in that study, the shredded polymers were not aged. Studying the recycling of aged nanocomposites containing nanoclay and olefins is important because there is evidence that photo-oxidation in such nanocomposites is increased, compared with neat olefins. Cleaning of systems that have processed nanomaterials is an important potential source of exposure.

9.8 Benefits of nanomaterials recycling

Recycling of nanomaterials can reduce the amount of waste that is sent to landfills and incinerators. Recycling can increase economic security by tapping a domestic source of materials, prevents pollution by reducing the need to collect new raw materials, saves energy, supports manufacturing conserves valuable resources, and helps to create jobs in the recycling and manufacturing industries. Recycling nanotechnology is a part of green technology that going in the direction to create a healthy environment.

9.9 Limitations of nanomaterials recycling

There are a number of limitations to the effective adoption of recycling processes. Recycling can involve high energy usage, hazards, high labor costs, and different practices by individuals and countries, which can hamper the efficient implementation of recycling plans. The biggest limit to recycling is that not all materials can be recycled, so materials can be recycled only a limited number of times, owing to degradation each time through the process. This degradation is referred to as downcycling. In addition, recycling poses a number of societal and ethical issues. E-waste recycling has led to electronic waste from developed countries being shipped to undeveloped countries for recycling. In many cases, this leads to low wages and terrible conditions for the workers who are involved in the recycling process and the release of toxins that are health risks for the individuals and their surrounding communities.

9.10 Conclusions

The impact of the risk by nanotechnology in the environment was extensively discussed in this chapter. To avoid the risks, researchers are classifying the waste generated from the nanomaterials and trying to determine the most effective processes to recycle them by taking various types of materials from metals, oxides, and polymers. How to handle the nanomaterials at the laboratory scale as well as at the industrial scale was discussed. Some of the larger consumer applications of nanowaste, such as concrete production, suspensions application, and low-cost sensors energy storage applications, were discussed. The benefits and limitations in the nanomaterials recycling process were discussed. It is hoped that these discussion will influence nanotechnology-related researchers to support green technology.

References

[1] T.A. Saleh, Trends in the sample preparation and analysis of nanomaterials as environmental contaminants, Trends Environ. Anal. Chem. 28 (2020) e00101. Available from: https://doi.org/10.1016/j.teac.2020.e00101.

[2] A. Behera, S. Aich, Characterization and properties of magnetron sputtered nanoscale NiTi thin film and the effect of annealing temperature, Surf. Interface Anal. 47 (2015) 805−814. Available from: https://doi.org/10.1002/sia.5777.

[3] C. Coll, D. Notter, F. Gottschalk, T. Sun, C. Som, B. Nowack, Probabilistic environmental risk assessment of five nanomaterials (nano-TiO_2, nano-Ag, nano-ZnO, CNT, and fullerenes, J. Nanotoxicol. 10 (4) (2016) 436−444. Available from: https://doi.org/10.3109/17435390.2015.1073812.

[4] A.T. Besha, Y. Liu, D.N. Bekele, Z. Dong, R. Naidu, G.N. Gebremariam, Sustainability and environmental ethics for the application of engineered nanoparticles, Environ. Sci. Policy 103 (2020) 85−98. Available from: https://doi.org/10.1016/j.envsci.2019.10.013.

[5] M.S. Bilgili, P. Agamuthu, A new issue in waste management: nanowaste, Waste Manage. Res. 37 (3) (2019) 197−198. Available from: https://doi.org/10.1177/0734242X19830044.

[6] C. Rodríguez-Ibarra, A. Déciga-Alcaraz, O. Ispanixtlahuatl-Meráz, E.I. Medina-Reyes, N.L. Delgado-Buenrostro, Y.I. Chirino, International landscape of limits and recommendations for occupational exposure to engineered nanomaterials, Toxicol. Lett. 322 (2020) 111−119. Available from: https://doi.org/10.1016/j.toxlet.2020.01.016.

[7] E. Osman, Nanofinished medical textiles and their potential impact to health and environment, in: A. Shukla (Ed.), Nanoparticles and Their Biomedical Applications, Springer, Singapore, 2020. Available from: https://doi.org/10.1007/978-981-15-0391-7_5.

[8] R. Singh, S. Singh, Nanomanipulation of consumer goods: effects on human health and environment, in: S. Singh, P. Maurya (Eds.), Nanotechnology in Modern Animal Biotechnology, Springer, Singapore, 2019. Available from: https://doi.org/10.1007/978-981-13-6004-6_7.

[9] S. Ranjan, N. Dasgupta, S. Singh, et al., Toxicity and regulations of food nanomaterials, Environ. Chem. Lett. 17 (2019) 929−944. Available from: https://doi.org/10.1007/s10311-018-00851-z.

[10] R.-Y. Liang, H.-F. Tu, X. Tan, Y.-S. Yeh, P. Ju Chueh, S.-M. Chuang, A gene signature for gold nanoparticle-exposed human cell lines, Toxicol. Res. 4 (2) (2015) 365−375. Available from: https://doi.org/10.1039/c4tx00181h.

[11] G. Lenz e Silva, C. Viana, D. Domingues, F. Vieira, Risk assessment and health, safety, and environmental management of carbon nanomaterials, InTechOpen (2019) 1−21. Available from: https://doi.org/10.5772/intechopen.85485.

[12] L. Măruțescu, M. Carmen Chifiriuc, C. Postolache, G.G. Pircalabioru, A. Bolocan, Chapter 17 — Nanoparticles' toxicity for humans and environment, in: A.M. Grumezescu (Ed.), Nanomaterials for Drug Delivery and Therapy, William Andrew Publishing, 2019, pp. 515—535. ISBN 9780128165058. Available from: https://doi.org/10.1016/B978-0-12-816505-8.00012-6.

[13] D. Tian, S.G. Das, J.M. Doshi, J. Peng, J. Lin, C. Xing, sHA 14—1, a stable and ROS-free antagonist against anti-apoptotic Bcl-2 proteins, bypasses drug resistances and synergizes cancer therapies in human leukemia cell, Cancer Lett. 259 (2) (2008) 198—208. Available from: https://doi.org/10.1016/j.canlet.2007.10.012.

[14] A. Baun, N.B. Hartmann, K. Grieger, et al., Ecotoxicity of engineered nanoparticles to aquatic invertebrates: a brief review and recommendations for future toxicity testing, Ecotoxicology 17 (2008) 387—395. Available from: https://doi.org/10.1007/s10646-008-0208-y.

[15] A.M. Genaidy, R. Sequeira, T. Tolaymat, J. Kohler, M. Rinder, An exploratory study of lead recovery in lead-acid battery lifecycle in United States market: an evidence-based approach, Sci. Total. Environ. 407 (1) (2008) 7—22. Available from: https://doi.org/10.1016/j.scitotenv.2008.07.043.

[16] E.P. Vejerano, Y. Ma, A.L. Holder, A. Pruden, S. Elankumaran, L.C. Marr, Toxicity of particulate matter from incineration of nanowaste, Environ. Sci. Nano 2 (2015) 143—154. Available from: https://doi.org/10.1039/C4EN00182F.

[17] S. Kotsilkov, E. Ivanov, N.K. Vitanov, Release of graphene and carbon nanotubes from biodegradable poly(lactic acid) films during degradation and combustion: risk associated with the end-of-life of nanocomposite food packaging materials, Materials 11 (2018) 2346.

[18] G.P. Nichols, Exploring the need for creating a standardized approach to managing nanowaste based on similar experiences from other wastes, Environ. Sci. Nano 3 (2016) 946—952. Available from: https://doi.org/10.1039/C6EN00214E.

[19] N. Musee, Nanotechnology risk assessment from a waste management perspective: are the current tools adequate? Hum. Exp. Toxicol. 30 (8) (2011) 820—835. Available from: https://doi.org/10.1177/0960327110384525.

[20] J. Jeevanandam, A. Barhoum, Y.S. Chan, A. Dufresne, M.K. Danquah, Review on nanoparticles and nanostructured materials: history, sources, toxicity and regulations, Beilstein J. Nanotechnol. 9 (2018) 1050—1074. Available from: https://doi.org/10.3762/bjnano.9.98.

[21] Z. Niu, N. Becknell, Y. Yu, et al., Anisotropic phase segregation and migration of Pt in nanocrystals en route to nanoframe catalysts, Nat. Mater. 15 (2016) 1188—1194. Available from: https://doi.org/10.1038/nmat4724.

[22] J. Park, Y.J. Sa, H. Baik, T. Kwon, S.H. Joo, K. Lee, Iridium-based multimetallic nanoframe@nanoframe structure: an efficient and robust electrocatalyst toward oxygen evolution reaction, ACS Nano 11 (6) (2017) 5500—5509. Available from: https://doi.org/10.1021/acsnano.7b00233.

[23] J.-L. Wang, J.-W. Liu, S.-H. Yu, Recycling valuable elements from the chemical synthesis process of nanomaterials: a sustainable view, ACS Mater. Lett. 1 (5) (2019) 541—548. Available from: https://doi.org/10.1021/acsmaterialslett.9b00283.

[24] A. Maynard, R. Aitken, T. Butz, et al., Safe handling of nanotechnology, Nature 444 (2006) 267—269. Available from: https://doi.org/10.1038/444267a.

[25] R.A. Hoerr, A. Gupta, M.J. Matuszewski, Developing practices for safe handling of nanoparticles and nanomaterials in a development-stage enterprise: a practical guide for research and development organizations, in: T. Webster (Ed.), Safety of Nanoparticles. Nanostructure Science and Technology, Springer, New York, NY, 2009. Available from: https://doi.org/10.1007/978-0-387-78608-7_1.

[26] https://www.ehs.washington.edu/system/files/resources/nanosafeguide.pdf (accessed 10.11.2020)

[27] G.H. Amoabediny, A. Naderi, J. Malakootikhah, M.K. Koohi, S.A. Mortazavi, M. Naderi, et al., Guidelines for safe handling, use and disposal of nanoparticles, J. Phys. Conf. Ser. 170 (2009) 012037. Available from: https://doi.org/10.1088/1742-6596/170/1/012037.

[28] V. Murashov, J. Howard, Risks to health care workers from nano-enabled medical products, J. Occup. Environ. Hyg. 12 (6) (2015) D75—D85. Available from: https://doi.org/10.1080/15459624.2015.1006641.

[29] G. Bystrzejewska-Piotrowska, J. Golimowski, P.L. Urban, Nanoparticles: their potential toxicity, waste and environmental management, Waste Manage. 29 (9) (2009) 2587−2595. Available from: https://doi.org/10.1016/j.wasman.2009.04.001.

[30] Y. Ding, T.A.J. Kuhlbusch, M. Van Tongeren, A. Sánchez Jiménez, I. Tuinman, R. Chen, et al., Airborne engineered nanomaterials in the workplace-a review of release and worker exposure during nanomaterial production and handling processes, J. Hazard. Mater. 322 (Part A) (2017) 17−28. Available from: https://doi.org/10.1016/j.jhazmat.2016.04.075.

[31] S. Yadav, Potentiality of earthworms as bioremediating agent for nanoparticles, in: M. Ghorbanpour, K. Manika, A. Varma (Eds.), Nanoscience and Plant−Soil Systems. Soil Biology, vol. 48, Springer, Cham, 2017. Available from: https://doi.org/10.1007/978-3-319-46835-8_8.

[32] R. Huang, K.-L. Huang, Z.-Y. Lin, J.-W. Wang, C. Lin, Y.-M. Kuo, Recovery of valuable metals from electroplating sludge with reducing additives via vitrification, J. Environ. Manag. 129 (2013) 586−592. Available from: https://doi.org/10.1016/j.jenvman.2013.08.019.

[33] K.J. Lee, S. Choi, S. Park, H.R. Moon, General recyclable redox-metallothermic reaction route to hierarchically porous carbon/metal composites, Chem. Mater. 28 (12) (2016) 4403−4408. Available from: https://doi.org/10.1021/acs.chemmater.6b01459.

[34] H. Lee, Y. Yoon, S. Park, et al., Selective dispersion of high purity semiconducting single-walled carbon nanotubes with regioregular poly(3-alkylthiophene)s, Nat. Commun. 2 (2011) 541. Available from: https://doi.org/10.1038/ncomms1545.

[35] X.B. Fang, Z.Q. Fang, P.K.E. Tsang, W. Cheng, X.M. Yan, L.C. Zheng, Selective adsorption of Cr (VI) from aqueous solution by EDA-Fe_3O_4 nanoparticles prepared from steel pickling waste liquor, Appl. Surf. Sci. 314 (2014) 655−662. Available from: https://doi.org/10.1016/j.apsusc.2014.06.191.

[36] B. Yameen, W.I. Choi, C. Vilos, A. Swami, J. Shi, O.C. Farokhzad, Insight into nanoparticle cellular uptake and intracellular targeting, J. Control. Release 190 (2014) 485−499. Available from: https://doi.org/10.1016/j.jconrel.2014.06.038.

[37] B. Chertok, B.A. Moffat, A.E. David, F. Yu, C. Bergemann, B.D. Ross, et al., Iron oxide nanoparticles as a drug delivery vehicle for MRI monitored magnetic targeting of brain tumors, Biomaterials 29 (4) (2008) 487−496. Available from: https://doi.org/10.1016/j.biomaterials.2007.08.050.

[38] S.K. Shahi, N. Kaur, A. Kaur, et al., Green synthesis of photoactive nanocrystalline anatase TiO_2 in recyclable and recoverable acidic ionic liquid [Bmim] HSO_4, J. Mater. Sci. 50 (2015) 2443−2450. Available from: https://doi.org/10.1007/s10853-014-8799-6.

[39] H. Du, W. Liu, M. Zhang, C. Si, X. Zhang, B. Li, Cellulose nanocrystals and cellulose nanofibrils based hydrogels for biomedical applications, Carbohydr. Polym. 209 (2019) 130−144. Available from: https://doi.org/10.1016/j.carbpol.2019.01.020.

[40] J. Yang, Z. Li, T. Guang, M. Hu, R. Cheng, R. Wang, et al., Green synthesis of high-performance $LiFePO_4$ nanocrystals in pure water, Green. Chem. 20 (2018) 5215−5223. Available from: https://doi.org/10.1039/C8GC02584C.

[41] J. Gong, J. Liu, X. Wen, Z. Jiang, X. Chen, E. Mijowska, et al., Upcycling waste polypropylene into graphene flakes on organically modified montmorillonite, Ind. Eng. Chem. Res. 53 (11) (2014) 4173−4181. Available from: https://doi.org/10.1021/ie4043246.

[42] P. Pati, S. McGinnis, P.J. Vikesland, Life cycle assessment of "green" nanoparticle synthesis methods, Environ. Eng. Sci. 31 (7) (2014) 410−420. Available from: https://doi.org/10.1089/ees.2013.0444.

[43] D. Wang, T. Xie, Y. Li, Nanocrystals: solution-based synthesis and applications as nanocatalysts, Nano Res. 2 (30) (2009) 46. Available from: https://doi.org/10.1007/s12274-009-9007-x.

[44] A. Zacco, L. Borgese, A. Gianoncelli, R.P.W.J. Struis, L.E. Depero, E. Bontempi, Review of fly ash inertisation treatments and recycling, Environ. Chem. Lett. (2014) 12:153−175, DOI 10.1007/s10311-014-0454-6

[45] R. Farzana, R. Rajarao, K. Hassan, P. Ranjan Behera, V. Sahajwalla, Thermal nanosizing: novel route to synthesize manganese oxide and zinc oxide nanoparticles simultaneously from spent Zn−C battery, J. Clean. Prod. 196 (2018) 478−488. Available from: https://doi.org/10.1016/j.jclepro.2018.06.055.

[46] Z. Wang, Z. Wu, N. Bramnik, S. Mitra, Fabrication high-performance flexible alkaline batteries implementing multiwalled carbon nanotubes and copolymer separator, Adv. Mater. 26 (6) (2014) 970−976. Available from: https://doi.org/10.1002/adma.201304020.

[47] L. Zhou, H. Zou, Y. Wang, Z. Le, Z. Liu, A.A. Adesina, Effect of potassium on thermogravimetric behavior and co-pyrolytic kinetics of wood biomass and low density polyethylene, Renew. Energy 102 (Part A) (2017) 134−141. Available from: https://doi.org/10.1016/j.renene.2016.10.028.

[48] G.S. Bajad, S.K. Tiwari, R.P. Vijayakumar, Synthesis and characterization of CNTs using polypropylene waste as precursor, Mater. Sci. Eng.: B 194 (2015) 68−77. Available from: https://doi.org/10.1016/j.mseb.2015.01.004.

[49] K.V. Manukyan, S. Rouvimov, E.E. Wolf, A.S. Mukasyan, Combustion synthesis of graphene materials, Carbon 62 (2013) 302−311. Available from: https://doi.org/10.1016/j.carbon.2013.06.014.

[50] J. Deng, Y. You, V. Sahajwalla, R.K. Joshi, Transforming waste into carbon-based nanomaterials, Carbon 96 (2016) 105−115. Available from: https://doi.org/10.1016/j.carbon.2015.09.033.

[51] K.K. Kefeni, T.A.M. Msagati, T.T.I. Nkambule, B.B. Mamba, Synthesis and application of hematite nanoparticles for acid mine drainage treatment, J. Environ. Chem. Eng. 6 (2) (2018) 1865−1874. Available from: https://doi.org/10.1016/j.jece.2018.02.037.

[52] X. Zhang, K. Wan, P. Subramanian, M. Xu, J. Luo, J. Fransaer, Electrochemical deposition of metal−organic framework films and their applications, J. Mater. Chem. A 8 (2020) 7569−7587. Available from: https://doi.org/10.1039/D0TA00406E.

[53] M.S. Mauter, M. Elimelech, Environmental applications of carbon-based nanomaterials, Environ. Sci. Technol. 42 (16) (2008) 5843−5859. Available from: https://doi.org/10.1021/es8006904.

[54] W. Liu, C. Weng, J. Zheng, X. Peng, J. Zhang, Z. Lin, Emerging investigator series: treatment and recycling of heavy metals from nanosludge, Environ. Sci. Nano 6 (2019) 1657−1673. Available from: https://doi.org/10.1039/C9EN00120D.

[55] W. Liu, F. Huang, Y. Liao, J. Zhang, G. Ren, Z. Zhuang, et al., Treatment of CrVI-containing $Mg(OH)_2$ nanowaste, Angew Chem Int Ed Engl. 47 (30) (2008) 5619−5622. Available from: https://doi.org/10.1002/anie.200800172.

[56] C. Donga, K.I.S. Mabape, S.B. Mishra, A.K. Mishra, Chapter 8 − Polymer-based engineering materials for removal of nanowastes from water, in: A.K. Mishra, H.M.D. Anawar, N. Drouiche (Eds.), Emerging and Nanomaterial Contaminants in Wastewater, Elsevier, 2019, pp. 217−243. Available from: https://doi.org/10.1016/B978-0-12-814673-6.00008-5.

[57] A. Campos, I. López, Current status and perspectives in nanowaste management, in: C. Hussain (Ed.), Handbook of Environmental Materials Management, Springer, Cham, 2019. Available from: https://doi.org/10.1007/978-3-319-73645-7_161.

[58] F. Xu, Review of analytical studies on TiO_2 nanoparticles and particle aggregation, coagulation, flocculation, sedimentation, stabilization, Chemosphere 212 (2018) 662−677. Available from: https://doi.org/10.1016/j.chemosphere.2018.08.108.

[59] C. Walden, W. Zhang, Biofilms vs activated sludge: considerations in metal and metal oxide nanoparticle removal from wastewater, Environ. Sci. Technol. 50 (16) (2016) 8417−8431. Available from: https://doi.org/10.1021/acs.est.6b01282.

[60] A. Patwa, A. Thiéry, F. Lombard, et al., Accumulation of nanoparticles in "jellyfish" mucus: a bio-inspired route to decontamination of nano-waste, Sci. Rep. 5 (2015) 11387. Available from: https://doi.org/10.1038/srep11387.

[61] S. Yuan, D. Peng, X. Hu, J. Gong, Bifunctional sensor of pentachlorophenol and copper ions based on nanostructured hybrid films of humic acid and exfoliated layered double hydroxide via a facile layer-by-layer assembly, Anal. Chim. Acta 785 (2013) 34−42. Available from: https://doi.org/10.1016/j.ac.2013.04.050.

[62] S. Luhar, I. Luhar, Potential application of E-wastes in construction industry: a review, Constr. Build. Mater. 203 (2019) 222−240. Available from: https://doi.org/10.1016/j.conbuildmat.2019.01.080.

[63] A. Mohajerani, L. Burnett, J.V. Smith, et al., Nanoparticles in construction materials and other applications, and implications of nanoparticle use, Mater. (Basel) 12 (19) (2019) 3052. Available from: https://doi.org/10.3390/ma12193052.

[64] A. Behera, S.S. Mohapatra, D.K. Verma, Nanomaterial: fundamental principle and application, Nanotechnology and Nanomaterial Applications in Food, Health and Biomedical Science, Apple Academic Press & CRC Press, 2019, p. 343. ISBN: 9781771887649.

[65] Z. Wang, Q. Yu, F. Gauvin, P. Feng, R. Qianping, H.J.H. Brouwers, Nanodispersed TiO_2 hydrosol modified Portland cement paste: the underlying role of hydration on self-cleaning mechanisms, Cem. Concr. Res. 136 (2020) 106156. Available from: https://doi.org/10.1016/j.cemconres.2020.106156.

[66] C. Nickel, J. Angelstorf, R. Bienert, et al., Dynamic light-scattering measurement comparability of nanomaterial suspensions, J. Nanopart. Res. 16 (2014) 2260. Available from: https://doi.org/10.1007/s11051-014-2260-2.

[67] Y. Liu, M. Tourbin, S. Lachaize, P. Guiraud, Silica nanoparticle separation from water by aggregation with $AlCl_3$, Ind. Eng. Chem. Res. 51 (4) (2012) 1853−1863. Available from: https://doi.org/10.1021/ie200672t.

[68] M.F. Nazar, S. Sakhawat Shah, J. Eastoe, A. Muhammad Khan, A. Shah, Separation and recycling of nanoparticles using cloud point extraction with non-ionic surfactant mixtures, J. Colloid Interface Sci. 363 (2) (2011) 490−496. Available from: https://doi.org/10.1016/j.jcis.2011.07.070.

[69] N.T.S. Phan, C.W. Jones, Highly accessible catalytic sites on recyclable organosilane-functionalized magnetic nanoparticles: an alternative to functionalized porous silica catalysts, J. Mol. Catal. A Chem. 253 (1−2) (2006) 123−131. Available from: https://doi.org/10.1016/j.molcata.2006.03.019.

[70] A. Anand, B. Unnikrishnan, S.-C. Wei, C.P. Chou, L.-Z. Zhang, C.-C. Huang, Graphene oxide and carbon dots as broad-spectrum antimicrobial agents-a minireview, Nanoscale Horiz. 4 (2019) 117−137. Available from: https://doi.org/10.1039/C8NH00174J.

[71] I. Beletskaya, V. Tyurin, Recyclable nanostructured catalytic systems in modern environmentally friendly organic synthesis, Molecules 15 (7) (2010) 4792−4814. Available from: https://doi.org/10.3390/molecules15074792.

[72] S. Navalon, A. Dhakshinamoorthy, M. Alvaro, M. Antonietti, H. García, Active sites on graphene-based materials as metal-free catalysts, Chem. Soc. Rev. 46 (2017) 4501−4529. Available from: https://doi.org/10.1039/C7CS00156H.

[73] Q. Zhou, J. Li, M. Wang, D. Zhao, Iron-based magnetic nanomaterials and their environmental applications, Crit. Rev. Environ. Sci. Technol. 46 (8) (2016) 783−826. Available from: https://doi.org/10.1080/10643389.2016.1160815.

[74] X.-Y. Jiao, L.-S. Li, S. Qin, Y. Zhang, K. Huang, L. Xu, The synthesis of fluorescent carbon dots from mango peel and their multiple applications, Colloids Surf. A Physicochem. Eng. Asp. 577 (2019) 306−314. Available from: https://doi.org/10.1016/j.colsurfa.2019.05.073.

[75] S.M. Abdelbasir, S.M. El-Sheikh, V.L. Morgan, H. Schmidt, L.M. Casso-Hartmann, D.C. Vanegas, et al., Graphene-anchored cuprous oxide nanoparticles from waste electric cables for electrochemical sensing, ACS Sustain. Chem. Eng. 6 (9) (2018) 12176−12186. Available from: https://doi.org/10.1021/acssuschemeng.8b02510.

[76] L. Yin, X. Wen, C. Du, J. Jiang, L. Wu, Y. Zhang, et al., Comparison of the abundance of microplastics between rural and urban areas: a case study from East Dongting Lake, Chemosphere 244 (2020) 125486. Available from: https://doi.org/10.1016/j.chemosphere.2019.125486.

[77] T. Dutta, K.-H. Kim, A. Deep, J.E. Szulejko, K. Vellingiri, S. Kumar, et al., Recovery of nanomaterials from battery and electronic wastes: a new paradigm of environmental waste management, Renew. Sustain. Energy Rev. 82 (Part 3) (2018) 3694−3704. Available from: https://doi.org/10.1016/j.rser.2017.10.094.

[78] M. Kovochich, C.C. Fung, R. Avanasi, et al., Review of techniques and studies characterizing the release of carbon nanotubes from nanocomposites: implications for exposure and human health risk assessment, J. Expo. Sci. Environ. Epidemiol. 28 (2018) 203−215. Available from: https://doi.org/10.1038/jes.2017.6.

[79] P.C. Raynor, J.I. Cebula, J.S. Spangenberger, B.A. Olson, J.M. Dasch, J.B. D'Arcy, Assessing potential nanoparticle release during nanocomposite shredding using direct-reading instruments, J. Occup. Environ. Hyg. 9 (2012) 1. Available from: https://doi.org/10.1080/15459624.2012.633061.

CHAPTER 10

Recycling of nanomaterials by solvent evaporation and extraction techniques

Haleema[1], Muhammad Usman Munir[2], Duy-Nam Phan[3] and Muhammad Qamar Khan[1]

[1]Nanotechnology Research Lab, Department of Textile & Clothing, Faculty of Engineering & Technology, National Textile University, Karachi Campus, Karachi, Pakistan
[2]Department of Materials Engineering, Kaunas University of Technology, Kaunas, Lithuania
[3]School of Textile — Leather and Fashion, Hanoi University of Science and Technology, Hanoi, Vietnam

10.1 Introduction

In recent years the field of nanoscience has grown significantly because of its many advanced applications and benefits to the society [1]. Nanotechnology offers effective opportunities in medical imaging, textile modification, scaffolds, materials construction, agriculture, aerogel coatings, paints and pigments, nanocomposites, biomedicine, in many more fields. As nanotechnology becomes part of our daily life, its presence in the environment poses challenges. As result, the nonutility research is gaining more attention. This field of nanotechnology has grown in volume and as in quantity as well in recent years [2].

Nanotechnology is expected to gain substantial market share among developing technologies. It was estimated that the worldwide nanotechnology industry will grow to reach USD $75.8 billion by 2020 [3]. This rapid growth offers numerous opportunities for economic growth and industrial development. Nanotechnology offers many new solutions to many past problems involving achieving economic sustainability and balanced growth. Specifically, the human population is expected to rise to 11.2 billion by 2100. According to statistical reports of allied market research, the nanomaterials market is expected to exceed USD $55 billion by 2022 from USD $14.7 billion in 2015. But many new challenges to human health, biodiversity, and ecosystems are presented by this field. Nano-size materials and particles that are usually used in commercial products and also in industries have been proven to be highly harmful when they are released into ecosystems or in the process of being recycled [4].

10.2 The importance of recycling in waste management

Recycling of waste materials is one element in strategies of waste minimization. These days, countries and international bodies are setting recycling objectives and are compelling measures to reach them [5].

Nanomaterials Recycling
DOI: https://doi.org/10.1016/B978-0-323-90982-2.00010-X

© 2022 Elsevier Inc.
All rights reserved.

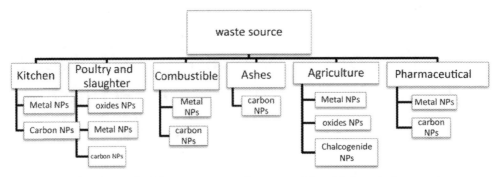

Figure 10.1 Flowchart of different groups of nanoparticles fabricated from various waste sources [6].

Today, products containing nanomaterials are being recycled along with their analogous products without nanomaterials. No separation or separate collection of products containing nanomaterials solely because of their nanomaterial content is known. Also, existing recycling techniques do not take into account the possible risks coming from waste containing nanomaterials (WCNM). The main concern is about possible nonspecific risks of WCNM. In the recycling processes are nanomaterials that might be released into the workstations' atmosphere or into the surroundings by the means of air, soil, or water [6].

10.2.1 Classification of wastes

Waste classification is important according to Lymer. Treatment, storage, and discarding of materials are very dependent on which type waste the materials are. Furthermore, waste classification helps in providing error-free landfill sites [7] (Fig. 10.1).

There are six categories of waste according to their initial source and the problems they create. The primary cause of all waste is industries and household (Table 10.1).

10.3 Nanomaterials in the environment

The environmental impact of engineered nanomaterials will be a serious problem in future resulting from the important growth of nanoparticles (NPs). They do not behave like bulk materials. They are more toxic and reactive in the environment. Because of their size, they can easily penetrate animal and plant cells and float in air, triggering unknown effects. The destination of NPs depends on their use [7,8]. For example, zinc oxide NPs may be present in sewage sludge, which is often spread on fields as a fertilizer, but CNTs are more likely to enter the waste stream from manufacturing processes, eventually ending up in landfills [8].

Table 10.1 The most common combinations of waste types and nanomaterial [8].

Source no.	Waste type	Common nanomaterials
1	Metal	Carbon nanotubes (CNTs), nano-SiO_2, nano-Ag
2	Plastic	CNTs, nano-SiO_2, nano-TiO_2, nanoclay
3	Waste electrical and electronic equipment	CNTs, nano-Ag, nano-SiO_2, nano-TiO_2, nano-ZnO
4	Textiles	CNTs, nano-Ag, nano-SiO_2, nano-TiO_2,
5	End of life vehicles	CNTs, nano-SiO_2, nano-TiO_2
6	Tires	CNTs, nano-SiO_2, nanoclay
7	Construction and demolition waste	CNTs, nano-SiO_2, nano-TiO_2, nano-ZnO, nanoclay, nano-CeO_2

In 2010 a team of public health experts listed the nanosilver in waste water as one of the main issues that could adversely affect the saving of biological diversity. The amount of waste containing residual silver NPs is increasing in proportion with the consumption of silver NPs in local products and in medical use. Silver NPs discharged in gray water may have a variety of outcomes, including being converted into ionic silver; combining with other ions, molecules, and groups; agglomerating; or remaining in nanomaterial form [9].

According to recent findings, synthetic nanowaste does not vanish even when exposed to harsh ecological conditions. For example, NPs of cerium oxide do not burn or change in the heat of a waste incineration plant. They remain intact in combustion residues or in the incineration system [10].

Moreover, NPs have ability to carry cadmium and other contaminants through air, water, and soil. NPs present potential environmental and human health risks as these particles can interact sturdily with biotic structures [4]. Nanostructures also present a problem for separation, recovery, and reuse of the particulate matter. Upto the present time, few strategies have been developed for recycling or reuse of nanomaterial. Recycling and reuse of engineered nanomaterial is challenging because to be practical, the methods have to be relatively cheap, simple, energy efficient, and fast [11,12] (Fig. 10.2).

10.3.1 Recovering nanomaterials from the environment

Even with recoverability built into nanoproducts, it is inevitable that some nanowaste will end up in waste streams. In recent years, researchers have proposed a number of ways to remove these NPs, which, owing to their small size, are incredibly difficult to filter out by conventional methods from the waste before it is released into the environment [4].

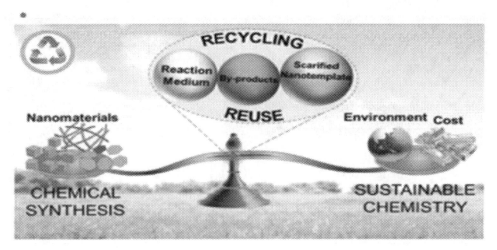

Figure 10.2 Recycling valuable materials from the chemical synthesis of nanomaterials [12].

Filtration of individual NPs is impossible. The recovery processes that are used for recovery and offer any possible commercial advantage involve high energy requirements and high cost. So it is important to find new approaches to achieve NPs' stability without affecting their functionality and properties by using low-energy methods and expensive equipment [13].

Special manufactured nanomaterials as nano-scale metals have high value, and there is strong interest among users and manufacturers in recovering certain nanomaterials for reuse. Nanomaterials may be declared hazardous waste, which could lead to a loss of valuable products because of lack of information about nanomaterials. They could exist in the waste stream recycling process in forms including pure manufactured nanomaterials (e.g., CNTs), nanobyproducts, liquid suspensions containing nanomaterials, items contaminated with nanomaterials (e.g., wipes), and solid matrices with integrated nanomaterials [5].

10.3.2 Recovering nanomaterials from products

Several research groups are attempting to find novel nanotechnology-enabled treatments for industrial waste water using recycled NPs that help to meet circularity and sustainability goals [3].

Given our limited understanding of how nanomaterials behave when released into the environment, most advice from expert committees suggests that manufacturers and retailers of nano-enhanced products should build plans for recovering and recycling the nanomaterials into the product lifecycle. Ideally, products should be designed so that NPs can be separated out and reused as easily as possible. As of yet, there is no specific legislation to control the release of NPs into the environment. Many conventional waste treatment processes (incineration, scrubbing/filtering etc.) do not decompose NPs, allowing their direct release into the environment. This means that new methods must be developed for recovering NPs from products at the end of their life [14] (Table 10.2).

Table 10.2 Environmental applications of waste nanoparticles [15].

Treatment	Nanoparticle	Source	Application
Waste water treatment and water remediation	Iron oxide NPs	Mill scale	Dye removal (adsorption)
	Magnetite (Fe_3O_4)	Iron ore tailings	Dye removal (adsorption)
	Graphene	Polyethylene terephthalate	Dye removal (adsorption)
	$CaCO_3$	Eggshells	Lead (Pb^{2+}) adsorption
	Porous aerogels	Paper, cotton textiles, and plastic bottles	Dye removal (adsorption)
	$NiFe_2O_4$/ ZnCuCr-LDH composite	Saccharin waste water	Dye removal (adsorption)
	Silica NPs (SiO_2 NPs)	Sugar cane waste ash	Dye removal (adsorption)
	Silica NPs ($MW-nSiO_2$, $nSiO_2$)	Corn husk waste	Dye removal (photocatalyst)
	Nanocomposite of ZnO and CuO	Printed circuit boards	Dye removal (photocatalyst)
	Metals-doped ZnO (M-ZnO)	Fabric filter dust	Nitrobenzene removal
	Zero-valent iron NPs	Pickling line of a steel plant	Detection of mercury (Hg^{2+})
Monitoring of pollutants in water	Carbon NPs	Pomelo peels	Nitrobenzene removal
	Nanocuprous oxide	Electrical waste	Detecting dopamine and mercury
Capture of air pollutants	Porous silica NPs	Rice husks	CO_2 capture

10.4 Nanomaterial recycling techniques

Effective and efficient NP recycling techniques are vital for the advancement of nanotechnology in healthcare and nanoscience industrial applications. Many NP separation and recycling methods are being used such as ultracentrifugation, filtration, chromatography, electrophoresis, the addition of antisolvent CO_2, temperature control, and selective precipitation. These separation methods are not continuous and involve multiple steps for preparation with minimum sample volume. However, alternative methods, such as application of magnetic fields, thermoresponsive materials, pH, and colloidal solvents, offer efficient and effective approaches for recycling nanomaterials without substantial cost, energy demand, and time consumption. Further studies on the intrinsic recyclability of nanomaterials are important in order to achieve efficient recycling. It is important to determine how chemical and mechanical properties of nanomaterials change once they are mixed with other product and to set guidelines regarding the reuse in such cases [16].

A reflection on how a proper product design could improve the disassembling of these materials would also help in identifying appropriate reuse and recycling options. Research has shown that the ionic liquid of 1-ethyl-3-methylimidazolium acetate could be recycled for the second synthesis of nanosheet-constructed Pd electrocatalyst or Ag particles via a centrifugation process [17]. A limited supply of the raw material may pose a substantial challenge to the development of gold-based nanotechnologies [18]. Therefore novel approaches for gold recovery from waste streams are being actively researched. Recently, Liu et al. 27 reported one approach to recovering gold using alpha-cyclodextrin (α-CD). This method involves the formation of a host–guest inclusion complex involving tetrabromoaurate anion $[AuBr_4]^-$, α-CD, and hexaaqua potassium cation $[K(OH_2)_6]^+$ followed by rapid coprecipitation at room temperature. This method has been reported to have high separation efficiencies ($>75\%$) and high recovery yields ($>90\%$), and unlike traditional methods for selective gold recovery, it does not involve the use of toxic cyanide or mercury [19].

The use of solvents as a method to recycle foamed polymers offers some advantages. Any insoluble contaminants can be removed by filtration, leaving the polymer clean for further reprocessing. The dissolution process also allows separation of plastics from other types of waste and polymers that are nonsoluble, depending on their chemical nature, in a recycling process known as selective dissolution [20].

10.4.1 Solvent evaporation method

The solvent evaporation method is a very popular and easy method for recycling NPs. It is based on the emulsification of an organic solution of polymer in an aqueous phase followed by evaporation of the solvent [21].

The process of emulsification and solvent evaporation involves two steps. The first step is emulsification of the polymer solution into an aqueous phase. During the second step, the polymer solvent is evaporated, inducing polymer precipitation as nanosphere. The NPs are collected by ultracentrifugation and washed with distilled water to remove stabilizer residue or any free drug and are then lyophilized for storage. A modification of this method is known as the high-pressure emulsification and solvent evaporation method This method involves preparation of an emulsion, which is then subjected to homogenization under high pressure followed by overall stirring to remove the organic solvent. The size can be controlled by adjusting the stirring rate, type, and amount of dispersing agent, viscosity of organic and aqueous phases, and temperature [22].

Several variations have been developed recently based on this technology. The solvent evaporation method has attracted the most attention because of its ease of use and scale-up and lower residual solvent potential compared to other processes. The systems of solvent evaporation method can be based on the nature of the external phase, either aqueous or nonaqueous; the incorporated mode of the core material in the organic solution of the polymer, either dissolved, dispersed, or emulsified; and the elimination procedure of the organic solvent by either evaporation or extraction [23].

10.4.1.1 Classification of solvent evaporation technique

The classification proposed by Aftabrouchad and Doelker is applicable to prepare the nanostructures using the solvent evaporation technique.

1. Solvent evaporation (emulsification–evaporation):
   a. Oil-in-water emulsion
   b. Multiple emulsions: water-in-oil-in-water
   c. No aqueous emulsions
2. Solvent extraction (emulsification–extraction) [24].

10.4.1.2 New modifications of solvent evaporation techniques

Several innovative modifications of emulsification solvent evaporation and extraction techniques have been developed, including water-in-oil-in-water-in-oil, water-in-oil-in-oil, water-in-oil-in-oil-in-oil, solid-in-oil in-water, and water-in-oil [25].

10.4.2 Solvent extraction method

Immiscible liquids (liquids that do not dissolve in one another) form layers when put together. This is because each liquid differs in polarity or orientation. The order of the phases, whether a particular liquid is on top or on the bottom, is determined by the liquids' density. Solvent extraction is the act of removing something or separating it [26].

10.4.2.1 Types of solvent extraction

A variety of procedures are used, including material liquid-liquid extraction, liquid-solid extraction, supercritical fluid extraction, and other special approaches. Liquid-liquid extraction is an extraction method that is implemented by using a liquid extracting medium to liquids, liquid solutions, or samples in solution [27].

One important factor regarding solvent selection is the lack of miscibility between them, such that they will form two distinct phases. Lower solubility of one solvent in another is also highly recommended. For various solvents, the miscibility is typically based on experimental results [28] (Fig. 10.3).

Solvent	Dielectric constant	Dipole moment (D)	Boiling point (°C)	Density (g/mL)	Solubility in water (% weight)	Solubility of water in solvent (% weight)
Hexane	1.89	0	68.7	0.66	0.01	—
Benzene	2.28	0	80.1	0.88	0.08	0.06
Cyclohexane	2.02	0	80.72	0.78	—	0.01
Toluene	2.33	0.31	110.62	0.87	0.05	0.06
m-Xylene	2.4	0.4	139	0.86	0.01	0.04
Chloroform	4.81	1.04	61.15	1.49	0.81	0.97
Carbon tetrachloride	2.2	0	77	1.58	0.08	0.01
Chlorobenzene	5.5	1.6	132	1.11	0.05	0.05
Diethyl ether	4.4	1.2	35	0.71	6.9	1.26
Diisopropyl ether	3.9	1.2	68	0.72	0.9	0.6
Dibutyl ether	3.1	1.2	142	0.77	0.3	0.19
Diamyl ether	3.1	—	187	0.78	—	—
Ethyl acetate	6.4	1.8	77	0.9	7.94	3.01
Methyl isobutyl ketone	13.1	—	116	0.8	1.7	1.9
Cyclohexanone	18.2	2.8	157	0.95	2.3	8

Figure 10.3 Physical properties of some common solvents used in extractions [29].

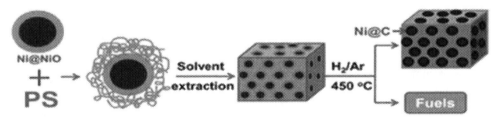

Figure 10.4 Synthesis of ferromagnetic core-shell nanoparticles [31].

10.5 Recycling of nanomaterials via solvent evaporation and extraction

10.5.1 Recycling of nanomaterials by solvent evaporation method

The liquid-liquid extraction of metals from spent batteries is a simple technique to recover high-purity products. Using this technique, some researchers have proposed using this technique to separate valuable metals from spent ZnMn. In these processes, the metal-loaded organic phase needs to be treated with HCl solution to back-extract the desired metals [30].

Nanomaterials such as polystyrene polymer nanocomposites can be recycled via the solvent extraction method to fabricate ferromagnetic core-shell NPs and simultaneously to produce liquid fuels and chemical radicals [31,32] (Fig. 10.4).

The NPs that are produced are found to have substantial catalytic effects on the pyrolysis of polystyrene and re more saturated than pure polystyrene, owing to catalytic hydrogenation of the Ni NPs [31].

Recently, ionic liquids have also shown great promise with regard to recyclability for the second synthesis of nanomaterials because of their thermal stability and no volatility during the synthesis process. Zhang et al. reported that iron oxide NPs with high quality could be prepared in the recycled 1-butyl-3-methylimidazolium bis(trifluoromethylsulfonyl)imide ionic liquid that was separated from the initially synthesized products and washing agent of hexane by evaporation process. Similarly, by using an evaporation technique, ionic liquid of 1-buty-3-methylimidazolium hydrogen sulfate for the green synthesis of photoactive nanocrystalline anatase TiO_2 can be recycled up to three cycles [33].

Further application of the recycled reaction medium can serve as an alternative way to decrease the cost and pollution during the synthesis process of nanomaterials when referring to their commercial production. Once the reaction media were recycled for the next synthesis, it would reduce the consumption of resources and decrease the cost of waste treatment. Conventional metals can also be recycled by solvent extraction method by nanowaste. A team of researchers worked on recovery of NPs of gold. The NPs are produce from nanowaste by supporting CTAb stabilized nanoemulsion and microemulsion. Phase separation of these emulsion induced by the addition of water. Wasters drive a better partitioning of the NPs in the upper co existing phase, along with redistribution of CTAb surfactant into the lower aqueous phase. The filtered NPs dried by solvent removal and diffuse them into another solvent such as octane [34] (Fig. 10.5).

Nanotemplates in the form of ions could be recycled to synthesize the corresponding nanowires. The valuable reagents could be reused for the secondary synthesis of nanocrystals with the same quality, and the solutes could be recycled [36] (Fig. 10.6).

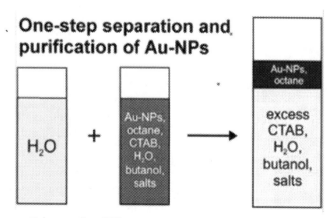

Figure 10.5 Nanoparticle recycling [35].

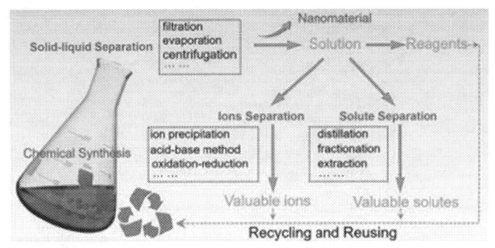

Figure 10.6 Recycling of nanoparticles via extraction method [32].

10.5.2 Recycling of waste by solvent extraction method

Batch- and pilot-scale experiments have demonstrated that solvent extraction can lead to energy conservation by removing the energy needed to move toxic materials to waste sites, reducing the energy needed to manufacture and refine raw metals previously discarded in waste sludge, and even reducing the energy needed to replace discharged waste water [37].

Solvent-based extraction approaches demonstrate a strong potential for extraction techniques to update and make acceptable closed-loop recycling solutions for plastic waste. To construct a suitable method of solvent-based polishing, solubility and diffusivity are key to estimating the efficiency of extraction. Also, it is important to use a simple classification of additives based on their physical characteristics [37].

10.5.3 Nanoparticle recovery using a microemulsion

In 2010 researchers from Bristol University published work on the separation of cadmium and zinc NPs using a special solvent. Solvent is a stable microemulsion of oil in water, which breaks down into two layers when heated. All of the NPs in the solution end up in one of the layers, allowing simple separation [38].

More than 60 million liters of 10%—15% copper-containing waste etchants (CCWEs) are disposed of annually from the manufacture of printed circuit board solutions. Copper chloride, hydrochloric acid, and water are mostly made of CCWEs. Although the dumping of etchants without adequate care posed an environmental issue, costly methods of isolation or storage were previously used; however, they were not successful or reliable. Therefore the resource recovery of these undesired waste etchants in the form of copper NPs will be economically and environmentally

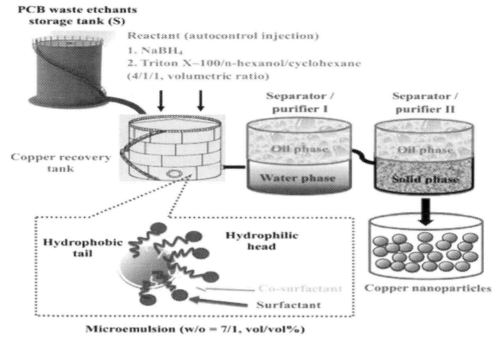

Figure 10.7 Recycling of nanoparticles by a microemulsion technique [39].

desirable. Experimentally, copper NPs have been recovered from CCWEs by microemulsion at controllable ambient temperature and pressure [39] (Fig. 10.7).

10.5.4 Nanoparticle recovery by cloud point extraction

A similar process was reported in 2011 by a research team from Pakistan. They used a technique called cloud point extraction (CPE) to separate gold and palladium NPs from an aqueous solution. The cloud point of an emulsion is the point at which the two phases are on the verge between mixing fully and forming two layers, causing clouding of the solution. In this team's method the NP solution is heated to this point and then centrifuged to fully separate the layers, extracting the NPs [40].

10.5.5 Recycling nanoparticles by employing a colloidal solvent

CPE is an efficient principle that has been developed recently for reversible regulation over NP stability. The use of a colloidal fluid that helps the context can allow colloid stability/instability transitions to isolate and recycle NPs (in place of a natural molecular solvent). This approach offers a number of advantages over traditional methods, such as low cost and the ability to avoid energy-intensive pressure changes or centrifugation cycles, potentially leading to significant economic and environmental benefits [31].

10.6 Potential opportunities for the recovery and reuse of nanowaste

In recent decades the key nanowaste-recycling processes that have been studied have been primarily based on traditional methods, such as centrifugation isolation or solvent evaporation. Recently, researchers have recommended a range of alternate recovery and separation methodologies focused on target NM products to efficiently and economically recover or isolate NMs from their waste streams. CPE, molecular antisolvents, or nanostructured colloidal solvents may be performed by using biological selective extraction procedures. These separation processes are efficient with greater cost savings and lower time and energy demand for recycling NMs. However, to set up efficient separation and recycling for NMs, the intrinsic properties of target nanomaterials must be taken into account [41].

10.7 Conclusion

The principle of this study is that energy consumption and waste prevention must be minimized. It is therefore essential to develop new approaches that enable clean and reversible separation and recovery processes [42]. Over recent decades, various studies have been dedicated to NP stability, and the majority of documented procedures involve extremely energetic methods, such as centrifugation or evaporation resolution. A few alternatives to these traditional methods of separation have recently been outlined. However. the application of magnetic fields, pH and thermo responsive materials, molecular antisolvents, or nanostructured colloidal solvents offers effective and efficient methodologies without excessive costs, time consumption, or energy requirements for recycling NPs. However, a lot of work needs to be done to create innovative approaches, such as mesoporous bioactive glasses applications or photoswitchable schemes, to understand the full potential of NPs in commercial applications [3,43].

References

[1] M. Thiruvengadam, G. Rajakumar, I.M. Chung, Nanotechnology: current uses and future applications in the food industry, 3 Biotech 8 (1) (2018) 74. Available from: https://doi.org/10.1007/s13205-018-1104-7.

[2] C. Buzea, I.I. Pacheco, K. Robbie, Nanomaterials and nanoparticles: sources and toxicity, Journal of Biomaterials and Bilogocal Interfaces 2 (4) (2007) MR17—MR71.

[3] T. Faunce, B. Kolodziejczyk, Nanowaste: Need for Disposal and Recycling Standards, 2017, pp. 1—12.

[4] M.E. Vance, T. Kuiken, E.P. Vejerano, S.P. McGinnis, M.F. Hochella, D.R. Hull, Nanotechnology in the real world: Redeveloping the nanomaterial consumer products inventory, Beilstein J. Nanotechnol. 6 (1) (2015) 1769—1780.

[5] A. Desa, N. Bayah Abd Kadir, F. Yusooff, A study on the knowledge, attitudes, awareness status and behaviour concerning solid waste management, Proc. Soc. Behav. Sci. 18 (2011) 643—648.

[6] S. Yamaguchi, ENV/EPOC/WPRPW (2013) 2/Final Unclassified: Recycling of Waste Containing Nanomaterials (WCNM), no. 2013, 2015, pp. 1—18.

[7] N. Ferronato, V. Torretta, Waste mismanagement in developing countries: A review of global issues, Int. J. Environ. Res. Public Health 16 (6) (2019) 1060.

[8] L. Andersen, F.M. Christensen, J.M. Nielsen, Nanomaterials in waste — issues and new knowledge, Project 1608, 2014.

[9] M.A. Perez, Effects of Silver Nanoparticles on Wastewater Treatment and *Escherichia Coli* Growth, The Florida State University Famu-Fsu Colege of Engineering, 2012.

[10] Jaiswal, 基因的改变 NIH public access, Bone 23 (1) (2014) 1−7.

[11] M.R. Barbara Karn, Nanotechnology and the environment: report of a national nanotechnology initiative workshop, Nanotechnology, 2003, p. 53.

[12] J.-L. Wang, J.-W. Liu, S.-H. Yu, Recycling valuable elements from the chemical synthesis process of nanomaterials: a sustainable view, ACS Mater. Lett. 1 (5) (2019) 541−548.

[13] J. Singh, T. Dutta, K.H. Kim, M. Rawat, P. Samddar, P. Kumar, "Green" synthesis of metals and their oxide nanoparticles: applications for environmental remediation, J. Nanobiotechnol. 16 (1) (2018) 1−24.

[14] O. Myakonkaya, C. Guibert, J. Eastoe, I. Grillo, Recovery of nanoparticles made easy, Langmuir 26 (6) (2010) 3794−3797.

[15] S.M. Abdelbasir, K.M. McCourt, C.M. Lee, D.C. Vanegas, Waste-derived nanoparticles: synthesis approaches, environmental applications, and sustainability considerations, Front. Chem. 8 (2020) 782. Available from: https://doi.org/10.3389/fchem.2020.00782.

[16] T. Salafi, K.K. Zeming, Y. Zhang, Advancements in microfluidics for nanoparticle separation, Lab. Chip 17 (2017) 11−33.

[17] U. Hamidah, T. Arakawa, Y.Y. Hng, A. Nakagawa-izumi, M. Kishino, Recycled ionic liquid 1-ethyl-3-methylimidazolium acetate pretreatment for enhancing enzymatic saccharification of softwood without cellulose regeneration, J. Wood Sci. 64 (2018) 149−156.

[18] M. Starr, K. Tran, Determinants of the physical demand for gold: evidence from panel data, World Econ. 31 (3) (2008) 416−436.

[19] P. Pati, S. McGinnis, P.J. Vikesland, Waste not want not: life cycle implications of gold recovery and recycling from nanowaste, Environ. Sci. Nano 3 (2016) 1133−1143. Available from: https://doi.org/10.1039/C6EN00181E.

[20] M.E. Grigore, Methods of recycling, properties and applications of recycled thermoplastic polymers, Recycling 2 (4) (2017) 1−11.

[21] R. Deshmukh, P. Wagh, J. Naik, Solvent evaporation and spray drying technique for micro- and nanospheres/particles preparation: a review, Dry Technol. 34 (15) (2016) 1758−1772.

[22] T. Urbaniak, W. Musiał, Influence of solvent evaporation technique parameters on diameter of submicron lamivudine-poly-ε-caprolactone conjugate particles, Nanomaterials (Basel) 9 (9) (2019) 1240. Available from: https://doi.org/10.3390/nano9091240.

[23] N.T. Hwisa, P. Katakam, B.R. Chandu, S.K. Adiki, Solvent evaporation techniques as promising advancement in microencapsulation, VRI Biol. Med. Chem. 1 (2013) 8−22.

[24] P.B. ODonnell, J.W. McGinity, Preparation of microspheres by the solvent evaporation technique, Adv. Drug Deliv. Rev. 28 (1) (1997) 25−42.

[25] M. Neeta, S. Mehta, P. Satija, Pandey, M. Dahiya, Solvent evaporation technique: An innovative approach to increase gastric retention, Int. J. Adv. Sci. Res. 1 (4) (2016) 60−67.

[26] D.B. Todd, Solvent extraction, in: H.C. Vogel, C.M. Todaro (Eds.), Fermentation and Biochemical Engineering Handbook — Principles, Process Design, and Equipment, third ed., Elsevier, 2014.

[27] S. Moldoveanu, V. David, Solvent extraction, Modern Sample Preparation for Chromatography, Elsevier, 2015, pp. 1−453.

[28] F. Xie, T.A. Zhang, D. Dreisinger, F. Doyle, A critical review on solvent extraction of rare earths from aqueous solutions, Miner. Eng. 56 (2014) 10−28. Available from: https://doi.org/10.1016/j.mineng.2013.10.021.

[29] A.K. Pabby, A.M. Sastre, Solvent extraction: principles and practices, Reference Module in Chemistry, Molecular Sciences and Chemical Engineering, 2018. Available from: http://doi.org/10.1016/B978-0-12-409547-2.14359-8.

[30] A.G. Chmielewski, T.S. Urbański, W. Migdał, Separation technologies for metals recovery from industrial wastes, Hydrometallurgy 45 (3) (1997) 333–344.

[31] M. Inamuddin, S. Thomas, R. Kumar Mishra, A.M. Asiri (Eds.), Sustainable Polymer Composites and Nanocomposites, Springer International Publishing, 2019. Available from: http://doi.org/10.1007/978-3-030-05399-4.

[32] J. Zhu, et al., Comprehensive and sustainable recycling of polymer nanocomposites, J. Mater. Chem. 21 (2011) 16239–16246.

[33] Q. Zhang, K. De Oliveira Vigier, S. Royer, F. Jérôme, Deep eutectic solvents: syntheses, properties and applications, Chem. Soc. Rev. 41 (2012) 7108–7146. Available from: https://doi.org/10.1039/C2CS35178A.

[34] P.P. Sheng, T.H. Etsell, Recovery of gold from computer circuit board scrap using aqua regia, Waste Manag. Res. 25 (4) (2007) 380–383. Available from: https://doi.org/10.1177/0734242X07076946.

[35] M.J. Hollamby, J. Eastoe, A. Chemelli, O. Glatter, S. Rogers, R.K. Heenan, et al., Separation and purification of nanoparticles in a single step, Langmuir 26 (10) (2010) 6989–6994. Available from: https://doi.org/10.1021/la904225k.

[36] V. Coman, B. Robotin, P. Ilea, Nickel recovery/removal from industrial wastes: a review, Resour. Conserv. Recycl. 73 (2013) 229–238.

[37] M.B. Mansur, Extração por solventes aplicada à recuperação de metais e da água a partir de resíduos industriais e efluentes líquidos, Rev. Esc. Minas 64 (1) (2011) 51–55.

[38] I. Capek, Preparation of metal nanoparticles in water-in-oil (w/o) microemulsions, Adv. Colloid Interface Sci. 110 (1–2) (2004) 49–74. Available from: https://doi.org/10.1016/j.cis.2004.02.003.

[39] N.V. Mdlovu, C.L. Chiang, K.S. Lin, R.C. Jeng, Recycling copper nanoparticles from printed circuit board waste etchants via a microemulsion process, J. Clean. Prod. 185 (2018) 781–796.

[40] S. Gumhold, R. Macleod, X. Wang, Feature extraction from point clouds, in: Proceedings of the 10th International Meshing Roundtable (2001) pp. 293–305.

[41] Z. Chen, A.M. Yadghar, L. Zhao, Z. Mi, A review of environmental effects and management of nanomaterials, Toxicol. Environ. Chem. 93 (6) (2011) 1227–1250. Available from: https://doi.org/10.1080/02772248.2011.580579.

[42] E. Kabir, V. Kumar, K.H. Kim, A.C.K. Yip, J.R. Sohn, Environmental impacts of nanomaterials, J. Environ. Manag. 225 (2018) 261–271. Available from: https://doi.org/10.1016/j.jenvman.2018.07.087.

[43] O. Myakonkaya, Z. Hu, M.F. Nazar, J. Eastoe, Recycling functional colloids and nanoparticles, Chem. A Eur. J. 16 (39) (2010) 11784–11790.

CHAPTER 11

Using pH/thermal responsive materials

Soheyl Mirzababaei[1], Kiyana Saeedian[2], Mona Navaei-Nigjeh[1] and Mohammad Abdollahi[1]

[1]Pharmaceutical Sciences Research Center, The Institute of Pharmaceutical Sciences (TIPS), and School of Pharmacy, Tehran University of Medical Sciences, Tehran, Iran
[2]Biotechnology Group, Faculty of Chemical Engineering, Tarbiat Modares University, Tehran, Iran

Abbreviations

APTES	(3-aminopropyl)triethoxysilane
HEMA	2-hydroxyethyl methacrylate
AAm	acrylamide
AA	acrylic acid
AN	acrylonitrile
ATPS	aqueous two-phase system
BMA	butyl methacrylate
CST	critical solution temperature
EAA	ethyl acrylic acid
EOPO	ethylene oxide-propylene oxide
Au NP	gold nanoparticle
HPEI	hyperbranched polyethylenimine
IBAm	isobutyramide
LCST	lower critical solution temperature
MAA	methacrylic acid
MMA	methyl methacrylate
MBA	N,N'-methylene-bis-acrylamide
DMAEMA	N,N-dimethylaminoethyl methacrylate
NP	nanoparticle
NVI	N-vinylimidazole
PBA	phenylboronic acid
PLA	polylactic acid
PEG	poly(ethylene glycol)
PLGA	poly(L-glutamic acid)
PLL	poly(L-lysine)
PEtOx	poly(N-ethyl oxazoline)
PNIPAM	poly(N-isopropylacrylamide)
PNVCL	poly(N-vinylcaprolactam)
PDEA	poly[(2-diethylamino)ethyl methacrylate]
PDPA	poly[(2-diisopropylamino)ethyl methacrylate]
PHA	polyhydroxyalkanoates
UCST	upper critical solution temperature

Nanomaterials Recycling
DOI: https://doi.org/10.1016/B978-0-323-90982-2.00011-1

© 2022 Elsevier Inc.
All rights reserved.

11.1 Introduction

Owing to nanotechnology advances, several nanostructures have gained commercial applications in fields such as in agriculture, biomedicine, water and waste water treatment, electronics, and catalytic reactions. This is because of their high specific surface area that, for example, could improve catalyst activity by enhancing mass transfer, adsorbent and drug carrier activity by increasing their capacity, and sensors performance by enhancing their selectivity. However, owing to their reduced size, their separation has always been a challenge, as it requires expensive and energy-consuming processes, such as centrifugation, nanofiltration, and supercritical solvent. Besides, the inefficiency of conventional methods could lead to leakage of nanoparticles (NPs) into the soil, water, and air, and in some cases, they may generate secondary wastes. Not only are they a threat to the fauna and flora of the contaminated region, but also they are detrimental to human health. As they accumulate in an organism, they can disrupt the homeostasis of that organism through mechanisms including the generation of reactive oxygen species and apoptosis [1,2]. Moreover, their synthesis mostly involves using organic solvents and chemicals that could have a negative environmental impact [3]. Accordingly, their application could have environmental and life-threatening consequences.

One method to tackle this problem is to develop recyclable nanomaterials. This could improve the ability to efficiently separate these structures with a more cost- and time-effective approach. Besides, enhancing their recyclability could reduce the need for industries that rely on nanomaterials. It will reduce both the environmental impact of their synthesis and the manufacturing cost of the industries using these nanomaterials in their process. Therefore recyclable nanomaterials could have ecological and economic advantages.

Developing stimuli-responsive materials is a fascinating approach to fabricate recyclable nanomaterials. As shown in Fig. 11.1, these materials undergo a reversible

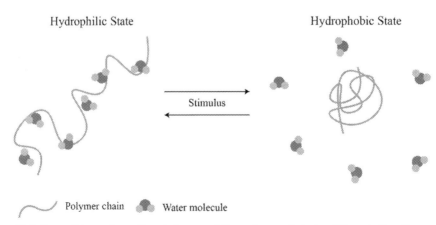

Figure 11.1 Reversible conformational change of the polymer chain by different stimuli.

conformational change in response to a stimulus, which could be a change in temperature or pH. This reversible change—from hydrophilic state to hydrophobic state and vice versa—could trigger the recovery process; while these materials are working under one condition, the stimulus could cause a change in the structure of the material that enables its recovery from the continuous phase [4]. In this chapter we focus on pH-responsive and thermoresponsive materials and their application to develop recyclable nanostructures. We start with a brief introduction to these two types of materials. Then we introduce common structures that have been generated from such materials. Finally, we discuss applications of these two types of stimuli-responsive materials to fabricate recyclable nanomaterials.

11.2 pH-responsive materials

As a subclass of stimuli-responsive materials, pH-responsive polymers are mostly polyelectrolytes consist of monomers with ionizable groups that can either accept (protonate) or donate (deprotonate) proton at a certain pH value [5]. In fact, a change in the pH of the environment could alter the degree of ionization of ionic and cationic functional groups. This phenomenon will be followed by a change in electrostatic repulsion between charged species and the expansion of the folded polymer chains, which will lead to a reversible change in the volume, solubility, surface activity, and configuration of the polymer [6,7]. The result of this reversible change could be swelling or deswelling of hydrogels, flocculation or resuspension of NPs, or formation and deformation of micelles and vesicles [8].

This property and the desired pH for this transition to happen (critical or transition pH) could be tuned through different approaches: using an ionizable moiety with a pK_a in the desirable pH range, adding and controlling the distribution of hydrophobic moieties in the polymer backbone to modify the pK_a of the polyelectrolyte and conformation in the uncharged state, and copolymerization with other ionizable and nonionizable polymers that have an impact on the repulsive forces and hydrophobic interactions [6,9]. Accordingly, this tunable property has enabled researchers to develop smart materials with applications in the fields of drug delivery, protein immobilization, sensors, surface decontamination, oil recovery, membrane technology, water treatment, and so on [10,11].

These smart materials are generally categorized into two subgroups: polyacids and polybases. Having weak acids in their structure, polyacids or polyanions release protons at pH values higher than their pK_a and gain an overall negative charge. There are some common functional groups in polyacids, including carboxylic, boronic, phosphonic, and sulfonic groups. Carboxyl group (COOH) deprotonates and converts to carboxylate (COO^-), while carboxylate receives protons at low pH, enabling the reversible transition of this group. Acrylic acid (AA), methacrylic acid (MAA), ethyl acrylic acid (EAA), maleic anhydride, and *N,N*-dimethylaminoethyl methacrylate (DMAEMA) are among the most frequently used monomers with the pendant carboxylic group [4,12]. Polymers with a sulfonic acid (SO_3H) functional group are another set of these polyanions, notably used for tuning the pK_a of styrene and vinyl-based particles. Poly(2-acrylamido-2-methylpropane sulfonic acid) and poly(4-styrene-sulfonic acid) are examples of sulfonate-containing polymers. Phosphonic acid—based polymers are another group of polyacids that are generally synthesized by treating natural or synthetic polymers with phosphonic acids, such as phosphorus-containing (meth)acrylate and vinyl phosphonic acid [4,8]. The last group of these materials has a boronic acid functional group that gives them properties such as glucose sensitivity and self-healing. Polymerization of phenylboronic acid (PBA)-containing monomers is the most common method to tailor these boronic acid—based materials [13].

The other subcategory of pH-responsive polymers, polycations or polybases, contains weak bases that accept proton at pH values lower than their pKb and become positively charged. Primary, secondary, and tertiary amines are the main ionizable functional groups in these polymers, and monomers such as vinylamine, pyridine, acrylamide (AAm), imidazole, pyrrolidine, and aminoethyl methacrylate are among the most frequently used monomers to synthesize these materials. Polymers including poly[(2-diisopropylamino)ethyl methacrylate] (PDPA), poly[(2-diethylamino)ethyl methacrylate] (PDEA), and PDMAEMA are examples of synthetic polybases. It is worth mentioning that chitosan and polypeptides are naturally derived polycations that exhibit such behavior [4,12].

11.3 Thermoresponsive materials

As their name reveals, an alteration in the conformational state of these polymers occurs when the temperature of the environment changes. The most important feature of these polymers, which is used for their characterization, is critical solution temperature (CST). Around this temperature the hydrophobic/hydrophilic interactions between the polymer chain and the solvent could reversibly change. A change in temperature within this specific range results in particular solubility behavior in aqueous solutions. Hence polymers undergo a sharp and reversible phase transition by collapsing or expanding the chains [4,14]. The physical states of the chains cause different responses of the polymer. The linear and solubilized chains form a monophasic medium, while the transition makes the polymer precipitate, and at this time, the formation of the two phases occurs [15].

On the basis of their behavior, these materials could be categorized into two groups: polymers having a lower critical solution temperature (LCST) and polymers having an upper critical solution temperature (UCST) [16]. In the former group, the polymer-solvent system is in a single-phase (isotropic) state at low temperature, and the polymer becomes insoluble at temperatures above CST, which leads to a two-phase system (anisotropic state). This could be described by the balance between enthalpy and entropy terms. In the single-phase condition below LCST, the enthalpy term, related to electrostatic attractions between the polymer chain and the water molecules, contributes to the solubility of the polymer chain and its hydrophilic, expanded state. As the temperature exceeds the LCST value, the entropy term dominates, leading to polymer precipitation and the collapsed form of the polymer chain due to the hydrophobic interactions [5,16,17]. Examples of these materials, also called negative temperature-sensitive polymers, are poly(N-isopropylacrylamide) (PNIPAM), poly(N-ethyl oxazoline) (PEtOx), and poly(N-vinylcaprolactam) (PNVCL) [4,16].

While the LCST polymers are studied and investigated extensively, there is limited research on another important class of thermoresponsive materials that demonstrate reduced solubility upon cooling [17]. Unlike LCST-type polymers, the polymer-solvent mixture is in a single phase at temperatures higher than the CST, and precipitation of the polymer occurs by reducing the temperature below the CST [4,18]. Some examples of UCSTs, which are also referred to as positive temperature-sensitive polymers, are poly (acrylic acid) (PAA), polyacrylamide (PAAm), and poly(acrylamide-co-butyl methacrylate) [16]. Despite the high potential of UCSTs, there are limited and few scattered publications regarding the important aspects of their phase transition, preparation, and stability, since achieving this behavior under physiological conditions is challenging [19,20].

This phase (or volume) transition is a reversible process. It can be applied in a pulsatile manner, allowing the polymer to be manipulated uniquely via on-demand or on-off switchable control by temperature [21,22]. The transition temperature depends on factors including molecular weight and salt concentration that can be changed by incorporating some moieties [4,21]. These additives can influence the volume phase transition position by altering solvent characteristics and hydrophilic/hydrophobic balance. Salts, surfactants, and cosolvents can be mentioned as the most critical of these additives. Also, the type of behavior could be affected by the solvent. For example, PNIPAM is well known for its LCST-type behavior; however, this polymer shows a UCST-type behavior in water-alcohol binary mixtures [23]. Therefore the transition temperature can be shifted in a wide range by manipulating the contributing factors, which provides us with this opportunity to adjust the transition temperature for the desired application [24].

11.4 Stimuli-responsive nanostructures

Similar to other types of nanomaterials, stimuli-responsive materials could be developed in different structures. Therefore a summary of some of the most common stimuli-responsive structures, including micelles, vesicles, core-shell NPs, hydrogels, and polymer brushes (illustrated in Fig. 11.2), and how these responsive properties affect these nanostructures is provided in this section.

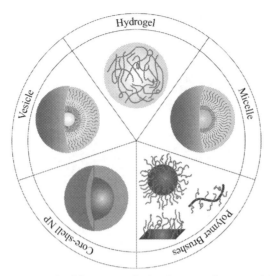

Figure 11.2 Common structures for fabrication of stimuli-responsive materials.

11.4.1 Micelle

Micelles are one of the most common nanostructures and in the case of stimuli-responsive materials are usually composed of self-assembled amphiphilic copolymers. Micellization occurs by aggregation of building blocks with the hydrophobic domain in the core and hydrophilic domain being exposed to the external environment [12]. Yang et al. synthesized a pH-sensitive micelle by self-assembly of triblock copolymer methyl poly(ethylene glycol) ether-b-poly(β-amino esters)-b-polylactic acid (MPEG-b-PBAE-b-PLA). MPEG and PLA are the outermost and inner blocks, respectively, while PBAE is the middle block, providing control over the swelling property of the system. At high pH, the size and zeta potential of the particle did not change; however, reducing the pH value below seven caused protonation of amine groups of the PBAE block. This phenomenon gave rise to a conformational change of the PBAE block from a globular to a stretched conformation, which led to swelling of the micelle [25]. By synthesizing an AB$_2$ Y-shaped miktoarm copolymer of poly(L-lysine) (PLL) and poly(L-glutamic acid) (PLGA), Rao et al. developed a micelle with switchable behavior. In this self-assembled micelle, helix PLGA serves as the core with coiled PLL at the outer layer under acidic conditions, while helix PLL is the core of the micelle, and coiled PLGA forms the outer layer at alkalin condition [26]. Shi's group synthesized a multilayer thermoresponsive polymeric micelle for supporting the gold NPs (Au NPs), which consist of a polystyrene core, a PAA/poly(4-vinylpyridine) (P4VP) shell, and a mixed corona of PEG and PNIPAM. The results showed that the PNIPAM chains collapsed onto the PAA/P4VP complex layer above the LCST and hindered the mass transfer of small molecules and kept Au NPs in the templates [27]. Akimoto et al. produced thermoresponsive micelles of poly(N-isopropylacrylamide-co-N,N-dimethylacrylamide)-b-poly(lactic acid) with the LCST of 39.4°C. The aggregation of micelles above the LCST increased their size from 20 to 600 nm. They reported that intracellular uptake of micelles inside cultured bovine carotid endothelial cells is significantly limited at a temperature below 37°C, and the aggregated micelles are internalized into the cells above the 42°C. Hence combined with a heating treatment, they can be used as a drug delivery system [28].

11.4.2 Vesicle

Vesicles are spherical structures with an aqueous core compartment and a bilayer outer membrane composed of phospholipids, nonionic surfactants, or amphiphilic copolymers. These structures provide the desired property by a structural change of the bilayer membrane that could be triggered by a stimulus such as pH, temperature, or light. Using carboxyl-terminated polydiacetylene vesicles, Jannah et al. developed a calorimetric urease sensor. With a change in pH of the solution the electrostatic repulsion of the deprotonated carboxyl group could cause conformational changes and consequently a colorimetric change in the vesicle. Accordingly, the catalytic activity of urease could be detected by a blue-to-red color transition, which is due to the increased pH by ammonia formation [29]. Li et al. also developed a polymersome with diblock copolypeptides of poly(2-(diisopropylamino)) ethyl L-aspartate block-poly(L-lysine) as the building block. The authors proposed this carrier for drug delivery to tumor tissue. Decreasing the pH value down to five led to the protonation of tertiary amine groups and disassembly of the vesicle and the eventual release of doxorubicin [30]. Li et al. benefited from changing the water temperature for self-assembly of poly[N-(3-aminopropyl) methacrylamide hydrochloride] to obtain a thermoresponsive vesicle system. These homogeneous vesicles, which are 280 nm in size, were formed by increasing the temperature of the solution from 25°C to 45°C [31]. McCoy et al. produced 30−40 nm thermoresponsive vesicles by using the small-angle scattering technique. The vesicles were formed spontaneously from the self-assembly of two surfactants in an aqueous mixture. The results showed that the vesicles collapse into 2−3 nm ellipsoidal micelles above an LCST of 37°C. These vesicles can encapsulate materials or substances, and the vesicle-micelle structural transition by heating from 25°C to 37°C plays a significant role in the release mechanism. Hence this system potentially can be used as responsive carriers for the controlled release of compounds such as drugs and pharmaceuticals [32].

11.4.3 Polymer brush

Tethering polymer chains to another polymer chain, planar surface, or a spherical structure yields one-, two-, or three-dimensional polymer brushes, respectively. This approach can alter the surface properties and functionality of the resultant materials, making them exciting candidates for many purposes, such as surface coatings, nano-lithography, drug delivery, and catalysis [33]. Mi et al. developed a polymer brush containing both quaternary amine and carboxylic groups. This copolymer can determine the charge of the surface; while the surface is neutral under alkaline and neutral conditions, the protonation of carboxylic groups led to a positively charged polymer chain under acidic conditions. This switching behavior enables adherence of tunable bacteria to the surface and, consequently, easier decontamination of the surface and analysis of the contaminant [34]. Fielding et al. grafted PDPA on a wafer that showed switching swelling behavior triggered by a pH change. Being exposed to humid HCl vapor, amine groups will be protonated, and increased hydrophilicity results in swelling that could be observed by the change in the brush color from brown to blue [35]. Liu et al. obtained a polymer brush from noncovalent interactions between a thermoresponsive hyperbranched polyethylenimine (HPEI) with isobutyramide groups (IBAm) and Au NPs. They reported that Au NPs could be used as thermoresponsive catalysts with LCSTs in a wide range, depending on the degree of substitution of the IBAm groups, the molecular weight of the core polymer, or the pH of the solution [36]. Bi et al. synthesized thermoresponsive amphiphilic linear-dendritic diblock copolymers with linear PNVCL as hydrophilic block and dendritic aromatic polyamide as a hydrophobic block. Because of their amphiphilic characteristics, these copolymers formed nanospherical micelles that could encapsulate hydrophobic drugs and showed the desired thermoresponsive drug release behavior [37].

11.4.4 Hydrogel

Hydrogels are water-swellable polymeric networks that can be achieved by physical or chemical crosslinking of hydrosols [38]. General characteristics of these materials, such as volume, water content, swelling property, and hydrophilicity, can be tuned by the degree of crosslinking [4]. Besides, by using stimuli-sensitive moieties as the building blocks, smart gels could be synthesized that undergo a phase transition in response to stimuli, including light, temperature, and pH [8]. To prevent the release of insulin and eventually its denaturation in the stomach, Qi et al. developed a pH-responsive hydrogel by grafting 2-acrylamido-2-methyl-1-propanesulfonic acid onto salecan. While protonation of sulfonic acid groups resulted in a collapsed state of the hydrogel under acidic conditions, its ionization in neutral conditions dissociates hydrogen bonds. This led to swelling of the hydrogel and release of insulin in places other than the gastric acidic environment [39]. Zhu et al. also developed multipolymer hydrogels containing

(9-phenanthryl)methyl methacrylate and (9-anthryl)methacrylate as a donor and acceptor pair of complementary nonradiative resonance energy transfer. Increased pH of the system led to swelling of the nanogel, which increased the distance between the donor and the acceptor pair, leading to an increased ratio of their photoluminescence intensity and a transition from blue to violet under ultraviolet light illumination. Therefore this nanogel could be used as a pH sensor [40]. By copolymerizing poly(glycidyl methacrylate) and PNIPAM, Jiang et al. synthesized a thermoresponsive Au NP—loaded hydrogel with an LCST of 50°C. The diameter of nanogels shrank from 280 to 113 nm by increasing the temperature to 75°C, and the researchers reported that the recoverable catalysts could be successfully used for the reduction of 4-nitrophenol [41]. By grafting PNIPAM on chitosan, Luckanagul et al. developed a thermoresponsive nano-sized hydrogel with an LCST close to body temperature (37°C). Owing to its biocompatibility and low toxicity, it could act as a smart platform for the controllable release of a drug, including lipophilic ones. The authors reported that nanogel successfully encapsulated curcumin and could regulate the release of the received dose of the drug [42].

11.4.5 Core-shell NP

Another possible structure for developing stimuli-responsive materials is the core-shell structure. In this case, one of the components—core or shell—usually fulfills the desirable function, while the other is made of a material that ensures stimuli-responsive behavior [4]. Also, s core-shell structure could be used to develop dual stimuli-responsive materials with core and shell responding to different stimuli. Xu et al. synthesized NPs with a poly(ethylene imine) dendrimer core and an acid-labile PEG shell. The dendritic structure was used as the drug carrier, while the shell improved the circulation of the particle. Besides, acidic pH triggered shell cleavage which was proposed as a method for controlled release [43]. Zhu et al. developed a core-shell structure based on poly(ethylene oxide)-b-polystyrene to prevent steel corrosion by the pH-controlled release of benzotriazole. While it was higher in acidic conditions, owing to the less protonated state of the poly(ethylene oxide) moieties, it was inhibited in basic conditions, owing to the collision of particles and formation of a sodium–rich layer around the particle [44]. Zhu et al. prepared thermoresponsive core-shell NPs through conjugating Au NPs as the core with a shell of thiol-terminated PNIPAM. The NP-containing solution was transparent, owing to the extended polymer chains below the LCST of 25°C. By increasing the solution temperature, PNIPAM chains went through a conformational change, exposed their hydrophobic surface to the solution, and intrachain coil-to-globule transitions and interchain self-association occurred, so the suspension turned opaque. This reversible thermal transition of PNIPAM-tethered Au NPs renders them a good candidate for fabricating smart liquid cell windows that block the solar heat by turning from transparent to opaque at high temperatures [45]. Lu et al. prepared core-shell nanogels with an L-proline-containing hydrophobic core and a thermoresponsive PNIPAM shell by the precipitation polymerization process. By increasing the temperature, the crosslinked chains of corona-shaped shell collapsed on the core and decreased the activity of the catalytic core. The recovery of this nanogel during several reuse cycles was promising, and it kept its high activity and enantioselectivity [46].

11.5 Applications in nanomaterials recycling

As was described earlier, stimuli-responsive materials could be used to develop recyclable nanomaterials with the stimuli triggering the recovery process. In this section we discuss the most common applications of these structures in the literature. More examples of these applications are summarized in Table 11.1.

11.5.1 Aqueous two-phase system

An aqueous two-phase system (ATPS) is formed as a result of polymer-polymer interaction or polymer-high ionic strength salt interaction that is incompatible [47]. It can be used as a liquid-liquid extraction technique in the downstream processing of several biomolecules, including enzymes, nucleic acids, viruses, and proteins. This is due to the mild operational condition of these systems compared to conventional liquid-liquid extraction techniques, which involve an excessive amount of organic solvents or high temperature and pressure. In addition, this method is cost-effective, scalable, and capable of steady-state operation [48,49]. However, the conventional chemicals used for ATPS formation lack the ability of recyclability which could affect the cost and environmental impact of the process. Therefore developing recyclable ATPS could render this method industrially applicable [50].

Stimuli-responsive polymers could be used to develop recyclable ATPS with the stimuli triggering the recycling process. He et al. developed an ATPS for the partitioning of porcine circovirus type 2 Cap protein, based on two copolymers: $P_{ADB4.99}$, which was a copolymer of AA, DMAEMA, and butyl methacrylate (BMA) as monomers and $P_{MDM7.08}$, which was synthesized using MAA, DMAEMA, and methyl methacrylate (MMA) as monomers (Fig. 11.3). After the formation of the ATPS with $P_{ADB4.99}$ and $P_{MDM7.08}$ enriched in the top and bottom phase, respectively, partitioning of the protein was optimized by tuning pH, temperature, and concentrations of salt and copolymers. Finally, recovery and phase separation were conducted by adjusting the pH of each phase to the isoelectric point (p_I) of the polymer that was enriched in the respective phase. Also, the addition of salt to the system improved the efficiency of phase recovery, as it results in weakened interactions and eventually precipitation of polymers [51]. Leong et al. used ethylene oxide−propylene oxide (EOPO) as a

Table 11.1 Examples of pH-responsive and thermoresponsive materials to develop recyclable nanomaterials.

Stimulus	Composition	Application	Noticeable results	References
pH	Ammonium-functionalized hollow polymer particles	Adsorbent for dye removal	• High selectivity of the adsorbent for acidic dyes due to the strong electrostatic attraction between the negative dye and positive ammonium group under acidic conditions • Regeneration of adsorbent under weak alkaline conditions due to electrostatic repulsion between carboxyl groups and acidic dye • 98% removal efficiency even after five cycles	[64]
	Chelating N-heterocyclic dicarbene palladium complex	Catalyst for Suzuki reaction	• Improved solubility, selectivity, and activity of the catalysts due to the presence of carboxyl groups • Reversible precipitation of the catalysts by acidifying the solution • Recovery and reuse of the catalyst at least four times with a small loss of activity	[65]
	Au NP-immobilized P4VP-g-cellulose nanocrystals	Catalyst for reduction of 4-nitrophenol	• Controlling the growth of Au NPs and improvement of their catalytic activity by P4VP brushes • Recovery of catalysts by their agglomeration above the pK_a of P4VP and a subsequent centrifugation step	[66]
	Pd-loaded PDEAEMA-g-UiO-66	Catalyst for Knoevenagel condensation reaction	• Stabilization of Pickering emulsion and occurrence of reaction at neutral condition by deprotonation of PDEAEMA • Enabling demulsification and accordingly recycling of catalysts by adjusting pH to 2	[67]

(Continued)

Table 11.1 (Continued)

Stimulus	Composition	Application	Noticeable results	References
	Au NP—loaded DMAEMA-g-starch NP	Catalyst for hydrogenation of p-nitroanisole	• Formation of a stable Pickering emulsion and progression of catalytic reaction at a pH value of nine or above • Demulsification and consequent partitioning of the catalyst and the product in two phases by reducing pH to two • Reusability of the catalyst at least for four cycles	[68]
	• P_{ADBA} composed of AA, DMAEMA, BMA, and allyl alcohol • P_{MDB} composed of MAA, DMAEMA, and BMA	ATPS for biosynthesis of cephalexin	• Significant reduction of penicillin acylase inhibition by cephalexin due to cephalexin partitioning to the P_{ADBA}-rich phase • Recycling copolymers in each phase by adjusting the pH to the p_I of the respective phase, resulting in a 90% recovery of polymers	[69]
	• P_{ADB} composed of AA, DMAEMA, BMA • P_{MDBH} composed of MMA, DMAEMA, and HEMA[a]	ATPS for extraction of tylosin	• Partitioning of tylosin in the P_{ADB}-rich phase due to the higher degree of carboxyl ionization of P_{ADB} compared to that of the P_{MDBH} • About 93% recovery of ATPS by adjusting the pH of each phase to its p_I	[70]
Temperature	2-Hydroxy-3-isopropoxypropyl starch/alginate composite hydrogel	Adsorbent for removal of heavy metals	• Facile shrinkage of the hydrogel at 35°C resulting in the reduction of the required HCl for adsorbent regeneration • More efficient and faster recycling compared to nonresponsive hydrogel	[71]
	Grapheneoxide—perylene bisimides-containing PNIPAM	Adsorbent for dye removal	• A hybrid structure with graphene and a polymeric part responsible for adsorption and thermoresponsive behavior, respectively • Facile separation of the adsorbent by heating and a following filtration step	[72]

Pd@graphene-g-PEtOx	Catalyst for Suzuki reaction	• Catalysts precipitation by adjusting the temperature above the LCST of the polymer and its recovery by hot filtration • An approximate conversion rate of 80% after recycling for five times	[73]
Lipase-immobilized poly(AAm-*co*-AN[b])	Catalyst for tributyrin hydrolysis	• Biocatalyst recovery by simply cooling the polypropylene solution to a temperature lower than the UCST of the carrier followed by decantation • 80% retained activity of immobilized enzyme after six cycles	[74]
• P_{VBAm} composed of NVCL, BMA, and AAm • P_N composed of NIPAM, 2.2'-azo-bis-isobutyronitrile	ATPS for ε-polylysine extraction	• ATPS formation and partitioning of the ε-polylysine to the P_N-rich phase • Recovery of each phase by raising the temperature above its LCST, resulting in polymer precipitation • Reusability of the recovered polymers in the next cycles of ATPS formation	[75]
• Poly(maleic anhydride modified β-cyclodextrin-*co*-NIPAM) • Dextran	ATPS for mandelic acid extraction	• Selectively separating (S)-enantiomer of mandelic acid by ATPS • Recovery of polymers by precipitating at a temperature above LCST followed by centrifugation • No obvious change in separation factor upto three cycles	[76]
• PEG-b-PNIPAM	ATPS for bromelain extraction	• Phase formation by addition of salt to the copolymer solution and partitioning of bromelain to the polymer-rich phase • Extraction of bromelain and recovery of the polymer by increasing temperature of the polymer-rich phase above its LCST	[77]

(*Continued*)

Table 11.1 (Continued)

Stimulus	Composition	Application	Noticeable results	References
pH and magnetic field	• Magnetic graphene oxide/poly (NVI[c]-co-AA) hydrogel	Adsorbent for dye removal	• Facile separation of the adsorbent by a magnetic field due to the presence of the magnetic graphene oxide • Repeated adsorption and desorption of both cationic and anionic dyes by adjusting the pH of the solution • 65% removal percentage after five cycles	[60]
	APTES[d]–coated magnetic NP	Adsorbent for oil separation	• Fast and easy separation of the adsorbent by an external magnetic field • Demulsification of oil droplets by NPs due to interfacial adsorption and electrostatic attraction under neutral and acidic conditions and their regeneration in weak alkaline condition • Recovery of adsorbent upto nine cycles	[78]
Temperature and magnetic field	Fe@SiO$_2$@PNIPAM-co-MAA core-shell NP	Adsorbent for phenolic compounds	• Contribution of hydrophobic/hydrophilic interactions, and electrostatic interactions to high adsorption rate • Easy recycling of the adsorbent due to its magnetic characteristics	[79]
	Malonyl-CoA synthetase—immobilized Fe$_3$O$_4$@ PNIPAM	Catalyst for productionof malonyl-CoA	• Controlling catalytic activity by the degree of polymer chain extension. • Full recovery of catalyst at temperatures above LCST using a magnetic field • High catalytic activity even after three cycles	[80]

[a]2-Hydroxyethyl methacrylate.
[b]Acrylonitrile.
[c]N-vinylimidazole.
[d](3-Aminopropyl)triethoxysilane.

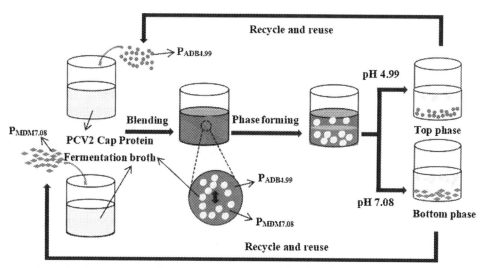

Figure 11.3 Application of stimuli-responsive materials for developing recoverable ATPS. Recovery was triggered by the pH-responsiveness of the polymers. *Reprinted with permission from J. He, J. Wan, T. Yang, X. Cao, L. Yang, Recyclable aqueous two-phase system based on two pH-responsive copolymers and its application to porcine circovirus type 2 Cap protein purification, J. Chromatogr. A 1555 (2018) 113−123. https://doi.org/10.1016/j.chroma.2018.04.032 [51], Copyright 2018 Elsevier.*

recyclable thermoseparating polymer to recover polyhydroxyalkanoates (PHAs) from a microbial culture with an ATPS method. They added the aqueous polymer solution to the PHA feedstock and heated them to reach EOPO's LCST, which finally induced thermoseparation and obtained two phases. The top phase was water-rich, including the PHAs, and the bottom phase consisted of concentrated polymers. The results demonstrated a high recovery yield and purification factor of PHAs. For the recycling studies, after removal of the top phase, a new feed was added to the recycled polymer, and the phase formation and PHA recovery process was performed again. They reported that recycling and reutilization of EOPO could be done at least two times with satisfying yield and PHA purity [52].

Recoverable ATPS could also be used in extractive bioconversion: an in situ method for the formation and recovery of biomolecules in one step. Based on citrate and a pH-responsible polymer, Chen et al. developed an ATPS to synthesize and purify cefprozil. Upon addition of substrate and the enzyme to the system, they were biased in the salt-containing phase (bottom phase), while cefprozil was partitioned to the polymer-enriched phase (top phase), which improved reaction efficiency by reducing the inhibitory effect of the product. Finally, cefprozil was recovered from the top phase by precipitating the polymer at its p_I, which led to a high-purity product and recycling of the polymer to form the ATPS repeatedly [47].

11.5.2 Catalysts

In the large-scale production of chemicals, catalysts facilitate the synthesis reaction, and their reusability significantly affects the productivity of the process. To improve their efficiency, they could be used in the form of nanocatalysts, as they increase the specific surface area of the system [53]. However, the separation and recycling of the nanocatalysts involve methods such as high-speed centrifugation that are expensive, time-consuming, and energy-consuming and require instrumentations that could become a serious problem as the size of the particles reduces. Besides, they often lead to the loss of catalyst during the separation process [54]. Therefore catalysts could be immobilized on smart materials to enable their simple separation and recovery from the reaction solution.

Dong et al. immobilized Pd NPs on P(AAm-co-AA) to achieve recyclability of this catalyst by tuning the pH of the environment (Fig. 11.4A). The hydrophilic nature of this copolymer at a weak basic condition improved mass transfer of the substrate, resulting in the higher catalytic activity of Pd NPs compared to that of the catalysts immobilized on the γ-Al$_2$O$_3$. In order to reuse the catalysts, centrifugation was required for Pd-γ-Al$_2$O$_3$, which led to the loss of approximately 8 wt.% of NPs and, accordingly, 10% of the catalytic activity. On the other hand, in the case of the P(AAm-co-AA) stabilizer, recovery of the catalyst was achieved by adjusting the pH to 3, which led to precipitation of the copolymer due to the interaction of amine and carboxyl groups. Not only did this mild recycling condition contribute to only 1%−2% of the catalyst loss and 3% of catalytic activity, but also it eliminated energy-consuming processes such as centrifugation [55]. Wang et al prepared temperature-regulated core-corona micelle of PNIPAM-b-P4VP with an LCST 32°C containing a P4VP-rich core and thermoresponsive PNIPAM block construct corona. Then they loaded Au NPs in the micelle and formed a thermoresponsive catalyst that could catalyze the hydrogenation of p-nitrophenol. They demonstrated that below LCST, diffusion of the reactants could be facilitated by the swollen PNIPAM corona, which increases the supply of the reactant to the surface of Au NPs. But above the LCST these chains became hydrophobic and collapsed, resulting in the formation of a barrier around the NPs. This switchable status decelerated the diffusion of reactants and consequently decreased the catalytic reaction rate. To recover the catalyst, the colloidal dispersion of micelle-supported Au NPs was first kept at 4°C; the, all the reactants were removed by dialyzing the colloidal dispersion against water at room temperature. The experiments showed that the catalyst kept its high stability and efficiency for at least three cycles [56].

Regarding catalytic reactions, there is another problem, which is the solvent being incompatible with the product, reactant, or catalyst. In these instances, biphasic catalytic reactions could enhance the rate and yield of the reaction, as they provide compartmentalization of chemicals [57]. To develop such a system, Pickering

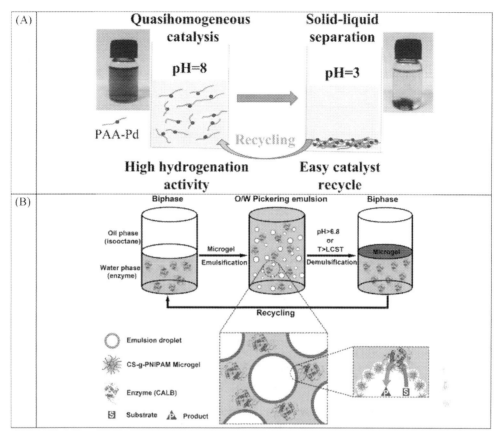

Figure 11.4 Application of stimuli-responsive materials for developing recyclable catalysts. (A) a recoverable pH-responsive catalyst. (B) A recoverable Pickering emulsion system to perform biphasic catalytic reaction at the interface of aquatic and organic phases. *Reprinted with permission from (A) Y. Dong, Y. Jin, J. Wang, J. Shu, M. Zhang, Pd nanoparticles stabilized by a simple pH-sensitive P(acrylamide-co-acrylic acid) copolymer: a recyclable and highly active catalyst system in aqueous medium, Chem. Eng. J. 324 (2017) 303–312. https://doi.org/10.1016/j.cej.2017.05.030 [55], Copyright 2017 Elsevier; (B) Y. Wang, L. Zhu, H. Zhang, H. Huang, L. Jiang, Formulation of pH and temperature dual-responsive Pickering emulsion stabilized by chitosan-based microgel for recyclable biocatalysis, Carbohydr. Polym. 241 (2020), 116373. https://doi.org/10.1016/j.carbpol.2020.116373 [59], Copyright 2020 Elsevier.*

emulsions are typically used that involve emulsions located at the interface of two phases and stabilized by NPs. Owing to the high stability of emulsions, their separation is usually energy-consuming, which renders these systems less favorable. Accordingly, attempts have been made to develop easily separable Pickering emulsions [58].

One approach to address this issue is to develop stimuli–responsive Pickering emulsion in which emulsification and demulsification are triggered by a stimulus, such as light, temperature, or pH. This property makes the recyclability of the system more comfortable and more effective. For example, Wang et al. grafted PNIPAM onto chitosan to develop thermoresponsive and pH-responsive microgels (Fig. 11.4B). The Pickering emulsion was formed by the addition of a microgel-containing water phase to isooctane and implementing homogenization. To assess its ability to develop a biphasic catalytic system, lipase was added to the water phase, and its substrate was incorporated into the oil phase. Compared to a conventional biphasic system, not only did the Pickering emulsion system improve the rate of the catalytic activity, but also it rendered the mixing process an unnecessary step, as the interfacial surface significantly improved. In addition, after completion of the reaction, demulsification occurred by raising the temperature above the LCST of PNIPAM, leading to three separate phases. After separating the oil phase containing the product, the two other phases, which were a water phase containing lipase and a microgel phase, were added to fresh solvent and reactant, enabling recovery of the catalyst. This recycling method resulted in reusability of the catalyst and the microgels with a conversion rate higher than 75% even after six cycles [59].

11.5.3 Adsorption

As a result of industrial development, several pollutants, such as pharmaceutical traces, dyes, heavy metals, pesticides, and emulsified oil, have been leaked into water that could be detrimental to aquatic life and human health. Accordingly, several methods have been proposed for waste water treatment, including biodegradation, oxidation, filtration, and adsorption. Among these methods, adsorption is one of the most interesting, owing to its simplicity, effectiveness, and ecofriendliness. However, after the adsorption process, labor-intensive and expensive downstream processes, including centrifugation and filtration, might be required to remove the adsorbent from the water. Accordingly, for this process to be economically favorable, a simple and cheap recovery process of the adsorbent is essential [60].

Zhu et al. prepared HPEI-based gel and employed it as a recyclable adsorbent to remove dyes from waste water (Fig. 11.5A). Through a crosslinking reaction between HPEI and N,N'-methylene-bis-acrylamide (MBA), HPEI-MBA gel was obtained that was treated with isobutyric anhydride to form HPEI-MBA-IBAm gel. The obtained gel had high capacity and performance in the adsorption/desorption of both single-dye and mixed systems of anionic and cationic dyes. Also, they conducted desorption experiments by changing the temperature and dried the recovered gel under a vacuum. Several adsorption-desorption processes were repeated, and results showed that HPEI-MBA-IBAm gel kept its adsorption efficiency through 11 cycles, which suggests that it is a promising thermoresponsive adsorbent for dye removal applications [61].

Aside from the recyclability of adsorbents, ease of separation is of significant importance for adsorbents to be commercially used. Accordingly, attempts have been made to develop dual stimuli-responsive materials for this purpose; one stimulus enables a simple separation, while the other triggers the recovery of the adsorbent. Wang et al.

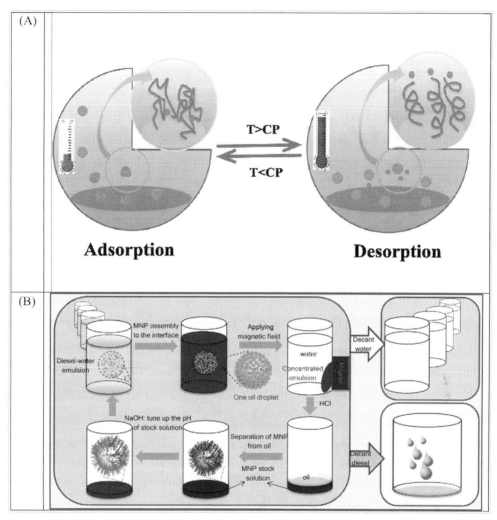

Figure 11.5 Application of stimuli-responsive materials for developing recoverable adsorbents. (A) A thermoresponsive adsorbent for heavy metal adsorption. (B) A dual-responsive nanomaterial for facile separation and recovery of adsorbent. *Reprinted with permission from (A) C. Zhu, Y. Xia, Y. Zai, Y. Dai, X. Liu, J. Bian, et al., Adsorption and desorption behaviors of HPEI and thermoresponsive HPEI based gels on anionic and cationic dyes, Chem. Eng. J. 369 (2019) 863–873. https://doi.org/10.1016/j.cej.2019.03.169 [61], Copyright 2019 Elsevier; (B) X. Wang, Y. Shi, R.W. Graff, D. Lee, H. Gao, Developing recyclable pH-responsive magnetic nanoparticles for oil-water separation, Polym. (Guildf.) 72 (2015) 361–367. https://doi.org/10.1016/j.polymer.2014.12.056 [62], Copyright 2020 Elsevier.*

developed hybrid NPs composed of a magnetic core and a shell made of PDMAEMA. (Fig. 11.5B). In a relatively neutral environment these NPs formed an oil-in-water Pickering emulsion that enables the collection of micro-size oil particles. Then particles were separated from the aqueous solution by utilizing an external magnetic field to achieve purified water. After that, HCl was added to the system to break the Pickering emulsion, and destabilization of the emulsion led to the separation of oil and magnetic NPs. Finally, NaOH treatment was used to disperse these core-shell structures into the aqueous solution so that they could be used again for oil separation [62]. Hayashi et al. prepared hybrid magnetite-thermoresponsive NPs to develop an environmentally friendly adsorbent for the removal of heavy metal pollutants. They synthesized magnetite NPs in the presence of thermoresponsive copolymer, P (NIPAM-*co*-AA) and formed well-dispersed NPs with high immobilized polymer loading. Results demonstrated successful control of Cu (II) adsorption and desorption based on the thermoresponsive conformational change of polymer with LCST properties. The adsorbent that was developed in this study could be recycled from the solution within 10 seconds by positioning a neodymium magnet close to the sample vial and could be used repeatedly while maintaining the adsorption capacity [63].

11.6 Conclusions

The booming development in nanomaterials and their industrial applications have raised concerns about their environmental impacts. Hazard compounds utilized in the synthesis of nanomaterials and accumulation of nanomaterials in the environment could be detrimental to living systems. Accordingly, several studies have been dedicated to methods and processes for the separation and recycling of nanomaterials. One interesting goal is developing recyclable materials that could reduce the energy and cost of downstream processes for nanomaterials separation. Therefore in this chapter we described pH-responsive and thermoresponsive materials, the most common geometries for their fabrication, and how they could be used to serve this purpose.

Despite all the advances in the field of stimuli-responsive materials, there are still issues that could be addressed. The most critical issue is the synthesis method of these materials. Most of them are based on environmentally harmful chemicals; therefore applying principles of green chemistry could be an answer to this problem. Moreover, developing simple methods for their fabrication could ensure their industrial applications. Another important aspect is more studies on developing dual-responsive materials, which could be significantly helpful in the field of the adsorption process. In most cases, the stimulus triggers the precipitation of the proposed pH-responsive or thermoresponsive nanomaterials, which might still require a following separation process that enables their reusability. In dual-responsive nanomaterials, one stimulus could enable the separation of the adsorbent-adsorbate complex from the solution whereas the other stimulus triggers the desorption process and recovery of the nanostructure to be used in the next cycles.

References

[1] T.A. Saleh, Nanomaterials: classification, properties, and environmental toxicities, Environ. Technol. Innov. 20 (2020) 101067. Available from: https://doi.org/10.1016/j.eti.2020.101067.

[2] L.M. Rossi, N.J.S. Costa, F.P. Silva, R. Wojcieszak, Magnetic nanomaterials in catalysis: advanced catalysts for magnetic separation and beyond, Green Chem. 16 (2014) 2906−2933. Available from: https://doi.org/10.1039/c4gc00164h.

[3] J.-L. Wang, J.-W. Liu, S.-H. Yu, Recycling valuable elements from the chemical synthesis process of nanomaterials: a sustainable view, ACS Mater. Lett. 1 (2019) 541−548. Available from: https://doi.org/10.1021/acsmaterialslett.9b00283.

[4] J. Zhao, V.E. Lee, R. Liu, R.D. Priestley, Responsive polymers as smart nanomaterials enable diverse applications, Annu. Rev. Chem. Biomol. Eng. 10 (2019) 361−382. Available from: https://doi.org/10.1146/annurev-chembioeng-060718-030155.

[5] J. Zhang, M. Zhang, K. Tang, F. Verpoort, T. Sun, Polymer-based stimuli-responsive recyclable catalytic systems for organic synthesis, Small 10 (2014) 32−46. Available from: https://doi.org/10.1002/smll.201300287.

[6] F. Reyes-Ortega, pH-Responsive Polymers: Properties, Synthesis and Applications, Woodhead Publishing Limited, 2014. Available from: https://doi.org/10.1533/9780857097026.1.45.

[7] A. Kawamura, T. Miyata, pH-Responsive polymer, in: S. Kobayashi, K. Müllen (Eds.), Encyclopedia of Polymeric Nanomaterials, Springer Berlin Heidelberg, Berlin, Heidelberg, 2014, pp. 1−9. Available from: https://doi.org/10.1007/978-3-642-36199-9_211-1.

[8] G. Kocak, C. Tuncer, V. Bütün, pH-Responsive polymers, Polym. Chem. 8 (2017) 144−176. Available from: https://doi.org/10.1039/c6py01872f.

[9] H. Tang, W. Zhao, J. Yu, Y. Li, C. Zhao, Recent development of pH-responsive polymers for cancer nanomedicine, Molecules (2019) 24. Available from: https://doi.org/10.3390/molecules24010004.

[10] S. Dai, P. Ravi, K.C. Tam, pH-Responsive polymers: synthesis, properties and applications, Soft Matter 4 (2008) 435−449. Available from: https://doi.org/10.1039/b714741d.

[11] W. Tao, J. Wang, W.J. Parak, O.C. Farokhzad, J. Shi, Nanobuffering of pH-responsive polymers: a known but sometimes overlooked phenomenon and its biological applications, ACS Nano. 13 (2019) 4876−4882. Available from: https://doi.org/10.1021/acsnano.9b01696.

[12] S. Bazban-Shotorbani, M.M. Hasani-Sadrabadi, A. Karkhaneh, V. Serpooshan, K.I. Jacob, A. Moshaverinia, et al., Revisiting structure-property relationship of pH-responsive polymers for drug delivery applications, J. Control. Rel. 253 (2017) 46−63. Available from: https://doi.org/10.1016/j.jconrel.2017.02.021.

[13] Y. Guan, Y. Zhang, Boronic acid-containing hydrogels: synthesis and their applications, Chem. Soc. Rev. 42 (2013) 8106−8121. Available from: https://doi.org/10.1039/c3cs60152h.

[14] Q. Zhang, C. Weber, U.S. Schubert, R. Hoogenboom, Thermoresponsive polymers with lower critical solution temperature: from fundamental aspects and measuring techniques to recommended turbidimetry conditions, Mater. Horiz. 4 (2017) 109−116. Available from: https://doi.org/10.1039/c7mh00016b.

[15] M.R. Aguilar, J. San Román, Smart polymers and their applications, Smart Polym. Their Appl. (2014) 1−568. Available from: https://doi.org/10.1533/9780857097026.

[16] A.K. Teotia, H. Sami, A. Kumar, Thermo-Responsive Polymers: Structure and Design of Smart Materials, Elsevier Ltd., 2015. Available from: https://doi.org/10.1016/B978-0-85709-713-2.00001-8.

[17] D. Roy, W.L.A. Brooks, B.S. Sumerlin, New directions in thermoresponsive polymers, Chem. Soc. Rev. 42 (2013) 7214−7243. Available from: https://doi.org/10.1039/c3cs35499g.

[18] P.A. Limadinata, A. Li, Z. Li, Temperature-responsive nanobiocatalysts with an upper critical solution temperature for high performance biotransformation and easy catalyst recycling: efficient hydrolysis of cellulose to glucose, Green Chem. 17 (2015) 1194−1203. Available from: https://doi.org/10.1039/c4gc01742k.

[19] J. Seuring, F.M. Bayer, K. Huber, S. Agarwal, Upper critical solution temperature of poly(N-acryloyl glycinamide) in water: a concealed property, Macromolecules 45 (2012) 374−384. Available from: https://doi.org/10.1021/ma202059t.

[20] F. Doberenz, K. Zeng, C. Willems, K. Zhang, T. Groth, Thermoresponsive polymers and their biomedical application in tissue engineering—a review, J. Mater. Chem. B 8 (2020) 607–628. Available from: https://doi.org/10.1039/c9tb02052g.

[21] M.R. Aguilar, J. San Román, Introduction to Smart Polymers and Their Applications, second ed., Elsevier Ltd., 2019. Available from: https://doi.org/10.1016/b978-0-08-102416-4.00001-6.

[22] Y.J. Kim, Y.T. Matsunaga, Thermo-responsive polymers and their application as smart biomaterials, J. Mater. Chem. B 5 (2017) 4307–4321. Available from: https://doi.org/10.1039/c7tb00157f.

[23] J. Niskanen, H. Tenhu, How to manipulate the upper critical solution temperature (UCST)? Polym. Chem. 8 (2017) 220–232. Available from: https://doi.org/10.1039/c6py01612j.

[24] D. Schmaljohann, Thermo- and pH-responsive polymers in drug delivery, Adv. Drug Deliv. Rev. 58 (2006) 1655–1670. Available from: https://doi.org/10.1016/j.addr.2006.09.020.

[25] C. Yang, Z. Xue, Y. Liu, J. Xiao, J. Chen, L. Zhang, et al., Delivery of anticancer drug using pH-sensitive micelles from triblock copolymer MPEG-b-PBAE-b-PLA, Mater. Sci. Eng. C 84 (2018) 254–262. Available from: https://doi.org/10.1016/j.msec.2017.12.003.

[26] J. Rao, Y. Zhang, J. Zhang, S. Liu, Facile preparation of well-defined AB2 Y-shaped miktoarm star polypeptide copolymer via the combination of ring-opening polymerization and click chemistry, Biomacromolecules 9 (2008) 2586–2593. Available from: https://doi.org/10.1021/bm800462q.

[27] D. Xiong, Z. Li, L. Zou, Z. He, Y. Liu, Y. An, et al., Modulating the catalytic activity of Au/micelles by tunable hydrophilic channels, J. Colloid Interface Sci. 341 (2010) 273–279. Available from: https://doi.org/10.1016/j.jcis.2009.09.045.

[28] J. Akimoto, M. Nakayama, K. Sakai, T. Okano, Thermally controlled intracellular uptake system of polymeric micelles possessing poly(N-isopropylacrylamide)-based outer coronas, Mol. Pharm. 7 (2010) 926–935. Available from: https://doi.org/10.1021/mp100021c.

[29] F. Jannah, J.M. Kim, pH-sensitive colorimetric polydiacetylene vesicles for urease sensing, Dye. Pigment. 169 (2019) 15–21. Available from: https://doi.org/10.1016/j.dyepig.2019.04.072.

[30] Z. Li, J. Li, J. Huang, J. Zhang, D. Cheng, X. Shuai, Synthesis and characterization of pH-responsive copolypeptides vesicles for siRNA and chemotherapeutic drug co-delivery, Macromol. Biosci. 15 (2015) 1497–1506. Available from: https://doi.org/10.1002/mabi.201500161.

[31] Y. Li, B.S. Lokitz, C.L. McCormick, Thermally responsive vesicles and their structural "locking" through polyelectrolyte complex formation, Angew. Chem. - Int. (Ed.) 45 (2006) 5792–5795. Available from: https://doi.org/10.1002/anie.200602168.

[32] T.M. McCoy, J.B. Marlow, A.J. Armstrong, A.J. Clulow, C.J. Garvey, M. Manohar, et al., Spontaneous self-assembly of thermoresponsive vesicles using a zwitterionic and an anionic surfactant, Biomacromolecules 21 (2020) 4569–4576. Available from: https://doi.org/10.1021/acs.biomac.0c00672.

[33] C. Feng, X. Huang, Polymer brushes: efficient synthesis and applications, Acc. Chem. Res. 51 (2018) 2314–2323. Available from: https://doi.org/10.1021/acs.accounts.8b00307.

[34] L. Mi, M.T. Bernards, G. Cheng, Q. Yu, S. Jiang, pH responsive properties of non-fouling mixed-charge polymer brushes based on quaternary amine and carboxylic acid monomers, Biomaterials 31 (2010) 2919–2925. Available from: https://doi.org/10.1016/j.biomaterials.2009.12.038.

[35] L.A. Fielding, S. Edmondson, S.P. Armes, Synthesis of pH-responsive tertiary amine methacrylate polymer brushes and their response to acidic vapour, J. Mater. Chem. 21 (2011) 11773–11780. Available from: https://doi.org/10.1039/c1jm11412c.

[36] X.Y. Liu, F. Cheng, Y. Liu, H.J. Liu, Y. Chen, Preparation and characterization of novel thermoresponsive gold nanoparticles and their responsive catalysis properties, J. Mater. Chem. 20 (2010) 360–368. Available from: https://doi.org/10.1039/b915313f.

[37] Y. Bi, C. Yan, L. Shao, Y. Wang, Y. Ma, G. Tang, Well-defined thermoresponsive dendritic polyamide/poly(N-vinylcaprolactam) block copolymers, J. Polym. Sci. Part A Polym Chem. 51 (2013) 3240–3250. Available from: https://doi.org/10.1002/pola.26716.

[38] C.K. Sudhakar, N. Upadhyay, A. Jain, A. Verma, R. Narayana Charyulu, S. Jain, Hydrogels-Promising Candidates for Tissue Engineering, Elsevier Inc, 2015. Available from: https://doi.org/10.1016/B978-0-323-32889-0.00005-4.

[39] X. Qi, W. Wei, J. Li, G. Zuo, X. Pan, T. Su, et al., Salecan-based pH-sensitive hydrogels for insulin delivery, Mol. Pharm. 14 (2) (2017) 431−440. Available from: https://doi.org/10.1021/acs.molpharmaceut.6b00875.

[40] M. Zhu, D. Lu, S. Wu, Q. Lian, W. Wang, A.H. Milani, et al., Responsive nanogel probe for ratiometric fluorescent sensing of pH and strain in hydrogels, ACS Macro Lett. 6 (2017) 1245−1250. Available from: https://doi.org/10.1021/acsmacrolett.7b00709.

[41] X. Jiang, D. Xiong, Y. An, P. Zheng, W. Zhang, L. Shi, Thermoresponsive hydrogel of poly(glycidyl methacrylate-co-N-isopropylacrylamide) as a nanoreactor of gold nanoparticles, J. Polym. Sci. Part A Polym Chem. 45 (2007) 2812−2819. Available from: https://doi.org/10.1002/pola.22037.

[42] J.A. Luckanagul, C. Pitakchatwong, P. Ratnatilaka Na Bhuket, C. Muangnoi, P. Rojsitthisak, S. Chirachanchai, et al., Chitosan-based polymer hybrids for thermo-responsive nanogel delivery of curcumin, Carbohydr. Polym. 181 (2018) 1119−1127. Available from: https://doi.org/10.1016/j.carbpol.2017.11.027.

[43] S. Xu, Y. Luo, R. Haag, Water-soluble pH-responsive dendritic core-shell nanocarriers for polar dyes based on poly(ethylene imine), Macromol. Biosci. 7 (2007) 968−974. Available from: https://doi.org/10.1002/mabi.200700066.

[44] Y. Zhu, Y. Ma, Q. Yu, J. Wei, J. Hu, Preparation of pH-sensitive core-shell organic corrosion inhibitor and its release behavior in simulated concrete pore solutions, Mater. Des. 119 (2017) 254−262. Available from: https://doi.org/10.1016/j.matdes.2017.01.063.

[45] M.Q. Zhu, L.Q. Wang, G.J. Exarhos, A.D.Q. Li, Thermosensitive gold nanoparticles, J. Am. Chem. Soc. 126 (2004) 2656−2657. Available from: https://doi.org/10.1021/ja038544z.

[46] A. Lu, D. Moatsou, I. Hands-Portman, D.A. Longbottom, R.K. O'Reilly, Recyclable L-proline functional nanoreactors with temperature-tuned activity based on core-shell nanogels, ACS Macro Lett. 3 (2014) 1235−1239. Available from: https://doi.org/10.1021/mz500704y.

[47] J. Chen, Z. Ding, H. Pan, X. Cao, Development of pH-responsive polymer and citrate aqueous two-phase system for extractive bioconversion of cefprozil, Talanta 174 (2017) 256−264. Available from: https://doi.org/10.1016/j.talanta.2017.05.027.

[48] A.L. Grilo, M.R. Aires-Barros, A.M. Azevedo, Partitioning in aqueous two-phase systems: fundamentals, applications and trends, Sep. Purif. Rev. 45 (2016) 68−80. Available from: https://doi.org/10.1080/15422119.2014.983128.

[49] M. Iqbal, Y. Tao, S. Xie, Y. Zhu, D. Chen, X. Wang, et al., Aqueous two-phase system (ATPS): an overview and advances in its applications, Biol. Proced. Online 18 (2016) 1−18. Available from: https://doi.org/10.1186/s12575-016-0048-8.

[50] P.L. Show, C.P. Tan, M.S. Anuar, A. Ariff, Y.A. Yusof, S.K. Chen, et al., Primary recovery of lipase derived from Burkholderia cenocepacia strain ST8 and recycling of phase components in an aqueous two-phase system, Biochem. Eng. J. 60 (2012) 74−80. Available from: https://doi.org/10.1016/j.bej.2011.10.005.

[51] J. He, J. Wan, T. Yang, X. Cao, L. Yang, Recyclable aqueous two-phase system based on two pH-responsive copolymers and its application to porcine circovirus type 2 Cap protein purification, J. Chromatogr. A 1555 (2018) 113−123. Available from: https://doi.org/10.1016/j.chroma.2018.04.032.

[52] Y.K. Leong, J.C.W. Lan, H.S. Loh, T.C. Ling, C.W. Ooi, P.L. Show, Cloud-point extraction of green-polymers from *Cupriavidus necator* lysate using thermoseparating-based aqueous two-phase extraction, J. Biosci. Bioeng. 123 (2017) 370−375. Available from: https://doi.org/10.1016/j.jbiosc.2016.09.007.

[53] M. Zhang, W. Zhang, Pd nanoparticles immobilized on pH-responsive and chelating nanospheres as an efficient and recyclable catalyst for suzuki reaction in water, J. Phys. Chem. C 112 (2008) 6245−6252. Available from: https://doi.org/10.1021/jp7121517.

[54] Y. Hao, Y. Liu, R. Yang, X. Zhang, J. Liu, H. Yang, A pH-responsive TiO_2-based Pickering emulsion system for in situ catalyst recycling, Chin. Chem. Lett. 29 (2018) 778−782. Available from: https://doi.org/10.1016/j.cclet.2018.01.010.

[55] Y. Dong, Y. Jin, J. Wang, J. Shu, M. Zhang, Pd nanoparticles stabilized by a simple pH-sensitive P(acrylamide-co-acrylic acid) copolymer: a recyclable and highly active catalyst system in aqueous medium, Chem. Eng. J. 324 (2017) 303−312. Available from: https://doi.org/10.1016/j.cej.2017.05.030.

[56] Y. Wang, G. Wei, W. Zhang, X. Jiang, P. Zheng, L. Shi, et al., Responsive catalysis of thermoresponsive micelle-supported gold nanoparticles, J. Mol. Catal. A Chem. 266 (2007) 233−238. Available from: https://doi.org/10.1016/j.molcata.2006.11.014.

[57] J. Tang, P.J. Quinlan, K.C. Tam, Stimuli-responsive Pickering emulsions: recent advances and potential applications, Soft Matter 11 (2015) 3512−3529. Available from: https://doi.org/10.1039/c5sm00247h.

[58] R. Luo, J. Dong, Y. Luo, pH-Responsive Pickering emulsion stabilized by polymer-coated silica nanoaggregates and applied to recyclable interfacial catalysis, RSC Adv. 10 (2020) 42423−42431. Available from: https://doi.org/10.1039/d0ra07957j.

[59] Y. Wang, L. Zhu, H. Zhang, H. Huang, L. Jiang, Formulation of pH and temperature dual-responsive Pickering emulsion stabilized by chitosan-based microgel for recyclable biocatalysis, Carbohydr. Polym. 241 (2020) 116373. Available from: https://doi.org/10.1016/j.carbpol.2020.116373.

[60] G. Yao, W. Bi, H. Liu, pH-Responsive magnetic graphene oxide/poly(NVI-co-AA) hydrogel as an easily recyclable adsorbent for cationic and anionic dyes, Colloids Surf. A Physicochem. Eng. Asp. 588 (2020) 124393. Available from: https://doi.org/10.1016/j.colsurfa.2019.124393.

[61] C. Zhu, Y. Xia, Y. Zai, Y. Dai, X. Liu, J. Bian, et al., Adsorption and desorption behaviors of HPEI and thermoresponsive HPEI based gels on anionic and cationic dyes, Chem. Eng. J. 369 (2019) 863−873. Available from: https://doi.org/10.1016/j.cej.2019.03.169.

[62] X. Wang, Y. Shi, R.W. Graff, D. Lee, H. Gao, Developing recyclable pH-responsive magnetic nanoparticles for oil-water separation, Polym. (Guildf.) 72 (2015) 361−367. Available from: https://doi.org/10.1016/j.polymer.2014.12.056.

[63] K. Hayashi, T. Matsuyama, J. Ida, A simple magnetite nanoparticle immobilized thermoresponsive polymer synthesis for heavy metal ion recovery, Powder Technol. 355 (2019) 183−190. Available from: https://doi.org/10.1016/j.powtec.2019.07.007.

[64] Y. Qin, L. Wang, C. Zhao, D. Chen, Y. Ma, W. Yang, Ammonium-functionalized hollow polymer particles as a pH-responsive adsorbent for selective removal of acid dye, ACS Appl. Mater. Interfaces 8 (2016) 16690−16698. Available from: https://doi.org/10.1021/acsami.6b04199.

[65] L. Li, J. Wang, C. Zhou, R. Wang, M. Hong, pH-Responsive chelating N-heterocyclic dicarbene palladium(II) complexes: recoverable precatalysts for Suzuki−Miyaura reaction in pure water, Green Chem. 13 (2011) 2071−2077. Available from: https://doi.org/10.1039/c1gc15312a.

[66] Z. Zhang, G. Sèbe, X. Wang, K.C. Tam, Gold nanoparticles stabilized by poly(4-vinylpyridine) grafted cellulose nanocrystals as efficient and recyclable catalysts, Carbohydr. Polym. 182 (2018) 61−68. Available from: https://doi.org/10.1016/j.carbpol.2017.10.094.

[67] W.L. Jiang, Q.J. Fu, B.J. Yao, L.G. Ding, C.X. Liu, Y. Bin Dong, Smart pH-responsive polymer-tethered and Pd NP-loaded NMOF as the Pickering interfacial catalyst for one-pot cascade biphasic reaction, ACS Appl. Mater. Interfaces 9 (2017) 36438−36446. Available from: https://doi.org/10.1021/acsami.7b12166.

[68] L. Qi, Z. Luo, X. Lu, Facile synthesis of starch-based nanoparticle stabilized Pickering emulsion: its pH-responsive behavior and application for recyclable catalysis, Green Chem. 20 (2018) 1538−1550. Available from: https://doi.org/10.1039/c8gc00143j.

[69] H. Pan, T. Yang, J. Chen, X. Cao, Biosynthesis of cefprozil in an aqueous two-phase system composed of pH-responsive copolymers and its crystallization analysis, Process. Biochem. 64 (2018) 124−129. Available from: https://doi.org/10.1016/j.procbio.2017.09.015.

[70] S. Bai, J.F. Wan, X.J. Cao, Partitioning of tylosin in recyclable aqueous two-phase systems based on two pH-responsive polymers, Process. Biochem. 87 (2019) 204−212. Available from: https://doi.org/10.1016/j.procbio.2019.09.020.

[71] M. Dai, Y. Liu, B. Ju, Y. Tian, Preparation of thermoresponsive alginate/starch ether composite hydrogel and its application to the removal of Cu(II) from aqueous solution, Bioresour. Technol. 294 (2019) 122192. Available from: https://doi.org/10.1016/j.biortech.2019.122192.

[72] L. Wang, L. Jiang, D. Su, C. Sun, M. Chen, K. Goh, et al., Non-covalent synthesis of thermoresponsive graphene oxide-perylene bisimides-containing poly(N-isopropylacrylamide) hybrid for organic pigment removal, J. Colloid Interface Sci. 430 (2014) 121−128. Available from: https://doi.org/10.1016/j.jcis.2014.05.031.

[73] A. Kozur, L. Burk, R. Thomann, P.J. Lutz, R. Mülhaupt, Graphene oxide grafted with polyoxazoline as thermoresponsive support for facile catalyst recycling by reversible thermal switching between dispersion and sedimentation, Polymer (Guildf.) 178 (2019) 121553. Available from: https://doi.org/10.1016/j.polymer.2019.121553.

[74] L. Lou, H. Qu, W. Yu, B. Wang, L. Ouyang, S. Liu, et al., Covalently immobilized lipase on a thermoresponsive polymer with an upper critical solution temperature as an efficient and recyclable asymmetric catalyst in aqueous media, ChemCatChem 10 (2018) 1166−1172. Available from: https://doi.org/10.1002/cctc.201701512.

[75] C. Xu, W. Dong, J. Wan, X. Cao, Synthesis of thermo-responsive polymers recycling aqueous two-phase systems and phase formation mechanism with partition of ε-polylysine, J. Chromatogr. A 1472 (2016) 44−54. Available from: https://doi.org/10.1016/j.chroma.2016.10.016.

[76] Z. Tan, F. Li, C. Zhao, Y. Teng, Y. Liu, Chiral separation of mandelic acid enantiomers using an aqueous two-phase system based on a thermo-sensitive polymer and dextran, Sep. Purif. Technol. 172 (2017) 382−387. Available from: https://doi.org/10.1016/j.seppur.2016.08.039.

[77] L. Wang, W. Li, Y. Liu, W. Zhi, J. Han, Y. Wang, et al., Green separation of bromelain in food sample with high retention of enzyme activity using recyclable aqueous two-phase system containing a new synthesized thermo-responsive copolymer and salt, Food Chem. 282 (2019) 48−57. Available from: https://doi.org/10.1016/j.foodchem.2019.01.005.

[78] T. Lü, S. Zhang, D. Qi, D. Zhang, G.F. Vance, H. Zhao, Synthesis of pH-sensitive and recyclable magnetic nanoparticles for efficient separation of emulsified oil from aqueous environments, Appl. Surf. Sci. 396 (2017) 1604−1612. Available from: https://doi.org/10.1016/j.apsusc.2016.11.223.

[79] J. Li, Q. Zhou, Y. Wu, Y. Yuan, Y. Liu, Investigation of nanoscale zerovalent iron-based magnetic and thermal dual-responsive composite materials for the removal and detection of phenols, Chemosphere 195 (2018) 472−482. Available from: https://doi.org/10.1016/j.chemosphere.2017.12.093.

[80] B.P. Krishnan, L.O. Prieto-López, S. Hoefgen, L. Xue, S. Wang, V. Valiante, et al., Thermomagneto-responsive smart biocatalysts for malonyl-coenzyme a synthesis, ACS Appl. Mater. Interfaces 12 (2020) 20982−20990. Available from: https://doi.org/10.1021/acsami.0c04344.

CHAPTER 12

Nanomaterials recycling standards

Arsalan Ahmed[1], Muhammad Fahad Arian[1] and Muhammad Qamar Khan[2]

[1]Department of Textile and Clothing, Faculty of Engineering & Technology, National Textile University Karachi Campus, Karachi, Pakistan
[2]Nanotechnology Research Lab, Department of Textile & Clothing, Faculty of Engineering & Technology, National Textile University, Karachi Campus, Karachi, Pakistan

12.1 Introduction

Nanomaterials (NMs) are nanosized particles (<100 nm) that remain in nature. These particles are available in various types of products that are available on the market. Most NMs are too small to be seen with the naked eye and sometimes cannot be seen by using the common laboratory microscopes [1−3].

A nanometer is one millionth of a millimeter, which is approximately 100,000 times smaller than the diameter of a human hair. NMs form naturally as the byproducts of the combustion or reactions of the products, and these products may be described as purposefully fabricated through engineering for specialized functions [4,5].

These materials normally have some physical and chemical properties that are different from those of their bulk form counterparts. Materials that are formed on such a small scale can be referred to as engineered NMs, which can have unique optical, magnetic, electrical, and other properties, which have potential for great impact on electronics, medicine, and other fields:

1. Nanotechnology (NT) can be used to design pharmaceuticals that can be target to the specific organs or cells in the human body, especially cancer cells, to create or increase the effectiveness of treatment.
2. NMs are also used in the cement industry, textile manufacturing, and other applications in which stronger and lighter-weight materials are required.
3. NMs can make electronics more efficient because of their good neutralizing effects in environmental remediation and cleanup to bind and neutralize toxins [1,4,6−8] (Fig. 12.1).

The use of NMs is increasing, and they are becoming very important components for NT. NMs are also significant in healthcare, electronics, cosmetics, and other areas [9]. As will be discussed later in the chapter, their physical and chemical properties are often different from those of bulk materials, which requires special attention in risk assessment. Therefore it is important to determine the risks to workers and consumers and the covering of the potential risks to the environment [10,11]. These properties to be studied include the physical and chemical shown in Table 12.1.

Nanomaterials Recycling
DOI: https://doi.org/10.1016/B978-0-323-90982-2.00012-3

© 2022 Elsevier Inc.
All rights reserved.

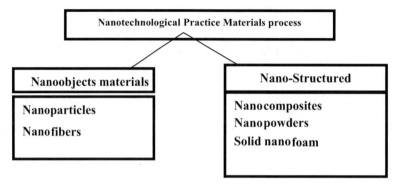

Figure 12.1 Illustrated nanotechnological practice materials.

Table 12.1 Physical and chemical properties of nanomaterials.

Physical properties	Chemical properties
1. The size, shape, specific surface area, and ratio of width and height of the material.	1. The molecular structure of the products.
2. Whether they stick together or not.	2. The composition, including the purity and whether there are known impurities or additives.
3. Depends on the distribution of the size of the numbers.	3. Whether it is held in a solid or liquid and or gas formation.
4. How smooth or bumpy the surface of the materials is.	4. The surface chemistry of the products.
5. The structure, including crystals, to avoid the structure and any crystal defects.	5. How the water attracts to the water molecules of the products in oils and fats [12–14].
6. How well they dissolve in the materials.	

Natural processes have produced nanoparticles (NPs), and modern science has recently learned how to synthesize a bewildering array of artificial materials with structures at the atomic scale [15]. The smallest NMs having a total of tens or hundreds of atoms and dimensions at the scale of nanometers; these are NPs. These NPs are comparable in size to viruses. The smallest NPs have dimensions of tens of the nanometers; by comparison, a human immunodeficiency virus particle is approximately 100 nm in diameter. Thus in the emerging science of NT, they might be called "nanoorganisms" [16,17]. Like viruses, some NPs can penetrate the (lungs) or skin and enter the circulatory system and lymphatic systems of humans and animals, reaching most of the tissues and organs of the body. Whenever there is a disturbing cellular process, it is plausible to make sure what the actual problem is and what factors led to the disease. The toxicity of each of these materials depends largely on the particular arrangement of the atom's quantities. To begin with, consider all of the possible variations, which will be happening in the form of the shape and chemistry of even the smallest NPs consisting of only tens of the atoms. Thus there are a huge number of distinct NMs with potentially very different physical and toxicological effects on the body. Asbestos is a good example of a toxic NM that causes lung cancer and other diseases. Asbestos exists in several forms with the slight variations in shape and chemistry, significantly varying of the toxicity of the asbestos products that are available [18–20].

Every person has been exposed to nanosize foreign particles, as we inhale them with every breath and consume them with every drink. Human skin and lungs and the gastrointestinal tract are is in the constant contact with the environment [21,22].

Skin is generally an effective barrier to the foreign substances; the lungs and gastrointestinal tract are more vulnerable parts of the human body. The skin, lungs, and gastrointestinal system are the three most likely points of entry for natural or anthropogenic NPs into the human body. Injections and implants are other possible routes of exposure, which are primarily limited to engineered materials that may affect to the body as a result of their small size. These NPs can be translocated from their entry points into the circulatory and lymphatic systems of the body and ultimately to body tissues and organs. Some NPs, depending on their composition and size, ad can produce irreversible damage to cells by oxidative stress or and organelle injury. In truth, every organism on earth continuously encounters nanosized entities. The vast majority of them cause little ill effect on the body and may go unnoticed, but occasionally an intruder will cause appreciable harm to the organism. The most advanced types of toxic intruders are viruses, which are composed of nuclear acid-based structures that allow them not only to interfere with the biological systems of the body but also to parasitically exploit cellular processes to replicate themselves [19,23].

Among the more benign viruses are the ones that cause the familiar symptoms of the common cold or flu, which occurring as a result of the operation of the immune systems of the body, whose nanosized constituents (chemicals and proteins) usually destroy and remove the viral invaders. However, a growing number of recent studies demonstrate that nanoorganisms and microorganisms may play a vital role in many chronic diseases in which infectious pathogens have not been suspected, diseases that were previously attributed only to genetic factors and lifestyle of the patient. Nanosized particles are created in countless physical processes from erosion to combustion with health risks ranging from lethal to benign. The industrial NP materials today constitute a tiny but significant pollution source that is literally laid to rest beneath much larger-sized natural sources, and the NPs have effects on the pollution incidental to other human activities, particularly automobile exhaust, which are very harmful for the human body [19,24].

12.2 Fundamental of nanoparticles

NPs are formed through the breaking down of larger particles or by controlling the adjoining the assembly processes. Natural phenomena and many human industrial and domestic activities, such as cooking food products, the manufacturing, and road and air transport, release NPs into the atmosphere [25–27].

In recent years, NPs that are intentionally used in large capacities and for use in advanced technologies and consumer products have become a new source of the exposure in the marketplace. At present, it is not clear how significantly human exposure to these types of NPs has increased in the workplace and or through the use of NT-based products [1,2,28].

There are two approaches to the manufacturing of NMs:
1. The top-down approach involves the breaking down of large pieces of material to generate NPs. This method is particularly suitable for making interconnected and integrated structures, such as in the electronic circulatory system.
2. In the bottom-up approach, single atoms and molecules are assembled into larger nanostructures. This is a very powerful method of creating identical structures with atomic precision, although to date, the human-made materials generated in this way are still much simpler than natural complex structures (Fig. 12.2).

12.3 Classification of nanoparticles

NPs can be classified into different types according to the size of the classes, such as morphology and physical and chemical properties. Some of them are carbon-based NPs, ceramic NPs, metal NPs, semiconductor NPs, polymeric NPs, and lipid-based NPs [30].

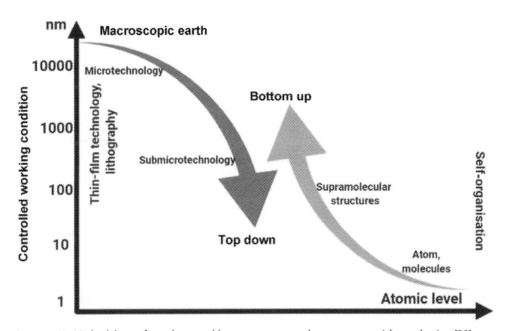

Figure 12.2 Methodology of top-down and bottom-up approaches to nanoparticle production [29].

12.3.1 Carbon-based nanoparticles

The two main carbon-based NPs are fullerenes and carbon nanotubes (CNTs). Fullerenes are the allotropes of the carbon having a structure of a hollow cage of 60 or more carbon atoms. The structure of C60 is called Buckminsterfullerene, and it would look like a hollow football. The carbon units in these structures have pentagonal and hexagonal arrangements. These are attractive for commercial applications because of their electrical conductivity, structure, high strength, and electron affinity.

CNTs are nothing but graphene sheets are rolled into tubes. These materials are used mainly for structural reinforcement, as they are 100 times stronger than the steel.

CNTs can be classified as single-walled carbon nanotubes or multiwalled carbon nanotubes. CNTs are unique in being thermally conductive along with the lengthwise nonconductive across the tube area [31,32].

12.3.2 Ceramic nanoparticles

The ceramic NPs are inorganic solids made up of oxides, carbides, carbonates, and phosphates. These NPs have high heat resistance and chemical inertness. They have applications in photocatalysis and photodegradation of dyes, drug delivery, and imaging [33].

Ceramic NPs can be used for drug delivery by controlling some of their characteristics, such as size, surface area, porosity, and surface to volume ratio [34]. These NPs have been used are effectively as a drug delivery system in treating bacterial infections, glaucoma, cancer, for example [33].

12.3.3 Metal nanoparticles

Metal NPs can be synthesized by chemical, electrochemical, and photochemical methods. In chemical methods the metal NPs are obtained by reducing the metal ion precursors in solution by chemical reducing agents. These have the ability to absorb small molecules and have high surface energy. These NPs have applications in research areas, detection and imaging of biomolecules, and environmental and bioanalytical applications [35,36].

12.3.4 Semiconductor nanoparticles

Semiconductor NPs have the properties between those of metals and nonmetals. They are used in the making of photocatalysts and electronics devices and in photooptics and for the separations of water-splitting applications. These metals are arranged by capacity in the periodic table in groups II−VI, III−V, or IV−VI [37].

12.3.5 Polymeric nanoparticles

Polymeric NPs are organic-based NPs in shapes that depend upon the method of preparation. They may be shaped like nanocapsules or nanospheres. A nanosphere particle has a matrix-like structure, whereas a nanocapsular particle has a core-shell morphology. In the former, the active compounds and the polymer are uniformly dispersed; in the latter, the active compounds are confined and surrounded by a polymer shell [38].

Some of the merits of polymeric NPs are controlled release, protection of drug molecules, and the ability to do combined therapy and the specific targeting segments. They have applications in drug delivery and diagnostics purposes. Drug deliveries with polymeric NPs are highly biodegradable and biocompatible formations [39,40].

12.3.6 Lipid-based nanoparticles

Lipid-based NPs are generally spherical in shape with a diameter ranging from 10 to 100 nm. They consist of a solid core made of lipid and a matrix containing the soluble lipophilic molecules in shapes [41]. The external core of these NPs is stabilized by surfactants and emulsifiers. These NPs are used in the biomedical field as drug carriers and for delivery and RNA release in cancer therapy [42] (Fig. 12.3).

12.4 Scope of nanotechnology

NT is considered the next step in the logical integration of technology-based science with other disciplines, including biology, chemistry, and physics. The Royal Society and Royal Academy of Engineering have defined *nanoscience* as the study of phenomena and manipulation of materials at the atomic, molecular, and macromolecular

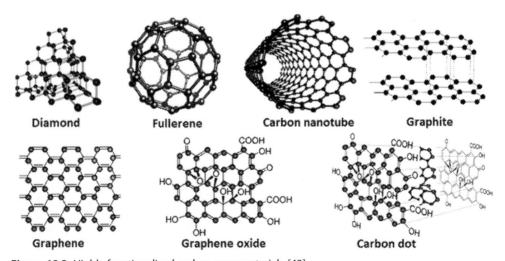

Figure 12.3 Highly functionalized carbon nanomaterials [43].

scales, while *NT* has been defined as the design, characterization, production, and application of structures devices and systems by controlling shape and size at the nano scale [43].

A current focus of NT is the building of devices at a microscopic scale or even the size of the molecules that could potentially benefit medicine, environmental protection, energy levels, and space exploration. In the last few years the term *NT* has become almost synonymous with objects that are innovative and highly promising [44]. The more sophisticated and precise description of NT would be the manipulation of the matter with at least one dimension of size from 1 to 100 nm. In recent years these NMs embody distinctive physical, chemical, and biological properties compared to their conventional counterparts, which will be the most likely the beneficial characteristics [45].

Researchers know that different NPs are evolving and NMs are increasingly seen in daily use and will be important parts of daily life in the form of cosmetics, food packaging, and drug delivery, biosensors, therapeutics systems, and so on. Unprecedented avenues for exposure of the environment and living beings to NPs are increasing day by day [46]. The increasing exposure to NMs makes it imperative to assess the toxic effect of NP-based materials. As far as the physical and chemical characteristics of NMs are concerned, the influence of properties of NPs are more important to evaluate than physicochemical properties of the NMs, including size, to measure its surface area to check whether its soluble effects are present [47]. To check whether chemical compositions are available, it is important to determine the shape of the substance, its agglomeration state, the shape of its crystal structure, the value of the surface energy, its surface charge, the morphology of the surface area, type of its surface coating, and the role of individual characteristic properties in imparting toxic manifestations. The materials are then checked and analyzed in an attempt to correlate their physicochemical properties with the toxicity of the NMs present in the materials [43].

These NPs may have toxic manifestations through diverse mechanisms and may cause and create allergies. They may also lead to fibrosis, organ failure, nephrotoxicity, hematological toxicities, neurotoxicity, herpetological toxicities, splendid toxicities, and pulmonary toxicities, among others [48,49].

12.5 Recycling of nanomaterials

The number of new commercial products containing NMs is growing. Many of these products are in the personal care, cosmetics, and sunscreen categories. The leftovers of other products, their packaging, and other products containing NMs, such as electronic equipment, textiles, and composite plastics, end their life cycle as municipal or industrial waste. They are disposed of by recycling, energy recovery, waste incineration, or landfilling. Recycling of NMs generally has a higher priority than incineration and/or landfill [50].

Currently these products are containing NMs are being recycling along with their analogous products without the NMs. No separation or separate collection of products containing surely NMs solely to their NM contents. The new existing recycling techniques do not allow them the possibility to the nano specific risks coming from waste containing from the NMs.

Recycling operations are not always helpful when it comes to NMs, as they are carried out by using the techniques that meet the general standards of an environmentally sound waste management process for products that do not containing NMs. These operations may not be enough to manage waste NPs in a safe and environmentally sound manner, which may give rise to other problems [51,52].

Because of the high value of the specially manufactured NMs that consist of nanoscaled metals, there is strong interest among manufacturers and users in recovering certain NMs for reuse or recharging of the products. Owing to the lack of information about NMs, the cases would normally be treated and declared as hazardous waste products, which could lead to the loss of valuable materials. It would be useful if the recycling process included the potential to separate NPs from the waste stream to recycle them. The majority of the testing processes are conventional forms of separation techniques, such as centrifugation or solvent evaporation, which demand high energy use. Alternative process methods, such as the application of magnetic fields, pH, and thermoresponsive materials of molecular antisolvents or nanostructures solvents, will provide effective and the efficient methods for the recycling of NPs without significant costs and large time or energy consumption. However, to set up the efficient recycling of NMs, further studies of their intrinsic recyclability properties, such as thermal, mechanical, and chemical properties, are needed. Information about how these characteristics may change once the NMs are mixed with other products is also needed, as well are guidelines regarding disassembly and reuse in some cases. Improvement in the disassembly of these types of the materials would also help in identifying the appropriate reuse and recycling process options. Other technological solutions for NMs make use of magnetically recoverable supports for the immobilization of gold NP catalysts, which will guarantees facile, clean, fast, and efficient separation of the catalyst at the end of the reaction cycle [50,52].

12.6 The importance of recycling in waste management

The importance of recycling waste is one of the elements in strategies of waste minimization or waste prevention and sustainable material management. Individual countries and international bodies are setting recycling goals and are taking the steps and measures to reach them. In the European Union (EU) several legal provisions oblige the member states to recycle municipal waste products and all household waste products [53].

The United States Environmental Protection Agency issued a statement that the proper regulation of household items, industrial manufacturing units, commercial solid waste products, and hazardous wastes would be the under the Resource Conservation and Recovery Act. Effective solid waste management is a cooperative effort that involves federal agencies, state departments, and regional and local entities [54].

According to the Organisation for Economic Co-operation and Development, the average recycling process rate, including composting, for municipal waste was 33% in 2011, with a range from <10% to 63%. In the EU the overall rate of material recovery (recycling plus composting) amounts to 42%, the range in individual countries is given as going from 2% to 70% [50,55].

In addition, there is a growing interest in the recycling process to recover metals and the minerals of secondary raw materials from the bottom ash (slag) of municipal solid waste incinerators. Based on the different sources, including the main waste streams in the community and industrial waste, the solid waste stream also includes the following:

1. Biowaste products
2. Food waste products
3. Plastic and glass (bottles)
4. Metal items
5. Paper and cardboard materials
6. Plastic materials such as bottles and jars
7. Leather and textile goods
8. Waste electrical and electronic items
9. All types of batteries
10. Wooden products
11. Construction demolition and wastes products
12. End-of-life vehicles
13. Rejected tires and accessories
14. Recycling of residues from waste incineration plants, including the recovery of metals from bottom ash by mechanical separation of fly ash by acid washing.

The processes in which nanowaste products streams are generated include production plants, distribution units, and handling materials as well as disposal. It is anticipated that the production and number of NM applications will increase the waste streams containing NMs will also increase. Along with accumulation of naturally occurring NPs, NMs are likely become more widespread in the atmosphere so an understanding of the fate and associated risks of NMs released from waste treatment operations such as recycling processes is particularly important to avoid inadequate management of these materials [50,55,56].

12.7 Uses of nanomaterials

Owing to the ability to generate materials in a particular way, NMs can play specific roles in uses across various industries from healthcare and cosmetics to environmental preservation and air purification.

The healthcare field utilizes NMs in different ways; one of the major uses is in drug delivery systems. There is one example of this process whereby NPs are being developed to assist in the transportation of chemotherapy drugs directly to cancerous growths as well as to deliver drugs to areas of damaged arteries that are in patients with cardiovascular disease. CNTs are also being developed to be used in processes such as the addition of antibodies to the nanotubes to act as bacteria sensors [57,58].

12.8 Uses of nanomaterials in consumer products

NPs can contribute as stronger, lighter, cleaner and "smarter" surfaces and systems. They are already being use in the manufacture of scratchproof eyeglasses, crack-resistant paints, antigraffiti coatings for walls, transparent sunscreens, stain-repellent fabrics, self-cleaning windows, and ceramic coatings for solar cells.

NTs are also being used to increase the safety of cars. NPs can improve the formation of tires, reducing the stopping distance in wet conditions. The stiffness of car bodies can also be improved by use of NPs to strengthen the steel, Moreover, ultrathin transparent coatings can be applied to displays or panes to avoid glare or condensation. In the future it may be possible to produce transparent car body parts [59].

NTs can also be applied in the processing of food and in food packaging, resulting in food safety. NMs may be used to place antimicrobial agents on coated films and to modify gas permeability as required for different lines of the products. NMs are also being used in biology and medicine in a wide variety of ways. For example, they are used in products for drug delivery and gene therapy, tissue engineering, and DNA probes and in the nanoscaled products known as biochips [60].

12.9 Nanomaterials and the industries they are used in

The use of NMs is prevalent in a wide range of industries and in consumer products. In the cosmetics industry, mineral NPs such as titanium oxide are used in sunscreens. Although bulk titanium oxide shows poor stability in conventional chemical ultraviolet (UV) protection, titanium oxide NPs are able to provide improved UV protection while having the added advantage of removing the cosmetically unappealing whitening associated with sunscreen.

The sports industry has been producing baseball bats that have been made lighter through the use of CNTs, thus improving their performance. Furthermore, the use of NMs in this industry can be identified in the use of the antimicrobial NT in items such as bath towels and room mats and to prevent illnesses caused by bacteria [61,62].

NMs have also been developed to be used in the military. One example is the use of mobile pigment NPs to produce a better formation of camouflage through the injection of the particles into the material of soldiers' uniforms. In addition, the military have been developing sensor systems that use NMs such as titanium dioxide that can detect biological agents.

Titanium dioxide NPs are also being used in coatings to form self-cleaning surfaces of products such as plastic garden chairs. This involves a sealed film of water created on the coating of NMs, and any dirt dissolves in the film, after which the next shower will remove the dirt, essentially cleaning the chairs.

12.10 Categories of wastage of nanomaterials

The increasing number of innovative NMs that are finding application in establishing the value and product chains are leading to an increased interest from industry, academia, governments, and stakeholders looking at life cycle assessment of NMs and how nanowaste is generated and handled. The Nanotechnologies Industry Association takes part in these discussions and brings forward the views of the NT industries.

There are three broad origins of waste from NMs with categorizations similar to those of conventional waste:

1. Waste generated during the production of NMs or NT- derived products.
2. Waste generated as a result of usage of NMs or NT-derived products (e.g., abrasion, degradation).
3. Waste generated through end-of-life activities (e.g., recycling, incineration, and landfill) [54,63].

12.11 Nanowaste ecotoxicology and treatment

Communities are reluctant to embrace new technologies when there is inadequate disclosure of the potential impacts on health or safety. Molecular genetics and NT are the two examples of such technologies that have generated extensive debate regarding the potential public and environmental health impacts. A far as NT is concerned, there has been potential for toxicity and there is a lack of characterization of data and management of nanowaste that needs to be addressed before community acceptance of the potential benefits of the NT can occur [50].

The human body and the ecosystem have been as the focus of the impacts of anthropogenic waste and most of the air pollution. There are few published studies that have investigated the impact on aquatic ecosystems, and they are limited primarily to the effects of C60, nanotubes, and titanium dioxide, with the crustacean *Daphnia magna* being the main test for the species. There are many new questions that might be raised about NMs. For example, how can exposure to the NMs is measured? What are the potential health benefits of NMs? What are the potential environmental effects of NMs? How well can we assess the risks from NMs? The answers to these questions are a faithful summary of the scientific opinion produced in 2006 by the Scientific Committee on Emerging and Newly Identified Health Risks (SCENIHR). This is the modified opinion (after public consultation) on the appropriateness of existing methodologies to assess the potential risks associated with engineered and adventitious products of nanotechnologies. SCENIHR concluded that NMs may have different (eco) toxicological properties than the substances in bulk form; therefore their risks need to be assessed on a case-by-case basis [54,64,65]. The main risks and benefits of NMs are shown in Table 12.2.

12.12 Waste generated during production

According to Živković, "The European Commission has been launched thematic studies exploring nanomaterials (NMs) and the EU environmental legislation in 2011, a report published in the accordance with the Coherence of Waste Legislation in the EU. The Consumer participation in this debate during the early stages of nanoproduct design may be help to the address information needs concerning the product life cycle requirements. The ratings of nano-products could be to introduce which exhibit in varying degrees of end-of-product, life practices and information provided for the local government recycling deposits. The Nano-manufacturers could be to engage to the consumers throughout this product life cycle in order to learn about their experiences, including the potential of misusing of these products whether deliberately or unintentionally" [54].

The current guidelines suggest that nanowaste should be classified as hazardous waste and should be, at a minimum, double bagged. It should be encompassed in a rigid, impermeable container, and it should preferably be bound within a solid matrix. However, nanoobject capture and containment must be considered as an overarching priority for operative nanowaste management development. Atmospheric dispersion of nanoobjects is occurring, and nanoobjects are being translocated to distant ecosystems as a result of deposition from surrounding research materials and nanomanufacturing facilities. To avoid potential damage to the ecosystem, the impacts of the parameters such as nanoobject size, configuration, and solubility are required for the assessment of each phase of the nanoproduct life cycle before routine disposal methods can be considered. Unless it is securely enclosed in an impermeable container or bound within a solid matrix, contaminant runoff and the atmospheric dispersion remain potential

Table 12.2 Nanomaterials risks and benefits [54].

Applications	Risk	Benefit
Nanocrystals harvest light in photovoltaic devices.	Light pollution in rural areas, opportunity cost to fossil fuel economies.	Green renewable energy new self-lighting displays for electronic devices.
An antimicrobial wound dressing contains nanocrystalline silver ion.	It has to release of antimicrobials into the environment, hazard to natural microbial systems.	Improved healing of wounds and reduced risk of infection.
Sunscreens containing titanium dioxide NMs are extremely effective at absorbing UV light.	Titanium hazard to intertidal organisms and sandy shore ecosystems.	Consumer preference for transparent but effective sunscreen. Potential decrease in skin cancer due to increased sunscreen use.
Metal NM supplements to fuels increase their burn efficiency.	Respiratory exposure to NMs in fuel exhausts. Long-range transport of particles in the atmosphere.	Less soot from diesel vehicles and aviation fuels. Economic savings for transport infrastructure on fuel costs. Reduced greenhouse gases. Less urban air pollution. Burn efficient.
Medical applications of hydroxyapatite and nanosilica applications in bone reconstruction.	Durability of particles eroded from the surface may cause pathology in other internal organs in the long term.	Structural repairs to teeth and bone using a natural material already in the body (no adverse immune response).
NMs used in the food packaging industry.	Unintended transfer of NM from the packaging to the food. Uncertain lifetime oral exposure risk.	Stronger, lighter packaging to protect soft foods. Antibacterial packaging to improve shelf life. Increased food safety.
Use of CNTs in sports equipment improves its strength and flexibility.	Life cycle analysis, what happens to their use? The materials end up as waste landfills.	Better product that lasts longer for the consumer. Reduced sports- related injuries.
Use of NMs as a catalyst in industrial processes such as coal liquefaction and producing gas.	Inadvertent incorporation of toxic catalysts in consumer products, waste disposal of catalytic converters to landfill.	Improved efficiency and economy of industrial and processes. Less industrial waste per ton of production.
Use of NMs using in the water filtration and purification industry.	Unintended waterborne exposure of wildlife to engineered NMs.	New sources of safe drinking water in poor regions of the world. More efficient water purification systems. Reduced exposure to waterborne pathogenic organisms and toxins.

12.13 Disposal and recycling of nanomaterials in waste

The disposal of NMs and the products in which they are used should be performed with particular care to ensure that the NMs do not pose a threat to human health and the environment in which they are being released. NMs that are considered to be hazardous materials in terms of toxicity or chemical reactivity should be neutralized, and other types of nanowaste should be recycled if possible. The nanowaste can be the result or byproduct of industrial or commercial processes, and because of the broad range of existing NMs, a single procedure for the disposal will not be suffice for all classes of NMs. Hence it is important to understand the properties of the specific nanowastes in question before developing effective disposal practices. In the developing safety measures, the disposal procedures necessary for handling nanowaste must be based on current knowledge and take into the account existing legislation. The disposal procedures must ensure that the waste is deactivated of its hazardous properties. This will depend on the type of the material and thermal, chemical, or physical processing of NT-containing possible waste and deactivation solutions [54,66].

12.14 Risks related to nanomaterials in waste

During the recycling of waste containing NMs, engineered NPs, like any other particles in the recycling process, can remain individually isolated or may form new, bigger piles. Potential exposure exists mainly in connection with free nanoobjects that are in the nano scale, that is, smaller than 100 nm in all three or in two spatial dimensions (e.g., NPs, nanofibers, and nanorods) [54].

1. NPs may penetrate biological barriers.
2. NPs may show intensified effects in case of substances in which the toxic properties are combined.
3. NPs may have increase due to the bioavailability of the products.
4. NPs may have different chemical and physical properties from the material available in the market.
5. NPs in some items, such as CNT and nanowires, may have effects in the lungs similar to those of asbestos fibers.
6. There may be a serious risk of dust explosions in all applications available in the market that contain inflammable powders or powdery substances.

Thus mainly possible exposures must be taken into the account in managing waste containing NMs, especially nanoobjects that are free or that are releasable in the recycling process [50,67].

12.15 Nanomaterials in recycling operations and potential exposure

NMs can potentially enter the body in three ways:
1. Through the skin.
2. Through the gastrointestinal tract.
3. Through the respiratory or inhalation process.

On the basis of current knowledge, the most important exposure route for NPs is through the inhalation process. It is unlikely that NPs would penetrate healthy skin, although it still is a good idea to wear protective gloves (e.g., nitrile gloves) in occupational handling NPs. To avoid exposure through the gastrointestinal tract, food and drinks should be never be ingested in the same laboratory or room where NPs are handled [54,68].

If they are inhaled, NPs have a high probability of being deposited in the lungs. How much is deposited depends on the particles' properties, such as size, volume, and physiological factors; the route into the lungs, such as oral or nasal inhalation; and whether inhalation occurred during exercise or while resting [69]. While the impaction and interception are factors that are more dominant for micro-sized particles, depositions of nanosized particle are more driven by diffusion in the particular at low air speed, such as in the alveolar sacs and surface charge. An important consideration in the deposition of particles in the respiratory system is the lung lining fluid, which is a complex mixture of lipids and the surfactant proteins, since with any depositing material it quickly becomes coated. Such coating gives the deposited particles a biological identity, known as the protein version, which is likely to play a role in the way particles interact with the lung's cells, such as the alveolar macrophages. Deposition in the deeper part of the lungs, the alveolar tract, is cleared by the macrophages, or engulfed particles are repositioned in the alveolar tract. However, if they fail in doing so, the particles can be translocated from their primary organ of entrance in the lungs to secondary organs via the circulatory system. If they are sited in the tracheobronchial tract, particles are cleared by the muscles that transport particles up to the pharynx, after which they are swallowed and thereby enter the gastrointestinal tract. Among other parameters, the translocation rate and fraction translocated to other organs depend on particle size, morphology, surface parameters such as composition, charge, and primary and secondary coatings with proteins, lipids, and functional groups.

Inhaled NPs can also be translocated along the sensory neuronal pathways to reach secondary organs and tissues such as the vascular endothelium process, heart, and brain tissues [70].

12.16 Nanotechnology is dangerous for humans

NT involves the manipulation of matter at nanoscale to produce new materials, structures, and devices. NMs are designed and manufactured at the nanoscale through NT. In commercial processes, products that incorporate NMs are growing rapidly, and these tiny products are increasingly found in the paint, fabrics, cosmetics products, wooden products, electronics, and sunscreen products. NMs have unique properties that make them different from chemical substances that are larger in size. NMs are providing the opportunities for the development of innovative products in the technologies and medicine [71].

Generally, the use of NT results in product attributes that are different from those of conventionally manufacturing products, and assessments of the safety and efficiency of the regulated products involving NT are necessary. The focus on the distinctive qualities and behaviors of NMs make them a special case. However, the regulatory bodies do not judge all products involving the application of NT. The agencies regulate NT products under their current purview and in accordance with the applicable legal standards.

The overarching purpose of NT is to create better circumstances for humans and the environment. Industries using NMs can help communities and the environment in general to solve the energy crisis and to diagnose and cure diseases. NT may provide cheap and easy water purification methods, cheap energy, faster computers, longer-lasting mobile phones, more efficient batteries, surgical and other implants, diagnostics procedures for diseases, better surgical tools, and targeted deliveries of medicines. The substances used in NT might pose a risk to the environment in terms of the need to make sure that handling of these new particles and materials are done safely. As a matter of fact, the things that makes NPs so interesting are their nanosize and the fact that they have properties, such as optical, chemical, magnetic, biological, electrical, and mechanical ones, that are completely different from the properties of the same materials in bulk. However, these new properties may cause problems if NPs are unintentionally released. Very few of these particles exist in nature, and we are as humans have not been exposed to them throughout the evolutionary process. Therefore we cannot be sure that our bodies have developed the defense mechanisms to deal with ongoing consequences in the future [54,72].

12.17 Possible dangers of nanotechnology

The minimum size of NMs and the way in which their surfaces are modified increase the chances that they will interact with biological systems. NMs vary in characteristics that make them attractive for applications in the pharmaceutical industry and other industries in which NMs may damage the human body and the ozone layer.

NPs are mostly dangerous for three main reasons:

1. NPs damages the lungs. The ultrafine particles from fuel burned in diesel machines, power plants, and incinerators are harmful and can cause the considerable damage to human lungs. The danger is due their size because they can get deep into the lungs and because they carry other chemicals, including metals and hydrocarbon particles that are mixed in the flames.
2. NPs can get into the body through the pores of the skin and go directly into the lungs and the digestive system. These conditions will create free radicals, which can cause damage to cell and DNA. There is also a concern that once NPs are in the bloodstream, they will be able to cross the blood-brain barrier and directly affect to the brain cells.
3. The human body develops a tolerance to substances through naturally occurring processes. When it comes into contact with new elements or molecules for which it has no natural immunity, it is more likely to find them the materials toxic.

The danger of contact with NPs is not just speculation [73]. As more research is undertaken into these conditions and the concerns are increasing, there are some recent findings that may clarify the scenario:

1. There are some NPs that cause lung damage in rats, and several studies have shown that CNTs, which are similar in shape to asbestos fibers, may cause mesothelioma in the lungs of rats.
2. Other NPs have been shown to lead to brain damage in fish and dogs.
3. A German study found clear evidence that if discrete nanometer diameter particles were deposited in the nasal region, they completely circumvented the blood−brain barrier and traveled up the olfactory nerve system and might go straight into the brain cells.
4. Inhaled CNTs can suppress the immune system by affecting the function of the T cells and a type of white blood cell that organizes the immune system to fight infections.

12.18 Conclusion

This chapter discussed NMs, generation of nanowaste from different applications disposal, and recycling of NMs. Top-down and bottom-up concepts of production can be applied to NMs with the main focus on NM recycling. Several approaches and applicable methods were also covered in this chapter for efficient functional performance of the NMs along with their respective recycling life. Overall, this chapter provides a comprehensive discussion of the recycling of NMs, the importance of nanowaste recycling standards, ecotoxicology and treatment techniques, NMs recycling operations, and the possible dangers of NT. This discussion should be beneficial for future researchers and technologists dealing with NMs.

References

[1] J. Jeevanandam, et al., Review on nanoparticles and nanostructured materials: history, sources, toxicity and regulations, Beilstein J. Nanotechnol. 9 (1) (2018) 1050–1074.

[2] M. Kamali, et al., Sustainability criteria for assessing nanotechnology applicability in industrial wastewater treatment: current status and future outlook, Environ. Int. 125 (2019) 261–276.

[3] A. Ahmed, et al., Preparation of PVDF-TrFE based electrospun nanofibers decorated with PEDOT-CNT/rGO composites for piezo-electric pressure sensor, J. Mater. Sci. Mater. Electron. 30 (15) (2019) 14007–14021.

[4] C. Binns, Introduction to Nanoscience and Nanotechnology, vol. 14, John Wiley & Sons, 2010.

[5] D.L. Schodek, P. Ferreira, M.F. Ashby, Nanomaterials, Nanotechnologies and Design: An Introduction for Engineers and Architects, Butterworth-Heinemann, 2009.

[6] A.S. Ali, Application of nanomaterials in environmental improvement, in: M. Sen (Ed.), Nanotechnology and Environment, IntechOpen, London, 2020.

[7] M.H. Fulekar, B. Pathak, R.K. Kale, Nanotechnology: perspective for environmental sustainability, in: M. Fulekar, B. Pathak, R. Kale (Eds.), Environment and Sustainable Development, Springer, New Delhi, 2013, pp. 87–114. Available from: https://doi.org/10.1007/978-81-322-1166-2_7.

[8] F.D. Guerra, et al., Nanotechnology for environmental remediation: materials and applications, Molecules 23 (7) (2018) 1760.

[9] Z. Abdullaeva, Nano- and Biomaterials: Compounds, Properties, Characterization, and Applications, John Wiley & Sons, 2017.

[10] A. Ehrmann, T.A. Nguyen, P.N. Tri, Nanosensors and Nanodevices for Smart Multifunctional Textiles, Elsevier, 2020.

[11] C. Sanchez, et al., Applications of advanced hybrid organic–inorganic nanomaterials: from laboratory to market, Chem. Soc. Rev. 40 (2) (2011) 696–753.

[12] G. Lofrano, G. Libralato, J. Brown, Nanotechnologies for Environmental Remediation, Springer, 2017.

[13] S.P. Patil, V.V. Burungale, Physical and chemical properties of nanomaterials, in: N.D. Thorat, J. Bauer (Eds.), Nanomedicines for Breast Cancer Theranostics, Elsevier, 2020, pp. 17–31. Available from: https://doi.org/10.1016/B978-0-12-820016-2.00002-1.

[14] S. Kumbar, S. Patil, C. Jarali, Chapter Three – Graphene in different extraction techniques, in: C.M. Hussain (Ed.), Analytical Applications of Graphene for Comprehensive Analytical Chemistry, vol. 91, Elsevier, 2020. Available from: https://doi.org/10.1016/bs.coac.2020.09.001.

[15] I. Khan, K. Saeed, I. Khan, Nanoparticles: properties, applications and toxicities, Arab. J. Chem. 12 (2017) 908.

[16] X. Xu, J. Jia, M. Guo, The most recent advances in the application of nano-structures/nano-materials for single-cell sampling, Front. Chem. 8 (2020) 718. Available from: https://doi.org/10.3389/fchem.2020.00718.

[17] C. Xu, et al., Waste-to-wealth: biowaste valorization into valuable bio (nano) materials, Chem. Soc. Rev. 48 (18) (2019) 4791–4822.

[18] T.A. Saleh, Nanomaterials for pharmaceuticals determination, Bioenergetics 5 (226) (2016) 2.

[19] C. Buzea, I.I. Pacheco, K. Robbie, Nanomaterials and nanoparticles: sources and toxicity, Biointerphases 2 (4) (2007) MR17–MR71.

[20] A.F. Gualtieri, et al., Structure model and toxicity of the product of biodissolution of chrysotile asbestos in the lungs, Chem. Res. Toxicol. 32 (10) (2019) 2063–2077.

[21] H.-R. Paur, et al., In-vitro cell exposure studies for the assessment of nanoparticle toxicity in the lung—a dialog between aerosol science and biology, J. Aerosol Sci. 42 (10) (2011) 668–692.

[22] W. Yao, et al., Human exposure to particles at the air-water interface: influence of water quality on indoor air quality from use of ultrasonic humidifiers, Environ. Int. 143 (2020) 105902.

[23] H. Meng, et al., Walking the line: the fate of nanomaterials at biological barriers, Biomaterials 174 (2018) 41–53.

[24] D. Herrera, et al., Biofilms around dental implants, in: Z. Artzi (Ed.), Bone Augmentation by Anatomical Region: Techniques and Decision-Making, Wiley, 2020, pp. 487–504.

[25] S. Panicker, et al., On demand release of ionic silver from gold-silver alloy nanoparticles: fundamental antibacterial mechanisms study, Mater. Today Chem. 16 (2020) 100237.

[26] H.-M. Ding, W.-D. Tian, Y.-Q. Ma, Designing nanoparticle translocation through membranes by computer simulations, ACS Nano 6 (2) (2012) 1230−1238.

[27] O. Adir, et al., Integrating artificial intelligence and nanotechnology for precision cancer medicine, Adv. Mater. 32 (13) (2020) 1901989.

[28] A. Dhasmana, et al., Nanoparticles: applications, toxicology and safety aspects, in: K. Kesari (Ed.), Perspectives in Environmental Toxicology. Environmental Science and Engineering, Springer, Cham, 2017, pp. 47−70. Available from: https://doi.org/10.1007/978-3-319-46248-6_3.

[29] P. Patel, et al., Cellular and molecular impact of green synthesized silver nanoparticles, in: S.M. Avramescu (Ed.), Engineered Nanomaterials − Health and Safety, IntechOpen, London, 2019.

[30] I. Ijaz, et al., Detail review on chemical, physical and green synthesis, classification, characterizations and applications of nanoparticles, Green. Chem. Lett. Rev. 13 (3) (2020) 223−245.

[31] M. Azizi-Lalabadi, et al., Carbon nanomaterials against pathogens; the antimicrobial activity of carbon nanotubes, graphene/graphene oxide, fullerenes, and their nanocomposites, Adv. Colloid Interface Sci. 284 (2020) 102250.

[32] S. Soleymani Eil Bakhtiari, et al., Polymethyl methacrylate-based bone cements containing carbon nanotubes and graphene oxide: an overview of physical, mechanical, and biological properties, Polymers 12 (7) (2020) 1469.

[33] R. Chaim, et al., Grain growth during spark plasma and flash sintering of ceramic nanoparticles: a review, J. Mater. Sci. 53 (5) (2018) 3087−3105.

[34] C. Thomas, S.P. Kumar Mishra, S. Talegaonkar, Ceramic nanoparticles: fabrication methods and applications in drug delivery, Curr. Pharm. Des. 21 (42) (2015) 6165−6188.

[35] A. Rana, K. Yadav, S. Jagadevan, A comprehensive review on green synthesis of nature-inspired metal nanoparticles: mechanism, application and toxicity, J. Clean. Prod. 272 (2020) 122880.

[36] M.J. Ndolomingo, N. Bingwa, R. Meijboom, Review of supported metal nanoparticles: synthesis methodologies, advantages and application as catalysts, J. Mater. Sci. 55 (2020) 1−47.

[37] M.K. Sahu, Semiconductor nanoparticles theory and applications, Int. J. Appl. Eng. Res. 14 (2) (2019) 491−494.

[38] M.B. Tahir, et al., Role of nanotechnology in photocatalysis, Ref. Module Mater. Sci. Mater. Eng. (2020). Available from: https://doi.org/10.1016/B978-0-12-815732-9.00006-1.

[39] J. Priya, et al., A review on polymeric nanoparticles: a promising novel drug delivery system, J. Glob. Pharma Technol. 10 (4) (2018) 10−17.

[40] B. Begines, et al., Polymeric nanoparticles for drug delivery: recent developments and future prospects, Nanomaterials 10 (7) (2020) 1403.

[41] M. Mirahadi, et al., A review on the role of lipid-based nanoparticles in medical diagnosis and imaging, Ther. Deliv. 9 (8) (2018) 557−569.

[42] L.M. Ickenstein, P. Garidel, Lipid-based nanoparticle formulations for small molecules and RNA drugs, Expert. Opin. Drug. Deliv. 16 (11) (2019) 1205−1226.

[43] Q.-L. Yan, et al., Highly energetic compositions based on functionalized carbon nanomaterials, Nanoscale 8 (9) (2016) 4799−4851.

[44] T.A. Saleh, Nanomaterials: classification, properties, and environmental toxicities, Environ. Technol. Innov. 20 (2020) 101067.

[45] X. Chen, J. Li, Superlubricity of carbon nanostructures, Carbon 158 (2020) 1−23.

[46] R. Rao, et al., Carbon nanotubes and related nanomaterials: critical advances and challenges for synthesis toward mainstream commercial applications, ACS Nano 12 (12) (2018) 11756−11784.

[47] P. Ganguly, A. Breen, S.C. Pillai, Toxicity of nanomaterials: exposure, pathways, assessment, and recent advances, ACS Biomater. Sci. Eng. 4 (7) (2018) 2237−2275.

[48] L. Ding, et al., Nanotoxicity: the toxicity research progress of metal and metal-containing nanoparticles, Mini Rev. Med. Chem. 15 (7) (2015) 529−542.

[49] H. Zazo, C.I. Colino, J.M. Lanao, Current applications of nanoparticles in infectious diseases, J. Control. Release 224 (2016) 86−102.

[50] T. Faunce, B. Kolodziejczyk, Nanowaste: need for disposal and recycling standards. G20 Insights, Policy Era: Agenda 2030, 2017.

[51] M.G. Wacker, A. Proykova, G.M.L. Santos, Dealing with nanosafety around the globe—regulation vs. innovation, Int. J. Pharm. 509 (1–2) (2016) 95–106.

[52] M. Cossutta, J. McKechnie, Environmental impacts and safety concerns of carbon nanomaterials, in: S. Kaneko, et al. (Eds.), Carbon Related Materials, Springer, Singapore, 2020, pp. 249–278. Available from: https://doi.org/10.1007/978-981-15-7610-2_11.

[53] C. Reynolds, et al., Consumption-stage food waste reduction interventions—what works and how to design better interventions, Food Policy 83 (2019) 7–27.

[54] D. Živković, et al., Nanomaterials environmental risks and recycling: actual issues, Recycling Sustain. Dev. 7 (1) (2014) 1–8.

[55] A.T. Besha, et al., Sustainability and environmental ethics for the application of engineered nanoparticles, Environ. Sci. Policy 103 (2020) 85–98.

[56] M.S. Bilgili, P. Agamuthu, A New Issue in Waste Management: Nanowaste, Sage Publications, London, 2019.

[57] B.A. Duguay, et al., The possible uses and challenges of nanomaterials in mast cell research, J. Immunology 204 (8) (2020) 2021–2032.

[58] U.A. Ashfaq, et al., Recent advances in nanoparticle-based targeted drug-delivery systems against cancer and role of tumor microenvironment, Crit. Rev. Ther. Drug. Carrier Syst. 34 (4) (2017) 317–353. Available from: https://doi.org/10.1615/CritRevTherDrugCarrierSyst.2017017845.

[59] M. Hamid Nazeer, Effects of nanoaprticles on reproductive health, 2017.

[60] S.K. Ameta, et al., Use of nanomaterials in food science, in: M. Ghorbanpour, P. Bhargava, A. Varma, D. Choudhary (Eds.), Biogenic Nano-Particles and their Use in Agro-Ecosystems, Springer, Singapore, 2020, pp. 457–488. Available from: https://doi.org/10.1007/978-981-15-2985-6_24.

[61] S. Kaul, et al., Role of nanotechnology in cosmeceuticals: a review of recent advances, Review J Pharm (Cairo) 2018 (2018) 3420204. Available from: https://doi.org/10.1155/2018/3420204.

[62] C. Kingston, et al., Release characteristics of selected carbon nanotube polymer composites, Carbon 68 (2014) 33–57.

[63] P.C. Ray, H. Yu, P.P. Fu, Toxicity and environmental risks of nanomaterials: challenges and future needs, J. Environ. Sci. Health Part. C 27 (1) (2009) 1–35.

[64] A. Baun, et al., Ecotoxicity of engineered nanoparticles to aquatic invertebrates: a brief review and recommendations for future toxicity testing, Ecotoxicology 17 (5) (2008) 387–395.

[65] K. Khosravi-Katuli, et al., Effects of nanoparticles in species of aquaculture interest, Environ. Sci. Pollut. Res. 24 (21) (2017) 17326–17346.

[66] A. Caballero-Guzman, T. Sun, B. Nowack, Flows of engineered nanomaterials through the recycling process in Switzerland, Waste Manage. 36 (2015) 33–43.

[67] M.P. Tsang, et al., Evaluating nanotechnology opportunities and risks through integration of life-cycle and risk assessment, Nat. Nanotechnol. 12 (8) (2017) 734.

[68] S. Kamali, S. Sanajou, M.N. Tazehzadeh, Nanomaterials in construction and their potential impacts on human health and the environment, Environ. Eng. Manage. J. (EEMJ) 18 (11) (2019) 2305–2318.

[69] H. Qiao, et al., The transport and deposition of nanoparticles in respiratory system by inhalation, J. Nanomater. 96 (2015) 1–8.

[70] T. Praphawatvet, J.I. Peters, R.O. Williams III, Inhaled nanoparticles—an updated review, Int. J. Pharm. 587 (2020) 119671. Available from: https://doi.org/10.1016/j.ijpharm.2020.119671.

[71] P.J. Borm, et al., The potential risks of nanomaterials: a review carried out for ECETOC, Part. Fibre Toxicol. 3 (1) (2006) 11.

[72] C. Coussens, L. Goldman, Implications of Nanotechnology for Environmental Health Research, National Academies Press, 2005.

[73] L. Handojo, et al., Application of nanoparticles in environmental cleanup: production, potential risks and solutions, in: R. Bharagava (Ed.), Emerging Eco-friendly Green Technologies for Wastewater Treatment. Microorganisms for Sustainability, vol. 18, Springer, Singapore, 2020, pp. 45–76. Available from: https://doi.org/10.1007/978-981-15-1390-9_3.

CHAPTER 13

Ionic liquids for nanomaterials recycling

Hani Nasser Abdelhamid[1,2]
[1]Advanced Multifunctional Materials Laboratory, Department of Chemistry, Faculty of Science, Assiut University, Assiut, Egypt
[2]Proteomics Laboratory for Clinical Research and Materials Science, Department of Chemistry, Assiut University, Assiut, Egypt

13.1 Introduction

Room-temperature ionic liquids (ILs) are liquid salts with melting points below $100°C$ and low vapor pressure or non-volatility under ambient conditions [1]. Molten salt, ethyl ammonium nitrate, was considered the first prepared ILs in 1914 with a melting point of $12°C$. ILs exhibit high thermal stability (to over $400°C$), offering non-volatile solvents, leading to reduced solvent emission compared to conventional volatile organic compounds. It exhibited a low-viscosity, tunable properties such as solubility, acidity, or basicity, long-range thermal stability, and very low corrosivity relative to mineral acids and bases [2]. There are several ILs, such as imidazolium-based ILs, including 1-butyl-3-methyl imidazolium bis(triflylmethylsulfonyl)imide ([BMIM][Tf$_2$N]). ILs are considered supramolecular fluids, owing to their applicability for forming extended hydrogen bond networks in the liquid state. This phenomenon is entropically driven, leading to spontaneous process, and is called the IL effect. Another type of ILs is known as polyionic liquids (P-ILs) [3]. Both types of ILs offer several advantages, making them promising as solvents or media for various reactions such as polymerization, alkylation, Beckmann rearrangement, esterification, acidic hydrolysis, carbonization, depolymerization, and extractions. They were have also been used for applications such as membrane [4] and surfactants [5].

Nanotechnology is the technology of dwarf particles [nanoparticles (NPs)] within a nanometer range dimension ($1 nm = 10^{-9} m$). NPs have unique properties compared to bulk materials of the same composition. They offer characteristic features of catalytic activity [6−11], artificial nanozymes [12−14], and antimicrobial activity [15,16]. They have been used for several application [17−22,23−55,56−79]. They have large surface areas and can be modified with inorganic reagents [80,81], organic molecules [82], and biomolecules [83]. The investment in nanotechnology is increasing exponentially and is expected to reach several trillion dollars in the future.

ILs and P-ILs were used for the synthesis of NPs [84,85]. They exhibit low interfacial tension and complexing ability, leading to controlled synthesis of nanomaterials [86]. ILs such as 1-pentyl-3-methyl-imidazolium bromide and 1-ethyl-3-methylimidazolium

Nanomaterials Recycling
DOI: https://doi.org/10.1016/B978-0-323-90982-2.00024-X

© 2022 Elsevier Inc.
All rights reserved.

tetrafluoroborate were used as templates for the synthesis of NPs preventing aggregation [87]. This property leads to an increase in the material's porosity and the average pore size [87]. The synthesis of NPs using ILs offered ligand-free NPs using hydrogen-driven chemical methods [88] and can be achieved via the simple hydrogenation and decomposition of zero-valent metal complexes or a physical method such as magnetron sputtering [88]. The synthesis of nanomaterials is still expensive. Therefore recycling is highly important [89].

13.2 Scope of ionic liquids

ILs have been implemented in basic and applied research. The number of publications on ILs has developed exponentially (Fig. 13.1). A literature survey indicates that ILs offered a valuable addition to science with their wide range of potential applications in the fields of chemistry, biology, and physics. Ethyl ammonium nitrate [EtNH$_3$][NO$_3$] with a melting point of 12°C was considered the first IL in the literature in 1914 [90]. In 1940, a research team at Rice Institute of Texas invented a salt consisting of the addition of alkyl pyridinium chloride and aluminum chloride. The resultant salt was liquid at room temperature. Since that date, many activities have been reported to explore the properties and applications of ILs.

1. The chemistry and structure of ILs
2. IL salts are a combination of organic cations (e.g., imidazolium, pyridinium, ammonium, and phosphonium) and organic or inorganic anions (e.g., CF$_3$COO, HSO$_4$, and Cl) (Fig. 13.2). ILs are green solvents.
3. ILs are liquids at room temperature over a wide range of temperatures.
4. They have low vapor pressure under ambient conditions (non-volatile).
5. ILs have excellent lubricating and hydraulic properties.

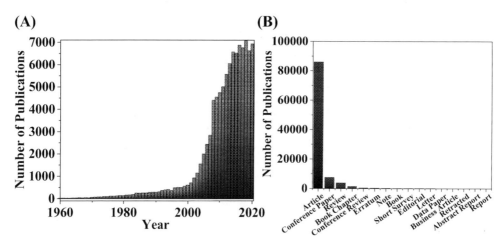

Figure 13.1 (A) Number of publications over the years. (B) Types of publications. *Data were extracted from Scopus (March 2021) using the search term "ionic liquids."*

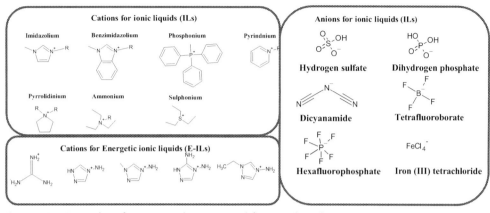

Figure 13.2 Examples of cations and anions used for ionic liquids.

6. ILs have tunable properties of acidity and basicity.
7. Most of the reported ILs are colorless and exhibit a nature of polar solvents.
8. ILs can uptake gases and control their release.
9. Most of the reported ILs exhibit a hydrophilic nature.
10. The viscosity of most of the reported ILs is very low.
11. ILs offer high solubility for many species, including supramolecular.
12. There are several ILs with a wide range of cations and anions.

ILs are claimed to be green solvents. As a rule of thumb, a liquid or solvent is called "green" if its production requires minimal environmental impacts over the entire life cycle of the compound [91]. The greenness of a solvent or liquid can be evaluated by using several factors, including the atom efficiency [92], and environmental factor [93]. The synthesis of ILs such as 1-methylimidazolium halides generated a stoichiometric amount of waste [94]. This was considered a green solvent. However, the purification steps are not included in the assessment. The synthesis of 1-methylimidazolium halides using conventional synthesis procedures and purification methods require excess amounts of the starting materials and organic solvents. The energy needed for the synthesis procedure should also be considered in the greenness assessment. Thus, modern methods such as microwave irradiation (MWI) and ultrasound-assisted reactions should be investigated. These methods require low energy, enhance the reaction rates, offer high selectivity of the desired product, and produce an extensive product quantity compared to traditional energy methods. The greenness of the IL production can also be improved via solvent-free synthesis procedures, which generate less harmful waste.

13.3 Synthesis of ionic liquids

The synthesis of ILs can be achieved via several methods (Fig. 13.3) [95]. ILs such as ethyl ammonium nitrate can be synthesized via the addition of concentrated nitric acid

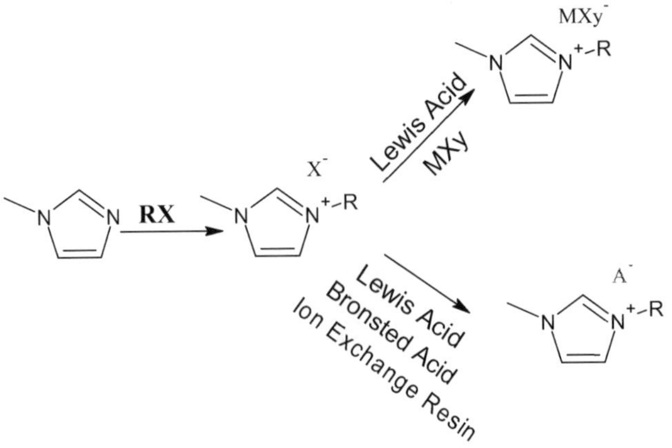

Figure 13.3 Synthesis procedure of ionic liquids using imidazole as an example.

with ethylamine. The undesirable aqueous layer can be removed via distillation. The synthesis of ILs can be achieved via a single step. The counter cation parts of ILs can be performed by using anion exchange reactions between halides and metal halides to form ILs with Lewis acidic properties. Ion exchange resin—based ILs can be applied to obtain Brønsted acidic ILs. ILs suffer from impurities such as volatile, halide, and cations. Thus, purification methods are usually required to ensure high purity.

13.4 Types of ILs

ILs can be classified into several types (Fig. 13.4). They can be classified as chiral ILs, bioionic liquids (Bio–ILs), P-ILs, energetic ionic liquids (E-ILs), neutral ionic liquids (N-ILs), protic ionic liquids (Pr-ILs), supported ionic liquids (S-ILs), basic ionic liquids (B-ILs), and metallic ionic liquids (M-ILs). Chiral ILs are ILs that contain chiral moieties. They are important for applications such as liquid chiral chromatography, stereoselective polymerization, synthesis of potential active chiral compounds, and liquid crystals. Bio–ILs are ILs without toxic moieties, such as alkyl imidazolium and alkyl benzimidazolium. They are biodegradable. P-ILs are polymer ILs with repeating motifs of monomeric units of ILs. E-ILs are ILs with energetic cations such as 4–amino–1–methyltriazolium, 4–amino–1–ethyltriazolium, 4–aminotriazolium, guanidinium, 4–amino–1–butyltriazolium, and 1,5–diaminotetrazolium. They offer several advantages compared to conventionally energetic compounds such as 2,4,6–trinitrotoluene (TNT), 1,3,5,7–tetranitro–1,3,5,7–tetraazocane (HMX), and 1,3,5–(tris–nitro)perhydro–1,3,5–triazine (RDX). They have high density, high thermal stability, negligible vapor pressure, and negligible vapor toxicity. They require a simple synthesis procedure and can be easily transported with high safety. Neutral ILs are ILs with weak electrostatic interactions between anionic and cations moieties. Acidic ILs contain moieties with acidic groups such as $COOH$, SO_3H, and PO_3H.

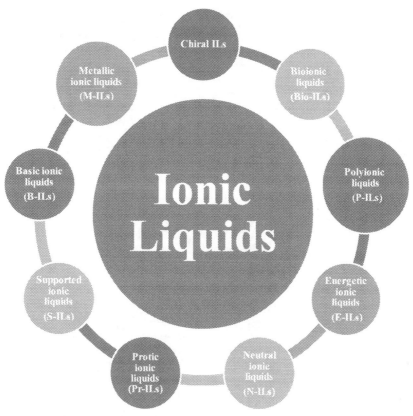

Figure 13.4 Types of ionic liquids.

13.5 Recycling ionic liquid

Recycling ILs is important. The recycling process may reduce the cost of the ILs. It is also vital for recovery materials to dissolve into ILs. The hydrophilic and hydrophobic properties of ILs play an essential role in recycling ILs [96]. Hydrophobic ILs can be easily extracted from water. This property enables the simple separation of water-soluble solute from the aqueous phase. The recovery of hydrophilic ILs is difficult compared to hydrophobic ILs.

Hydrophilic ILs can be recovered by using methods such as supercritical CO_2, membrane technology, such as nanofiltration membranes (STARME^{MTM}) [97], distillation, extraction, adsorption, induced phase separation, crystallization, and physical force fields (Table 13.1). The advantages and disadvantages of each method are tabulated in Table 13.1. Extraction is usually used for nonvolatile products. On the other side, distillation is helpful for volatile solutes. Both methods are the most common methods for the recycling and reuse of ILs. Advanced technologies, such as

Table 13.1 Summary of methods used for recovery and recycling of ionic liquids.

Technique	Example	Advantages	Disadvantages
Distillation		Simple, rapid, and robust method	Energy consuming, Partial decomposition
Extraction	Solvent extraction	Simple, no complex equipment, controlled recovery, selectivity, and flexibility	Emulsion formation, not efficient, loss of compounds, a complicated, laborious, preconcentration step required
	CO_2 extraction	Green process, high extraction quality, efficient, selective, minimized product degradation, eliminates solvent residues	High cost, technical skill required
Adsorption	Adsorption/ desorption process	Robust, relatively easy to operate	Requires equilibrium adsorption and desorption data; desorption solvent
	Chromatography	Robust, easy to operate, efficient, nondisruptive, selective	Equipment complexity, preconcentration step required
Induced phase separation	Salting-out process	Simple, effective, inexpensive	Environmental problems of high inorganic salt
	Carbon dioxide	Green process, CO_2 can be recycled, minimized degradation	High cost, technical skill required
	Temperature	Simple, ease of operation, less energy consumption	Restricted to some unique ILs system
Membrane-based process	Nanofiltration	Simple, less energy and solvent demand	Relatively low flux and recovery yield
	Electrodialysis	Low energy consumption, osmotic pressure not a limiting factor	Elaborate control required, material for membranes and stack is important
	Reverse osmosis	Low energy requirement, compact and lower space requirement, modular design	Requires pretreatment of mixture, limited by osmotic pressure

(Continued)

Table 13.1 (Continued)

Technique	Example	Advantages	Disadvantages
	Pervaporation	Effective for concentrating ILs, highly selective, scalable for quantitatively recovering volatile solutes from nonvolatile solvent	Requires large membrane area
	Vacuum membrane distillation	Effective for concentrating ILs, highly selective, scalable for quantitatively recovering volatile solutes or solvents directly from nonvolatile solvent	Requires large membrane area, may face fouling
Crystallization	Solution crystallization	High purity	Energy consuming
	Melt crystallization	High purity, simple	Requires controlled temperature
	Pressure-induced crystallization		Requires controlled pressure
Physical methods	Magnetic separation	Simple, low energy consumption	Applicable for ILs that respond to a magnetic field
	Centrifugation		Less effective, low processing rate

membrane-based processes, including filtration, pervaporation, reverse osmosis, and electrodialysis, are promising for recovering ILs (Table 13.1). Adsorption such as ion exchange was also reported. The adsorption of ILs onto Amberlite IR 120Na resin was found to be pseudo-second-order adsorption [98]. The adsorption of ILs using an adsorbent is practical and offers high performance in terms of adsorbate amount and cost [99]. A combination of more than one method can also be used [100].

The vacuum membrane distillation technique uses a membrane such as polyacrylonitrile-based hydrophobic membrane to separate 1-butyl-3-methylimidazolium chloride ([Bmim]Cl) at high concentration (≥ 20 wt.%) [101]. The effects of membrane properties such as elemental composition, zeta potential, and roughness were investigated for the same IL ([Bmim]Cl) [102]. The authors reported that the repulsion was due to

Lifshitz—van der Waals and electrostatic interaction components. On the other side, the attraction was due to the acid-base nteraction component. The membrane's roughness decreased the interaction energy barrier and increased the fouling risk [102]. The high negative zeta potential of membranes enhanced the recovery of ILs via non-covalent interactions such as electrostatic interactions [103].

13.6 The applications of ionic liquids for recycling

ILs have been applied for the recovery of drugs, NPs, and biopolymers such as cellulose. The applications of ILs are promising. ILs offer high efficiency and green technology for recovery. They can be applied for extraction techniques such as microextraction. IL-assisted microextraction can be used for the analysis of pesticide residues in samples such as fruits and vegetables [104]. ILs such as tetrabutylammonium chloride, 1-butyl-3-methylimidazolium chloride, and benzyl dimethyl(2-hydroxyethyl) ammonium chloride have been used for the recovery of non-steroidal anti-inflammatory drugs such as ibuprofen, naproxen, and ketoprofen [105].

ILs have been advanced the synthesis of nanomaterials [106,107]. They can be used for the synthesis procedure in various forms such as micelles, vesicles, gels, emulsions, and microemulsions [107]. ILs improved the synthesis of carbon-based nanomaterials. ILs- carbon nanomaterial hybrids show unique properties due to both components' synergistic effects [108]. ILs were used as precursors to prepare carbon materials via the carbonization at ambient pressure. They offered carbon-based nanomaterials with multiple heteroatoms, such as N, P, and S. They were used for biomass-derived carbon-based nanomaterials such as carbon dots. They can be used for the synthesis of nanocomposite based on carbon nanomaterials. Metal NPs such as ruthenium or rhodium supported on chemically derived graphene surfaces were synthesized via the decomposition of metal carbonyl precursors in ILs 1-butyl-3-methylimidazolium tetrafluoroborate [109]. The process was performed by using MWI. The synthesis procedure offered small and uniform particle sizes (2.2 ± 0.4 nm for Ru and 2.8 ± 0.5 nm for Rh) [109]. ILs such as 1-ethyl-3-methylimidazolium bis(trifluoromethylsulfonyl)amide ($[emim][NTf_2]$) improved the contact angles of graphene, leading to the high performance of the materials [110]. A nanocomposite of iridium (Ir) and graphene was synthesized in 1-butyl-3-methyl-imidazolium tetrafluoroborate ([BMIm][BF_4]) via the decomposition of $Ir_4(CO)_{12}$ using MWI or electron-beam (e-beam) irradiation (IBA Rhodotron accelerator) [111]. Carbon nanomaterials such as graphene oxide [112] and graphene [113] can be covalently modified by using ILs. ILs improved the synthesis procedure and offered uniform particle size for the synthesized NPs. They can be considered green synthesis methods. They can also be applied to the synthesis of several categories of inorganic, organic, and hybrid materials.

ILs improved the synthesis procedure of metallic NPs. An IL $[BMIM][PF_6]$ was used for the synthesis of single-crystalline gold dendrites via the reaction between a zinc plate and a solution of HAuCl₄ [114]. Metallic NPs such as Ru NPs and Ni NPs were synthesized via the reduction and decomposition of solution of organometallic complexes [Ru(COD)(2-methylallyl)₂] and [Ni(COD)₂] (COD = 1,5-cyclooctadiene) in the presence of imidazolium ILs [115]. This method requires no reducing agents [115]. The synthesis of metallic NPs using ILs can be used for the synthesis of several morphologies such as spherical [86], dendrites [114], nanorods [116], and hierarchical gold-platinum nanodendrites [117]. ILs offered good solvation properties via the ions' charged moieties leading to controlled morphology of the synthesis NPs [118]. The synthetic NP can be easily extracted and recycled using ILs.

Metal hydroxide and oxide NPs can be synthesized and recycled using ILs. ILs were used to synthesize a family of amorphous transition metal hydroxides, such as FeNi hydroxide [119]. The method is a simple, one-step protocol and offers an ultrasmall size (2−3 nm) [119]. Mn_5O_8 nanoplates with bivalent properties were synthesized via a microwave using ILs [120]. ILs offered several interactions, such $\pi-\pi$ stacking between imidazolium rings of 1-butyl-3-methyl-imidazolium chloride and hydrogen bonds between Cl anions and hydrohausmannite. These interactions improved the synthesis of bivalent Mn_5O_8 nanoplates. MnO_2 can be synthesized via phase and morphology transformation of MnO_2 [121]. The use of ILs tunes the wettability of metal oxides such as TiO_2 [122]. ILs assist and control the morphology of inorganic nanocrystals [123] and offer the morphology of hierarchically olive-like Al_2O_3 and CeO_2/Al_2O_3 [124] and SnO microflowers [125]. ILs were used for the synthesis of magnetic NPs [126]. A freshly made or recycled 1-butyl-3-methylimidazolium bis(triflylmethylsulfonyl)imide ([BMIM][Tf₂N]) was applied for the synthesis of iron oxide (Fe_3O_4) magnetic NPs [127]. The synthesis involves the dissolving of iron pentacarbonyl ($Fe(CO)_5$) in [BMIM][Tf₂N] followed by thermal decomposition. The synthesized Fe_3O_4 NPs can be separated out simply from the imidazolium-based ILs via an external magnet. The material can also be synthesized by using a freshly made and 60% (v/v) recycled [BMIM][Tf₂N] IL [127]. ILs can be used as a sacrificial precursor. IL n-octylammonium hypophosphite was used as a source for phosphorus to synthesize nickel metaphosphate ($Ni_2P_4O_{12}$) [128]. It acted as both reactant and solvent to synthesize $Ni_2P_4O_{12}$ via a one-step calcination approach [128].

Recycling cellulose by using ILs is promising. The process is called IonoSolv. The recovery of cellulose can be performed for pure cellulose films [129], hemicellulose [130], biomass [130,131], hemicellulose fraction of wheat straw into pentoses [132], and river red gum [133]. ILs such as 1-butyl-3methylimidazolium chloride (BmimCl) were applied to recover cellulose with recovery rates of 95.0 wt.% [129]. 1-Allyl-3-methylimidazolium chloride ([Amim][Cl]) was used to recycle cellulose from waste papers [134]. The regenerated cellulose was converted to carbon dots for bioimaging applications [134]. 1-Butyl-3-methylimidazolium chloride ([BMIM]Cl) was used for the recovery of cellulose nanocrystals from cotton fibers [135]. The extraction method can be achieved in two stages: swelling treatment and hydrolysis using acid conditions [135]. The method involves hydrolysis under mild conditions, using only 1 wt.% of H_2SO_4. This procedure involves 60 times less acid than the traditional extraction method based on concentrated H_2SO_4 [135]. BMImCl was used for the extraction of cellulose nanocrystals from cotton gin motes [136]. ILs can be implemented in different extraction methods, such as acid and ultrasonic [137].

Applications of ILs as solvent (IonoSolv) enabled the separation of cellulose-rich material from river red gum [133]. Pyridinium-based IL offers a one-pot deconstruction and conversion of lignocellulose into reducing sugars [138]. ILs enhanced the bioconversion for the pretreatment of lignocellulosic biomass [139]. 1-Ethyl-3-methylimidazolium acetate ([C2C1Im][OAc]) IL showed separation factors as high as 1500 [140]. The method provided recovery of >99.9 wt.% [C2C1Im][OAc] from aqueous solution (≤ 20 wt.% IL) with recyclability over five times [140].

ILs were applied for other materials such as porous materials [141], mesoporous organic polymers [142], and proteins [143]. ILs such as imidazolium provide hydrophobic or hydrophilic regions (polar and nonpolar regions) [144,145]. They produce a protective layer on the NP's surface, preventing aggregation and agglomeration due to steric and electronic protection [146].

13.7 Conclusions

The applications of ILs in the preparation and recovery of several inorganic and hybrid materials are promising. ILs can be generally applied for several materials, such as metals, nonmetal elements, silicas, metal oxides, chalcogenides, and porous materials. ILs offer high recyclability and can be cost-effective after optimization, making them attractive and receiving a growing number of scientists. ILs can be considered green solvents. The methods that implement ILs are sustainable and could be effective in comparison to conventional methods. Future research will explore the synthesis of cheap ILs, making use of cost-effective methods.

References

[1] P. Wasserscheid, W. Keim, Ionic liquids-new "solutions" for transition metal catalysis, Angew. Chem. Int. Ed. 39 (2000) 3772−3789. Available from: https://doi.org/10.1002/1521-3773 (20001103)39:21 < 3772::AID-ANIE3772 > 3.0.CO;2-5.

[2] T. Welton, Room-temperature ionic liquids. Solvents for synthesis and catalysis, Chem. Rev. 99 (1999) 2071−2084. Available from: https://doi.org/10.1021/cr980032t.

[3] W. Qian, J. Texter, F. Yan, Frontiers in poly(ionic liquid)s: syntheses and applications, Chem. Soc. Rev. 46 (2017) 1124−1159. Available from: https://doi.org/10.1039/C6CS00620E.

[4] X. Yan, S. Anguille, M. Bendahan, P. Moulin, Ionic liquids combined with membrane separation processes: a review, Sep. Purif. Technol. 222 (2019) 230−253. Available from: https://doi.org/10.1016/j.seppur.2019.03.103.

[5] A. Bera, J. Agarwal, M. Shah, S. Shah, R.K. Vij, Recent advances in ionic liquids as alternative to surfactants/chemicals for application in upstream oil industry, J. Ind. Eng. Chem. 82 (2020) 17−30. Available from: https://doi.org/10.1016/j.jiec.2019.10.033.

[6] H.N. Abdelhamid, M.N. Goda, A.E.-A.A. Said, Selective dehydrogenation of isopropanol on carbonized metal—organic frameworks, Nano-Struct. Nano-Objects 24 (2020) 100605. Available from: https://doi.org/10.1016/j.nanoso.2020.100605.

[7] A.A. Kassem, H.N. Abdelhamid, D.M. Fouad, S.A. Ibrahim, Hydrogenation reduction of dyes using metal-organic framework-derived CuO@C, Micropor. Mesopor. Mater. 305 (2020) 110340. Available from: https://doi.org/10.1016/j.micromeso.2020.110340.

[8] A.A. Kassem, H.N. Abdelhamid, D.M. Fouad, S.A. Ibrahim, Catalytic reduction of 4-nitrophenol using copper terephthalate frameworks and CuO@C composite, J. Environ. Chem. Eng. (2020) 104401. Available from: https://doi.org/10.1016/j.jece.2020.104401.

[9] H.N. Abdelhamid, High performance and ultrafast reduction of 4-nitrophenol using metal-organic frameworks, J. Environ. Chem. Eng. 9 (2021) 104404. Available from: https://doi.org/10.1016/j.jece.2020.104404.

[10] H.N. Abdelhamid, UiO-66 as a catalyst for hydrogen production via the hydrolysis of sodium borohydride, Dalton Trans. 49 (2020) 10851—10857. Available from: https://doi.org/10.1039/D0DT01688H.

[11] H.N. Abdelhamid, Hierarchical porous ZIF-8 for hydrogen production via the hydrolysis of sodium borohydride, Dalton Trans. 49 (2020) 4416—4424. Available from: https://doi.org/10.1039/D0DT00145G.

[12] H.N. Abdelhamid, W. Sharmoukh, Intrinsic catalase-mimicking MOFzyme for sensitive detection of hydrogen peroxide and ferric ions, Microchem. J. 163 (2021) 105873. Available from: https://doi.org/10.1016/j.microc.2020.105873.

[13] H.N. Abdelhamid, G.A.-E. Mahmoud, W. Sharmoukh, Correction: a cerium-based MOFzyme with multi-enzyme-like activity for the disruption and inhibition of fungal recolonization, J. Mater. Chem. B 8 (2020) 7557. Available from: https://doi.org/10.1039/D0TB90139C.

[14] H.N. Abdelhamid, G.A.-E. Mahmoud, W. Sharmouk, A cerium-based MOFzyme with multi-enzyme-like activity for the disruption and inhibition of fungal recolonization, J. Mater. Chem. B 8 (2020) 7548—7556. Available from: https://doi.org/10.1039/D0TB00894J.

[15] M.S. Yousef, H.N. Abdelhamid, M. Hidalgo, R. Fathy, L. Gómez-Gascón, J. Dorado, Antimicrobial activity of silver-carbon nanoparticles on the bacterial flora of bull semen, Theriogenology 161 (2021) 219—227. Available from: https://doi.org/10.1016/j.theriogenology.2020.12.006.

[16] S. Kumaran, H.N. Abdelhamid, N. Hasan, H.-F. Wu, Cytotoxicity of palladium nanoparticles against *Aspergillus niger*, Nanosci. Nanotechnol. Asia 10 (2020) 80—85. Available from: https://doi.org/10.2174/2210681208666180904113754.

[17] H.-F. Wu, J. Gopal, H.N. Abdelhamid, N. Hasan, Quantum dot applications endowing novelty to analytical proteomics, Proteomics 12 (2012) 2949—2961. Available from: https://doi.org/10.1002/pmic.201200295.

[18] H.N. Abdelhamid, H.-F. Wu, A method to detect metal-drug complexes and their interactions with pathogenic bacteria via graphene nanosheet assist laser desorption/ionization mass spectrometry and biosensors, Anal. Chim. Acta 751 (2012) 94—104. Available from: https://doi.org/10.1016/j.ac.2012.09.012.

[19] H.N. Abdelhamid, S. Kumaran, H.-F. Wu, One-pot synthesis of $CuFeO_2$ nanoparticles capped with glycerol and proteomic analysis of their nanocytotoxicity against fungi, RSC Adv. 6 (2016) 97629—97635. Available from: https://doi.org/10.1039/C6RA13396G.

[20] H.N. Abdelhamid, H.-F. Wu, Selective biosensing of Staphylococcus aureus using chitosan quantum dots, Spectrochim. Acta A Mol. Biomol. Spectrosc. 188 (2018) 50—56. Available from: https://doi.org/10.1016/j.saa.2017.06.047.

[21] H.N. Abdelhamid, Lanthanide Metal-Organic Frameworks and Hierarchical Porous Zeolitic Imidazolate Frameworks: Synthesis, Properties, and Applications, Stockholm University, Faculty of Science, Stockholm, 2017.

[22] H.N. Abdelhamid, H.-F. Wu, Synthesis and multifunctional applications of quantum nanobeads for label-free and selective metal chemosensing, RSC Adv. 5 (2015) 50494—50504. Available from: https://doi.org/10.1039/C5RA07069D.

[23] M. Zulfajri, H.N. Abdelhamid, S. Sudewi, S. Dayalan, A. Rasool, A. Habib, et al., Plant part-derived carbon dots for biosensing, Biosensors 10 (2020) 68. Available from: https://doi.org/10.3390/bios10060068.

[24] H.N. Abdelhamid, H.-F. Wu, Multifunctional graphene magnetic nanosheet decorated with chitosan for highly sensitive detection of pathogenic bacteria, J. Mater. Chem. B 1 (2013) 3950−3961. Available from: https://doi.org/10.1039/c3tb20413h.

[25] H.N. Abdelhamid, H.-F. Wu, Proteomics analysis of the mode of antibacterial action of nanoparticles and their interactions with proteins, TRAC Trend Anal. Chem. 65 (2015) 30−46. Available from: https://doi.org/10.1016/j.trac.2014.09.010.

[26] H.N. Abdelhamid, H.-F. Wu, Probing the interactions of chitosan capped CdS quantum dots with pathogenic bacteria and their biosensing application, J. Mater. Chem. B 1 (2013) 6094−6106. Available from: https://doi.org/10.1039/c3tb21020k.

[27] M. Manikandan, H. Nasser Abdelhamid, A. Talib, H.-F. Wu, Facile synthesis of gold nanohexagons on graphene templates in Raman spectroscopy for biosensing cancer and cancer stem cells, Biosens. Bioelectron. 55 (2014) 180−186. Available from: https://doi.org/10.1016/j.bios.2013.11.037.

[28] J. Gopal, H.N. Abdelhamid, P.-Y. Hua, H.-F. Wu, Chitosan nanomagnets for effective extraction and sensitive mass spectrometric detection of pathogenic bacterial endotoxin from human urine, J. Mater. Chem. B 1 (2013) 2463. Available from: https://doi.org/10.1039/c3tb20079e.

[29] H.N. Abdelhamid, H.-F. Wu, Gold nanoparticles assisted laser desorption/ionization mass spectrometry and applications: from simple molecules to intact cells, Anal. Bioanal. Chem. 408 (2016) 4485−4502. Available from: https://doi.org/10.1007/s00216-016-9374-6.

[30] M.L. Bhaisare, H.N. Abdelhamid, B.-S. Wu, H.-F. Wu, Rapid and direct MALDI-MS identification of pathogenic bacteria from blood using ionic liquid-modified magnetic nanoparticles (Fe_3O_4@SiO_2), J. Mater. Chem. B 2 (2014) 4671−4683. Available from: https://doi.org/10.1039/C4TB00528G.

[31] H.N. Abdelhamid, H.-F. Wu, Facile synthesis of nano silver ferrite ($AgFeO_2$) modified with chitosan applied for biothiol separation, Mater. Sci. Eng. C 45 (2014) 438−445. Available from: https://doi.org/10.1016/j.msec.2014.08.071.

[32] P.-Y. Hua, M. Manikandan, H.N. Abdelhamid, H.-F. Wu, Graphene nanoflakes as an efficient ionizing matrix for MALDI-MS based lipidomics of cancer cells and cancer stem cells, J. Mater. Chem. B 2 (2014) 7334−7343.

[33] H.N. Abdelhamid, H.-F. Wu, Polymer dots for quantifying the total hydrophobic pathogenic lysates in a single drop, Colloid Surface B 115 (2014) 51−60. Available from: https://doi.org/10.1016/j.colsurfb.2013.11.013.

[34] R. Sekar, S.K. Kailasa, H.N. Abdelhamid, Y.-C. Chen, H.-F. Wu, Electrospray ionization tandem mass spectrometric studies of copper and iron complexes with tobramycin, Int. J. Mass Spectrom. 338 (2013) 23−29. Available from: https://doi.org/10.1016/j.ijms.2012.12.001.

[35] M. Shahnawaz Khan, H.N. Abdelhamid, H.-F. Wu, Near infrared (NIR) laser mediated surface activation of graphene oxide nanoflakes for efficient antibacterial, antifungal and wound healing treatment, Colloid Surface B 127C (2015) 281−291. Available from: https://doi.org/10.1016/j.colsurfb.2014.12.049.

[36] H.N. Abdelhamid, Ionic liquids for mass spectrometry: matrices, separation and microextraction, TRAC Trend Anal. Chem. 77 (2015) 122−138. Available from: https://doi.org/10.1016/j.trac.2015.12.007.

[37] H.N. Abdelhamid, M.S. Khan, H.F. Wu, Graphene oxide as a nanocarrier for gramicidin (GOGD) for high antibacterial performance, RSC Adv. 4 (2014) 50035−50046. Available from: https://doi.org/10.1039/c4ra07250b.

[38] B.-S. Wu, H.N. Abdelhamid, H.-F. Wu, Synthesis and antibacterial activities of graphene decorated with stannous dioxide, RSC Adv. 4 (2014) 3722. Available from: https://doi.org/10.1039/c3ra43992e.

[39] H.N. Abdelhamid, A. Talib, H.-F. Wu, Facile synthesis of water soluble silver ferrite ($AgFeO_2$) nanoparticles and their biological application as antibacterial agents, RSC Adv. 5 (2015) 34594−34602. Available from: https://doi.org/10.1039/C4RA14461A.

[40] M. Dowaidar, H.N. Abdelhamid, M. Hällbrink, X. Zou, Ü. Langel, Graphene oxide nanosheets in complex with cell penetrating peptides for oligonucleotides delivery, Biochim. Biophys. Acta Gen. Subj. 1861 (2017) 2334−2341. Available from: https://doi.org/10.1016/j.bbagen.2017.07.002.

[41] M. Dowaidar, H.N. Abdelhamid, M. Hällbrink, K. Kurrikoff, K. Freimann, X. Zou, et al., Magnetic nanoparticle assisted self-assembly of cell penetrating peptides-oligonucleotides complexes for gene delivery, Sci. Rep. 7 (2017) 9159. Available from: https://doi.org/10.1038/s41598-017-09803-z.

[42] H.N. Abdelhamid, A. Talib, H.F. Wu, One pot synthesis of gold — carbon dots nanocomposite and its application for cytosensing of metals for cancer cells, Talanta 166 (2017) 357−363. Available from: https://doi.org/10.1016/j.talanta.2016.11.030.

[43] H.N. Abdelhamid, Laser assisted synthesis, imaging and cancer therapy of magnetic nanoparticles, Mater. Focus 5 (2016) 305−323. Available from: https://doi.org/10.1166/mat.2016.1336.

[44] H.N. Abdelhamid, H.M. El-Bery, A.A. Metwally, M. Elshazly, R.M. Hathout, Synthesis of CdS-modified chitosan quantum dots for the drug delivery of Sesamol, Carbohydr. Polym. 214 (2019) 90−99. Available from: https://doi.org/10.1016/j.carbpol.2019.03.024.

[45] R.K. Keservani, A.K. Sharma (Eds.), Nanoparticulate Drug Delivery Systems, first ed., Apple Academic Press, Boca Raton, FL, 2019. Available from: https://doi.org/10.1201/9781351137263.

[46] H.N. Abdelhamid, H.-F. Wu, Nanoparticles advance drug delivery for cancer cells, in: R.K. Keservani, A.K. Sharma (Eds.), Nanoparticulate Drug Delivery Systems, first ed., Apple Academic Press, Boca Raton, FL, 2019, pp. 121−150. ISBN 978-1-77188-695-6.

[47] H.N. Abdelhamid, J. Gopal, H.-F.F. Wu, Synthesis and application of ionic liquid matrices (ILMs) for effective pathogenic bacteria analysis in matrix assisted laser desorption/ionization (MALDI-MS), Anal. Chim. Acta 767 (2013) 104−111. Available from: https://doi.org/10.1016/j.ac.2012.12.054.

[48] M. Dowaidar, H. Nasser Abdelhamid, M. Hällbrink, Ü. Langel, X. Zou, Chitosan enhances gene delivery of oligonucleotide complexes with magnetic nanoparticles−cell-penetrating peptide, J. Biomater. Appl. 33 (2018) 392−401. Available from: https://doi.org/10.1177/0885328218796623.

[49] H.N. Abdelhamid, M. Dowaidar, M. Hällbrink, Ü. Langel, Cell penetrating peptides-hierarchical porous zeolitic imidazolate frameworks nanoparticles: an efficient gene delivery platform, SSRN Electron. J. (2019). Available from: https://doi.org/10.2139/ssrn.3435895.

[50] H.N. Abdelhamid, M. Dowaidar, Ü. Langel, Carbonized chitosan encapsulated hierarchical porous zeolitic imidazolate frameworks nanoparticles for gene delivery, Micropor. Mesopor. Mater. 302 (2020) 110200. Available from: https://doi.org/10.1016/j.micromeso.2020.110200.

[51] H.N. Abdelhamid, M. Dowaidar, M. Hällbrink, Ü. Langel, Gene delivery using cell penetrating peptides-zeolitic imidazolate frameworks, Micropor. Mesopor. Mater. 300 (2020) 110173. Available from: https://doi.org/10.1016/j.micromeso.2020.110173.

[52] K.H. Hussein, H.N. Abdelhamid, X. Zou, H.-M. Woo, Ultrasonicated graphene oxide enhances bone and skin wound regeneration, Mater. Sci. Eng. C 94 (2019) 484−492. Available from: https://doi.org/10.1016/j.msec.2018.09.051.

[53] H.N. Abdelhamid, A.M. El-Zohry, J. Cong, T. Thersleff, M. Karlsson, L. Kloo, et al., Towards implementing hierarchical porous zeolitic imidazolate frameworks in dye-sensitized solar cells, R. Soc. Open Sci. 6 (2019) 190723. Available from: https://doi.org/10.1098/rsos.190723.

[54] M.N. Goda, H.N. Abdelhamid, A.E.-A.A. Said, Zirconium oxide sulfate-carbon ($ZrOSO_4$@C) derived from carbonized UiO-66 for selective production of dimethyl ether, ACS Appl. Mater. Interfaces 12 (2020) 646−653. Available from: https://doi.org/10.1021/acsami.9b17520.

[55] A.A. Kassem, H.N. Abdelhamid, D.M. Fouad, S.A. Ibrahim, Metal-organic frameworks (MOFs) and MOFs-derived CuO@C for hydrogen generation from sodium borohydride, Int. J. Hydrog. Energy 44 (2019) 31230−31238. Available from: https://doi.org/10.1016/j.ijhydene.2019.10.047.

[56] T. Van Cleve, G. Wang, M. Mooney, C.F. Cetinbas, N. Kariuki, J. Park, et al., Tailoring electrode microstructure via ink content to enable improved rated power performance for platinum cobalt/high surface area carbon based polymer electrolyte fuel cells, J. Power Sources 482 (2021) 228889. Available from: https://doi.org/10.1016/j.jpowsour.2020.228889.

[57] Y. Yang, K. Shen, J. Lin, Y. Zhou, Q. Liu, C. Hang, et al., A Zn-MOF constructed from electron-rich π-conjugated ligands with an interpenetrated graphene-like net as an efficient nitroaromatic sensor, RSC Adv. 6 (2016) 45475−45481. Available from: https://doi.org/10.1039/C6RA00524A.

[58] H.N. Abdelhamid, Salts induced formation of hierarchical porous ZIF-8 and their applications for CO_2 sorption and hydrogen generation via $NaBH_4$ hydrolysis, Macromol. Chem. Phys. 221 (2020) 2000031. Available from: https://doi.org/10.1002/macp.202000031.

[59] H.N. Abdelhamid, Zinc hydroxide nitrate nanosheets conversion into hierarchical zeolitic imidazolate frameworks nanocomposite and their application for CO_2 sorption, Mater. Today Chem. 15 (2020) 100222. Available from: https://doi.org/10.1016/j.mtchem.2019.100222.

[60] Q. Yao, A. Bermejo Gómez, J. Su, V. Pascanu, Y. Yun, H. Zheng, et al., Series of highly stable isoreticular lanthanide metal—organic frameworks with expanding pore size and tunable luminescent properties, Chem. Mater. 27 (2015) 5332—5339. Available from: https://doi.org/10.1021/acs.chemmater.5b01711.

[61] H.N. Abdelhamid, M.L. Bhaisare, H.-F. Wu, Ceria nanocubic-ultrasonication assisted dispersive liquid-liquid microextraction coupled with matrix assisted laser desorption/ionization mass spectrometry for pathogenic bacteria analysis, Talanta 120 (2014) 208—217. Available from: https://doi.org/10.1016/j.talanta.2013.11.078.

[62] A.S. Etman, H.N. Abdelhamid, Y. Yuan, L. Wang, X. Zou, J. Sun, Facile water-based strategy for synthesizing MoO_{3-x} nanosheets: efficient visible light photocatalysts for dye degradation, ACS Omega 3 (2018) 2193—2201. Available from: https://doi.org/10.1021/acsomega.8b00012.

[63] R.M. Ashour, H.N. Abdelhamid, A.F. Abdel-Magied, A.A. Abdel-khalek, M. Ali, A. Uheida, et al., Rare earth ions adsorption onto graphene oxide nanosheets, Solvent Extr. Ion Exch. 35 (2017) 91—103. Available from: https://doi.org/10.1080/07366299.2017.1287509.

[64] H.N. Abdelhamid, X. Zou, Template-free and room temperature synthesis of hierarchical porous zeolitic imidazolate framework nanoparticles and their dye and CO_2 sorption, Green Chem. 20 (2018) 1074—1084. Available from: https://doi.org/10.1039/C7GC03805D.

[65] H.N. Abdelhamid, H.-F. Wu, Synthesis of a highly dispersive sinapinic acid@graphene oxide (SA@GO) and its applications as a novel surface assisted laser desorption/ionization mass spectrometry for proteomics and pathogenic bacteria biosensing, Analyst 140 (2015) 1555—1565.

[66] H.N. Abdelhamid, Delafossite nanoparticle as new functional materials: advances in energy, nanomedicine and environmental applications, Mater. Sci. Forum 832 (2015) 28—53. Available from: https://doi.org/10.4028/http://www.scientific.net/MSF.832.28.

[67] H. Nasser Abdelhamid, H.F. Wu, Furoic and mefenamic acids as new matrices for matrix assisted laser desorption/ionization-(MALDI)-mass spectrometry, Talanta 115 (2013) 442—450. Available from: https://doi.org/10.1016/j.talanta.2013.05.050.

[68] H. Nasser Abdelhamid, B.-S. Wu, H.-F. Wu, Graphene coated silica applied for high ionization matrix assisted laser desorption/ionization mass spectrometry: a novel approach for environmental and biomolecule analysis, Talanta 126 (2014) 27—37. Available from: https://doi.org/10.1016/j.talanta.2014.03.016.

[69] H.N. Abdelhamid, M.S. Khan, H.-F.F. Wu, Design, characterization and applications of new ionic liquid matrices for multifunctional analysis of biomolecules: a novel strategy for pathogenic bacteria biosensing, Anal. Chim. Acta 823 (2014) 51—60. Available from: https://doi.org/10.1016/j.ac.2014.03.026.

[70] J. Gopal, H.N. Abdelhamid, J.H. Huang, H.F. Wu, Nondestructive detection of the freshness of fruits and vegetables using gold and silver nanoparticle mediated graphene enhanced Raman spectroscopy, Sensors Actuat. B Chem. 224 (2016) 413—424. Available from: https://doi.org/10.1016/j.snb.2015.08.123.

[71] L. Valencia, H.N. Abdelhamid, Nanocellulose leaf-like zeolitic imidazolate framework (ZIF-L) foams for selective capture of carbon dioxide, Carbohydr. Polym. 213 (2019) 338—345. Available from: https://doi.org/10.1016/j.carbpol.2019.03.011.

[72] A.F. Abdel-Magied, H.N. Abdelhamid, R.M. Ashour, X. Zou, K. Forsberg, Hierarchical porous zeolitic imidazolate frameworks nanoparticles for efficient adsorption of rare-earth elements, Micropor. Mesopor. Mater. 278 (2019) 175—184. Available from: https://doi.org/10.1016/j.micromeso.2018.11.022.

[73] H.N. Abdelhamid, Ionic liquids matrices for laser assisted desorption/ionization mass spectrometry, Mass Spectrom. Purif. Tech. 1 (2015) 109—119. Available from: https://doi.org/10.4172/2469-9861.1000109.

[74] M.N. Iqbal, A.F. Abdel-Magied, H.N. Abdelhamid, P. Olsén, A. Shatskiy, X. Zou, et al., Mesoporous ruthenium oxide: a heterogeneous catalyst for water oxidation, ACS Sustain. Chem. Eng. 5 (2017) 9651—9656. Available from: https://doi.org/10.1021/acssuschemeng.7b02845.

[75] H.N. Abdelhamid, M. Wilk-Kozubek, A.M. El-Zohry, A. Bermejo Gómez, A. Valiente, B. Martín-Matute, et al., Luminescence properties of a family of lanthanide metal-organic frameworks, Micropor. Mesopor. Mater. 279 (2019) 400–406. Available from: https://doi.org/10.1016/j.micromeso.2019.01.024.

[76] H.N. Abdelhamid, Physicochemical properties of proteomic ionic liquids matrices for MALDI-MS, J Data Mining Genomics Proteomics 7 (2016) 2. Available from: https://doi.org/10.4172/2153-0602.1000189.

[77] H.N. Abdelhamid, Organic matrices, ionic liquids, and organic matrices@nanoparticles assisted laser desorption/ionization mass spectrometry, TRAC Trend. Anal. Chem. 89 (2017) 68–98. Available from: https://doi.org/10.1016/j.trac.2017.01.012.

[78] H.N. Abdelhamid, Nanoparticle assisted laser desorption/ionization mass spectrometry for small molecule analytes, Microchim. Acta 185 (2018) 200. Available from: https://doi.org/10.1007/s00604-018-2687-8.

[79] H.N. Abdelhamid, A. Bermejo-Gómez, B. Martín-Matute, X. Zou, A water-stable lanthanide metal-organic framework for fluorimetric detection of ferric ions and tryptophan, Microchim. Acta 184 (2017) 3363–3371. Available from: https://doi.org/10.1007/s00604-017-2306-0.

[80] H.N. Abdelhamid, Dehydrogenation of sodium borohydride using cobalt embedded zeolitic imidazolate frameworks, J. Solid State Chem. 297 (2021) 122034. Available from: https://doi.org/10.1016/j.jssc.2021.122034.

[81] H.N. Abdelhamid, A review on hydrogen generation from the hydrolysis of sodium borohydride, Int. J. Hydrog. Energy 46 (2021) 726–765. Available from: https://doi.org/10.1016/j.ijhydene.2020.09.186.

[82] H.N. Abdelhamid, Dye encapsulated hierarchical porous zeolitic imidazolate frameworks for carbon dioxide adsorption, J. Environ. Chem. Eng. 8 (2020) 104008. Available from: https://doi.org/10.1016/j.jece.2020.104008.

[83] H.N. Abdelhamid, Biointerface between ZIF-8 and biomolecules and their applications, Biointerface Res. Appl. Chem. 11 (2021) 8283–8297. Available from: https://doi.org/10.33263/BRIAC111.82838297.

[84] M. Gao, J. Yuan, M. Antonietti, Ionic liquids and poly(ionic liquid)s for morphosynthesis of inorganic materials, Chem. A Eur. J. 23 (2017) 5391–5403. Available from: https://doi.org/10.1002/chem.201604191.

[85] S.K. Singh, A.W. Savoy, Ionic liquids synthesis and applications: an overview, J. Mol. Liq. 297 (2020) 112038. Available from: https://doi.org/10.1016/j.molliq.2019.112038.

[86] L.L. Lazarus, C.T. Riche, B.C. Marin, M. Gupta, N. Malmstadt, R.L. Brutchey, Two-phase microfluidic droplet flows of ionic liquids for the synthesis of gold and silver nanoparticles, ACS Appl. Mater. Interfaces 4 (2012) 3077–3083. Available from: https://doi.org/10.1021/am3004413.

[87] J. Olchowka, T. Tailliez, L. Bourgeois, M.A. Dourges, L. Guerlou-Demourgues, Ionic liquids to monitor the nano-structuration and the surface functionalization of material electrodes: a proof of concept applied to cobalt oxyhydroxide, Nanoscale Adv. 1 (2019) 2240–2249. Available from: https://doi.org/10.1039/C9NA00171A.

[88] I. Qadir, N. Simon, D. Rivillo, J. Dupont, MNP catalysis in ionic liquids, Recent Advances in Nanoparticle Catalysis, Springer, 2020, pp. 107–128. Available from: https://doi.org/10.1007/978-3-030-45823-2_4.

[89] J.-L. Wang, J.-W. Liu, S.-H. Yu, Recycling valuable elements from the chemical synthesis process of nanomaterials: a sustainable view, ACS Mater. Lett. 1 (2019) 541–548. Available from: https://doi.org/10.1021/acsmaterialslett.9b00283.

[90] S. Sugden, H. Wilkins, CLXVII.—The parachor and chemical constitution. Part XII. Fused metals and salts, J. Chem. Soc. (1929) 1291–1298. Available from: https://doi.org/10.1039/JR9290001291.

[91] P.G. Jessop, Fundamental properties and practical applications of ionic liquids: concluding remarks, Faraday Discuss. 206 (2018) 587–601. Available from: https://doi.org/10.1039/C7FD90090B.

[92] B.M. Trost, The atom economy—a search for synthetic efficiency, Science 254 (1991) 1471–1477. Available from: https://doi.org/10.1126/science.1962206.

[93] R.A. Sheldon, The E factor: fifteen years on, Green Chem. 9 (2007) 1273—1283. Available from: https://doi.org/10.1039/b713736m.

[94] M. Deetlefs, K.R. Seddon, Assessing the greenness of some typical laboratory ionic liquid preparations, Green Chem. 12 (2010) 17—30. Available from: https://doi.org/10.1039/B915049H.

[95] R. Ratti, Ionic liquids: synthesis and applications in catalysis, Adv. Chem. 2014 (2014) 1—16. Available from: https://doi.org/10.1155/2014/729842.

[96] B. Wu, W. Liu, Y. Zhang, H. Wang, Do we understand the recyclability of ionic liquids? Chem. A Eur. J. 15 (2009) 1804—1810. Available from: https://doi.org/10.1002/chem.200801509.

[97] H.-T. Wong, Y.H. See-Toh, F.C. Ferreira, R. Crook, A.G. Livingston, Organic solvent nanofiltration in asymmetric hydrogenation: enhancement of enantioselectivity and catalyst stability by ionic liquids, Chem. Commun. (2006) 2063—2065. Available from: https://doi.org/10.1039/b602184k.

[98] H. Sui, J. Zhou, G. Ma, Y. Niu, J. Cheng, L. He, et al., Removal of ionic liquids from oil sands processing solution by ion-exchange resin, Appl. Sci. 8 (2018) 1611. Available from: https://doi.org/10.3390/app8091611.

[99] F. Yu, Y. Zhou, B. Gao, H. Qiao, Y. Li, E. Wang, et al., Effective removal of ionic liquid using modified biochar and its biological effects, J. Taiwan Inst. Chem. Eng. 67 (2016) 318—324. Available from: https://doi.org/10.1016/j.jtice.2016.07.038.

[100] J. Zhou, H. Sui, Z. Jia, Z. Yang, L. He, X. Li, Recovery and purification of ionic liquids from solutions: a review, RSC Adv. 8 (2018) 32832—32864. Available from: https://doi.org/10.1039/C8RA06384B.

[101] H. Wu, F. Shen, J. Wang, J. Luo, L. Liu, R. Khan, et al., Separation and concentration of ionic liquid aqueous solution by vacuum membrane distillation, J. Membr. Sci. 518 (2016) 216—228. Available from: https://doi.org/10.1016/j.memsci.2016.07.017.

[102] H. Wu, F. Shen, J. Wang, Y. Wan, Membrane fouling in vacuum membrane distillation for ionic liquid recycling: interaction energy analysis with the XDLVO approach, J. Membr. Sci. 550 (2018) 436—447. Available from: https://doi.org/10.1016/j.memsci.2018.01.018.

[103] A.M. Avram, P. Ahmadiannamini, A. Vu, X. Qian, A. Sengupta, S.R. Wickramasinghe, Polyelectrolyte multilayer modified nanofiltration membranes for the recovery of ionic liquid from dilute aqueous solutions, J. Appl. Polym. Sci. 134 (2017) 45349. Available from: https://doi.org/10.1002/app.45349.

[104] L.B. Abdulra'uf, A. Lawal, A.Y. Sirhan, G.H. Tan, Review of ionic liquids in microextraction analysis of pesticide residues in fruit and vegetable samples, Chromatographia 83 (2020) 11—33. Available from: https://doi.org/10.1007/s10337-019-03818-6.

[105] F.A. e Silva, M. Caban, M. Kholany, P. Stepnowski, J.A.P. Coutinho, S.P.M. Ventura, Recovery of nonsteroidal anti-inflammatory drugs from wastes using ionic-liquid-based three-phase partitioning systems, ACS Sustain. Chem. Eng. 6 (2018) 4574—4585. Available from: https://doi.org/10.1021/acssuschemeng.7b03216.

[106] Z. Ma, J. Yu, S. Dai, Preparation of inorganic materials using ionic liquids, Adv. Mater. 22 (2010) 261—285. Available from: https://doi.org/10.1002/adma.200900603.

[107] X. Kang, X. Sun, B. Han, Synthesis of functional nanomaterials in ionic liquids, Adv. Mater. 28 (2016) 1011—1030. Available from: https://doi.org/10.1002/adma.201502924.

[108] M. Tunckol, J. Durand, P. Serp, Carbon nanomaterial—ionic liquid hybrids, Carbon N. Y. 50 (2012) 4303—4334. Available from: https://doi.org/10.1016/j.carbon.2012.05.017.

[109] D. Marquardt, C. Vollmer, R. Thomann, P. Steurer, R. Mülhaupt, E. Redel, et al., The use of microwave irradiation for the easy synthesis of graphene-supported transition metal nanoparticles in ionic liquids, Carbon N. Y. 49 (2011) 1326—1332. Available from: https://doi.org/10.1016/j.carbon.2010.09.066.

[110] T. Kinoshita, S. Maruyama, Y. Matsumoto, Ionic liquid wettability of CVD-grown graphene on $Cu/\alpha\text{-}Al_2O_3(0\ 0\ 0\ 1)$ characterized by in situ contact angle measurement in a vacuum, Chem. Phys. Lett. 735 (2019) 136781. Available from: https://doi.org/10.1016/j.cplett.2019.136781.

[111] R.M. Esteban, K. Schütte, P. Brandt, D. Marquardt, H. Meyer, F. Beckert, et al., Iridium@graphene composite nanomaterials synthesized in ionic liquid as re-usable catalysts for solvent-free hydrogenation of benzene and cyclohexene, Nano-Struct. Nano-Objects 2 (2015) 11—18. Available from: https://doi.org/10.1016/j.nanoso.2015.07.001.

[112] C. Liu, S. Qiu, P. Du, H. Zhao, L. Wang, An ionic liquid—graphene oxide hybrid nanomaterial: synthesis and anticorrosive applications, Nanoscale 10 (2018) 8115—8124. Available from: https://doi.org/10.1039/C8NR01890A.

[113] R. Gusain, H.P. Mungse, N. Kumar, T.R. Ravindran, R. Pandian, H. Sugimura, et al., Covalently attached graphene—ionic liquid hybrid nanomaterials: synthesis, characterization and tribological application, J. Mater. Chem. A 4 (2016) 926—937. Available from: https://doi.org/10.1039/C5TA08640J.

[114] Y. Qin, Y. Song, N. Sun, N. Zhao, M. Li, L. Qi, Ionic liquid-assisted growth of single-crystalline dendritic gold nanostructures with a threefold symmetry, Chem. Mater. 20 (2008) 3965—3972. Available from: https://doi.org/10.1021/cm8002386.

[115] M.H.G. Prechtl, P.S. Campbell, J.D. Scholten, G.B. Fraser, G. Machado, C.C. Santini, et al., Imidazolium ionic liquids as promoters and stabilising agents for the preparation of metal(0) nanoparticles by reduction and decomposition of organometallic complexes, Nanoscale 2 (2010) 2601. Available from: https://doi.org/10.1039/c0nr00574f.

[116] X. Bai, Y. Gao, H. Liu, L. Zheng, Synthesis of amphiphilic ionic liquids terminated gold nanorods and their superior catalytic activity for the reduction of nitro compounds, J. Phys. Chem. C 113 (2009) 17730—17736. Available from: https://doi.org/10.1021/jp906378d.

[117] P. Song, L. Liu, J.-J. Feng, J. Yuan, A.-J. Wang, Q.-Q. Xu, Poly(ionic liquid) assisted synthesis of hierarchical gold-platinum alloy nanodendrites with high electrocatalytic properties for ethylene glycol oxidation and oxygen reduction reactions, Int. J. Hydrog. Energy 41 (2016) 14058—14067. Available from: https://doi.org/10.1016/j.ijhydene.2016.06.245.

[118] A.S. Pensado, A.A.H. Pádua, Solvation and stabilization of metallic nanoparticles in ionic liquids, Angew. Chem. Int. (Ed.) 50 (2011) 8683—8687. Available from: https://doi.org/10.1002/anie.201103096.

[119] Y. Cao, S. Guo, C. Yu, J. Zhang, X. Pan, G. Li, Ionic liquid-assisted one-step preparation of ultrafine amorphous metallic hydroxide nanoparticles for the highly efficient oxygen evolution reaction, J. Mater. Chem. A 8 (2020) 15767—15773. Available from: https://doi.org/10.1039/D0TA00434K.

[120] L. Chen, T. Zhang, H. Cheng, R.M. Richards, Z. Qi, A microwave assisted ionic liquid route to prepare bivalent Mn_5O_8 nanoplates for 5-hydroxymethylfurfural oxidation, Nanoscale 12 (2020) 17902—17914. Available from: https://doi.org/10.1039/D0NR04738D.

[121] G. Yan, Y. Lian, Y. Gu, C. Yang, H. Sun, Q. Mu, et al., Phase and morphology transformation of MNO_2 induced by ionic liquids toward efficient water oxidation, ACS Catal. 8 (2018) 10137—10147. Available from: https://doi.org/10.1021/acscatal.8b02203.

[122] S. Maruyama, Y. Matsumoto, Intrinsic nature of interfacial interactions between ionic liquids and rutile TiO_2 single crystal surfaces studied by in situ contact angle measurement in a vacuum, J. Phys. Chem. C 119 (2015) 17755—17761. Available from: https://doi.org/10.1021/acs.jpcc.5b05276.

[123] K. Qi, W. Zheng, Morphology-controlled synthesis of inorganic nanocrystals by ionic liquid assistance, Curr. Opin. Green Sustain. Chem. 5 (2017) 17—23. Available from: https://doi.org/10.1016/j.cogsc.2017.03.011.

[124] X. Yang, X. Lei, S. Tang, Ionic liquid-assisted synthesis of hierarchically olive-like Al_2O_3 and CeO_2/Al_2O_3 with enhanced catalytic activity for Orange II degradation, Ceram. Int. 47 (2021) 5446—5455. Available from: https://doi.org/10.1016/j.ceramint.2020.10.126.

[125] B. Qin, H. Zhang, T. Diemant, D. Geiger, R. Raccichini, R.J. Behm, et al., Ultrafast ionic liquid-assisted microwave synthesis of sno microflowers and their superior sodium-ion storage performance, ACS Appl. Mater. Interfaces 9 (2017) 26797—26804. Available from: https://doi.org/10.1021/acsami.7b06230.

[126] R. Gupta, M. Yadav, R. Gaur, G. Arora, P. Yadav, R.K. Sharma, Magnetically supported ionic liquids: a sustainable catalytic route for organic transformations, Mater. Horizons 7 (2020) 3097—3130. Available from: https://doi.org/10.1039/D0MH01088J.

[127] Y. Wang, S. Maksimuk, R. Shen, H. Yang, Synthesis of iron oxide nanoparticles using a freshly-made or recycled imidazolium-based ionic liquid, Green Chem. 9 (2007) 1051. Available from: https://doi.org/10.1039/b618933d.

[128] H. Ying, C. Zhang, T. Chen, X. Zhao, Z. Li, J. Hao, A new phosphonium-based ionic liquid to synthesize nickel metaphosphate for hydrogen evolution reaction, Nanotechnology 31 (2020) 505402. Available from: https://doi.org/10.1088/1361-6528/abb508.

[129] L. Li, Y. Zhang, Y. Sun, S. Sun, G. Shen, P. Zhao, et al., Manufacturing pure cellulose films by recycling ionic liquids as plasticizers, Green Chem. 22 (2020) 3835−3841. Available from: https://doi.org/10.1039/D0GC00046A.

[130] A.M. da Costa Lopes, R.M. èukasik, Separation and recovery of a hemicellulose-derived sugar produced from the hydrolysis of biomass by an acidic ionic liquid, ChemSusChem 11 (2018) 1099−1107. Available from: https://doi.org/10.1002/cssc.201702231.

[131] A.M. Asim, M. Uroos, S. Naz, M. Sultan, G. Griffin, N. Muhammad, et al., Acidic ionic liquids: promising and cost-effective solvents for processing of lignocellulosic biomass, J. Mol. Liq. 287 (2019) 110943. Available from: https://doi.org/10.1016/j.molliq.2019.110943.

[132] A.M. da Costa Lopes, R.M.G. Lins, R.A. Rebelo, R.M. èukasik, Biorefinery approach for lignocellulosic biomass valorisation with an acidic ionic liquid, Green Chem. 20 (2018) 4043−4057. Available from: https://doi.org/10.1039/C8GC01763H.

[133] P. Halder, S. Kundu, S. Patel, M. Ramezani, R. Parthasarathy, K. Shah, A comparison of ionic liquids and organic solvents on the separation of cellulose-rich material from river red gum, BioEnergy Res. 12 (2019) 275−291. Available from: https://doi.org/10.1007/s12155-019-09967-8.

[134] Y. Jeong, K. Moon, S. Jeong, W.-G. Koh, K. Lee, Converting waste papers to fluorescent carbon dots in the recycling process without loss of ionic liquids and bioimaging applications, ACS Sustain. Chem. Eng. 6 (2018) 4510−4515. Available from: https://doi.org/10.1021/acssuschemeng.8b00353.

[135] J. Lazko, T. Sénéchal, N. Landercy, L. Dangreau, J.-M. Raquez, P. Dubois, Well defined thermostable cellulose nanocrystals via two-step ionic liquid swelling-hydrolysis extraction, Cellulose 21 (2014) 4195−4207. Available from: https://doi.org/10.1007/s10570-014-0417-x.

[136] J.H. Jordan, M.W. Easson, B.D. Condon, Cellulose hydrolysis using ionic liquids and inorganic acids under dilute conditions: morphological comparison of nanocellulose, RSC Adv. 10 (2020) 39413−39424. Available from: https://doi.org/10.1039/D0RA05976E.

[137] Z. Pang, P. Wang, C. Dong, Ultrasonic pretreatment of cellulose in ionic liquid for efficient preparation of cellulose nanocrystals, Cellulose 25 (2018) 7053−7064. Available from: https://doi.org/10.1007/s10570-018-2070-2.

[138] S. Naz, M. Uroos, A.M. Asim, N. Muhammad, F.U. Shah, One-pot deconstruction and conversion of lignocellulose into reducing sugars by pyridinium-based ionic liquid−metal salt system, Front. Chem. 8 (2020) 236. Available from: https://doi.org/10.3389/fchem.2020.00236.

[139] J. Zhang, D. Zou, S. Singh, G. Cheng, Recent developments in ionic liquid pretreatment of lignocellulosic biomass for enhanced bioconversion, Sustain. Energy Fuels (2021). Available from: https://doi.org/10.1039/D0SE01802C.

[140] J. Sun, J. Shi, N.V.S.N. Murthy Konda, D. Campos, D. Liu, S. Nemser, et al., Efficient dehydration and recovery of ionic liquid after lignocellulosic processing using pervaporation, Biotechnol. Biofuels 10 (2017) 154. Available from: https://doi.org/10.1186/s13068-017-0842-9.

[141] C. Fan, Y. Liang, H. Dong, J. Yang, G. Tang, W. Zhang, et al., Guanidinium ionic liquid-controlled synthesis of zeolitic imidazolate framework for improving its adsorption property, Sci. Total Environ. 640−641 (2018) 163−173. Available from: https://doi.org/10.1016/j.scitotenv.2018.05.282.

[142] X. Yu, Z. Yang, H. Zhang, B. Yu, Y. Zhao, Z. Liu, et al., Ionic liquid/H_2O-mediated synthesis of mesoporous organic polymers and their application in methylation of amines, Chem. Commun. 53 (2017) 5962−5965. Available from: https://doi.org/10.1039/C7CC01910F.

[143] S.Y. Lee, I. Khoiroh, C.W. Ooi, T.C. Ling, P.L. Show, Recent advances in protein extraction using ionic liquid-based aqueous two-phase systems, Sep. Purif. Rev. 46 (2017) 291−304. Available from: https://doi.org/10.1080/15422119.2017.1279628.

[144] U. Schröder, J.D. Wadhawan, R.G. Compton, F. Marken, P.A.Z. Suarez, C.S. Consorti, et al., Water-induced accelerated ion diffusion: voltammetric studies in 1-methyl-3-[2,6-(S)-dimethyloc-ten-2-yl]imidazolium tetrafluoroborate, 1-butyl-3-methylimidazolium tetrafluoroborate and hexa-fluorophosphate ionic liquids, New J. Chem. 24 (2000) 1009−1015. Available from: https://doi.org/10.1039/b007172m.

[145] J.N.A. Canongia Lopes, A.A.H. Pádua, Nanostructural organization in ionic liquids, J. Phys. Chem. B 110 (2006) 3330−3335. Available from: https://doi.org/10.1021/jp056006y.

[146] J. Dupont, J.D. Scholten, On the structural and surface properties of transition-metal nanoparticles in ionic liquids, Chem. Soc. Rev. 39 (2010) 1780. Available from: https://doi.org/10.1039/b822551f.

SECTION III

Properties of recycled nanomaterials

CHAPTER 14

Techniques used to study the physiochemical properties of recycled nanomaterials

Vojislav Stanić
Laboratory of Radiation and Environmental Protection, Vinča Institute of Nuclear Sciences, University of Belgrade, Belgrade, Serbia

14.1 Introduction

Nanomaterials are used in a wide range of commercial products, including electronic components, textiles, cosmetics, and various biomedical materials. The introduction of nanomaterials into commercial products requires an assessment of their safety and an understanding of their impact on the environment and human health as well as the possibility of their recycling. The impact of synthetic nanomaterials on human health and the environment is still relatively little known. The nanometer size of the particles is a serious problem with respect to the separation and recycling of commercial products. Various physicochemical processes used in the recycling process can affect the characteristics of recycled nanoparticles: chemical composition, crystal properties, size, morphology, surface properties, and other characteristics. Table 14.1 shows some of the most commonly used nondestructive physiochemical techniques for the characterization of inorganic recycled nanomaterials. Based on these methods, an optimal technological recycling process from starting material to final product with satisfactory physicochemical characteristics can be found.

14.1.1 X-ray diffraction method

X-ray diffraction (XRD), is a method for characterization of nanomaterials and primarily serves to determine the structure of materials, qualitative identification of crystal phases, quantitative phase analysis, sample purity, crystallinity, size, shape and orientation of crystallites in a sample [1,2]. Depending on the type of sample being tested, XRD analysis can be divided into diffraction methods used on single-crystal samples (Laue method, rotating crystal method) and diffraction methods used on polycrystalline samples (crystal powder method, Debye-Scherrer method). Powdered XRD is a nondestructive method and is most widely used as an experimental method in nanomaterial testing. In diffraction analysis, X-rays of small wavelengths in the range from

Nanomaterials Recycling
DOI: https://doi.org/10.1016/B978-0-323-90982-2.00013-5

© 2022 Elsevier Inc.
All rights reserved.

a few to 0.1Å are used, which corresponds to photon energies from 1 to 120 keV. The choice of the appropriate X-ray radiation for material testing depends on the parameters of the crystal lattice, the distance between the crystallographic planes, the energy to be introduced into the sample, and so on. The powder XRD method uses X-ray radiation, which is monochromatic. The most common characteristic radiations are given in Table 14.2. In practice, CuKα 0.154056 nm monochromatic radiation is mostly used.

Each crystal can be viewed as a series of parallel and equidistant planes of atoms, which are located at a distance d. If a beam of monochromatic X-rays of wavelength λ passes through the crystal, most of the rays will pass through the crystal, while only a small part will bounce off the crystal planes in which the atoms are arranged. The reflected rays interfere with each other, as a result of which their intensity weakens or increases. XRD on equidistant planes in the crystal is shown in Fig. 14.1.

Table 14.1 Applicability of nondestructive analytical techniques to providing specific information on recycled nanoparticles.

Technique	Main information derived
X-Ray Diffraction method	Crystal structure, composition, crystalline grain size
Infrared Spectroscopy	Functional chemical groups, surface composition, structure of compounds
Raman Spectroscopy	Functional chemical groups, surface composition, structure of compounds
X-ray photoelectron spectroscopy	Electronic structure, elemental composition, oxidation states
Auger Electron Spectroscopy	Chemical composition of the surface of particles
X-Ray Fluorescence Spectroscopy	Chemical elemental composition
Scanning Electron Microscopy	Size of particles, morphology
Transmission Electron Microscopy	Size of particles, morphology, crystal structure of compounds
Atomic Force Microscopy	Size of particles, morphology
Electron Probe Micro Analysis	Chemical elemental composition

Table 14.2 Most commonly used X-rays for material testing.

X-rays	AgKα	MoKα	CuKα	CoKα	FeKα	CrKα
λ (nm)	0.0540	0.0711	0.154056	0.17889	0.1937	0.2291

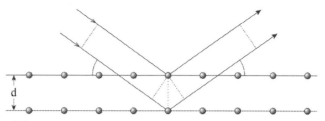

Figure 14.1 Bragg diffraction.

Constructive interference will occur only if the reflected rays are in phase, that is, if the Bragg equation is satisfied:

$$n\lambda = 2d\sin\theta, n = 1, 2, 3 \tag{14.1}$$

where n is the diffraction order, λ is the wavelength of the X-rays, d is the distance between crystallographic planes, and θ is the angle between the incident air and the plane from which the air is reflected.

If the incident X-ray beam falls on the crystal at an angle different from the Brag angle θ, there is destructive interference, that is, mutual cancellation of air.

Qualitative XRD analysis is based on the fact that each crystalline compound has characteristic reflections that are determined by the position of diffraction maxima, intensities, and their profile [full width at half maximum (FWHM) factor]. The identification of a crystalline compound is usually performed by comparing the diffractogram of an unknown sample with a catalog of standard data, which can be found in the International Center for Diffraction Data database. Quantitative XRD analysis is based on the fact that the intensity of diffraction maxima is proportional to the phase fraction in the analyzed sample. The position of reflective maxima in the diffractogram depends on the conditions given by Bragg's law (Eq. 14.1), while their intensity depends on a number of different factors related to the content of the unit cell and the characteristics of the sample itself. In examining crystalline materials, defined sharp reflections are obtained, in contrast to nanoparticles and amorphous substances, in which broad lines are obtained. The width of the diffraction maxima is influenced by the size of the crystallites, microvoltages, various types of crystal defects, packing errors in the crystal lattice, and characteristics of the XRD instrument. Based on the fact that the width of the diffraction maximum depends on the size of the crystallite on which the diffraction is performed, it is possible to determine the mean size of the crystallite, D, from Scherer's formula:

$$D = \frac{k\lambda}{\beta\cos\theta} \tag{14.2}$$

where λ is the wavelength of the X-ray, β is the width of the diffraction maximum at half its height, θ is the incident angle of the X-ray, and k is a constant that depends on the degree of symmetry of the crystal. This equation is suitable for the characterization of nanomaterials. The size of the crystallites has a significant influence on the appearance of the diffraction maxima; the smaller the crystallites, the more homogeneous is their powder distribution, while the width of the diffraction maxima is higher and the resolution, that is, the separation of reflections is weaker. XRD is the most common method for determining the crystal structure and purity of recycled nanomaterials [3–10].

Electronic waste often contains toxic substances and poses a danger to the environment [11]. The principal methods for recycling electronic scrap efficiently are the separation of parts of electronic components of various devices and the application of a chemical method involving mechanical treatment (milling) followed by pyrolysis or hydrometallurgy [12,13]. Milling technology has been widely used in waste treatment. During the milling process, there may be a reduction in particle size, in which specific surface area and crystal-phase structure can be changed. The low-temperature recycling processes lead to a highly efficient separation of nanoparticles. Fig. 14.2 shows the XRD analysis of the mixture obtained by grinding electronic waste, the printed circuit boards, using a cryomill process [3]. The results showed the presence of ceramic, polymer, glass, and metal in the ground waste. Average particle dimensions from powder were calculated from FWHM of different peaks and were in the range of 20–150 nm.

Nanomaterials are present in various biological materials. Their isolation can be used to obtain products of great economic importance. Biological apatites can be used in the remediation of soil and water, owing to their ability to retain a variety of ionic species, especially heavy metals [14]. Apatit IITM, developed by Wright and Conca [15] for the purpose of stabilizing toxic metals and radionuclides, is a commercially available biogenic form of apatite derived from fish bones. Biological apatites from bones are a nanometer in size, weakly crystalline, almost amorphous. The crystals have the appearance of needles, thin plates, or leaves [16]. Apatites are mainly isolated from bones by grinding, in which then the organic component is removed by thermal treatment [14]. The diffraction patterns of raw fish bones are broad, indicating that the hydroxyapatite (HAP) component is poorly crystallized (Fig. 14.3A2–4). Thermal

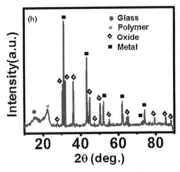

Figure 14.2 X-ray diffraction diffractogram of the printed circuit board milled for 30 min in the cryomill process [3].

Techniques used to study the physiochemical properties of recycled nanomaterials 295

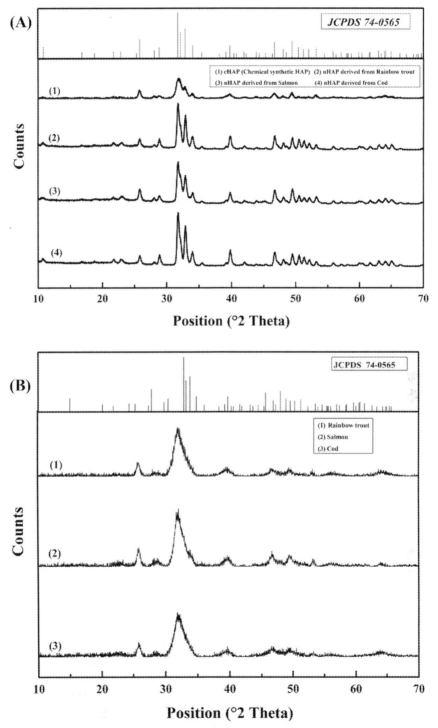

Figure 14.3 X-ray diffraction pattern of hydroxyapatite. (A1) Chemical synthetic hydroxyapatite, (A2—4) Hydroxyapatite produced from fish bones by thermal heating. (B) Hydroxyapatite in raw bones [17].

14.2 Fourier transform infrared spectroscopy

Fourier transform infrared spectroscopy (FT-IR) is one of the most commonly used methods for the characterization of organic and inorganic nanoparticle materials [18,19]. It is used to identify chemical groups, to solve problems with the structure of molecules, and to detect traces of impurities. The information obtained by FT-IR spectrum analysis alone is not always sufficient to resolve compound structure or identification but must be combined with data from other assays, such as XRD, NMR, and Raman spectroscopy.

FT-IR spectroscopy is based on measuring the absorption (or reflection) of radiation from the range of $1-1000\,\mu m$. The infrared (IR) part of the electromagnetic spectrum of radiation can be divided into three areas: the near-IR region, which extends from 0.7 to 2.5 μm (12,800−5000/cm, 3.81014−1.21014 Hz); the middle or fundamental region, which extends from 2.5 to 50 μm (4000−200/cm), and the far region, which includes wavelengths between 50 and 1000 μm (200−10/cm). Absorption bands in near-IR may be significant for organic compounds containing functional groups O—H, N—H, and C—H. Recently, this area has become important for hydrogen bond testing and for quantitative analysis. The most frequently used region for chemical analysis is the mid-IR region. The far-IR region is used to determine the structure of inorganic and organometallic compounds [20,21].

For IR radiation to be absorbed, the vibration frequency of the chemical bond must be equal to the radiation frequency, and the chemical bond must have the properties of an electric dipole. The frequency of active vibration in the IR spectrum is given by the following expression:

$$\nu = \frac{1}{2\pi}\sqrt{\frac{k}{\mu}} \tag{14.3}$$

where k is the force constant and μ is the reduced mass.

Each absorption maximum in the IR spectrum is characterized by its position, intensity, and shape (width and sometimes multiplicity). The position of the absorption band depends on the type of vibration of the chemical bond, that is, on its strength and mass of atoms in the bond. There are the following types of vibrations: stretching, which corresponds to higher values of the wave number ($\nu = 1/\lambda$), and bending, which corresponds to lower wave numbers. The intensity of the band depends on the magnitude of the change in the dipole moment.

It often happens that the number of absorption bands in the IR spectrum is less than the number of possible vibrations. This can happen if the vibration is inactive in the IR spectrum because it is not accompanied by a change in the dipole moment. Some vibrations can be degenerated (absorbed at the same frequency), owing to the symmetry of molecules or very low vibration frequencies involving heavy atoms (e.g., iodine), so they are often outside the measuring range of the device. Some bands are not detected, owing to low intensity.

The identification of absorption bands is performed on the basis of correlation tables or atlases, in which the frequencies of a large number of groups in different compounds are collected [22,23]. The position of the absorption bands can be influenced by interactions of atoms in the molecule, intermolecular interactions, and changes in the physical state of molecules. In organic molecules the factors that affect the position of the absorption maxima of individual functional groups are electronic effects (induced, resonant, field effects, and hybridization), hydrogen bonds, and bond angles (ring size). In the spectra of solids, in contrast to liquids and gases, there may appear bands as a result of a decrease in the symmetry of molecules or ions in the crystal field or as a consequence of interaction with other molecules or ions in the same crystal cell. In addition, bands can appear from the vibration of the crystal lattice. On this occasion the strips may split into doublets. Inorganic compounds have a relatively poor IR spectrum, in which identification is performed on the basis of characteristic salt anion bands or ligand bands of complex compounds. Many simple metal oxides do not absorb in the region 4000−650/cm, while oxides with more oxygen atoms show characteristic absorptions in a given area [20]. IR spectroscopy is a nondestructive method, and it is relatively easy to obtain a spectrum of samples in all three aggregate states (gas, liquid, and solid). Many inorganic compounds and a number of organic compounds can be tested in the solid state, either in the form of lozenges (KBr) or as a suspension of viscous oils (Nudzol). In cases in which the samples are strongly absorbed in the IR region of the spectrum, reflection techniques are applied: attenuated total reflection (ATR), internal reflectance spectroscopy, and diffuse reflectance spectroscopy. ATR is one of the most commonly used IR techniques and requires little or no sample preparation. IR spectroscopy is a qualitative and quantitative analysis because IR spectra are a characteristic feature of each compound [24]. Quantitative FT-IR analyses are based on measuring the intensity of the selected absorption maximum of the compound to be determined. The height of the absorption maximum is most often used, although the measurement of its surface gives more accurate results.

Pircheraghi et al. [9] reported the fabrication of a polyethylene separator for lead-acid batteries from recycled and fresh silica. Reclaimed silica from the spent lead-acid battery separator was obtained by pyrolysis. Fig. 14.4 shows the FTIR analysis of fabricated separators from fresh and recycled silica, that is, samples FS1.0 and RS1.0. Small differences in

298 Nanomaterials Recycling

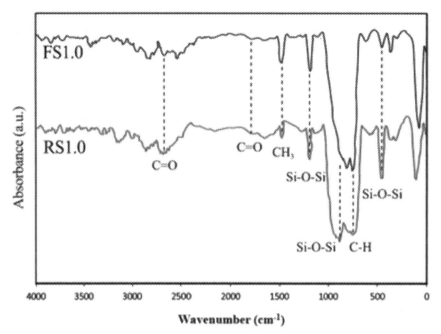

Figure 14.4 Fourier transform infrared spectroscopy spectrum of FS1.0 and RS1.0 as representative separator samples [9]. This figure shows as if it is taken by atomic force microscope. Please check.

the spectrum can be observed in the results as the smaller size of the recycled silica particles and the presence of some impurities in it compared to the fresh silica.

14.3 Raman spectroscopy

Raman spectroscopy is based on the inelastic scattering of electromagnetic radiation due to its interaction with the vibrational modes of the molecule. Indian physicist C.V. Raman (1928) noticed that in the spectrum of scattered radiation, in addition to the intense line that has the same frequency as the incident radiation, symmetrical lines of lower and higher frequencies of much lower intensity appear. The part of the scattering that does not change frequency when interacting with molecules is called Rayleigh scattering. The component of Raman radiation of lower frequency, v^-_R, is known as Stokes radiation, and the component of higher frequency, v^+_R, as anti-Stokes radiation. The intensity of Stokes radiation is significantly higher than the intensity of anti-Stokes. The magnitude of the change in frequency, $\Delta v = v_0 - v^{\pm}_R$, is called the Raman shift. It is characteristic of each molecular species and does not depend on incident radiation. In the Raman spectrum, only those oscillations of molecules that change the polarizability of molecules can appear. The basic parameters that determine the appearance of the Raman spectrum are the strength and type of bond between the atoms in the molecule, the symmetry of the molecules as a whole, and the symmetry of the oscillations that give the bands in the spectrum. In the case that the frequency (energy) of the excitation radiation is close to the energy of the electronic transitions into the molecule, the radiation is resonantly scattered, and resonant Raman spectra are formed. Raman spectroscopy most often uses high-intensity laser radiation with wavelengths in the visible and UV ranges and, less often, X-rays. Molecules that have a center of oscillation symmetry that show bands in the IR spectrum will be inactive in the Raman spectrum and vice versa. For less symmetric and asymmetric molecules this is not the case.

In the process of recycling nanomaterials, Raman spectroscopy can be useful for identifying molecules and crystal structure and determining the size and shape of nanoparticles [10,24].

Schauerman et al. [10] reported about recycling single-walled carbon nanotube (SWCNT) anodes from lithium ion batteries. Purification of the SWCNT material occurs through a series of acid and thermal treatments. During the using of a SWCNT/lithium battery, a portion of the lithium atoms react with the nonaqueous solvent at the electrode surface and form the solid electrolyte interphase (SEI) layer. Ultimately, the SEI layer can continue to grow until it negatively affects the performance of the battery. The position of the Raman peak for end-of-life single-wall carbon nanotubes (EOL-SWCNTs) is shifted to the right, owing to the existence of the SEI layer (Fig. 14.5). The positions of the pure SWCNTs and acid-treated EOL-SWCNTs (HCl acid and acid reflux) peaks are similar, indicating that the SEI layers were removed in the recycling process.

14.4 X-ray photoelectron spectroscopy

X-ray photoelectron spectroscopy (XPS) is also called electron spectroscopy for chemical analysis based on the photoelectric effect. The XPS method is based on the irradiation of a sample with X-rays and on the measurement of kinetic energy (velocity) and the number of electrons emitted. Al Kα (1486.6 eV), Mg Kα (1253.6 eV), and Ti Kα (4510.9 eV) radiation are/is most commonly used for material testing. Owing to the action of X-rays, electrons can be ejected by electron orbitals; their kinetic energy (E_K) is discrete and is determined by the binding energy of the orbitals (E_B) according to the following equation:

$$E_K = h\nu - E_B - E_W \tag{14.4}$$

where $h\nu$ is the energy of the applied X-rays and E_W is the work function of the electron.

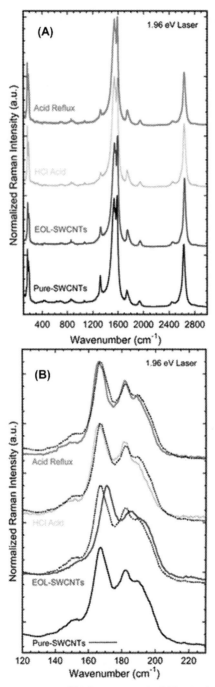

Figure 14.5 Normalized Raman spectra of high-purity end-of-life-single-wall carbon nanotubes and high-purity recycled single-wall carbon nanotubes [10].

XPS is one of the best methods for examining the surface structure of materials, elemental compositions, oxidative state of elements, uniformity of elements, determination of empirical formula of pure materials, and detection of impurities that contaminated the surface. The detection limits for most chemical elements are of the order of parts per million. XPS detects all elements with an ordinal number greater than Li. H and He cannot be detected because the diameter of their orbitals is so small that it reduces the probability of X-ray interaction with electrons in their orbitals to almost zero.

The maximum depth of analysis for an XPS measurement is about 10 nm. XPS is successfully used for the analysis of nanomaterials: inorganic and organic compounds, metals and their alloys, polymers, and biological materials [25]. The method is very useful in the analysis of the surface chemistry of the material in its basic state and/or after some treatment, such as recycling.

The binding energies of the internal electrons of each chemical element are unique, so they are used for their direct identification (qualitative analysis). In addition to the stated obtained spectrum with tabular values and in Internet databases, it is possible to identify a chemical element [26,27]. If the test element is part of a chemical compound, its XPS spectrum changes as a result of chemical shift.

The chemical shift can be explained by the effective charge potential change on an atom due to the action of another type of atom in a chemical bond or in the immediate vicinity. Owing to the mutual interaction of different atoms, the energies of their electronic orbitals change. For example, when an atom is bonded to an oxygen atom, which has higher electronegativity, a charge transfer to the oxygen and the effective charge of the former becomes positive, thus increasing the binding energy of its electrons. By contrast, the binding energy of the oxygen atom with higher electronegativity is decreased. Therefore XPS peaks of metal oxides typically show a shift of the metal to higher binding energies, where the increase in shift is proportional with the increase of valence state of the metal atom. The lines of the XPS spectrum are named after the orbital from which the electron was ejected (e.g., 1s, 2s, and 2p).

If a chemical element builds multiple bonds in a material, the resulting peaks are split into several secondary peaks that sometimes join. Deconvolution of the obtained peak on its components is usually performed by using software, in which each component corresponds to a certain type of chemical bond. The relative concentration of elements in the examined sample can be determined by comparing the relative intensity ratios of XPS lines in the photoelectron spectrum. XPS spectra are usually given with a number of detected electrons (sometimes per unit time) as a function of the binding energy.

$PM_{2.5}$ particles are present in the air with a size (diameter) generally less than 2.5 μm and may contain a significant amount of nanoparticles. High concentrations of $PM_{2.5}$ can have a very complex and detrimental effect on human health [28]. On the other hand, many technological solutions clearly reduce the emission of motor vehicle nanoparticles, such as catalysts, low-sulfur fuels, and diesel particulate filters.

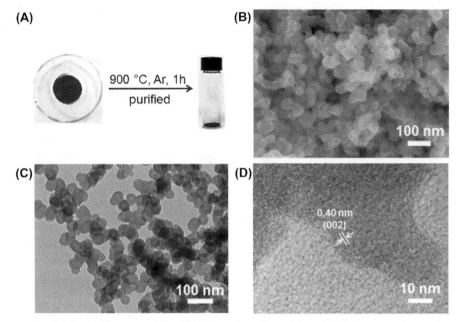

Figure 14.6 (A) Optical images of the PM$_{2.5}$ sample collected by a PTFE filter membrane in an air sampler and the product of PM-carbon nanoparticles after thermal annealing and purification. (B) Scanning electron microscopy image. (C) Transmission electron microscopy image. (D) HRTEM image of as-obtained PM-carbon nanoparticles [29].

Zhu et al. [29] reported that nitrogen-doped carbon nanoparticles (PM-CNPs) derived from diesel engines PM$_{2.5}$ emissions can serve as excellent electrode material for supercapacitors and efficient metal-free electrocatalysts toward oxygen reduction reaction. Morphologic analysis by scanning electron microscopy (SEM) and transmission electron microscopy (TEM) has shown that the PM-CNPs consist of carbon nanoparticles with diameters distributed between 25 and 40 nm, forming necklace-like networks (Fig. 14.6).

The XRD pattern (Fig. 14.7A) possesses a sharp peak around 25 degrees and another weak peak around 44 degrees, which are characteristic reflections of the (002) and (100) planes of carbon materials. The Raman spectrum of PM-CNPs (Fig. 14.7B) shows the D band at 1358/cm (related to defects) and the G band at 1693/cm (related to the crystalline graphite). The chemical states and atomic ratios of C, O, and N elements existing in the PM$_{2.5}$ derived carbon nanoparticles were analyzed by XPS. The C 1s signal comprises four major peaks at 284.6, 286.1, 287.8, and 289.1 eV, corresponding to C—C, C—O, C=O, and O—C=O bonds, respectively (Fig. 14.7C). The N 1s spectrum in Fig. 14.2D can be straightened into two peaks: pyrrolic nitrogen at 400.5 eV and pyridinic nitrogen at 398.7 eV.

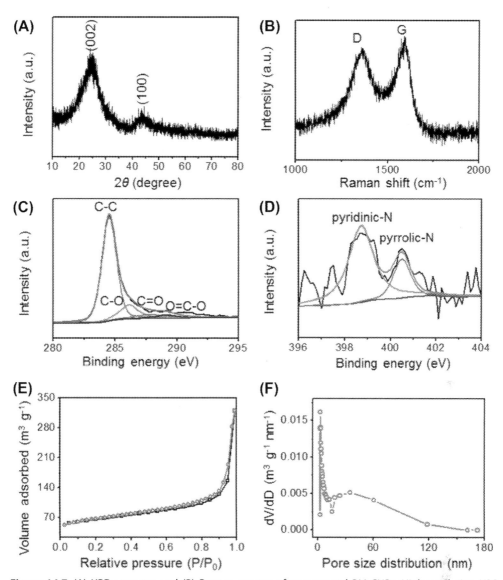

Figure 14.7 (A) XRD patterns and (B) Raman spectra of as-prepared PM-CNPs. High-resolution XPS spectra of PM-CNPs at (C) C 1s and (D) N 1s regions. (E) Nitrogen adsorption-desorption isotherms and (F) BJH desorption pore size distribution of PM-CNPs [29].

14.5 Auger electron spectroscopy

Auger electron spectroscopy (AES) is used to examine the chemical composition of the surface of solid materials. It is very useful for studying surface adsorption and reaction. The basis of this method is the Auger effect, which is based on the appearance of

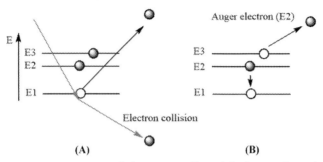

Figure 14.8 Schematic representation of the Auger effect. (A) The incident electron creates an electron gap at the 1s level. (B) An electron from the 2s level fills the 1s hole, while the transition energy is transferred to the 2p electron being emitted.

electron quantized energy in the mass of secondary electrons when the substance is irradiated with X-rays or high-energy electrons. When an atom is excited by external radiation or a beam of a high-energy electron (2–50 keV), an electron is ejected from the internal electronic levels. The resulting electronic gap is filled by an electron with some of the higher electronic levels. The energy difference between the electron levels will be manifested in the emission of a quantum of X-rays or may be coupled with the emission of an electron with some other higher energy level, which then represents the Auger electron (Fig. 14.8).

The emitted electron will have the kinetic energy E_A of the electron beam (Eq. 14.5):

$$E_A = E_1 - E_2 - E_3' - \Delta\phi \tag{14.5}$$

where E_A is the energy of the Auger electron; E_1, E_2, and E_3' are the energy of the electron in the basic electronic levels, the first higher energy level, and the second higher level, respectively; and $\Delta\phi$ represents the output work for the tested material. E_3' is not equal to the energy of the corresponding level of the neutral atom; because of ionization, there was a certain redistribution of energies by electronic levels. Since Auger electrons have a very low energy, only those in the thin surface layer of the test sample can leave the sample.

The marking of Auger electrons is performed on the basis of the electronic levels that participate in their formation, such as KLL and LMM. Hydrogen and helium cannot be detected by the AES method because they do not have enough electrons. As the ordinal number increases, so does the number of potential Auger crossings. The most common electronic transitions are between close energy levels. Elements with ordinal numbers 3–14 are predominantly analyzed with CLL transitions, while elements with a higher ordinal number (15–39) are mainly analyzed with LMM transitions. Elements with ordinal numbers from 40 to 79 are mostly analyzed with MNN transitions, while even more difficult elements are analyzed with the help of NVV transitions. Electrons in valence energy levels can also fill gaps or be emitted during KVV transition.

The AES spectrometer consists of a high-vacuum chamber, electron cannon, and ion cannon for cleaning the sample surface, an electron energy analyzer and components for spectrum registration. The ion cannon, in addition to its role of cleaning the surface of the sample from the products formed by the action of the electron beam, can also remove the surface layers and thus prepare the sample for the analysis of deeper layers. The quality of AES is based on the fact that Auger electrons are emitted with quantized energies, characteristic of chemical elements. The difficulties that arise in qualitative analysis are the proximity of lines originating from different chemical elements and the existence of a chemical shift (Ag4). Quantitative analysis is based on the line intensity of the spectra of Auger electrons. AES is very sensitive to surface elements; it is very useful for studying surface adsorption and reaction [30−32]. During recycling, adsorption of ions from the solution or their diffusion from the melt into the surface layers of recycled nanoparticles can occur. Various other chemical processes can take place on the surface of nanoparticles, such as oxidation and carbon dioxide binding [33]. Fig. 14.9 shows the LMM Auger energy of Zn exposed in various oxygen exposures.

The Zn LMM Auger energy of pure zinc is 991.6 eV, and that of ZnO is 987.7 eV. When zinc atoms on the are surface exposed to oxygen, the Auger peak Zn LMM shifts to a lower energy. Zinc atoms react with oxygen and form ZnO, and the zinc atom loses two electrons and becomes a positive ion. Owing to zinc oxidation, the binding energy of the zinc atom increases a little. Fig. 14.9 shows that the shift is not completely toward ZnO, which indicates that only Zn atoms are oxidized on the surface.

14.6 X-ray fluorescence analysis

X-ray fluorescence (XRF) spectroscopy is a method that allows rapid and multielemental qualitative and quantitative analysis of recycled nanoparticles [9]. The method is nondestructive, it has high precision and accuracy, and detection can be achieved at the level of parts per million or a few tenths of a percent. The basic principle of this method is based on the ability of atoms to absorb X-rays, which results in the excitation of electrons (ionization) from the inner part of the electron shell, thus creating electronic cavities. Electrons from the outer shells fill these cavities with the emission of fluorescent X-rays, which is characteristic of each chemical element. The disadvantage of the method is the existence of an interelemental influence that requires correction, and elements with an ordinal number $Z < 9$ (F) cannot be determined [34]. Based on the emitted radiation, qualitative and quantitative information on the elements present in the examined sample was obtained. Qualitative analysis is based on the determination of wavelength (WD XRF) or radiated energy (ED XRF), with the

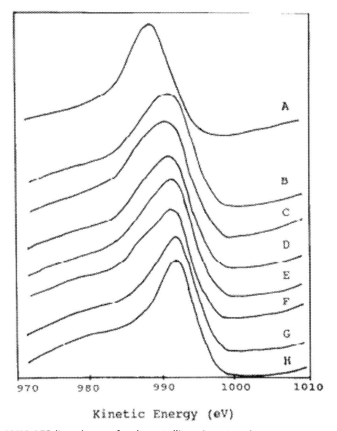

Figure 14.9 Zn LMM AES line shape of polycrystalline zinc at various oxygen exposures. (A) 2110. (B) 3000 L. (C) 2500 L. (D) 2000 L. (E) 1000 L. (F) 750 L. (G) 300 L. (H) zinc [33].

comparison of standard spectra for individual elements. It is necessary to identify all peaks in the spectrum, owing to the appearance of false peaks that can, for example, originate from coherent and incoherent X-rays. Quantitative analysis is based on measuring the integral peak intensity or peak height. The intensity of XRF radiation, in addition to the concentration of the analyte, depends on many factors, such as sample thickness, radiation absorption, and various geometric factors. Owing to the absorption of primary radiation and the absorption of fluorescent radiation, only a thin surface layer of material is analyzed by the described method.

14.7 Scanning electron microscopy

A scanning electron microscope is a device that provides information on sample structure, particle size, sample morphology, surface texture, and chemical composition using a high-energy electron beam [35]. SEM produces an image of a sample by scanning its surface with a very precisely focused beam of high-energy electrons. The magnification at SEM go upto 500,000 times, with a resolution of upto 0.6 nm. The power of SEM decomposition depends on the diameter of the electron beam and the current. The electron beam is partially disintegrated, owing to the interaction with the sample, which leads to an increase in its diameter. The electron beam current must be strong enough to show an image of each point of the sample on the screen. The electron beam interacts with the electrons in the sample, producing secondary electrons, backscattered electrons, and characteristic X-rays. SEM produces several different signals, which can be used to create the final image of the sample: secondary electrons, backscattered electrons, X-rays, electron current, light, and transmitted electrons. Secondary electrons are low-energy, so only those electrons that come out at or near the sample surface can reach the detector. Therefore the image of the sample obtained from the secondary electrons represents the true image of the surface. The amount of secondary electrons depends on the relief (slope of the surface) from which these electrons come and is used to determine the topography of the surface. Backscattered electrons are formed during elastic scattering, and their energies are close to the energies of incident electrons and can originate from a depth of $1-2$ μm. The image obtained with backscattered electrons represents, to some extent, the depth of the sample. The number of backscattered electrons depends on the ordinal number of atoms present in the sample and is often used together with the generated X-rays for qualitative chemical analysis of the tested sample. Secondary electron detectors are present in all SEM devices. Samples for SEM imaging must be nonvolatile and electrically conductive.

The process of recycling batteries combines two stages: a mechanical recycling process (physical process) and a metallurgy processes [36]. Materials such as electrodes (anode, cathode), electrolyte, and separator can be obtained after the proper mechanical conditioning treatment. Further recycling depends on the physicochemical characteristics of each component. The separator is a permeable membrane that prevents electrical contact in between the electrodes.

Because of the importance of recycling waste materials in the industry, an especially considerable amount of waste separators obtained from a lead-acid battery module built from polyethylene and silica nanoparticles (SNPs) was recycled through a pyrolysis process (at 700°C) [37]. A temperature of 700°C is often used to remove (pyrolyze) plastic components from separators and other battery parts [38]. The SNPs used in battery separators were nano-sized. FESEM images (Fig. 14.10C) showed very useful information about the state of SNPs after the pyrolysis process at 700°C. Particle size ranges from 30 nm in tiny particles to 200 nm in agglomerated ones. This shows that there were initial stages of sintering and partial sticking of the particles. According to the results of XRF (Fig. 14.10A), the purity of the obtained SNP was 96%, showing

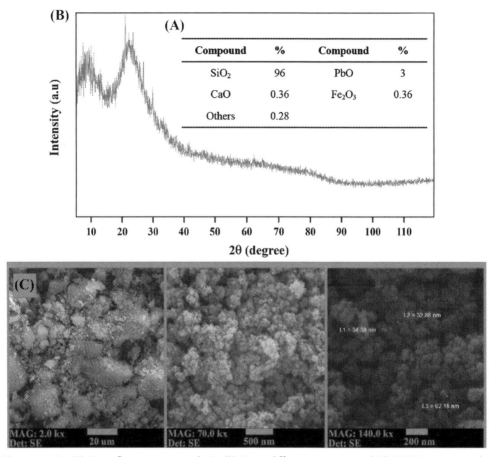

Figure 14.10 (A) X-ray fluorescence analysis, (B) X-ray diffraction curve, and (C) FESEM micrographs of the recycled silica nanoparticles from spent lead-acid battery separators [37].

that the recycling process was successful. A small amount of PbO (3 wt.%) was seen in the product. The XRD (Fig. 14.10B) confirmed an amorphous structure of silica (cristobalite) that remained stable during pyrolysis.

14.8 Transmission electron microscopy

TEM provides the most powerful magnification and can also be used to study the particle size, sample morphology, surface texture, crystal structures, and chemical composition of a sample [39,40]. The most modern devices achieve below the atomic resolution of 0.08 nm, i.e. magnification of 150,000,000×. The principle of operation of TEM is based on the fact that a beam of high-energy electrons is passed through a thin layer of samples, usually below 100 nm in thickness. The TEM consists of three systems. The first contains an electron-emitting source, in which the electrons are accelerated to an energy in the range 20−1000 keV, and a condenser system, which produces a beam of electrons with a desired diameter and focuses the beam onto the object. The second is the image-producing system, consisting of a movable specimen stage, the objective lens, and intermediate and projector lenses. The beam strikes the specimen, and the scattered or unscattered electrons are directed by a series of electromagnetic lenses and are projected onto the screen, obtaining electron diffraction, an amplitude-contrast image, a phase-contrast image, or a shaded image of different shading intensity, depending on the density of unscattered electrons. The third consists of an image capture system typically containing a fluorescent screen for viewing and focusing an image and a digital camera for permanent recordings. All parts of the microscope through which the electron beam passes are in a very high vacuum because the particles in space could stop or slow down the movement of electrons. The formation of contrast in TEM is based on the elastic interaction of incident electrons and atoms of the tested sample. The mass-thickness contrast depends on the mass of the atoms in the sample and the thickness of the sample. The probability of electronic deflection from the initial paths increases with the increase of the charge, that is, the mass of the atom. Areas containing heavier atoms will be darker than those containing lighter atoms. The number of scattered electrons is higher in thicker samples. The contrast caused by the crystal structure of the sample is called the diffraction contrast. Images on a TEM microscope can be obtained in two ways: in a light field and in a dark field. The first method is usually applied, in which only electrons that have passed through the sample without deflection participate in the formation of the image. Dense regions become darker, while regions of lower density (also in terms of crystal lattice or atomic number) or without a pattern in the path of the electron become brighter. This method represents a two-dimensional projection of the sample. The dark field mode uses electrons that bounce off the crystallographic layers sample. In this mode the image is formed so that the parts without the pattern remain dark. Electron diffraction obtained during TEM imaging provides important crystallographic data for the identification of a crystal material.

The waste recycling process can be monitored on the basis of TEM images. Fig. 14.11 shows the protocol for low-temperature recycling of electronic wastes such as printed circuit boards [3].

14.9 Atomic force microscopy

Atomic force microscopy (AFM) is a high-resolution type of scanning microscopy, which has a resolution of 0.1 nm in the sample plane (x, y) and 0.001 nm in the z-axis for hard and flat surfaces and 0.75−5 nm for soft materials (e.g., polymers, biological materials) [41,42]. It can be used to test the size, shape, structure, and

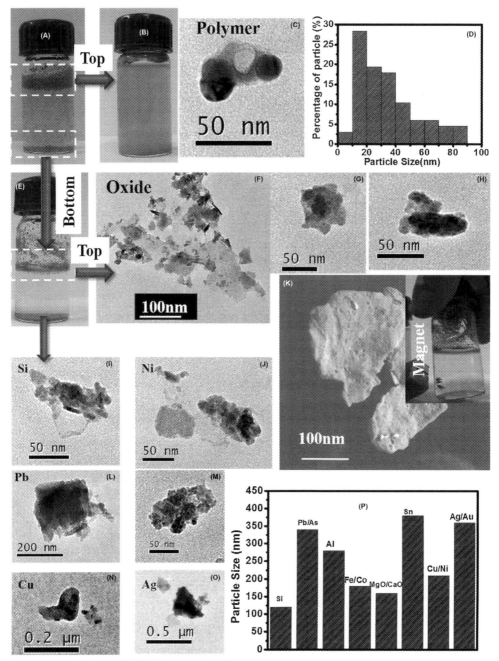

Figure 14.11 Digital image of the powder produced in milling and dissolved in water (A). The top part after mixing with distilled water and sonication for 5 min (B). Transmission electron microscopy images of the colloid particle (C). Particles distribution of the colloids in accordance with several transmission electron microscopy images (D). The powder collected from the bottom of the water is further dissolved in water, showing two separate layers (E). Transmission electron microscopy image of top oxide layer (F). High magnification of image of two individual oxide particles of lead oxide and silicon oxide (G, H). (I). Transmission electron microscopy images of individual metallic particles collected from bottom layer of the solution (J, L, O). Separation of magnetic particles with the help of a magnet (K). The mean particle size of different metals and oxides with the help of EDS attached to transmission electron microscopy (P) [3].

aggregation of nanoparticles. AFM gives a three-dimensional surface profile and does not require prior sample preparation, conductive sample surface, or a special vacuum environment, so measurements can be performed in atmospheric conditions. AFM is based on measuring the force between the test surface and the needle (tip) of the microscope. The measuring probe is a console of microscopic dimensions with a sharp tip of less than 10 nm in diameter that slides on the sample surface, and its bending is measured optically (by using laser interferometry) or electrically (by piezoelectric methods when the console is made of a piezoelectric material such as quartz). The console has a certain resonant oscillation frequency (50—150 kHz). The change in oscillation frequency of the console is proportional to the force acting between the tip of the needle and the sample for the reconstruction of the surface image. The dominant forces of interactions in AFM are Van der Waals forces, while the magnetic and electrostatic forces occur at greater distances from the sample. The topography of the sample surface was obtained as a result of changes in the amplitude, phase, or oscillation frequency of the console.

There are three types of sample surface recording modes: contact, noncontact, and tapping modes. In the contact mode, the tip of the needle is always in contact with the surface of the sample, and its topography is obtained as a result of monitoring the change in the oscillation frequency of the console. The total force of interaction between the needle tip and the sample surface is always repulsive and is maintained constant during scanning. The deflection of the probe needle is proportional to the force acting on its tip, according to Hooke's law (Eq. 14.6):

$$F = -kx \qquad (14.6)$$

where F is force, k is the elastic constant, and x is bending of the holder.

In the noncontact mode, the tip of the needle is not in contact with the surface of the sample but is located 50—150 Å above it. Then an attractive Van der Waals force acts on the needle, causing the console to bend. Van der Waals forces are weaker than the forces in the contact mode. The contactless mode of operation is suitable for the characterization of biological materials and polymers. In the tapping mode, the tip of the needle is occasionally in contact with the sample surface.

Clay minerals, because of their fine particle size (often nano-sized), extensive surface area, layer structure, and surface charge, are the subject of a substantial amount of research in science and technology in the field of heterogeneous catalysis. Many organic reactions are catalyzed by nanoclay or modified nanoclay materials [43]. Clay catalysts can be easily recycled from the reaction mixture without a significant loss of activity. Tajbakhsh et al. [44] reported that sulfonated nanoclay (Cloisite 30B) can be used efficiently in the catalytic synthesis of quinoxaline derivatives and is easily recycled. The yield of the catalytic reaction is 85%—100%. After completing the reaction, the catalyst remains insoluble and is recycled by rinsing with dichloromethane and drying at 120°C for 2 h (reused in subsequent catalytic cycle). The catalytic activity of the catalyst in terms of yields slightly decreased with the increase of catalytic cycles. The AFM images (Fig. 14.12) of the recycled catalyst after the first catalytic (Fig. 14.12D,E) cycle, in comparison with primary catalyst (Fig. 14.12A,B), showed that there was a small aggregation of clay nanoparticles and a small increase average size, 40 nm. The decrease in the activity of the catalyst is related to the resulting morphological changes.

14.10 Electron probe microanalysis

Electron probe microanalysis is used to qualitatively and quantitatively determine the chemical composition of the test sample. The instrumentation required for electronic microanalysis, known as the electronic microprobe, is almost identical to the instrumentation present in SEM, so microanalysis is most often performed in SEM to which appropriate detectors have been attached. In the first phase, a focused electron beam from the SEM is used to select a portion of the sample surface for analysis; in the second phase, it is used to excite X-ray emission. The incident beam of electrons reacts with the material, part of the electron dissipates elastically, and part of the electron loses its energy in contact with the atom of the material, emitting characteristic X-rays and inhibitory X-rays. The resulting primary X-ray on its way through the sample can further excite the atom, so part of its energy is spent on secondary X-ray radiation (X-ray fluorescence, primary and characteristic). Some atoms in the material deoxidize by emitting characteristic Auger electrons. For elemental analysis, only the characteristic X-ray radiation is important, while continuously it is only a nuisance. Qualitative analysis is established on the identification of the chemical element based on the measurement of the wavelength of the characteristic X-ray emission radiation. Quantitative analysis is based on comparing the line intensities of the observed element in the sample and in the standard.

Depending on the method of X-ray detection, two methods of electronic microanalysis are distinguished: energy dispersive spectroscopy (EDS/EDX) and wave dispersion spectroscopy (WDS/WDX).

EDS/EDX analysis is based on the measurement of X-ray energy, which is a characteristic of each element from which it is emitted. The advantage of EDS is the short analysis time, in which the complete spectrum of chemical elements can be analyzed. The main disadvantages of the EDS technique are poor energy resolution and lower accuracy of quantitative analysis.

In WDS/WDX analysis, X-rays are separated according to their wavelength, which is measured by the principles of XRD on a crystal in accordance with Bragg's law. For one crystal and one crystal position, X-rays of only one wavelength can be detected. If a larger number of wavelengths are to be detected, the

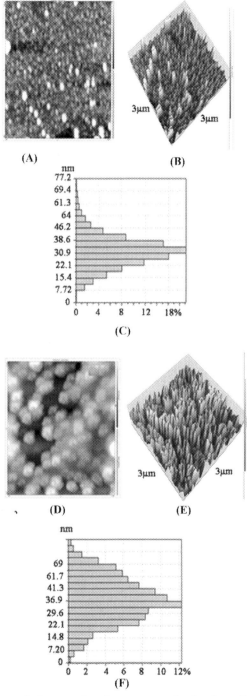

Figure 14.12 AFM images of phase and surface topography for the sulfonated nanoclay catalyst (A, B) and the recycled catalyst (D, E) and images of the distribution of particle size for the catalyst (C) and for the recycled catalyst (F) [44].

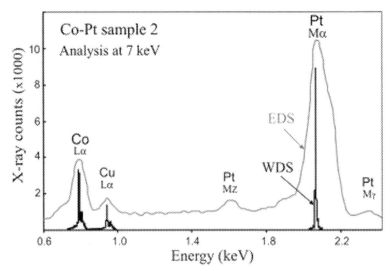

Figure 14.13 The energy dispersive spectroscopy spectrum acquired at 7 keV from a Co-Pt sample. Superimposed wave dispersion spectroscopy peaks reveal superior resolution and peak-to-background ratios compared to energy dispersive spectroscopy [45].

instrument allows the position of the crystals and detectors to be moved in a circular path. Each spectrometer is equipped with several diffraction crystals with different crystallographic constants to cover the whole spectrum of X-rays of interest for microelectronic analysis. The WDS method is suitable for the detection of lighter chemical elements. In WDS analysis the overlap of peaks of similar energies is significantly less pronounced than in EDS analysis. Fig. 14.13 shows the X-ray spectrum of the same sample of Co-Pt thin film, recorded by the EDS method and the WDS method [45].

Acknowledgments

This work is supported by the Ministry of Education, Science and Technological Development of the Republic of Serbia.

References

[1] C.F. Holder, R.E. Schaak, Tutorial on powder X-ray diffraction for characterizing nanoscale materials, ACS Nano 13 (2019) 7359–7365.
[2] V. Stanić, Dj Janaćković, S. Dimitrijević, S.B. Tanasković, M. Mitrić, M.S. Pavlović, et al., Synthesis of antimicrobial monophase silver-doped hydroxyapatite nanopowders for bone tissue engineering, Appl. Surf. Sci. 257 (2011) 4510–4518.
[3] C.S. Tiwary, S. Kishore, R. Vasireddi, D.R. Mahapatra, P.M. Ajayan, K. Chattopadhyay, Electronic waste recycling via cryo-milling and nanoparticle beneficiation, Mater. Today 20 (2017) 67–73.

[4] V.K. Booramurthy, R. Kasimani, D. Subramanian, S. Pandian, Production of biodiesel from tannery waste using a stable and recyclable nano-catalyst: an optimization and kinetic study, Fuel 260 (2020) 116373.

[5] B. Niu, Z. Xu, Innovating e-waste recycling: from waste multi-layer ceramic capacitors to Nb\\Pb codoped and ag-Pd-Sn-Ni loaded $BaTiO_3$ nano-photocatalyst through one-step ball milling process, Sustain. Mater. Technol. 21 (2019) e00101.

[6] A. Deep, A.L. Sharma, G.C. Mohanta, P. Kumar, K.-H. Kim, A facile chemical route for recovery of high quality zinc oxide nanoparticles from spent alkaline batteries, Waste Manage 51 (2016) 190−195.

[7] L.-H. Zhang, S.-S. Wu, Y. Wan, Y.-F. Huo, Y.-C. Luo, M.-Y. Yang, et al., Mn_3O_4/carbon nanotube nanocomposites recycled from waste alkaline $Zn-MnO_2$ batteries as high-performance energy materials, Rare Metals 36 (2017) 442−448.

[8] C. Gan, C. Zhang, P. Liu, Y. Liu, W. Wen, B. Liu, et al., Polymeric carbon encapsulated Si nanoparticles from waste Si as a battery anode with enhanced electrochemical properties, Electrochim. Acta 307 (2019) 107−117.

[9] G. Pircheraghi, M. Nowrouzi, Sh Nemati, Fabrication of polyethylene separator for lead-acid batteries from waste and recycled silica and investigation of its performance, J. Clean. Prod. 250 (2020) 119535.

[10] C.M. Schauerman, M.J. Ganter, G. Gaustad, C.W. Babbitt, R.P. Raffaelle, B.J. Landi, Recycling single-wall carbon nanotube anodes from lithium ion batteries, J. Mater. Chem. 22 (2012) 12008.

[11] Q. Liu, J. Cao, K.Q. Li, X.H. Miao, G. Li, F.Y. Fan, et al., Chromosomal aberrations and DNA damage in human populations exposed to the processing of electronics waste, Environ. Sci. Pollut. Res. 16 (2009) 329−338.

[12] B. Ghosh, M.K. Ghosh, P. Parhi, P.S. Mukherjee, B.K. Mishr, Waste printed circuit boards recycling: an extensive assessment of current status, J. Clean. Prod. 94 (2015) 5−19.

[13] J. Cui, L. Zhang, Metallurgical recovery of metals from electronic waste: a review, J. Hazard. Mater. 158 (2008) 228−256.

[14] V.Đ. Stanić, B.K. Adnađević, S.I. Dimitrijević, S.D. Dimović, M.N. Mitrić, B.B. Zmejkovski, et al., Synthesis of fluorapatite nanopowders by a surfactant-assisted microwave method under isothermal conditions, Nucl. Technol. Radiat. Prot. 33 (2018) 180−187.

[15] J.V. Wright, J.L. Conca, United States Patent #6217775 Stabilisation and Remediation of Soil, Water and Wastes Contaminated with Metals Using Fish Bones and Hard Parts, United States Patent Office, Alexandria, VA, 2000.

[16] S.V. Dorozhkin, Nanodimensional and nanocrystalline apatites and other calcium orthophosphates in biomedical engineering, biology and medicine, Materials 2 (2009) 1975−2045.

[17] P. Shi, M. Liu, F. Fan, C. Yu, W. Lu, M. Du, Characterization of natural hydroxyapatite originated from fish bone and its biocompatibility with osteoblasts, Mater. Sci. Eng. C. 90 (2018) 706−712.

[18] C. Baudota, C.M. Tanb, J.C. Konga, FTIR spectroscopy as a tool for nano-material characterization, Infrared Phys. Technol. 53 (2010) 434−438.

[19] G. Vucković, S. Tanasković, Z.M. Miodragović, V. Stanić, High-spin binuclear Co(II) complexes with a pendant octaazamaclocycle and carboxylates, J. Serb. Chem. Soc. 72 (2007) 1295−1308.

[20] E. Kendix, G. Moscardi, R. Mazzeo, P. Baraldi, S. Prati, E. Joseph, et al., Far infrared and Raman spectroscopy analysis of inorganic pigments, J. Raman Spectrosc. 39 (2008) 1104−1112.

[21] J.-M. Ortega, F. Glotin, R. Prazeres, X. Li, R. Gref, Far infrared micro-spectroscopy: an innovative method to detect individual metal−organic framework particles, Appl. Opt. 56 (2017) 6663−6667.

[22] K. Nakamoto (Ed.), Infrared and Raman Spectra of Inorganic and Coordination Compounds: Part A: Theory and Applications in Inorganic Chemistry, sixth ed., John Wiley & Sons, 2008. Available from: https://doi.org/10.1002/9780470405840.

[23] G. Socrates (Ed.), Infrared and Raman Characteristic Group Frequencies Tables and Charts, John Wiley & Sons Ltd, Chichester, 2004, pp. 323−326.

[24] A.O. Odeh, Oualitative and quantitative ATR-FTIR analysis and its application to coal char of different ranks, J. Fuel Chem. Technol. 43 (2015) 129−137.

[25] S. Hofmann (Ed.), Auger- and X-Ray Photoelectron Spectroscopy in Materials Science, Springer-Verlag, Berlin Heidelberg, 2013.

[26] D. Briggs, Handbook of X-ray Photoelectron Spectroscopy C.D. Wanger, W.M. Riggs, L.E. Davis, J.F. Moulder and G.E. Muilenberg Perkin-Elmer Corp., Physical Electronics Division, Eden Prairie, Minnesota, USA, 1979. 190 pp. $195, Surf. Interface. Anal. 3 (4) (1981). Available from: https://doi.org/10.1002/sia.740030412.

[27] C.D. Wagner, A.V. Naumkin, A. Kraut-Vass, J.W. Allison, C.J. Powell, J.R. Rumble, NIST X-Ray Photoelectron Spectroscopy Database, SRD 20, Version 3.5, National Institute of Standards and Technology, Gaithersburg, 2008.

[28] T. Rönkkö, H. Timonen, Overview of sources and characteristics of nanoparticles in urban traffic-influenced areas, J. Alzheimers Dis. 72 (2019) 15−28.

[29] G. Zhu, T. Chen, Y. Hu, L. Ma, R. Chen, H. Lv, et al., Recycling PM2.5 carbon nanoparticles generated by diesel vehicles for supercapacitors and oxygen reduction reaction, Nano. Energy 33 (2017) 229−237.

[30] Q. Cheng, C. Li, V. Pavlinek, P. Saha, H. Wang, Surface-modified antibacterial TiO_2/Ag^+ nanoparticles: preparation and properties, Appl. Surf. Sci. 252 (2006) 4154−4160.

[31] A.R. Gheisi, C. Neygandhi, A.K. Sternig, E. Carrasco, H. Marbach, D. Thomele, et al., O_2 adsorption dependent photoluminescence emission from metal oxide nanoparticles, Phys. Chem. Chem. Phys. 16 (2014) 23922−23929.

[32] R. Ramamoorthy, M.K. Kennedy, H. Nienhaus, A. Lorke, F.E. Kruis, H. Fissan, Surface oxidation of monodisperse SnO_x nanoparticles, Sens. Actuat. B Chem. 88 (2003) 281−285.

[33] Z. Yongfa, S. Yangming, A study of the oxygen adsorption and initial oxidation on polycrystalline zinc by AES line shapes and EELS, Surf. Sci. 275 (1992) 357−364.

[34] M. Todorović, P. Djurdjević, V. Antonijević, Optičke Metode Instrumentalne Analize, Hemijski fakultet, Beograd, 1997.

[35] V. Stanić, S. Dimitrijević, D.G. Antonović, B.M. Jokić, S.P. Zec, S.T. Tanasković, et al., Synthesis of fluorine substituted hydroxyapatite nanopowders and application of the central composite design for determination of its antimicrobial effects, Appl. Surf. Sci. 290 (2014) 346−352.

[36] L. Yun, D. Linh, L. Shui, X. Peng, A. Garg, M.L.P. Le, et al., Metallurgical and mechanical methods for recycling of lithium-ion battery pack for electric vehicles, Resour. Conserv. Recycl. 136 (2018) 198−208.

[37] Sh Nemati, G. Pircheragh, Fabrication of a form-stable phase change material with green fatty acid and recycled silica nanoparticles from spent lead-acid battery separators with enhanced thermal conductivity, Thermochim. Acta 693 (2020) 178781.

[38] H.-J. Kim, T.N.V. Krishna, K. Zeb, V. Rajangam, C.V.V. Muralee Gopi, S. Sambasivam, et al., A comprehensive review of li-ion battery materials and their recycling techniques, Electronics 9 (2020) 1161.

[39] Y. Leng (Ed.), Materials Characterization. Introduction to Microscopic and Spectroscopic Methods, Wiley-VCH Verlag GmbH & Co, KgaA, 2013.

[40] V. Stanić, S. Dimitrijević, J. Antić-Stanković, M. Mitrić, B. Jokić, I.B. Plećaš, et al., Synthesis, characterization and antimicrobial activity of copper and zinc-doped hydroxyapatite nanopowders, Appl. Surf. Sci. 256 (2010) 6083−6089.

[41] V. Jokanović, Instrumentalne metode ključ razumevanja nanotehnologije i nanomedicine, Inženjerska akademija Srbije, Institut za nuklearne nauke Vinča, Beograd, 2014.

[42] V. Stanić, A.S. Radosavljević-Mihajlović, V. Živković-Radovanović, B. Nastasijević, M. Marinović-Cincović, J.P. Marković, et al., Synthesis, structural characterisation and antibacterial activity of Ag^+-doped fluorapatite nanomaterials prepared by neutralization method, Appl. Surf. Sci. 337 (2015) 72−80.

[43] B.K.G. Theng (Ed.), Clay Mineral Catalysis of Organic Reactions, CRC Press, 2018.

[44] M. Tajbakhsh, M. Bazzar, S.F. Ramzanian, M. Tajbakhsh, Sulfonated nanoClay minerals as a recyclable eco-friendly catalyst for the synthesis of quinoxaline derivatives in green media, Appl. Clay Sci. 88−89 (2014) 178−185.

[45] Z. Samardžija, K.Ž. Rožman, S. Kobe, Determination of the composition of Co−Pt thin films with quantitative electron-probe microanalysis, Mater. Charact. 60 (2009) 1241−1247.

CHAPTER 15

Mechanical properties of recycled nanomaterials

S. Behnam Hosseini
Department of Wood and Paper Science and Technology, Faculty of Natural Resources, University of Tehran, Karaj, Iran

15.1 Introduction

Nanomaterials (NMs) are defined as particles with one or more dimensions in the nanometer range, taking 100 nm as an arbitrary limit [1]. NMs can play a key role in enhancing the overall biomass to biohydrogen production process, owing to their unique properties [2,3], including high electroconductivity, larger surface area, and a high surface-to-volume ratio, and can catalyze each step of cellulosic biohydrogen production technology [4].

A list of established and/or newly emerging fields of industrial applications for NMs includes, but is not limited to, chemically inert additives (e.g., carbon, titania, and silica polymer fillers), catalysts, antimicrobial additives, pigments, UV filters, liquid crystal display technology (widely used in the screens of mobile phones, and desktop and laptop computers, tablets, notebooks, and high-definition televisions), biomedical applications (e.g., drug delivery, magnetic resonance imaging), membranes, and fuel cells [5]. Several types of NMs, such as iron, nickel, copper, gold, silver, and titanium, have been well demonstrated to improve biohydrogen production using various biological routes such as biophotolysis, dark fermentation, and photofermentation [4,6]. NMs, especially iron and nickel, play a significant role by acting as a cofactor on the active site of hydrogenase and nitrogenase enzymes and thus improve the biohydrogen production yield significantly [4,6,7]. NMs have been extensively investigated and applied for water treatment since their disclosure [8−12]. Moreover, by employing nanomaterials at pre-treatment step of lignocellulosic biomass may improve the removal of lignin thereby enhances the yield of sugar and also speed up the entire process [13]. Further, nanomaterials can enhance the production, influence the thermal and pH stability of cellulose enzymes for efficient hydrolysis i.e. for cellulose to sugar conversion process [14-16]. Employment of nanomaterial for the pretreatment of lignocellulosic biomass may enables to carry out the pre-treatment at much less extreme conditions, minimize the use of chemicals, fasten up the process, enhances the yield of reducing sugar released and thus may perhaps makes the overall process cost-effective [13,17,18]. Implementation of NM for biohydrogen production not only improves the yield and productivity of the biohydrogen but also enhances the reusability

Nanomaterials Recycling
DOI: https://doi.org/10.1016/B978-0-323-90982-2.00014-7

© 2022 Elsevier Inc.
All rights reserved.

of enzymes, enables the retention of the activity after multiple cycles, enables faster completion of reactions, and makes the process more cost-effective and sustainable for biohydrogen production [19].

According to shape or morphology, NMs comprise nanoparticles (NPs), nanotubes (NTs), nanofibers, nanowires, and nanomembranes (NMBs). NMs can be divided into inorganic NMs and organic molecule-based NMs. Inorganic NMs include carbon-based NMs (CNMs) [e.g., graphene, graphene oxide, carbon nanotubes (CNTs)], boron nitride-based NMs, silicon-based NPs, and transition metal/metal oxide/metal sulfide NPs (e.g., the most studied Fe(0) NPs and TiO_2 NPs). Meanwhile, organic molecule-based NMs include small organic molecule-supported NPs, organic polymer-supported NMs, metal-organic frameworks, and organic molecule-based NMBs [20].

Each category of NMs or NPs can be divided into more subcategories. For instance, the synthesis and application of CNMs have attracted extensive interest because they have unique and tunable physical, chemical, and electronic properties [11,12,21−24]. CNMs are usually classified as single-walled carbon nanotubes (SWCNTs) and multiwalled carbon nanotubes (MWCNTs) and their derivatives, advanced two-dimensional; graphene-based materials (i.e., graphene, graphene oxide, and related modification derivatives), and the C60 fullerene. Because of their large specific surface area, high surface-free energy, porosity, and activity, CNMs have been widely used as effective adsorbents for the remediation of waste water [20].

The usefulness of NMs, which has been reported by various researchers, is due to their large surface area and surface-free energy, small size, active atomicity, and reactivity. The high surface area to volume ratio of NMs immensely improves its absorption efficiency. Further, surface modification is a powerful tool that improves the properties of a material through imparting the desired functional groups on its surface without compromising the bulk properties [25−30].

Among NMs, TiO_2 NPs have received considerable attention because of their favorable properties; they are nontoxic, odorless, nonirritating, and thermal-stable [31,32], as well as having low cost, ready availability, wide acceptance by the market, and large-scale production in the industry. It is now common to apply TiO_2 NPs to solve a variety of environmental problems, such as odor elimination of drinking water and degradation of oil in surface water and of harmful organic contaminants in air or water [33].

15.2 Recycling of nanomaterials

15.2.1 Recycling definitions

Recycling is defined as any recovery operation in which waste materials are reprocessed into products, materials, or substances whether, for the original or other purposes [34]. Recycling also provides the opportunity to reduce energy usage by using reusable or recyclable materials [35]. In the field of recycling, an important topic is the quality of the recycled material, particularly to avoid reinsertion of contaminated inputs into the production chain [36].

Typically, recycling technologies can be divided into four categories according to their approaches: primary, secondary, tertiary, and quaternary. The primary approach involves recycling of industrial scrap; the secondary approach includes mechanical processing of a postconsumer product; tertiary recycling involves processes such as pyrolysis and hydrolysis that convert the plastic wastes into basic chemicals, monomers, or fuel; and the quaternary recycling approach involves the burning of the fibrous solid waste for generation of heat [37–39]. Mechanical recycling offers the following advantages over chemical recycling: a lower processing cost, lower global warming potential, less nonrenewable energy use, and less acidification and eutrophication [40].

NMs can be released into the environment throughout their entire life cycle [41]. The life cycle is generally understood to be the lifespan covering production, use, and disposal of a material, chemical, or product [42]. As the number of nanotechnology applications grows, more nanoproducts will enter the waste stream during the end-of-life (EoL) stage of their life cycle. That influx raises concerns related to the appropriate management of waste. At EoL a nanoproduct may enter the recycling system, be incinerated, or be landfilled [36].

The effects of recycling on nanoprocesses can be divided into three types: (1) occupational health effects in relation to the recycling processes themselves, (2) environmental impacts related to the residues generated or treated during the recycling processes (incineration, landfill, or sewage treatment), and (3) introduction of recovered NMs into products containing recycled NMs [43,44]. Also, the effectiveness of NM recycling and recovery processes depends on the form of nanoproduct-based technology and the types and class of NMs and matrices containing NMs (pure NMs, nanobyproducts, liquid suspension containing NMs, other contaminants with NMs, and solid matrices with integrated NMs) [45]. However, NMs also have certain limitations, including high cost [46], poor efficiency of recycling, and toxicity [47].

The primary tested nanowaste-recycling processes in the past decades have been based mainly on conventional techniques, such as separation by centrifugation or solvent evaporation. Recently, to effectively and economically recover or separate NMs from their waste streams, researchers have recommended a number of alternative recovery and separation methodologies that are based on the properties of the target NMs, including magnetic fields, pH, and thermoresponsivity, or use selective extraction procedures, including biological techniques, cold point extraction, molecular antisolvents, or nanostructured colloidal solvents. These separation processes provide effective separation processes for recycling NMs with more cost reduction and lower time and energy demand. However, to set up efficient separation and recycling for NMs, the intrinsic properties of the target NMs, such as mechanical, thermal, and chemical properties, are needed [45]. Table 15.1 provides examples of selected nanowaste streams with possible recycling procedures [45].

Table 15.1 Selected nanowaste streams and current recycling procedures for given nanomaterials [45].

Type of waste	Existing NMs	Recycling procedure	Possible sources of nanoexposure during the recycling process
Metal waste (scrap)	In coatings: metal oxides, CNTs, SiO_2	Shredding, smelting.	Shredding if engineered NMs can be set free from coatings. Smelting: Engineered NMs that are not destroyed in the melting process, insufficient exhaust gas purification.
Paper and cardboard	Carbon black (from ink), TiO_2 (except in special papers, TiO_2 is not in nanoform)	Pulping, de-inking (wet processes).	Dust from collection and transport. Aerosols of ink from pulping and de-inking.
Plastic	CNT, SiO_2, TiO_2	Collect and sorting or separate collection (e.g., for PET bottles). Mechanical recycling: shredding, washing, and regranulation. Feedstock recycling: depolymerization and cracking (for basic chemicals).	Shredding and regranulation if nanosubstance is set free. Feedstock (chemical) recycling: nanosubstance that is not destroyed in the process may be emitted or end up in the cracking residues (tar). Problem of dispersion of NMs to regranulated plastics.
Textiles	CNTs, Ag NPs	Collect, reuse, sort, prepare for reuse, shredding to get fiber NMs.	Shredding if nanosubstances are set free.
Waste electronic and electrical equipment	Carbon black (in plastic and in toners), CNTs (in electronic devices and in plastic housings), nano-iron oxide, ZnO, SiO_2, Ag (in coatings)	Collecting, dismantling, sorting by hand, shredding, and separating of the fractions, processing of fractions (nonmagnetic metals, iron, glass, plastics, etc.), further processing of the components (metal melting, material recovery of iron and noniron metals, extraction of metals from circuit boards).	Any step of the procedure, depending on the NM-containing component and on the specific type of NMs.

(Continued)

Table 15.1 (Continued)

Batteries	Electrodes with CNTs or nanophosphate ($nLiFePO_4$)	Collect, sorting. Mechanical, chemical, and/or thermal treatment. Various procedures, for example, BATREC (Switzerland) for alkali and mercury batteries; Battery Solutions (United States), Toxco (United States), for lithium batteries; INMETCO (United States) for Ni–Cd batteries.	In principle during mechanical, chemical, or thermal treatment, dependent on the process and on the type of battery with NMs.
Construction and demolition wastes	CNTs, SiO_2, TiO_2, Fe_2O_3, Cu, Ag NPs	Reuse of components, sorting of fractions (wood, concrete, brick, metal, etc.), metal recycling, secondary building materials, incineration and landfill	During destruction of buildings (dust emissions), shredding, grinding if nanosubstances are set free. Problem of dispersion of NMs fractions of recycled material.
EoL vehicles	CNTs, SiO_2, TiO_2 (in plastics, coatings, and paints)	Dismantling for reusable parts (including tires), removal of hazardous components (e.g., batteries), shredding and separation of fractions, metals go to smelting and refining, glass is recycled or landfilled, nonmetallic shredder residues for incineration or landfill.	Shredding and sorting of fractions, smelting of metals (NMs from coatings), disposal of nonmetallic shredder fraction. Modern cars contain electronic components that are normally not removed before shredding; this is a possible source of NM emissions.

(*Continued*)

Table 15.1 (Continued)

Tires	Carbon black, silica; there are indications that future developments will include others, for example, CNTs, nanoclay (SiO_2), or organic nanopolymers.	Collect, store (danger of ignition), refurbish, and reuse; shredding of metal, reuse of rubber for downcycled products or for energy recovery. 27% of tires worldwide were recycled in 2005, up from 6% in 1995; landfilling dropped from 62% to 22% in the same time period.	In principle, when shredding, actual tires contain NMs that are bound to the rubber matrix.
Recycling of residues from waste incineration	NMs from nanowaste in the municipal waste that are not destroyed or evaporated may stay in the bottom ash.	Separation of metal bottom ash from municipal solid waste (MSW); ~ 220 kg of bottom ash is produced in incinerating 1 ton of MSW; these contain metal residues (iron, aluminum, copper, and even gold) from MSW.	The most efficient recovery of metals from bottom ash is done with dry ash, with dust generation. Nanosubstances can be emitted during pouring, sieving, and mechanical and magnetic separation.

15.2.1.1 Environmental aspects

Recently, increased awareness of the environment and organized waste management have contributed to enhancing the image of recycling as an important instrument for solving the problem of waste materials [48]. With a better understanding of the recycled products, technological advancement in recycling technologies, and the societal pressure to reduce the environmental pollution load, many types of wastes and agroresidues are being recycled or reused for making valued processes and products [49]. With the advancement of recycling technology, it has been possible to recycle or reuse many important wastes and agroresidues for the same or other valued processes and products, thus also reducing the environmental load [49].

NMs can be synthesized in various architectures, such as NPs, NTs, nanowires, nanofibers, and nanosheets [50]. NPs display unique properties, owing to their high specific surface area and reactivity, allowing for detection and removal of diverse targets, including chemicals, pollutant gases (SO_2, CO, NO_x, etc.), organic pollutants, and biological substrates (e.g., bacteria, virus, antibiotics) from different media, such as natural waters, waste water, and air [12].

NMs can be designed for high selectivity against target pollutants by tailoring their uptake mechanism. Silver, iron, gold, titanium oxide, iron oxide, and zinc oxide are some of the common nano-scale metals and metal oxides that are used for such purposes [50]. For instance, gold NPs are useful for removing contaminants, such as chlorinated organic compounds, pesticides, and mercury, from water [50]. A very high proportion of machining waste is generated with titanium components [51]; hence recycling of the titanium swarf into a useful commodity at an affordable price is a strong motivation for converting swarf into powders using milling processes. Recently, Shial et al. [52] recycled titanium machining swarf into micron-sized nanocrystalline powders.

15.2.1.2 Economic aspects

From a logistical point of view, the recycling process is less economically viable, given the low weight-to-volume ratio and the complex heterogeneity of mixed waste, which implies an investment in transport, storage, and sorting facilities. Therefore chemical recycling is a preferable option for complex and contaminated wastes [53]. It is also possible to use the recycled materials along with virgin material to reduce costs. The construction industry is one the fastest-growing sectors, and to meet the demand, an enormous amount of construction materials is required, resulting in exhaustion of natural resources. Almost 15% of the total energy of a building life cycle is consumed during the generation of virgin construction materials [54], and part of this energy consumption can be decreased by the use of recycled materials in the construction industry [55]. Polymer materials are a suitable example, in which blending virgin and recycled material at various ratios via an extrusion process can be a cost-effective way

324　Nanomaterials Recycling

to increase reuse of recycled materials. Advances in this regard were found by using polypropylene (PP), polyethyleneterephtalate (PET), and polystyrene [56]; acrylonitrile butadiene styrene, polylactic acid (PLA), and high-impact polystyrene [57]; and PP and tires wastes [58].

The reuse and recyclability of NMs such as graphene-based NMs need to be further investigated from an economic point of view [59]. For example, the development of cost-effective CNTs to increase their commercial use has been an obstacle. In addition, the recycling or reuse of CNTs from water or waste water after adsorption of toxic metals is difficult, which increases the risk of secondary pollution. Hence the development of new materials or modification of the existing materials with high adsorption capacity is required, which would allow the effective removal of toxic metals with ease of separation and reusability [47]. Recycling efficiency is an area for improvement in order to meet the requirements of industrial production [60].

15.3 Mechanical properties of recycled nanomaterials

15.3.1 Recycled nanoclay

Nanoclays have been reported to improve thermal stability, resistance to fire, and mechanical properties of several polymers, including polyethylene (PE), PP, nylon 6, and PLA [61−68].

Sánchez et al. evaluated the effect of recycled four NMs: layered silicate modified nanoclay (Nanoclay1), calcium carbonate ($CaCO_3$), silver (Ag), and zinc oxide (ZnO) on visual appearance, material quality, and mechanical properties of three polymer matrices: PE, PP, and PET (Table 15.2) [69]. The results of incorporation of recycled Nanoclay1 and PE into recycled polyethylene matrix showed no significant changes in elongation at break or tear strength. Nevertheless, reductions in tensile modulus and tensile strength were visible, although values were in the same range, and therefore no restrictions for current applications of recycled PE are initially expected (Fig. 15.1) [69]. Sánchez et al. also investigated PET matrix containing three levels of recycled Nanoclay1 and PET, and the results of the mechanical properties do not show significant reductions in tensile modulus, tensile strength, or elongation at break. In fact, increases in this last parameter could be observed in S19 and S20 in comparison with the blank (S12) (Fig. 15.1) [69].

15.3.2 Recycled nano-$CaCO_3$

The effect of recycled nano-$CaCO_3$ on recycled PE matrix was studied by Sánchez et al., and the results indicated no significant changes in tensile modulus, elongation at break, or tear strength [69]. Nevertheless, a reduction in tensile strength was identified between blank (S1) and PE-$CaCO_3$ samples (S5−S7) (Fig. 15.1) [69].

Table 15.2 Composition of the mixtures of polymers in each of the recycled samples [69].

	Sample	Conventional recycled pellets	Nanoreinforced film	Nanomaterial in the mixture
Recycled PE samples	S1–Blank	100% PE	—	—
	S2	99% PE	1% PE + Nanoclay1	0.04% Nanoclay1
	S3	95% PE	5% PE + Nanoclay1	0.2% Nanoclay1
	S4	80% PE	20% PE + Nanoclay1	0.8% Nanoclay1
	S5	99% PE	1% PE + $CaCO_3$	0.04% $CaCO_3$
	S6	95% PE	5% PE + $CaCO_3$	0.2% $CaCO_3$
	S7	80% PE	20% PE + $CaCO_3$	0.8% $CaCO_3$
Recycled PP samples	S8–Blank	100% PP	—	—
	S9	99% PP	1% PP + Ag	0.04% Ag
	S10	95% PP	5% PP + Ag	0.2% Ag
	S11	80% PP	20% PP + Ag	0.8% Ag
Recycled PET samples	S12–Blank	100% PET	—	—
	S13	99% PET	1% PET + ZnO	0.04% ZnO
	S14	95% PET	5% PET + ZnO	0.2% ZnO
	S15	80% PET	20% PET + ZnO	0.8% ZnO
	S16	99% PET	1% PET + Ag	0.04% Ag
	S17	95% PET	5% PET + Ag	0.2% Ag
	S18	80% PET	20% PET + Ag	0.8% Ag
	S19	99% PET	1% PET + Nanoclay1	0.04% Nanoclay1
	S20	95% PET	5% PET + Nanoclay1	0.2% Nanoclay1
	S21	80% PET	20% PET + Nanoclay1	0.8% Nanoclay1

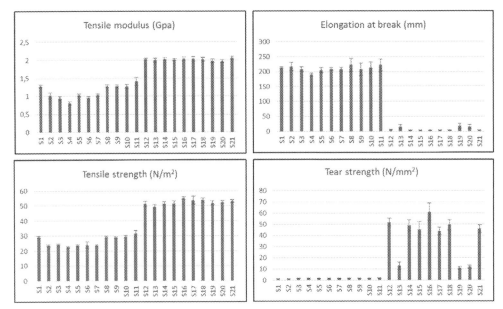

Figure 15.1 Mechanical properties of dog bone shapes of polyethylene and polypropylene and films of polyethyleneterephtalate. [69].

15.3.3 Recycled nanosilver

The addition of recycled nano-Ag into PP and PET matrixes and their influence were investigated by Sánchez et al. [69]. The results of the study revealed that incorporation of recycled nano-Ag into both PP and PET matrixes (S9—S11 and S16—S18, respectively) led to no significant changes in mechanical properties, including tensile modulus, tensile strength, and tear strength. By contrast, recycled nano-Ag + PET showed decrease in elongation at break compared to blank sample (S12), which was due to slight degradation of the polymer (Fig. 15.1) [69].

15.3.4 Recycled nano-zinc oxide

The effect of recycled nano-ZnO and PET on mechanical properties of recycled PET matrix showed a decrease in tear strength and elongation at break due to a slight degradation of the PET polymer [69]. Nevertheless, the variations observed in these mechanical properties were not initially in ranges to restrict the use of S13, S14, and S15 in the same applications of the blank material (S12). Finally, no significant variations could be observed in the tensile modulus and tensile strength (Fig. 15.1) [69].

15.3.5 Recycled carbon nanotubes

CNTs are an excellent combination of outstanding mechanical (20 GPa tensile strength and almost 1 TPa modulus) [70] and electrical properties; in other words, they are excellent candidates as fillers in multifunctional nanocomposites [71–79]. CNTs can further be categorized according to the number of graphene layers as SWCNTs and MWCNTs consisting of two or more layers of graphene sheets that form concentric cylinders [80]. A number of investigations have been conducted to resolve these issues, such as the use of chemical processing methods to wash the CNTs out of the matrix and mechanical methods to grind the whole composite and use it as a new filler material for other products [81]. The recycling of CNT composites depends on the properties of the matrix. However, because the scrap composites may come from different sources, mechanical processing methods for recycling of CNT composites work better than chemical methods [81]. Zhang et al. studied how recycling affects the nature of CNT-filled PP composites and how exposures to the recycling process affect workers. The results show that recycling has very little effect on the mechanical properties, especially the tensile properties, thus indicating the potential for reuse [79].

15.3.6 Recycled nano-SiO$_2$

Nano-SiO$_2$ (NS) particles are widely used in engineering because of their characteristics, namely, smaller particle sizes, larger specific surface area, and higher activity [82]. Currently, polymer matrix composites, obtained by the dispersion of micro-sized and/or nano-sized fillers, extracted from residues of mineral activity or agribusiness are a category of materials of great interest, even for durable applications that require more rigorous technical performances, particularly in terms of mechanical properties and stability [83,84].

Rice husk contains nearly 20 wt.% of silica, which is present in hydrated amorphous form. Currently, rice husk and sugarcane bagasse are widely used in the energy cogeneration process, and as a result of its burning, tons of ashes are produced. These ashes can be used directly as reinforcement of polymer or processed for obtaining "green silica," in the form of microparticles and NPs for addition in polymeric materials [85]. In this study, rice husk ashes were added to a copolyester-starch blend by melt extrusion, using a twin-screw extruder, to improve the mechanical and morphological behaviors of this biodegradable blend [85]. The results showed a gain of crystallinity of the polymeric blend, changes in the morphological behaviors, and a significant gain in tensile strength at break and Young's modulus (Table 15.3). From the results of this study, it can be inferred that rice husk ashes, a renewable and low-cost agroresource and a widely available raw material from energy cogeneration processes, are a good candidate for the reinforcement of biodegradable plastics in order to obtain composite materials with advanced mechanical behaviors for several applications [85].

The incorporation of sugar bagasse ashes obtained from the burning of sugarcane bagasse to produce energy in the sugar and bioethanol industries into PBT resulted in no important visual changes on the surface of the cryofractured PBT (Table 15.4).

Table 15.3 Mechanical test results of polybutylene adipate coterephthalate (PBAT)-starch blend and PBAT-starch-ash composite [86].

Properties	Neat copolyes ter-starch blend	Composite
Tensile strength at break (MPa)	6.64 ± 0.44	9.96 ± 0.7
Elongation at break (%)	83.22 ± 12	11.12 ± 0.5
Young's modulus (MPa)	35.35 ± 0.2	99.9 ± 2.7
Notched izod impact (J/m)	a	35.62 ± 2.5

[a]PBAT-starch blend is a nonbreakable extremely flexible material.

Table 15.4 Mechanical and thermomechanical results for the neat polybutylene terephthalate (PBT) and PBT-ash composite [86].

Test	Neat PBT	PBT-ash composites	Variation (%)
HDT (°C)	55.4	75.7	+36.6
Tensile strength at break (MPa)	38.0	56.1	+47.6
Elongation at break (%)	161.6	16.2	−90.0
Flexural strength (MPa)	74.2	86.4	+16.4
Flexural modulus (GPa)	2.4	2.9	+20.8

The addition of the ashes in this study showed that it is possible to get interesting property gains by using waste from renewable sources instead of traditional ones [87].

15.3.7 Recycled lignocellulose nanofibers

Cellulose is one of the most important, abundant, renewable, and biodegradable natural polymers and exists in the several plant biomasses, such as wood, cotton, hemp, straw, sugarcane bagasse, and other plant-based materials. It has a wide range of applications in the form of fiber, paper, films, and polymers. The utilization of this natural biomass for processing of novel material applications has recently attracted global interest, owing to its ecological and renewable characteristics [88]. Cellulose nanocrystals feature an attractive combination of properties, such as biocompatibility, large specific surface area and aspect ratio, high elastic modulus, high thermal stability, and excellent optical transparency [89], which have been exploited to improve the properties of other biopolymer matrices, such as PLA [90,91], polyhydroxyalkanoates [92,93], polyisoprene [94], and pea starch [95]. Besides their utilization as nanofillers, cellulose nanocrystals themselves can be used to produce high–barrier films [96], although this approach has not been fully explored to date. According to the common definitions, cellulose nanofibers (CNFs) are aggregates of cellulose chains 5−100 nm in width and with an estimated length of a few micrometers; they includes both crystalline and amorphous regions [97,98]. CNF is typically produced by using various mechanical methods, such as homogenization, microfluidization, microgrinding, refining, or cryo-crushing, or in combination with chemical and enzymatic pretreatments [98−100].

Many studies of CNF are related to the production and application of pure CNF, while the use of recycled impure lignocellulosic raw materials containing lignin, hemicelluloses, extractives, and so on for the production of lignocellulose nanofibers (LCNF) may have the potential for higher yields, less production energy, and lower costs for many appropriate end products. In addition, rather than economical reasons, LCNF may be more prospective as a result of environmental issues, as purification and bleaching processes are not included for the production of LCNF [101−103]. As a recycled paper, which can be reinforced with nanofibers, old corrugated containers (OCCs) are one of the most significant types of waste papers for recycling. This type of pulp is usually recycled repeatedly and therefore possesses weak strength properties, for which the application of LCNF may be helpful. The effect of LCNF addition and its interaction with cationic starch (CS)-colloidal silica NP solution (NS) system on tensile strength are presented in Fig. 15.2. Individually, 3% LCNF and CS-NS were able to improve tensile strength by 22%−41%, respectively, in comparison to control pulp (23.6 N/mg). This result highlighted the point that even nanofiber produced from OCCs that contained some materials other than cellulose, such as lignin, hemicellulose, extractives, and even minerals, could improve the tensile strength of the handsheet [104]. The mechanical fibrillation process causes permanent changes in the cellulose fiber structure, and it increases the bonding ability of cellulose by modifying the morphology and reducing the size of the fibers [105,106]. Thus some previous studies have shown that CNF can notably improve the mechanical properties of paper

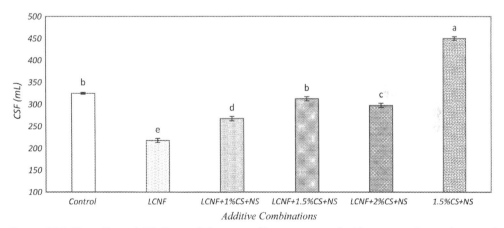

Figure 15.2 The effect of 3% lignocellulose nanofibers content of old corrugated containers and cationic starch-nanoparticle solution system on the tensile index. The bars that do not share the same lowercase letter were found to be significantly different from each other (confidence level: 95%) [104].

or board. It has been found that tensile strength and elastic modulus can be improved significantly [107–113].

15.3.8 Recycled nanoalumina

Nanostructural alumina has a large surface area, thermal stability, conductivity, mechanical strength, stiffness, inertness to most acids and alkalis, adsorption capacity, wear resistance, oxidation, electrical insulation, and nontoxicity. It has potential applications in industrial fields [114–117].

Sufian et al. investigated the performance of PP composites reinforced with alpha-alumina particles that were recovered from aluminum dross [118]. Fig. 15.3 shows the tensile modulus and strength of the neat PP and PP-α-Al$_2$O$_3$ composites as a function of particle content. The moduli were observed to have continuously increased with the increased concentration of α-Al$_2$O$_3$ particles, suggesting the reinforcement effect of rigid filler in the PP matrix. From Fig. 15.3A the dependence on the particle size can also be noticed. The microcomposites and nanocomposites exhibited huge variations in the moduli values across all compositions. Additionally, treatment with the maleic anhydride grafted polypropylene (MAPP) coupling agent also contributed to the improvement of the composites' modulus. As demonstrated in Fig. 15.3A, an increment of about twofold occurred after the neat PP was added with 5 wt.% of treated NPs. This finding suggests that better particle dispersion resulted from good interfacial adhesion between the PP matrix and nano-α-Al$_2$O$_3$ particles. As a consequence, there was an efficient transfer of stress between the matrix and the filler. As for the effect of filler content, the linear increment in the moduli ceased when the α-Al$_2$O$_3$ content went beyond 5 wt.%. Just like the moduli's result, the presence of treated nano-sized α-Al$_2$O$_3$ particles improved the tensile strength more effectively than did

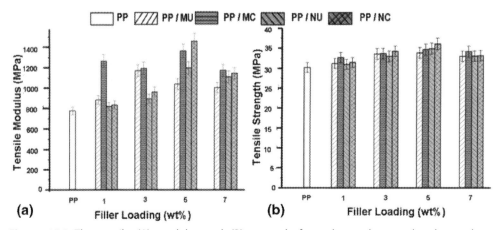

Figure 15.3 The tensile (A) modulus and (B) strength for polypropylene and polypropylene-α-Al$_2$O$_3$ nanocomposites [118].

Mechanical properties of recycled nanomaterials   331

Table 15.5 Comparison of tensile properties of nanocomposites augmented with 5 wt.% of treated Al_2O_3 [118].

Nanocomposite sample	Coupling agent	Modulus (MPa)	Tensile strength (MPa)	References
PP-α-Al_2O_3 (commercial)	TiO_2 (0.02, 0.04, 0.06, 0.08, 0.10 wt.%)	1290	33.02	Mirjalili et al. [120]
PP-Al_2O_3 (commercial)	MAPP (10 wt.%)	1400	~ 34.8	Pérez et al. [123]
PP-α-Al_2O_3 (recovered from waste)	MAPP (2 wt.%)	1425	36	Sufian et al. [118]

the micron-sized particles (Fig. 15.3B). Similarly, an optimum tensile strength was observed to occur at 5 wt.%, at which the enhancement was about 20% of the neat PP. However, it began to decrease rapidly when the filler content was increased beyond the optimal level. In accordance with many previous studies [119−124], the higher particle content formed large particle clusters as a result of agglomerated particles, which eventually led to a weakened tensile property [118]. Table 15.5 shows that the tensile properties of PP reinforced with 5 wt.% treated α-Al_2O_3 particles in this work exhibited a better performance in comparison to the results reported in works by Mirjalili et al. [120] and Pérez et al. [123]. Hence it can be emphasized that the alpha-alumina particles that were retrieved from the industrial waste showed mechanical properties comparable to those of commercial Al_2O_3 particles [118].

15.4 Conclusions

Properties that are specific to engineered NMs make them very interesting to include in a variety of consumer products. This use inevitably results in release of engineered NMs into technical or environmental compartments. As the number of nanotechnology applications grows, more nanoproducts will enter the waste stream during the EoL stage of their life cycle. Recently, increased awareness of the environment and organized waste management have contributed to enhancing the image of recycling as an important instrument for solving the problem of waste materials, and there are many chemical and mechanical methods to recycle NMs. The results showed that recycled and reused NMs have a wide range of effects on mechanical properties, including improvement, no significant change, and decrease. This study also indicated that it is possible to get interesting mechanical properties by recycling and reusing waste NMs instead of using virgin ones, depending on the NM, the matrix, and the desired mechanical properties.

References

[1] M. Auffan, J. Rose, J. Bottero, G.V. Lowry, J. Jolivet, M.R. Wiesner, Towards a definition of inorganic nanoparticles from an environmental, health and safety perspective, Nat. Nanotechnol. 4 (2009) 634–641.

[2] G. Kumar, T. Mathimani, E.R. Rene, A. Pugazhendhi, Application of nanotechnology in dark fermentation for enhanced biohydrogen production using inorganic nanoparticles, Int. J. Hydrogen Energy 44 (2019) 13106–13113.

[3] A. Pugazhendhi, S. Shobana, D.D. Nguyen, J.R. Banu, P. Sivagurunathan, S.W. Chang, et al., Application of nanotechnology (nanoparticles) in dark fermentative hydrogen production, Int. J. Hydrogen Energy 44 (2019) 1431–1440.

[4] M. Taherdanak, H. Zilouei, K. Karimi, Investigating the effects of iron and nickel nanoparticles on dark hydrogen fermentation from starch using central composite design, Int. J. Hydrogen Energy 40 (38) (2015) 12956–12963.

[5] W. Stark, P. Stoessel, W. Wohlleben, A. Hafner, Industrial applications of nanoparticles, Chem. Soc. Rev. 44 (2015) 5793–5805.

[6] H.N. Lin, B.B. Hu, M.J. Zhu, Enhanced hydrogen production and sugar accumulation from spent mushroom compost by *Clostridium thermocellum* supplemented with PEG8000 and JFC-E, Int. J. Hydrogen Energy 41 (4) (2016) 2383–2390.

[7] R. Lin, J. Cheng, L. Ding, W. Song, M. Liu, J. Zhou, et al., Enhanced dark hydrogen fermentation by addition of ferric oxide nanoparticles using Enterobacter aerogenes, Bioresour. Technol. 207 (2016) 213–219.

[8] M.S. Mauter, M. Elimelech, Environmental applications of carbon-based nanomaterials, Environ. Sci. Technol. 42 (2008) 5843–5859.

[9] N. Savage, M.S. Diallo, Nanomaterials and water purification: opportunities and challenges, J. Nanopart. Res. 7 (2005) 331–342.

[10] T. PradeepAnshup, Noble metal nanoparticles for water purification: a critical review, Thin Solid. Films 517 (2009) 6441–6478.

[11] S.C. Smith, D.F. Rodrigues, Carbon-based nanomaterials for removal of chemical and biological contaminants from water: a review of mechanisms and applications, Carbon (2015) 91122–91143.

[12] M.M. Khin, A.S. Nair, V.J. Babu, R. Murugana, S. Ramakrishna, A review on nanomaterials for environmental remediation, Energy Environ. Sci. 5 (2012) 8075–8109.

[13] Z. Wei, G. Zeng, F. Huang, M. Kosa, D. Huang, A.J. Ragauskas, Bioconversion of oxygen-pretreated Kraft lignin to microbial lipid with oleaginous *Rhodococcus opacus* DSM 1069, Green. Chem. 17 (5) (2015) 2784–2789.

[14] N. Srivastava, M. Srivastava, A. Manikanta, P. Singh, P.W. Ramteke, P.K. Mishra, Nanomaterials for biofuel production using lignocellulosic waste, Environ. Chem. Lett. 15 (2) (2017) 179–184.

[15] M.R. Ladole, J.S. Mevada, A.B. Pandit, Ultrasonic hyperactivation of cellulose immobilized on magnetic nanoparticles, Bioresour. Technol. 239 (2017) 117–126.

[16] S. Bilal, L. Ali, A.L. Khan, R. Shahzad, S. Asaf, M. Imran, et al., Endophytic fungus *Paecilomyces formosus* LHL10 produces sester-terpenoid YW3548 and cyclic peptide that inhibit urease and α-glucosidase enzyme activities, Arch. Microbiol. 200 (10) (2018) 1493–1502.

[17] R. Amin, A. Khorshidi, A.F. Shojaei, S. Rezaei, M.A. Faramarzi, Immobilization of laccase on modified $Fe_3O_4@$ $SiO_2@$ Kit-6 magnetite nanoparticles for enhanced delignification of olive pomace bio-waste, Int. J. Biol. Macromol. 114 (2018) 106–113.

[18] M.J. Khalid, A. Waqas, I. Nawaz, Synergistic effect of alkaline pretreatment and magnetite nanoparticle application on biogas production from rice straw, Bioresour. Technol. 275 (2019) 288–296.

[19] N. Srivastava, M. Srivastava, P.K. Mishra, M. Adnan Kausar, M. Saeed, V.K. Gupta, et al., Advances in nanomaterials induced biohydrogen production using waste biomass, Bioresour. Technol. 307 (2020) 123094. Available from: https://doi.org/10.1016/j.biortech.2020.123094.

[20] F. Lu, D. Astruc, Nanocatalysts and other nanomaterials for water remediation from organic pollutants, Coord. Chem. Rev. 408 (2020) 213180–213210.

[21] K. Yang, L. Zhu, B. Xing, Adsorption of polycyclic aromatic hydrocarbons by carbon nanomaterials, Environ. Sci. Technol. 40 (2006) 1855–1861.

[22] H. Yi, D. Huang, L. Qin, G. Zeng, C. Lai, M. Cheng, et al., Selective prepared carbon nanomaterials for advanced photocatalytic application in environmental pollutant treatment and hydrogen production, Appl. Catal. B: Environ. 239 (2018) 408−424.

[23] R. Hu, T. Furukawa, Y. Gong, L. Chen, X. Wang, X. Tian, et al., Tailoring of Cu@graphitic carbon nanostructures enables the selective detection of copper ions and highly efficient catalysis of organic pollutants, Adv. Mater. Interfaces 5 (2018) 1800551.

[24] G. Wang, X. Nie, X. Ji, X. Quan, S. Chen, H. Wang, et al., Enhanced heterogeneous activation of peroxymonosulfate by Co and N codoped porous carbon for degradation of organic pollutants: the synergism between Co and N, Environ. Sci. Nano 6 (2019) 399−410.

[25] L.M. Pandey, S.K. Pattanayek, Hybrid surface from self-assembled layer and its effect on protein adsorption, Appl. Surf. Sci. 257 (2011) 4731−4737.

[26] A. Hasan, L. Pandey, 6 − Self-assembled monolayers in biomaterials, in: R. Narayan (Ed.), Nanobiomaterials, Woodhead Publishing, 2017, pp. 137−178. Available from: https://doi.org/10.1016/B978-0-08-100716-7.00007-6.

[27] L.M. Pandey, Effect of solid surface with self assembled monolayers on adsorption of proteins, Dissertation, 2012.

[28] A. Hasan, M. Pandey, Polymers, surface-modified polymers, and self assembled monolayers as surface-modifying agents for biomaterials, Polym. Technol. Eng. 54 (2015) 1358−1378.

[29] A. Hasan, G. Waibhaw, L.M. Pandey, Conformational and organizational insights into serum proteins during competitive adsorption on self-assembled monolayers, Langmuir 34 (2018) 8178−8194.

[30] A. Hasan, V. Saxena, L.M. Pandey, Surface functionalization of Ti_6Al_4V via selfassembled monolayers for improved protein adsorption and fibroblast adhesion, Langmuir 34 (2018) 3494−3506.

[31] S. Kodama, H. Nakaya, M. Anpo, Y. Kubokawa, A common factor determining the features of the photocatalytic hydrogenation and isomerization of alkenes over Ti−Ti oxides, Bull. Chem. Soc. Jpn. 58 (1985) 3645−3646.

[32] F. Fujishima, TiO_2 photocatalysis fundamentals and applications, A Revolut. Clean. Technol. (1999) 14−21.

[33] H. Zhu, J. Orthman, J.-Y. Li, J.-C. Zhao, G. Churchman, E. Vansant, Novel composites of TiO_2 (anatase) and silicate nanoparticles, Chem. Mater. 14 (2002) 5037−5044.

[34] EU, 2008. Directive 2008/98/EC of the European Parliament and of the Council of 19 November 2008 on Waste and Repealing Certain Directives***. European Parliament.

[35] T. Alexander, B. Mainela, D. Laura, B. Irina, Mechanical and environmental performances of concrete using recycled materials, Procedia Manuf. 32 (2019) 253−258.

[36] A. Caballero-Guzman, T. Sun, B. Nowack, Flows of engineered nanomaterials through the recycling process in Switzerland, Waste Manage. 36 (2014) 33−43.

[37] Y. Wang, Fiber and textile waste utilization, Waste Biomass Valor 1 (2010) 135−143. Available from: https://doi.org/10.1007/s12649-009-9005-y.

[38] J. Scheirs, Polymer Recycling, Science Technology and Applications, Wiley, New York, 1998.

[39] Y. Wang (Ed.), Recycling in Textiles, Woodhead Publishing, Cambridge, 2006.

[40] L. Shen, E. Worrell, M.K. Patel, Open-loop recycling: a LCA case study of PET bottle-to-fibre recycling, Resour. Conserv. Recycl. 55 (2010) 34−52. Available from: https://doi.org/10.1016/j.resconrec.2010.06.014.

[41] F. Gottschalk, B. Nowack, The release of engineered nanomaterials to the environment, J. Environ. Monit. 13 (2011) 1145−1155.

[42] C. Som, M. Berges, Q. Chaudhry, M. Dusinska, T.F. Fernandes, S.I. Olsen, et al., The importance of life cycle concepts for the development of safe nanoproducts, Toxicology 269 (2010) 160−169.

[43] B. Nowack, Is anything out there? What life cycle perspectives of nano-products can tell us about nanoparticles in the environment, Nano Today 4 (2009) 11−12. Available from: https://doi.org/10.1016/j.nantod.2008.10.001.

[44] B. Mrowiec, Directions and possibilities of the safe nanowaste management, CHEMIK 70 (2016) 593−596.

[45] S.A. Younis, E.M. El-Fawal, P. Serp, Nano-wastes and the environment: potential challenges and opportunities of nano-waste management paradigm for greener nanotechnologies, in: C. Hussain

(Ed.), Handbook of Environmental Materials Management, Springer, Cham, 2018, pp. 1–72. Available from: https://doi.org/10.1007/978-3-319-58538-3_53-1.

[46] J. Wang, Q. Zhang, X. Shao, J. Ma, G. Tian, Properties of magnetic carbon nanomaterials and application in removal organic dyes, Chemosphere 207 (2018) 377–384.

[47] A. Jawed, V. Saxena, L.M. Pandey, Engineered nanomaterials and their surface functionalization for the bremoval of heavy metals: a review, J. Water Process. Eng. 33 (2020) 101009.

[48] Y. Huang, R.N. Bird, O. Heidrich, A review of the use of recycled solid waste materials in asphalt pavements, Resour. Conserv. Recycl. 52 (2007) 58–73.

[49] K.K. Samanta, S. Basak, S.K. Chattopadhyay, Recycled fibrous and nonfibrous biomass for value-added textile and nontextile applications, in: S. Muthu (Ed.), Environmental Implications of Recycling and Recycled Products. Environmental Footprints and Eco-design of Products and Processes, Springer, Singapore, 2015, pp. 167–212. Available from: https://doi.org/10.1007/978-981-287-643-0_8.

[50] T. Dutta, K.-H. Kim, A. Deep, J.E. Szulejko, K. Vellingiri, S. Kumar, et al., Recovery of nanomaterials from battery and electronic wastes: a new paradigm of environmental waste management, Renew. Sustain. Energy Rev. 82 (Part 3) (2018) 3694–3704. Available from: https://doi.org/10.1016/j.rser.2017.10.094.

[51] P. Luo, et al., Analysis of microstructure and strengthening in pure titanium recycled from machining chips by equal channel angular pressing using electron backscatter diffraction, Mater. Sci. Eng. A 538 (2012) 252–258.

[52] S.R. Shial, M. Masanta, D. Chaira, Recycling of waste Ti machining chips by planetary milling: generation of Ti powder and development of in situ TiC reinforced Ti-TiC composite powder mixture, Powder Technol. 329 (2018) 232–240.

[53] K. Ragaert, L. Delva, K. Van Geem, Mechanical and chemical recycling of solid plastic waste, Waste Manage 69 (2017) 24–58. Available from: https://doi.org/10.1016/j.wasman.2017.07.044.

[54] K. Adalberth, Energy use during the life cycle of single-unit dwellings: examples, Build. Environ. 32 (4) (1997) 321–329.

[55] W. Gao, T. Ariyama, T. Ojima, A. Meier, Energy impacts of recycling disassembly material in residential buildings, Energy Build. 33 (6) (2001) 553–562.

[56] N.E. Zander, M. Gillan, Z. Burckhard, F. Gardea, Recycled polypropylene blends as novel 3D printing materials, Addit. Manuf. 25 (2019) 122–130. Available from: https://doi.org/10.1016/j.addma.2018.11.009.

[57] R. Singh, R. Kumar, I. Farina, F. Colangelo, L. Feo, F. Fraternali, Multi-material additive manufacturing of sustainable innovative materials and structures, Polym. (Basel) 11 (2019) 62. Available from: https://doi.org/10.3390/polym11010062.

[58] J. Domingues, T. Marques, A. Mateus, P. Carreira, C. Malça, An additive manufacturing solution to produce big green parts from tires and recycled plastics, Procedia Manuf. 12 (2017) 242–248. Available from: https://doi.org/10.1016/j.promfg.2017.08.028.

[59] L. Xu, J. Wang, The application of graphene-based materials for the removal of heavy metals and radionuclides from water and wastewater, Crit. Rev. Environ. Sci. Technol. 47 (2017) 1042–1105.

[60] F.A. Cruz Sanchez, H. Boudaoud, M. Camargo, J.M. Pearce, Plastic recycling in additive manufacturing: a systematic literature review and opportunities for the circular economy, J. Clean. Prod. 264 (2020) 121602. Available from: https://doi.org/10.1016/j.jclepro.2020.121602.

[61] C. Silvestre, D. Duraccio, S. Cimmino, Food packaging based on polymer nanomaterials, Prog. Polym. Sci. 36 (12) (2011) 1766–1782.

[62] T.V. Duncan, Applications of nanotechnology in food packaging and food safety: barrier materials, antimicrobials and sensors, J. Colloid Interface Sci. 363 (1) (2011) 1–24.

[63] J.M. Lagaron, A. López-Rubio, Nanotechnology for bioplastics: opportunities, challenges and strategies, Trends Food Sci. Technol. 22 (11) (2011) 611–617.

[64] S. Sinha Ray, M. Okamoto, Polymer/layered silicate nanocomposites: a review from preparation to processing, Prog. Polym. Sci. 28 (11) (2003) 1539–1641.

[65] M. Baniassadi, A. Laachachi, F. Hassouna, F. Addiego, R. Muller, H. Garmestani, et al., Mechanical and thermal behavior of nanoclay based polymer nanocomposites using statistical homogenization approach, Compos. Sci. Technol. 71 (16) (2011) 1930–1935.

[66] T. Hanemann, Vinga Szabó, D., Polymer-nanoparticle composites: from synthesis to modern applications, Materials 3 (6) (2010) 3468–3517.

[67] Z. Akbari, T. Ghomashchi, S. Moghadam, Improvement in food packaging industry with biobased nanocomposites, Int. J. Food Eng. 3 (4) (2007) 1556–3758.

[68] Z. Wang, G. Xie, X. Wang, L. Guicum, Z. Zhang, Rheology enhancement of polycarbonate/calcium carbonate nanocomposites prepared by meltcompounding, Mater. Lett. 60 (8) (2006) 1035–1038.

[69] C. Sánchez, M. Hortal, C. Aliaga, A. Devis, V.A. Cloquell-Ballester, Recyclability assessment of nano-reinforced plastic packaging, Waste Manage. 34 (12) (2014) 2647–2655. Available from: https://doi.org/10.1016/j.wasman.2014.08.006.

[70] D. Qian, E. Dickey, R. Andrews, T. Rantell, Load transfer and deformation mechanisms in carbon nanotube-polystyrene composites, Appl. Phys. Lett. 76 (2000) 2868. Available from: https://doi.org/10.1063/1.126500.

[71] A. Carrillo, J.A. Swartz, J.M. Gamba, R.S. Kane, N. Chakrapani, B. Wei, et al., Noncovalent functionalization of graphite and carbon nanotubes with polymer multilayers and gold nanoparticles, Nano Lett. 3 (10) (2003) 1437–1440. Available from: https://doi.org/10.1021/nl034376x.

[72] A.B. Dalton, H.J. Byrne, J.N. Coleman, S. Curran, A.P. Davey, B. McCarthy, et al., Optical absorption and fluorescence of a multi-walled nanotube-polymer composite, Synth. Met. 102 (1-3) (1999) 1176–1177. Available from: https://doi.org/10.1016/S0379-6779(98)01067-4.

[73] M. Alvaro, P. Atienzar, J.L. Bourdelande, H. Garcia, Photochemistry of single wall carbon nanotubes embedded in a mesoporous silica matrix, Chem. Commun. (2002) 3004–3005. Available from: https://doi.org/10.1039/B209225P.

[74] H. Ago, T. Kugler, F. Cacialli, K. Petritsch, R.H. Friend, W.R. Salaneck, et al., Workfunction of purified and oxidised carbon nanotubes, Synth. Met. 103 (1-3) (1999) 2494–2495. Available from: https://doi.org/10.1016/S0379-6779(98)01062-5.

[75] R. Martel, T. Schmidt, H.R. Shea, T. Hertel, P. Avouris, Single- and multi-wall carbon nanotube field-effect transistors, Appl. Phys. Lett. 73 (1998) 2447. Available from: https://doi.org/10.1063/1.122477.

[76] P.M. Ajayan, Nanotubes from carbon, Chem. Rev. 99 (7) (1999) 1787–1800. Available from: https://doi.org/10.1021/cr970102g.

[77] S.J. Tans, M.H. Devoret, H. Dal, A. Thess, R.E. Smalley, L.J. Geerligs, et al., Individual single-wall carbon nanotubes as quantum wires, Nature 386 (1997) 474–477. Available from: https://doi.org/10.1038/386474a0.

[78] L.C. Venema, J.W.G. Wildoer, J.W. Janssen, S.J. Tans, H.L.J.T. Tuinstra, L.P. Kouwenhoven, et al., Imaging electron wave functions of quantized energy levels in carbon nanotubes, Science 283 (5398) (1999) 52–55. Available from: https://doi.org/10.1126/science.283.5398.52.

[79] J. Zhang, A. Panwar, D. Bello, T. Jozokos, J.A. Isaacs, C. Barry, et al., The effects of recycling on the properties of carbon nanotube-filled polypropylene composites and worker exposures, Env. Sci: Nano 3 (2) (2016) 409–417.

[80] K. Pyrzynska, Nanomaterials in speciation analysis of metals and metalloids, Talanta 212 (2020) 120784. Available from: https://doi.org/10.1016/j.talanta.2020.120784.

[81] J. Bhadra, N. Al-Thani, A. Abdulkareem, 11 - Recycling of polymer-polymer composites, in: R.K. Mishra, S. Thomas, N. Kalarikkal (Eds.), Micro and Nano Fibrillar Composites (MFCs and NFCs) from Polymer Blends, Woodhead Publishing, 2017, pp. 263–277. Available from: http://doi.org/10.1016/B978-0-08-101991-7.00011-X.

[82] Y. Wang, P. Hughes, H. Niu, Y. Fan, A new method to improve the properties of recycled aggregate concrete: composite addition of basalt fiber and nano-silica, J. Clean. Prod. 236 (2019) 117602.

[83] O. Güven, S.N. Monteiro, E.A.B. Moura, J.W. Drelich, Re-emerging field of lignocellulosic fiber – polymer composites and ionizing radiation technology in their formulation, Polym. Rev. 1 (6) (2016) 1560.

[84] M. Kotal, A.K. Bhowmick, Polymer nanocomposites from modified clays: recent advances and challenges, Prog. Polym. Sci. 51 (2015) 127–187.

[85] E.H. de Oliveira, V.A. Silva, R.R. Oliveira, A.S. Teran, A.V.A. Castillo, J. Harada, F.R.V. Diaz, E. A.B. Moura, Investigation on mechanical and morphological behaviours of copolyester/starch blend reinforced with rice husk ash, in: J.S. Carpenter, C. Bai, J.-Y. Hwang, S. Ikhmayies, B. Li, S.N.

Monteiro, Z. Peng, M. Zhang (Eds.), Characterization of Minerals, Metals, and Materials. The Minerals, Metals & Materials Society, Wiley, Hoboken, pp. 491–498. Available from: https://doi.org/10.1002/9781118888056.ch57.

[86] E.A.B. Moura, The potential of micro- and nano-sized fillers extracted from agroindustry residues as reinforcements of thermoplastic-based biocomposites—a review, in: S. Ikhmayies, J. Li, C. Vieira, J. Margem (Deceased), F. de Oliveira Braga (Eds.), Green Materials Engineering. The Minerals, Metals & Materials Series, Springer, Cham, 2019, pp. 89–100. Available from: https://doi.org/10.1007/978-3-030-10383-5_10.

[87] A.V. Ortiz, C.A. Pozenato, M.N. Sartori, R.R. Oliveira, M.A. Scarpin, E.A.B. Moura, Thermal andmorphological behavior of PBT/sugarcane bagasse ash composite, J. Nanostruct. Polym. Nanocompos. 8 (2012) 78–81.

[88] J. Li, X. Wei, Q. Wang, J. Chen, G. Chang, L. Kong, et al., Homogeneous isolation of nanocellulose from sugarcane bagasse by high pressure homogenization, Carbohydr. Polym. 90 (2012) 1609–1613.

[89] A. Dufresne, Comparing the mechanical properties of high performances polymer nanocomposites from biological sources, J. Nanosci. Nanotechnol. 6 (2006) 322–330.

[90] E. Fortunati, F. Luzi, D. Puglia, R. Petrucci, J. Kenny, L. Torre, Processing of PLA nanocomposites with cellulose nanocrystals extracted from *Posidonia oceanica* waste: innovative reuse of coastal plant, Ind. Crop. Prod. 67 (2015) 439–447.

[91] E. Lizundia, et al., PLLA-grafted cellulose nanocrystals: role of the CNC content and grafting on the PLA bionanocomposite film properties, Carbohydr. Polym. 142 (2016) 105–113.

[92] M.P. Arrieta, E. Fortunati, F. Dominici, E. Rayón, J. López, J.M. Kenny, PLA-PHB/cellulose based films: mechanical, barrier and disintegration properties, Polym. Degrad. Stab. 107 (2014) 139–149.

[93] I.T. Seoane, E. Fortunati, D. Puglia, V.P. Cyras, L.B. Manfredi, Development and characterization of bionanocomposites based on poly (3-hydroxybutyrate) and cellulose nanocrystals for packaging applications, Polym. Int. 65 (2016) 1046–1053.

[94] G. Siqueira, H. Abdillahi, J. Bras, A. Dufresne, High reinforcing capability cellulose nanocrystals extracted from *Syngonanthus nitens* (Capim Dourado), Cellulose 17 (2010) 289–298.

[95] X. Cao, Y. Chen, P.R. Chang, M. Stumborg, M.A. Huneault, Green composites reinforced with hemp nanocrystals in plasticized starch, J. Appl. Polym. Sci. 109 (2008) 3804–3810.

[96] M. Martínez-Sanz, A. Lopez-Rubio, J.M. Lagaron, Highbarrier coated bacterial cellulose nanowhiskers films with reduced moisture sensitivity, Carbohydr. Polym. 98 (2013) 1072–1082.

[97] B. Wang, M. Sain, The effect of chemically coated nanofiber reinforcement on biopolymer based nanocomposites, BioResources 2 (2007) 371–388.

[98] A. Ferrer, E. Quintana, I. Filpponen, et al., Effect of residual lignin and heteropolysaccharides in nanofibrillar cellulose and nanopaper from wood fibers, Cellulose 19 (2012) 2179–2193.

[99] F.W. Brodin, O.W. Gregersen, K. Syverud, Cellulose nanofibrils: challenges and possibilities as a paper additive or coating material—a review, Nord. Pulp Pap. Res. J. 29 (2014) 156–166.

[100] H.P.S.A. Khalil, Y. Davoudpour, M.N. Islam, et al., Production and modification of nanofibrillated cellulose using various mechanical processes: a review, Carbohydr. Polym. 99 (2014) 649–665.

[101] I. Solala, A. Volperts, A. Andersone, et al., Mechanoradical formation and its effects on birch kraft pulp during the preparation of nanofibrillated cellulose with Masuko refining, Holzforschung 66 (2012) 477–483. Available from: https://doi.org/10.1515/hf.2011.183.

[102] Y. Jiang, X. Liu, Q. Yang, et al., Effects of residual lignin on mechanical defibrillation process of cellulosic fiber for producing lignocellulose nanofibrils, Cellulose 25 (2018) 6479–6494.

[103] H. Yousefi, V. Azari, A. Khazaeian, Direct mechanical production of wood nanofibers from raw wood microparticles with no chemical treatment, Ind. Crop. Prod. 115 (2018) 26–31.

[104] S.M. Yousefhashemi, A. Khosravani, H. Yousefi, Isolation of lignocellulose nanofiber from recycled old corrugated container and its interaction with cationic starch–nanosilica combination to make paperboard, Cellulose 26 (2019) 7207–7221.

[105] S. Kamel, Nanotechnology and its applications in lignocellulosic composites, a mini review, Express Polym. Lett. 1 (2007) 546–575.

[106] V. da Costa Correia, V. dos Santos, M. Sain, et al., Grinding process for the production of nanofibrillated cellulose based on unbleached and bleached bamboo organosolv pulp, Cellulose 23 (2016) 2971–2987.

[107] Ø. Eriksen, K. Syverud, Ø. Gregersen, The use of microfibrillated cellulose produced from kraft pulp as strength enhancer in TMP paper, Nord. Pulp Pap. Res. J. 23 (2008) 299–304.

[108] C. Hii, Ø.W. Gregersen, G. Chinga-Carrasco, Ø. Eriksen, The effect of MFC on the pressability and paper properties of TMP and GCC based sheets, Nord. Pulp Pap. Res. J. 27 (2012) 388.

[109] H. Sehaqui, Q. Zhou, L.A. Berglund, Nanofibrillated cellulose for enhancement of strength in high-density paper structures, Nord. Pulp Pap. Res. J. 28 (2013) 182–189.

[110] I. González, F. Vilaseca, M. Alcalá, et al., Effect of the combination of biobeating and NFC on the physico-mechanical properties of paper, Cellulose 20 (2013) 1425–1435.

[111] K. Missoum, F. Martoïa, M.N. Belgacem, J. Bras, Effect of chemically modified nanofibrillated cellulose addition on the properties of fiber-based materials, Ind. Crop. Prod. 48 (2013) 98–105.

[112] S.R. Djafari Petroudy, K. Syverud, G. Chinga-Carrasco, et al., Effects of bagasse microfibrillated cellulose and cationic polyacrylamide on key properties of bagasse paper, Carbohydr. Polym. 99 (2014) 311–318.

[113] M. Hietala, A. Ämmälä, J. Silvennoinen, H. Liimatainen, Fluting medium strengthened by periodate–chlorite oxidized nanofibrillated celluloses, Cellulose 23 (2016) 427–437.

[114] A.A. Dubrovskiy, D.A. Balaev, K.A. Shaykhutdinov, O.A. Bayukov, O.N. Pletnev, S.S. Yakushkin, et al., Size effects in the 19 magnetic properties of ε-Fe_2O_3 nanoparticles, J. Appl. Phys. 118 (2015) 213901–213917.

[115] N. Kamboj, M. Aghayan, F. Rubio-Marcos, K. Nazaretyan, M.A. Rodríguez, S. Kharatyan, et al., Nanostructural evolution in mesoporous networks using in situ high-speed temperature scanner, Ceram. Int. 44 (2018) 12265–12272.

[116] K. El-Boubbou, Magnetic iron oxide nanoparticles as drug carriers: clinical relevance, Nanomedicine 13 (2018) 953–971.

[117] S. Said, S. Mikhail, M. Riad, Recent progress in preparations and applications of meso-porous alumina, Mater. Sci. Ene. Technol. 2 (2019) 288–297.

[118] A.S. Sufian, N. Samat, M.Y. Meor Sulaiman, W. Paulus, Alumina recovery from industrial waste: study on the thermal, tensile and wear properties of polypropylene/alumina nanocomposites, Int. J. Precis. Eng. Manuf.-Green Technol. 7 (2020) 163–172. Available from: https://doi.org/10.1007/s40684-019-00135-z.

[119] N. Samat, C.D. Marini, M.A. Maritho, F.A. Sabaruddin, Tensile and impact properties of polypropylene/microcrystalline cellulose treated with different coupling agents, Comput. Interface 20 (7) (2013) 497–506.

[120] F. Mirjalili, L. Chuah, E. Salahi, Mechanical and morphological properties of polypropylene/nano α-Al_2O_3 composites, Sci. World J. 2014 (2014) 1–12.

[121] P. Kakde, S.N. Paul, Effect of alumina nanoparticle addition on the mechanical and wear behaviour of reinforced acrylonitrile butadiene styrene polymer, Int. J. Adv. Res. Sci. Eng. Technol. 5 (6) (2014) 185–192.

[122] N.I. Zulkifli, N. Samat, H. Anuar, N. Zainuddin, Mechanical properties and failure modes of recycled polypropylene/microcrystalline cellulose composites, Mater. Des. 69 (2015) 114–123.

[123] E. Pérez, V. Alvarez, C.J. Pérez, C. Bernal, A comparative study of the effect of different rigid fillers on the fracture and failure behavior of polypropylene based composites, Comp: Part. B 52 (2013) 72–83.

[124] M.H.M. Hamdan, J.P. Siregar, M.R.M. Rejab, D. Bachtiar, J. Jamiluddin, C. Tezara, Effect of maleated anhydride on mechanical properties of rice husk filler reinforced PLA matrix polymer composite, Int. J. Precis. Eng. Manuf.-Green Technol. 6 (1) (2019) 113–124.

SECTION IV

Applications of recycled nanomaterials

CHAPTER 16

Industrial scale up applications of nanomaterials recycling

Ajit Behera[1] and Suman Chatterjee[2]

[1]Department of Metallurgical & Materials Engineering, National Institute of Technology, Rourkela, India
[2]Department of Mechanical Engineering, National Institute of Technology, Rourkela, India

16.1 Why industrial-scale recycling is require for nanomaterials

In the past three decades, nanotechnology has gained much attention as it enables the restructuring and redesign of materials at the nano scale (1−100 nm) to manipulate material properties for industrial applications. Nanomaterials (NMs), which are essentially nano-sized materials with at least one dimension falling in the nano-scale range, are the products of nanotechnology [1]. The penetration of nanotechnology in industries such as manufacturing, construction, robotics, communication, energy, materials, agriculture, healthcare, and medicine has led to a new era of growth in the fields of science and technology. With the increase in implementation of nanotechnology and incorporation of NMs into consumer products, environmental concerns over generation of nanowastes and nanopollutants have emerged. Nanowastes are defined as collectable wastes that contain NMs, which includes byproduct wastes, nanocontaminated wastes, and end-of-life nanoproducts generated from the production or use of nanoproducts [2]. By contrast, nanopollutants are NMs released from nanoproducts in a diffused manner with the potential of unexpected release into the environment. Both nanowastes and nanopollution are unwanted in the environment. However, nanowastes are more quantifiable because of their release as point sources, in contrast to nanopollutants, for which the pathways of nanoparticle release are relatively dispersed [3]. Therefore from a waste management perspective, nanowastes are potentially major contributors of nanoparticle release into the environment. Use of NMs in cosmetics, health and fitness products, fuel additives, catalysts, and nanochips ultimately releases the constituent nanoparticles into air, water streams, and/or soil [4,5]. Table 16.1 shows the sources of major of nanowastes and nanopollutants that are released into the environment [6−8]. Not all NMs possesses hazardous properties. In fact, studies performed on the same types of NMs are in disagreement; some studies show their biocompatibility, while others prove their potentially hazardous nature [e.g., carbon nanotubes (CNTs)]. The potential risks of these materials also depend on their solubility, size, shape, and agglomeration, among other physicochemical parameters (e.g., crystallinity, redox potential). Few of the initial

Nanomaterials Recycling
DOI: https://doi.org/10.1016/B978-0-323-90982-2.00015-9

© 2022 Elsevier Inc.
All rights reserved.

342 Nanomaterials Recycling

Table 16.1 Various large-scale industrial nanoproducts and their effect on the environment.

Industrial nanoproducts	NM ingredient	Risk to environment
Automobile parts	Fullerenes	Medium
	Single-walled CNT	Medium
	Multiwalled CNT	Medium
	Nanoclay	Low
Coatings and paints	TiO_2	Low
	ZnO	Medium
	Ag	Low
	Al_2O_3	Low
	MgO	Low
	CeO_2	Low
Cosmetics (sunscreen products)	ZnO	Medium
	TiO_2	Low
	Fullerenes	High
	Dendrimer	Medium
Personal care	Ag	Medium
	Fullerenes	High
	Fe_2O_3	Medium
	TiO_2	Low
Food and beverages	ZnO	Medium
	TiO_2	Low
	Fullerenes	High
	Dendrimer	Medium
Electronics (lithium ion batteries, etc.)	Lithium and lithium oxide	High
	Sodium and magnesium oxide	Medium
	CNT	Medium
Textiles (antiultraviolet textiles)	Single-walled CNT	Low
	Multiwalled CNT	Low

concerns over nanowastes were mentioned in an early report by the Royal Society and Royal Academy of Engineering in 2004 outlining the various nanoexposure pathways [9]. These byproducts are highly active in the atmosphere at nanometer scale [10]. These byproducts are actively functioning at nano levels, which makes nanobyproducts as a major concern. These nanobyproducts can easily float in the atmosphere and be diluted in water. This makes them more dangerous for living beings, including plants, as they might easily penetrates living cells [11]. One study by Mueller and Nowack [12] estimated that 5%—90% of nanoparticles (for silver, TiO_2, and carbon) are released from nanoproducts during application, disposal, dissolution, and recycling of these products. Routes of human exposure include ingestion from food and dietary supplements; direct skin contact through such products as cosmetics, sunscreen, and textiles; and inhalation of airborne nanopollutants. The surface coating, shape, size, surface area, and chemical composition of NMs can dictate their relative toxicity among humans, which can vary from one product to another. Additionally, direct entry of such nanowastes into traditional waste water and industrial effluent treatment plants may reduce existing treatment efficiency, posing new challenges in waste management [13].

Treating byproducts of NMs or nanotechnology as nanowaste by academia, industry, and government institutions has remarkably opened pathways for NM recovery and recycling from wastes on an industrial scale [14]. NM recycling refers to effective recycling of nanobyproducts or NMs from nanowastes by recovering either parent material aggregates or NM itself. It was estimated that roughly 160,000–280,000 tons/year of global NM produced ended up in landfills, 20,000–80,000 tons/year were discharged into soils, and 260–21,000 tons/year were released into water bodies and atmosphere in 2010. These estimates were based on probable material flows and estimates during production, manufacturing, product usage, and end-of-life release from waste water treatment and incineration [15]. The study included seven major NMs: silica, zinc oxide (ZnO), aluminum oxide (Al_2O_3), CNTs, TiO_2, nanoclays (ceramic NMs), and iron-based NMs [16]. A study by Giese et al. [17] presents a conservative estimate of 59,000 tons/year of NMs (limited to nano-Ag, ceric oxide (CeO_2), and silica) in 2017 which is expected to increase upto approximately 250,000 tons/year in 2050. Moreover, existing waste treatment mechanisms are not designed to recover or retain nanowastes released into the waste streams [18]. Given the potential toxicity of NMs upon release to environment along with the estimated increase in NM applications in industry and consumer products, there is an increasing need for NM-specific recycling for sustainable growth of the nanotechnology industry. The benefits of upscaling NM recycling are twofold: It can reduce the environmental risks and health hazards associated with NM industry, and successful recovery and recycling of NMs can curtail NM-manufacturing processes [19].

16.2 Market scenario of recycled nanomaterials

The rate of global demand for NMs increases more when we look at the market perception of existing materials and future demand for new materials and applications. Currently, CNTs along with other NMs, such as quantum dots, dendrimers, and nanoclays, are growing at the fastest pace. The global NMs market size was valued at USD $5.5 billion in 2016 and increased to USD $8.5 billion in 2019 [20]. Based on type of materials, NMs market is divided into carbon-based NMs, metal-based NMs, metal and nonmetal oxides, chemicals, and polymers. Metal- and metal oxide–based NMs represent a considerable share, owing to the growth of applications in the production of microcircuits, sensors, and passivation coatings [21]. The carbon-based NM market is estimated to hold a noticeable share, owing to their uncompetitive thermal and electrical conductivity and mechanical strength properties. Again, the NMs market

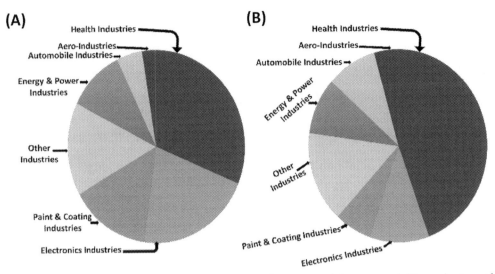

Figure 16.1 Diagram showing (A) the percentage of nanomaterial usage and (B) production of nanowaste in different industries in 2019 according to market share.

is segregated on the basis of their applications in electrical equipment, electronics, chemical products, medical instruments, aviation components, paints and coatings, adhesives and sealants, polymers, and composites. Rapid developments in healthcare technology, growth in the medical diagnostics industry, and various advantages of medicinal imaging applications are anticipated to drive the market [22]. Furthermore, increased focus on research concerning nanotechnology and rising government spending on biotechnology and pharmaceutical research and development are expected to augment the growth of the NMs market. In chemical industries, oxide materials, aerogels, and sol-gels find wide use on account of catalytic and high surface area with high absorption behavior.

Key players in the world NM market have endorsed various business strategies, such as capacity and business-expansion, partnerships, mergers, acquisitions, and application development to expand the global competitive market [23]. The market players profiled here are NMs manufacturers who either use NMs internally in some applications or supply them to other end users in the industry. The data are shown in Fig. 16.1.

16.3 Nanowaste generation from various industries and practices

Nanowastes generated through various processes can be classified as carbon-based nanobyproducts, metal-based nanobyproducts, polymer-based nanobyproducts, and nanocomposite byproducts, depending on the parent NM (Fig. 16.2). Release of nanobyproducts may occur during the product life cycle, usage, or end-of-life disposal [24]. The production also

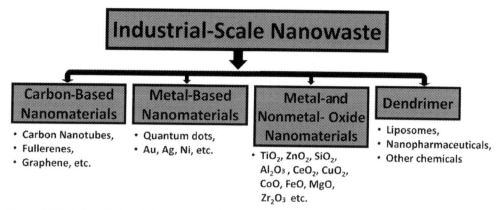

Figure 16.2 Various industrial nanomaterials byproducts.

depends on the circumstances of manufacturing, its nature at the end of life, and use of product in the specific environment. Synthesis and manufacturing of NMs is the primary process of release of nanowastes as excluded material from the NM size selection phase [25]. Processes such as material generation, drilling, cutting, rolling, and thermal heat addition processes that change the material's structure create the greatest possibility of release of nanowaste.

16.3.1 Medical industries' nanowaste

Healthcare industries occupy the largest position to use NMs product as well as producing respective wastes. Currently, applications of antibacterial, biocidal, and antifungal NM-assisted coating can be found everywhere. CNTs, iron oxide, gold, silver, and copper nanoparticles, quantum dots, nanogels, dendrimers, and liposomes are commonly used NM in medical industries [26,27]. Silver NMs have the most diverse range of applications among metal NMs. This is substantiated by the fact that there are currently more than 1000 products containing Ag NMs. Ag NMs are incorporated into textiles for antistain and adding antibacterial properties to medical textiles, in medicine and healthcare as topical ointments, in cell imaging and cancer therapy, as catalysts in fuel cells and hydrogen production, in environmental applications for water and waste water disinfection, and in activated carbon filters to reduce microbial growth [28]. Because of their benign nature and imparting of antibacterial or antifungal properties by the release of Ag ions from Ag NMs, these products are widely advertised and have greater market value than other NMs. Interestingly, Ag NMs released from consumer products diffuse into their surrounding environment as Ag^+ ions and are less likely to do so in the form of particulate Ag NMs. Washing of conventional Ag-based and/or nano-Ag-based textiles releases silver in ionic form (Ag^+) and particulate Ag [29]. Although incorporated into a wide variety of products, the overall production of Ag NMs is much lower in comparison to TiO_2 and SiO_2 NMs. Use of nanoparticles and nanodevices in medicinal applications and drug delivery is a growing rapidly scientific area [30]. Magnetic nanoparticles are used to quantify several parameters in polarized cells, in molecular biology studies, for diagnostic purposes, for cancer therapy, for transporting chemical factors in drug targeting, for drug biomanipulation, for labeling biomolecules to evaluate physiological features of the cell, as contrasting agents in imaging, and in biochemical analysis. These products generates huge amounts of waste after a predefined working period [31].

16.3.2 Electronic industries' nanowaste

Electronic industries generate more than 14 million tons of electronic waste (e-waste) per year. Electronic waste includes precious metals (Ag, Au, etc.), heavy metals, rare earth elements, glass, organic compounds (polychlorinated biphenyls, polyvinyl chloride, etc.), and composite materials [32]. Most e-waste is toxic to human health and the environment. Nano-scaled carbon fiber composite is the most wanted material among all e-materials because of its high strength-to-weight ratio, even more than that of steel. Carbon fibers can be cabled together into fiber bundles and framed in a resin to form a carbon composite sheet that can be used in portable electric supply in vehicles, airplanes, and building panels [33]. Carbon fiber in electric-assisted equipment facilitates lower fuel consumption with an increase in efficiency [34]. Thus carbon-fiber consistently meet the demands for structural carbon composites. Working with the carbon nanocomposite, such as grinding or cutting sheets, can release fine particulates that could potentially cause lung irritation, but this is more likely to occur in an occupational setting as opposed to a landfill, although exposure to high heat and to prolonged ultraviolet light can fracture the matrix and possibly contribute to the release of fibers and resin. Once the carbon fiber is embedded in a matrix to create the composite material, no carbon fiber can be extracted. Typically, the practice in many industries is to order large sheets of the composite and punch holes for structures and dispose of the excess carbon composite. This fabrication procedure generates waste, and the company pays for a large composite sheet and may end up using only 50% or less of it and scrapping the rest. Even though e-wastes are different in many regards, they have been recognized as international health and environmental threat [35]. One of the leading solutions for managing e-waste has been recycling. Although e-waste is generated within multiple industries and countries, universal practices such as responsible recycling practices standards and the waste from electrical and electronic equipment directive have created consistent and uniform standards for e-waste recycling that transcend individual communities of practice. Creating a solution to the problem of carbon fiber composite scrap was driven much more by economics than by anything else. Since the toxicity of carbon fiber is virtually a nonstarter compared with byproducts of electronics, it was the fact that so much carbon composite was being wasted and shipped off to landfills that really forced industry into acting [36].

16.3.3 Energy-harvesting industries' nanowaste

Nano-TiO_2 is the most widely and commonly produced NM in energy harvesting industries, owing to its physicochemical properties of absorption of ultraviolet light and high refractive index [37]. These properties have diversified the use of TiO_2 in solar photovoltaic cells, photocatalytic degradation of contaminants, and use in common consumer products such as sunscreens and additives in paint coatings. In 2019 light energy harvesting accounted for the largest share of the energy-harvesting market. The growth of light energy harvesting is attributed to the increasing number of companies involved in the production of solar products for building automation, consumer electronics, and security applications. The raw materials used in this industries such as silicon wafers, PV cells, polysilicon, ingots, Al, and Ag that are used to build solar cells. Solar cells, transducers, and the like consist of thin film of various materials that enters the environment when it crosses its fatigue limit and becomes waste [38]. Little research has yet taken place on nanowaste produced by thermoelectric energy harvesting, photoelectric energy harvesting, mechanical energy harvesting, radio wave energy harvesting, and regenerative energy harvesting. Energy harvesting of industrial waste is a focus of interest nowadays.

16.3.4 Construction industries' nanowaste

New building materials based on NMs will develop and will influence the construction sector. Nanotechnology has the potential to make construction faster, cheaper, safer, and more varied. If nanotechnology is implemented in the construction of homes and infrastructure, such structures will be stronger. If buildings are stronger, then fewer of them will require reconstruction, and less waste will be produced in long term. Waste can be used for producing nanopowder or other nanoproducts, which, by using new nanotechnologies, allow obtaining a new generation of cement-based materials that will be more durable, with higher mechanical properties, or even with desired properties, such as electrical conductivity as well as temperature. Nanotechnology in construction involves using nanoparticles, especially alumina and silica. If cement with nano-size particles can be manufactured and processed, it will open up a large number of opportunities in the fields of ceramics, high-strength composites, and electronic applications. Nanosilica (SiO_2) can be introduced into conventional concrete as part of the normal mix [39]. The construction sector utilizes a significant amount of nanosilica (after nano-TiO_2) as concrete additives, additives for wall paints, and additives to road marking materials. Being incorporated into buildings and structures, $<5\%$ of nanosilica used in the construction sector ends up in the waste management system. Also, approximately $20\%-30\%$ of the construction materials are typically recycled by the same industry, thereby reducing the release of nanosilica to the environment. Additional silica nanowaste release occurs from disposal, weathering, and abrasion processes, which are nonpoint sources of silica nanowastes, or from waste that ends up in the landfills. However, one of the advances made by the study of concrete at the nano scale is that particle packing in concrete can be improved by using nanosilica, which leads to a densifying of the microstructure and nanostructure, resulting in improved mechanical properties. Nanosilica addition to cement-based materials can also control the degradation of the fundamental calcium–silicate hydrate reaction of concrete caused by calcium leaching in water and block water penetration and therefore lead to improvements in durability. Around 40% of nano-TiO_2 is recycled, 36% is put into landfills, and approximately 20% becomes waste that diffuses into the environment. Most nano-TiO_2 in the construction industry is used as additives for wall paints and road markings to provide scratch resistance. Because of the application of nano-TiO_2 onto secondary surfaces, paint-based TiO_2 nanowastes end up in landfills and/or waste incineration plants. Nano-TiO_2 is also integrated into ceramics and filters, which end up in waste incineration plants, and in plastics, which are half incinerated and half recycled. Nano-TiO_2 is also incorporated into textiles, which release upto 4% of TiO_2 NM, typically during product use and washing processes, hence ending up indirectly in waste waters. The use of nanotechnology in steel helps to improve the physical properties of steel. Fatigue, the structural failure of steel, occurs as a result of cyclic loading. Current steel designs are based on the reduction in allowable stress, service life, or regular inspection regime. This has a significant impact on the life cycle costs of structures and limits the effective use of resources. Stress risers are responsible for initiating cracks from which fatigue failure results. The addition of copper nanoparticles reduces the surface unevenness of steel, which then limits the number of stress risers and hence fatigue cracking. Advancements in this technology through the use of nanoparticles would lead to increased safety, less need for regular inspection, and more efficient materials that are free from fatigue issues for construction. Steel cables can be strengthened by using CNTs. Stronger cables reduce the costs and period of construction, especially in suspension bridges, as the cables are run from end to end of the span. The use of vanadium and molybdenum nanoparticles improves the delayed fracture problems associated with high-strength bolts. This reduces the effects of hydrogen embrittlement and improves the steel microstructure by reducing the effects of the intergranular cementite phase. The addition of nanoparticles in construction can be reused from other recovered nanowaste from other industries [40].

16.3.5 Other industries' nanowaste

Both the paper and food packaging industries generate nanowaste materials. The highest chance of contamination is of plastic containers with various NMs in the respective environment. For example, a plastic bottle for a detergent that contained NMs in liquid suspension would then be contaminated with some residue of NMs. Pretreatments of recycled wastes, such as shredding, may serve to liberate NMs from host materials and introduce them into effluents following washing processes. Nowadays, these industries channel the waste for recycling by household products. Many consumer products containing NMs will be disposed of in municipal waste, so controlling the entry of nanowaste into waste streams targeted for recycling would be a challenging task. The removal of packaging materials contaminated with nanowastes from packaging waste streams targeted for recycling would then have implications for the achievement of overall recycling goals for those materials [41].

16.4 Potential nanobyproduct materials, their recovery, and recycling

Different industrial processes and practices generate various nano-scale derivatives from the same material.

16.4.1 Carbon-based nanobyproducts

Carbon is a ubiquitous element that is present in all living matter and is released from natural events occurring on the planet. Carbon byproducts come from various sources in nature and during production and manufacturing of these raw materials. Therefore carbon nanobyproducts are produced in enormous amount by forest fires, volcanic eruptions, soot, and fossil fuel combustion derived from pyrogenic and petrogenic sources. On the other end, engineered carbon NMs such as CNTs (single-walled/ multiwalled) fullerenes, graphenes, and nanofibers that are absorbed into water and waste water treatment. Most of the carbon-based NMs are used in sensors and electronics because of their excellent thermal, optical, and structural properties [42]. Fiber-like carbon derivatives are potentially health hazardous, but tightly agglomerated particle-like bundles of CNTs did not cause an inflammatory response when present in the peritoneal cavity of mice. The sources from which nanobyproducts can be easily obtained are leaves, petals, asphalt, urea, sugar mills, coal tar pitch, raw coal, vehicles, carbonization of paper, food waste, and glycoside hydrolase producers. By tracking the product life cycle, it is possible to trace the release conditions of carbon nanoproducts. The literature suggests that little research work has been conducted to understand the potential reasons for the release of the carbon-based nanobyproducts. Generally, the release of carbon-based nanobyproducts from polymer composites depends on two important scenarios: (1) release during high-energy manufacturing processes and environmental exposures occurring from grinding, cutting, sanding, and drilling in dry and wet machining conditions and (2) release of carbon byproducts from bound matrices as a result of low-energy processes, such as consumer use and environmental degradation from ultraviolet light and weathering [43]. During manufacturing of engineered carbon nanoparticle/nanotube (CNP/CNT), chemical vapor deposition is the prime location of release of significant carbon nanowastes followed by engineered CNP/CNT preparation. An approximate release of 0.9–1250 g carbon soot per gram of product was noted during single-walled CNT synthesis. A substantial portion ($\sim 50\%$) of engineered CNP/CNTs in electronics and energy storage devices are recycled and end up in landfills as solid carbon nanowaste. Carbon nanowastes from paints, plastics, and sporting equipment either end up in waste incineration plants and are burned or undergo unaccounted release during use. Roughly 6%–5.5% of CNTs are released after treatment from wastewater treatment plants and solid waste incineration, approximately 15% ends up in the soil, and approximately 1.5% is disposed to the air [44].

Recycled carbons fibers might cannot compete with virgin carbon fibers in terms of performance, but they would be viable in terms of weight savings, cost, and synthesis of complex shape, making them acceptable for specific automotive and parts industries in a circular system. Finding an effective recovery method that can produce good-quality fibers in mild conditions and their reuse for new composites is still a part of ongoing research. The pyrolysis process produces short, discontinuous, fluffy fibers that have undergone various processes. These remanufacturing processes were introduced in 2011 [45]. The reuse mainly depends on the physical structure and mechanical properties of the recycled fibers. Recycled discontinuous, three-dimensional-oriented fibers have ability to manufacture new structure components as technology demonstrators, including secondary components for the automotive industry, interior components of aircraft, and tooling. In general, recovered fibers after removal of residual char show acceptable mechanical properties in comparison to virgin material, especially in the case of recovered carbon fibers. However, further technical developments in remanufacturing processes are needed, especially in fiber alignment, fiber content in new composite and low fiber damage during synthesis. A known example of the reuse of byproducts generated in the production of NMs is found in the use of low-purity C60 fullerene soot in the lubricants industry [46].

16.4.2 Metal nanobyproducts

From aviation industries to medical industries, each and every piece of equipment is associated with a metal nanoproduct. The use of metal nanobyproducts is increasing at a great rate. Retaining and disposing of active metal nanoparticles is highly challenging, as the kinetics associated with NMs is rapid. NMs are thermodynamically metastable and lie in the region of high-energy local minima. Hence they are prone to attack and undergo transformation with other elements in the environment. Metal-based NMs include quantum dots, nanogold, nanosilver, and oxides with metal bases [47]. Metal-based waste NMs are a focus of the biomedical and pharmaceutical industries. Apart from catalysis, potential applications of metal nanoparticles are well known in other fields, including pigments, electronic and magnetic materials, and drug delivery. Metal nanoparticles such as Mg extracted from various sources during processing act as strong explosives when they come into contact with oxygen. Metal nanoparticles are highly reactive, and they inherently interact with impurities as well. NMs are usually considered harmful as they can penetrate the cells of the dermis [48]. Toxicity of NMs also appears to be high because of their high surface area and enhanced surface activity. NMs have been shown to cause irritation and have been indicated to be carcinogenic. If they are inhaled, their low mass entraps them inside lungs, and they cannot be expelled out of body. Their interaction with liver and blood could also prove to be harmful, although this aspect is still being debated. No hard-and-fast safe disposal policies have evolved for metal nanoparticles. Issues of their toxicity are still under question, and results of exposure experiments are not available. Hence the uncertainty associated with the effects of NMs is yet to be assessed to develop disposal policies [49].

16.4.3 Metal oxide and nonmetal oxide nanobyproducts

Nanoparticles of titanium dioxide (TiO_2) are used in sunscreens, paints, and electronic circuits. TiO_2 can penetrate diseased or damaged skin in significant amounts and enter the bloodstream, where it can affect the central nervous system, resulting in permanent damage through intranasal instillation and neuroinflammation in the brain. Moreover, TiO_2 dust is carcinogenic to humans. TiO_2 waste, when present in the environment, can kill beneficial soil microbes and bacteria, completely changing the ecosystem balance. Nanoparticles of metal oxides, including TiO_2, ZnO, and tungsten oxide (WO_3), are used in water purification techniques because of their superior abilities to improve the chemical and biological properties of water [50]. Similarly, Al_2O_3, B_2O_3, CaO, Fe_2O_3, K_2O, MgO, MnO, Na_2O, and P_2O_5 are in a young stage to recover from various sources and be reused. Sewage sludge, e-wastes, and construction wastes such as cementitious materials contain SiO_2-based nanoparticles. Zou et al. [51] presented an interesting study of recovering high purity nano-SiO_2 from sewage sludge by acid washing sludge carbon at high temperatures (500°C—1000°C). Raw sewage sludge was acid washed using HCl for 2 hours and washed further with deionized water to result in acid-washed sludge carbon. The sludge carbon was further treated with NaOH to form sodium silicates. The authors found that acid-washed sludge carbon consisted primarily of crystalline SiO_2 (quartz) and some amounts of hydroxyl water. Further, a sol-gel synthesis method was used to synthesize nano-SiO_2, which can be further used in products. A similar acid-washing procedure (with hydrogen peroxide and sodium hydroxide) was utilized by Kauldhar and Yadav to extract silica from paddy straw agricultural waste [52]. Their method resulted in extraction of silica and lignin from the rice straw, which was then filtered to give sodium silicate, which was further titrated to form 15—20-nm spherical and uniform SiO_2 nanoparticles.

Another study by Sapra et al., was able to recover silica nanoparticles from waste photovoltaic modules by chemical and thermal treatment. This method used an organic solvent (e.g., toluene) to dissolve the photovoltaic module and remove the glass layer. The solar cell was further heated, chemically treated by nitric acid, and constantly stirred to recover silicon and synthesize $10-40$-nm SiO_2 nanoparticles. The yield for this process was greater than 99.99% [53]. Nano-SiO_2 can also be recovered from fly ash (a byproduct of pulverized coal combustion) that has $30\%-60\%$ silica content in addition to alumina. The method of recovery and formation of nano-SiO_2 involves dissolving fly ash in a strong base (sodium or potassium hydroxide) solution to form silicate salts. This is followed by treating sodium silicate with hydrochloric or sulfuric acid by using a sol-gel technique to result in nano-SiO_2. This chemical treatment procedure of extracting nano-SiO_2 has been also found to be successful in recovering nano-SiO_2 from glass wastes such as bottles. The alkali fusion method of using base and heat to result in sodium silicate followed by acid treatment has also been applied to waste powder from the photonic and semiconductor industry to result in silica nanoparticles. Nano-SiO_2 can also be obtained from thermal heat treatment of wastes containing silica. Temperatures above $1500°C$, are used to form carbon nanoparticles from rubber tire waste, which are mixed with ground glass waste to form $10-150$-nm silica carbide (SiC) nanoparticles. These methods of recovering nano-SiO_2 from waste economically and simply open pathways for SiO_2-based NM recycling by incorporating nano-SiO_2 into other products such as concrete structures, rubber, ceramics, and membrane-based applications [54].

Tin plate electroplating sludge contains nano-SnO_2 as waste. Recent studies have developed novel approaches to nano-SnO_2 recovery from tin plate electroplating sludge. Sn is an important material in solder, tin plating, ceramic glazes, and glass. Therefore a large amount of Sn is disposed as waste in industry. It is estimated that Sn might run out in 20 years or less at current global consumption rates. Because of the presence of Sn in waste and the importance and limited quantity of Sn, the recycling of this material from sludge is an urgent and nonnegligible task. Many metal-recycling techniques have been developed, including leaching solvent extraction, electrolysis, ion exchange, membrane separation, and microbiological methods, but the disadvantages of these methods lie in the high costs of operation. The problem with nanowaste in sludge is a difficult one to solve because the complexity and diversity of the nanowaste make it hard to develop a universal method for managing waste. A specific treatment procedure for specific types of sludge is required with the following goals: First, metals in the sludge could be transformed into reusable materials after recycling expensive NMs; second, the strategy should be economically feasible in a real system. The major components in the sludge are nanophase or amorphous metal oxides or hydroxides. Metal recovery is frequently hampered by the difficulties in the precipitation of the highly suspended small-sized metal-hydroxide particles. A feasible approach for treating sludge is to artificially induce the fast crystal growth of the amorphous or nanophase form into larger forms to make the precipitation easier and to reduce their adsorption capacities. This method was designed as a strategy for recycling SnO_2 from tin plate electroplating sludge that mainly contains amorphous Sn and Fe compounds [55].

16.4.4 Polymer nanobyproducts

Polymer NMs are found particularly in the areas of packaging and transport. The focus of disposal of these nanopolymers, which are not biodegradable, is on recycling and reuse. Thermosetting and thermoplastic polymer nanowaste can be used as a strengthening filler material in most flexible materials. One of the drawbacks of using polymer fillers is their low thermal resistance. For instance, the synthetic biopolymer polylactic acid, which is frequently used as a food-packaging material, exhibits low performance due to low heat distortion temperature and low resistance to extreme heat and humidity. Nevertheless, weighing the advantages of using nanopolymer fillers as packaging materials, especially to cater for demands in society for sustainability and environmental safety [56].

16.5 Nanowaste recycling processes

In terms of the ease of recycling wastes containing NMs and their possible effects on recycling processes, knowledge remains limited. Some commentators have raised concerns that the mechanics of the recycling process, including crushing, cutting and grinding, may serve to liberate significant amounts of NMs. These processing byproducts and nanowaste are recycled in many way as shown in Fig. 16.3.

16.5.1 Mechanical recycling

Mechanical processing involves crushing and shredding followed by further grinding and milling downto 10—50 mm of fine particles. The key principle of recycling carbon fiber—reinforced polymer involves the separation of carbon fibers from the resin matrix. Recovered materials or flakes are separated into resin-rich powders and carbon fibers of different lengths still embedded in the resin matrix. Resin-rich powders find applications as fillers in bulk as well as sheet-melding composites. While recovered fiber-rich fractions find reuse in composites, their cost and quality are key factors affecting efficient reuse. Cutting mills were found to produce longer as well as more homogeneous fiber lengths as compared to hammer mills. Direct reforming without grinding has also been used in some cases. Owing to significant impairment in mechanical properties and poor bonding between the fibrous fraction and resin, recycled materials are still being used in small concentrations only. Electrodynamic fragmentation has been used to shred carbon fiber—reinforced thermoplastics. Milling

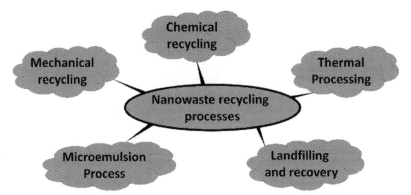

Figure 16.3 Major nanowaste processing and recycling.

dust and chips produced during milling operations have found applications as fillers in the production of thermoplastic granulates with increased tensile strength and overall rigidity. Degradation in carbon fiber quality is a key issue affecting mechanical processing. Mechanical attrition of any type of NMs' structural waste can be converted into uniform particle sizes that can be used mostly as filler for recycling purpose [57].

16.5.2 Chemical recycling

Gold nanoparticles are currently among the most studied industrial NMs. Their applications include electronics and photonic devices, catalysts, and sensors, with a particular emphasis in lateral flow immunoassays. The last application, related to the medical and veterinary industries, is expected to increase by more than four times in the next five years. In many research and industrial laboratories, Au NPs are produced by chemical methods that allow obtaining them with different morphologies (e.g., spheres, rods, triangles, and bipyramids,) and sizes. All these processes enable a huge production and consumption of Au NPs with the consequent generation of nanowaste. Since Au(III) salts (the precursor used to produce Au NPs) are expensive resources and their price depends on the world's gold market, a simple methodology that allows laboratory-scale gold recovery and its transformation into new Au NPs would present a significant profit in terms of laboratory's resources. Moreover, it would help to diminish the amount of nanowastes that are produced and must be treated. Additionally, this procedure could pave the way toward the recovery of other wastes that contain gold, at both laboratory and industrial scales. The recovery of gold from bulk solid wastes, as either Au(0) or Au(III), is well established at both industrial and laboratory scales [58]. One of the most widespread strategies is based on hydrometallurgical methods that involve the use of toxic leaching solutions (including cyanide, halides, thiosulfate, thiourea, thiocyanate, etc.).

The steps of the process of separating gold from laboratory nanowaste were as follows: (1) The Au NPs suspension volume was measured with a graduated test tube, and solid

NaCl was added until a final concentration of ~1M was reached (58 g of NaCl per liter of nanowaste solution). (2) The mixture was stirred manually to ensure NaCl dissolution. (3) After 12–24 hours, a sediment of Au NPs formed spontaneously, resulting in a black solid and a supernatant. To reduce the volume, the resulting supernatant was manually discarded in the adequate waste container. More than 90% of supernatant volume was removed in this step. (4) Then the black solid suspension was transferred to centrifugation tubes and was centrifuged for 10 minutes at 1000 RPM. The supernatant was discarded, and the solid was resuspensed with water; the same procedure was repeated three times. Alternatively, owing to the coalescence of the Au NPs into microparticles by the addition of NaCl, the black solid suspension can be filtered and washed. The final black solid represents the recovered gold, which was employed in the next step, as an aqueous suspension. Preparation of gold solution from a recovered gold:$HAuCl_4$ (aq) aqueous $HAuCl_4$ solution from recovered gold was prepared by employing a mixture of hydrogen peroxide and hydrochloric acid. Typically, 500 µL of recovered gold suspension was added to 25 mL of a solution of H_2O_2 (6M) and HCl (1M). The solid dissolved almost immediately, and the solution became light yellow. Then the solution was heated under reflux for 15 minutes to decompose the excess H_2O_2 [59].

16.5.3 Thermal processing

A variety of thermal processes, including pyrolysis (heating in the absence of air), fluidized bed and microwave-assisted pyrolysis, and combustion (heating in the presence of oxygen), have been used for recycling nanowaste. During incineration and once released from the waste matrix, NMs can be destroyed or can undergo a number of transformation processes that will influence their properties. If the matrix is combustible, NMs have higher chances of being liberated and being present in the gas phases. The combustion temperature and melting and boiling points of NMs affect the distribution of NMs between the solid and gaseous phases and determine whether the NMs are destroyed as a result of complete combustion. Thus it can be speculated that CNTs undergo complete combustion, and silver could be expected to enter the gas phase, whereas NMs such as ZnO, TiO_2, and cerium dioxide are likely to end up in the bottom ash. The chemical composition, size, and oxidation state of the NMs are other determinants of the fate of NMs during incineration. For example, if the temperature is high enough, reduced particles (e.g., aluminum) may undergo combustion to an extent that depends on their size and aggregation state. Conversely, particles that are already oxidized and have high melting points (e.g., cerium dioxide) may exit the combustion zone essentially unchanged. NMs that escape destruction during incineration may be captured by the flue gas treatment system and end up in the fly ash or other residues (e.g., bottom ash). Inorganic NMs will mostly end up in bottom ash, possibly in some aggregated and more stable form as a result of chemical reactions with other compounds [60]. Nonnegligible amounts of NMs, especially metal, can be expected to be found in fly ash and other flue gas cleaning residues. The few studies on the incineration of waste containing NMs indicate that CNTs (and possibly other organic NMs) are likely to be destroyed during combustion if the temperatures are sufficiently high at all times. However, complete destruction cannot be ensured. CNTs that are not destroyed are likely to end up in bottom ash. NMs that are not destroyed can undergo various transformation processes. They may react with other substances to form new particles, as in the case of silver in biosolids, which is transformed to silver-sulfur species during incineration. Larger nanoparticles may decompose into smaller particles, and NMs may agglomerate to form bigger particles, thereby losing their nanoform. Temperatures upto 1550°C are used for recycling of carbon fibers for steel-making applications. Devolatilized resin produces CO_2, H_2, and CH_4 gases and an oily fraction as well as char that may be deposited on fiber surfaces. Key advantages of pyrolysis (temperature range: 400°C−1550°C) include retention of mechanical properties, recovery of carbon fibers, and the absence of chemicals. Some of the detrimental factors include the deposition of chars on fiber surfaces and hazardous emissions due to thermal degradation of resin. While higher amounts of energy may be consumed during pyrolysis, depending on operating temperatures, this process does not cause the local area contamination that is typically associated with mechanical and physical processing of waste [61].

16.5.4 Landfilling treatment

If waste containing NMs is landfilled, such as nonrecyclable construction waste and shredder waste, there is a potential for leaching of the NMs from the waste to the leachate. Any organic acids in the leachate are shown to reduce the agglomeration of CNT. During landfilling, NMs released from the waste matrix can enter landfill leachate, which forms when water infiltrates in landfills and passes through the waste. Leachate is loaded with different organic and inorganic compounds and tends to be deposited at the bottom of the landfill. In engineered landfills, liners are placed at the bottom and on the sides of the landfill to prevent the flow of leachate out of the landfill. In a particular period, leachate is collected and treated before discharge to the environment. Landfill treatment also includes biological treatment, mechanical treatment by ultrafiltration, treatment with active carbon filters, electrochemical treatment including electrocoagulation by various proprietary technologies, and reverse osmosis membrane filtration using disc tube module technology [62].

16.5.5 Microemulsion process

In 2010 researchers from Bristol University published work on the separation of cadmium and zinc nanoparticles by using a special solvent [63]. The solvent is a stable microemulsion of oil in water, which breaks down into two layers when heated. All of the NPs in the solution end up in one of the layers, allowing simple separation. Composting and anaerobic digestion are the most commonly used biological processes for treatment of solid organic waste originating from industries and municipal or urban activities. Among urban organic waste streams, the important ones are household organic waste, garden and park waste, and sludge from waste water treatment plants. During both composting and anaerobic digestion, inorganic elements are conserved through the process and appear in the final outputs. Inorganic elements are not destroyed during biological waste treatment processes but are likely to undergo transformation processes, such as complexation and aggregation with other compounds. For example, silver and ZnO are transformed during biological treatment into different and more stable forms. In some cases, the presence of nanoparticles had positive effects on specific soil properties. For example, the use of iron oxide was reported to enhance the availability of nutrients in saline sodic soils [64].

16.5.6 Other methods of recycling

To date, several conventional techniques have been developed for remediation of soluble radionuclides, including chemical precipitation, evaporation, adsorption, ion exchange, membrane separation, cloud point extraction, and separation and removal of heavy metals from used materials such as batteries [65,66]. Conventional metal recovery methods, such as solvent extraction, ion exchange, electrolysis, plasma technology, and microbiological methods, offer strategies to recover high-value metals and rare earth elements from nanowaste. However, the diversity and complexity of nanowaste make it difficult to develop universally applicable methods for waste management. Thermoreversible liquid-liquid phase transition and cloud point extraction also hold promise for the successful separation and recovery of critical, high-value, and resource-limited materials from nanowaste. One such critical raw material is gold. The gold contained within the precipitated superstructure can be isolated by $Na_2S_2O_5$ reduction. The percent recoveries of gold samples were 59.6%−77.4%. Citrate-reduced Au NPs can be synthesized by using this recovered gold. However, the Au NPs were colloidally unstable and coalesced soon after synthesis, presumably owing to the as yet unidentified impurities in the recovered gold. Second, the dissolution of freshly recovered gold in aqua regia and subsequent boil-off of HNO_3 is necessary only for closed-loop recycling, in which the recovered gold is used for synthesizing new Au NPs. The dissolution in aqua regia followed by HNO_3 boil-off may not be necessary for an open-loop recycle scenario, where the objective is to simply recover the gold and not use it for Au NP synthesis. The energy footprint of this boiling step can be substantially reduced by first dissolving several batches of recovered gold in aqua regia and then boiling off HNO_3 in a single step. This study focused on

16.6 Summary and future perspectives

This chapter provided a brief discussion of how nanowastes can pollute various parts of the environment. Currently, nanowaste is becoming a serious area to be explored. Not all the sources of nanowaste generation have been identified. Nanowaste industrial treatment is still in its nascent stage. Most focus on the balancing of costs versus scientific and administrative benefits. However, increasing research and funding from government and private sources has greatly influenced the market valuation for waste recycling.

References

[1] W. Hannah, P.B. Thompson, Nanotechnology, risk and the environment: a review, J. Environ. Monit. 10 (2008) 291–300. Available from: https://doi.org/10.1039/B718127M.

[2] A. Boldrin, S.F. Hansen, A. Baun, N.I.B. Hartmann, T.F. Astrup, Environmental exposure assessment framework for nanoparticles in solid waste, J. Nanopart. Res. 16 (2014) 2394. Available from: https://doi.org/10.1007/s11051-014-2394-2.

[3] M. Naghdi, S. Metahni, Y. Ouarda, S.K. Brar, R.K. Das, M. Cledon, Instrumental approach toward understanding nano-pollutants, Nanotechnol. Environ. Eng. 2 (2017) 3. Available from: https://doi.org/10.1007/s41204-017-0015-x.

[4] I.S. Yunus, Harwin, A. Kurniawan, D. Adityawarman, A. Indarto, Nanotechnologies in water and air pollution treatment, Environ. Technol. Rev. 1 (1) (2012) 136–148. Available from: https://doi.org/10.1080/21622515.2012.733966.

[5] H. Pastrana, A. Avila, C.S.J. Tsai, Nanomaterials in cosmetic products: the challenges with regard to current legal frameworks and consumer exposure, Nanoethics 12 (2018) 123–137. Available from: https://doi.org/10.1007/s11569-018-0317-x.

[6] M.C. Coelho, G. Torrão, N. Emami, J. Gŕcio, Nanotechnology in automotive industry: research strategy and trends for the future-small objects, big impacts, J. Nanosci. Nanotechnol. 12 (8) (2012) 6621–6630. Available from: https://doi.org/10.1166/jnn.2012.4573.

[7] see: https://en.wikipedia.org/wiki/Pollution_from_nanomaterials (accessed 10.07.20).

[8] J. Saravanan, R. Karthickraja, J. Vignesh, Nanowaste, Int. J. Civil Eng. Technol. 8 (12) (2017) 483–491. Available from: http://iaeme.com/Home/issue/IJCIET?Volume = 8&Issue = 12.

[9] S. Dalton-Brown, Global ethics and nanotechnology: a comparison of the nanoethics environments of the EU and China, Nanoethics 6 (2012) 137–150. Available from: https://doi.org/10.1007/s11569-012-0146-2.

[10] J. Jeevanandam, A. Barhoum, Y.S. Chan, A. Dufresne, M.K. Danquah, Review on nanoparticles and nanostructured materials: history, sources, toxicity and regulations, Beilstein J. Nanotechnol. 9 (2018) 1050–1074. Available from: https://doi.org/10.3762/bjnano.9.98.

[11] R. Gupta, H. Xie, Nanoparticles in daily life: applications, toxicity and regulations, J. Environ. Pathol. Toxicol. Oncol. 37 (3) (2018) 209–230. Available from: https://doi.org/10.1615/JEnvironPatholToxicolOncol.2018026009.

[12] N.C. Mueller, B. Nowack, Exposure modeling of engineered nanoparticles in the environment, Environ. Sci. Technol. 42 (12) (2008) 4447–4453. Available from: https://doi.org/10.1021/es7029637.

[13] A. Behera, S.S. Mohapatra, D.K. Verma, Nanomaterials: fundamental principle and application, Nanotechnology and Nanomaterial Applications in Food, Health and Biomedical Science, Apple Academic Press, 2019, p. 32. Available from: https://doi.org/10.1201/9780429425660.

[14] P. Samaddar, Y. Sik Ok, K.-H. Kim, E.E. Kwon, D.C.W. Tsang, Synthesis of nanomaterials from various wastes and their new age applications, J. Clean. Prod. 197 (Part 1) (2018) 1190−1209. Available from: https://doi.org/10.1016/j.jclepro.2018.06.262.

[15] A.L. Holder, E.P. Vejerano, X. Zhou, L.C. Marr, Nanomaterial disposal by incineration, Environ. Sci. Process. Impacts 15 (2013) 1652−1664. Available from: https://doi.org/10.1039/C3EM00224A.

[16] M.-S. Choi, T. Park, W.-J. Kim, J. Hur, High-performance ultraviolet photodetector based on a zinc oxide nanoparticle@single-walled carbon nanotube heterojunction hybrid film, Nanomaterials 10 (2020) 395. Available from: https://doi.org/10.3390/nano10020395.

[17] B. Giese, F. Klaessig, B. Park, et al., Risks, release and concentrations of engineered nanomaterial in the environment, Sci. Rep. 8 (2018) 1565. Available from: https://doi.org/10.1038/s41598-018-19275-4.

[18] N. Ferronato, V. Torretta, Waste mismanagement in developing countries: a review of global issues, Int. J. Environ. Res. Public Health 16 (6) (2019) 1060. Available from: https://doi.org/10.3390/ijerph16061060.

[19] A. Caballero-Guzman, T. Sun, B. Nowack, Flows of engineered nanomaterials through the recycling process in Switzerland, Waste Manage. 36 (2015) 33−43. Available from: https://doi.org/10.1016/j.wasman.2014.11.006.

[20] see: < https://www.grandviewresearch.com/industry-analysis/nanotechnology-and-nanomaterials-market > (accessed 15.07.20).

[21] see: < https://www.alliedmarketresearch.com/nanotechnology-market > (accessed 15.07.20).

[22] see: < https://www.grandviewresearch.com/industry/nanoparticles > (accessed 15.07.20).

[23] see: < https://www.gminsights.com/sitemap > (accessed 15.07.20).

[24] D. Singh, G.A. Sotiriou, F. Zhang, J. Mead, D. Bello, W. Wohlleben, et al., End-of-life thermal decomposition of nano enabled polymers: effect of nanofiller loading and polymer matrix on by-products, Environ. Sci. Nano 3 (2016) 1293−1305. Available from: https://doi.org/10.1039/C6EN00252H.

[25] I. Khan, K. Saeed, I. Khan, Nanoparticles: properties, applications and toxicities, Arab. J. Chem. 12 (7) (2019) 908−931. Available from: https://doi.org/10.1016/j.arabjc.2017.05.011.

[26] O.V. Salata, Applications of nanoparticles in biology and medicine, J. Nanobiotechnol. 2 (2004) 3. Available from: https://doi.org/10.1186/1477-3155-2-3.

[27] A. Behera, S. Patel, M. Priyadarshini, Chapter 7 − Fiber-reinforced metal matrix nanocomposites, Available from: https://doi.org/10.1016/B978-0-12-819904-6.00007-4in: B. Han, S. Sharma, T.A. Nguyen, L. Longbiao, K. Subrahmanya Bhat (Eds.), Fiber-Reinforced Nanocomposites: Fundamentals and Applications, Elsevier, 2020, pp. 147−156.

[28] L. Sbardella, J. Comas, A. Fenu, I. Rodriguez-Roda, M. Weemaes, Advanced biological activated carbon filter for removing pharmaceutically active compounds from treated wastewater, Sci. Total Environ. 636 (2018) 519−529. Available from: https://doi.org/10.1016/j.scitotenv.2018.04.214.

[29] D.M. Mitrano, E. Rimmele, A. Wichser, R. Erni, M. Height, B. Nowack, Presence of nanoparticles in wash water from conventional silver and nano-silver textiles, ACS Nano 8 (7) (2014) 7208−7219. Available from: https://doi.org/10.1021/nn502228w.

[30] J.K. Patra, G. Das, L.F. Fraceto, E.V.R. Campos, M. del Pilar Rodriguez-Torres, L.S. Acosta-Torres, et al., Nano based drug delivery systems: recent developments and future prospects, J. Nanobiotechnol. 16 (2018) 71. Available from: https://doi.org/10.1186/s12951-018-0392-8.

[31] S.K. Kohli, S. Bali, R. Tejpal, V. Bhalla, V. Verma, R. Bhardwaj, et al., In-situ localization and biochemical analysis of bio-molecules reveals Pb-stress amelioration in Brassica juncea L. by co-application of 24-epibrassinolide and salicylic acid, Sci. Rep. 9 (2019) 3524. Available from: https://doi.org/10.1038/s41598-019-39712-2.

[32] L. Wei, S. Wang, Q. Zuo, S. Liang, S. Shen, C. Zhao, Nano-hydroxyapatite alleviates the detrimental effects of heavy metals on plant growth and soil microbes in e-waste-contaminated soil, Environ. Sci. Process. Impacts 18 (2016) 760−767. Available from: https://doi.org/10.1039/C6EM00121A.

[33] K. Drechsler, M. Heine, P. Mitschang, W. Baur, U. Gruber, L. Fischer, et al., Carbon fiber reinforced composites, Ullmann'S Encyclopedia of Industrial Chemistry, Wiley-VCH Verlag GmbH & Co, 2009. Available from: https://doi.org/10.1002/14356007.m05_m02.

[34] C.A. Lucier, B.J. Gareau, Electronic waste recycling and disposal: an overview, in: H. El-Din Saleh (Ed.), Assessment and Management of Radioactive and Electronic Wastes. IntechOpen. DOI: 10.5772/intechopen.85983. Available from: https://www.intechopen.com/books/assessment-and-management-of-radioactive-and-electronic-wastes/electronic-waste-recycling-and-disposal-an-overview.

[35] see: < https://www.greenbiz.com/article/electronic-waste-collection-conundrum > (accessed 15.07.20).

[36] K. Wang, J. Qian, L. Liu, Understanding environmental pollutions of informal e-waste clustering in global south via multi-scalar regulatory frameworks: a case study of Guiyu Town, China, Int. J. Environ. Res. Public Health 17 (2020) 2802. Available from: https://doi.org/10.3390/ijerph17082802.

[37] I.D. Ibrahim, E.R. Sadiku, T. Jamiru, Y. Hamam, Y. Alaylie, A.A. Eze, Prospects of nanostructured composite materials for energy harvesting and storage, J. King Saud Univ. Sci. 32 (1) (2020) 758—764. Available from: https://doi.org/10.1016/j.jksus.2019.01.006.

[38] R.J. Honda, V. Keene, L. Daniels, S.L. Walker, Removal of TiO_2 nanoparticles during primary water treatment: role of coagulant type, dose, and nanoparticle concentration, Environ. Eng. Sci. 31 (3) (2014) 127—134. Available from: https://doi.org/10.1089/ees.2013.0269.

[39] A.M. Said, M.S. Zeidan, M.T. Bassuoni, Y. Tian, Properties of concrete incorporating nano-silica, Constr. Build. Mater. 36 (2012) 838—844. Available from: https://doi.org/10.1016/j.conbuildmat.2012.06.044.

[40] N. Musee, Nanowastes, the environment: potential new waste management paradigm, Environ. Int. 37 (1) (2011) 112—128. Available from: https://doi.org/10.1016/j.envint.2010.08.005.

[41] see: < https://www.nanopartikel.info/en/nanoinfo/cross-cutting/2518-nanomaterials-in-waste > (accessed 18.07.20).

[42] Y. Yang, B. Chen, J. Hower, M. Schindler, C. Winkler, J. Brandt, et al., Discovery and ramifications of incidental Magnéli phase generation and release from industrial coal-burning, Nat. Commun. 8 (2017) 194. Available from: https://doi.org/10.1038/s41467-017-00276-2.

[43] T. Nguyen, E.J. Petersen, B. Pellegrin, J.M. Gorham, T. Lam, M. Zhao, et al., Impact of UV irradiation on multiwall carbon nanotubes in nanocomposites: formation of entangled surface layer and mechanisms of release resistance, Carbon N. Y. 116 (2017) 191—200. Available from: https://doi.org/10.1016/j.carbon.2017.01.097.

[44] R. Das, B.F. Leo, F. Murphy, The toxic truth about carbon nanotubes in water purification: a perspective view, Nanoscale Res. Lett. 13 (2018) 183. Available from: https://doi.org/10.1186/s11671-018-2589-z.

[45] S. Pimenta, S.T. Pinho, Recycling carbon fibre reinforced polymers for structural applications: technology review and market outlook, Waste Manage. 31 (2011) 378—392. Available from: https://doi.org/10.1016/j.wasman.2010.09.019.

[46] P.J.A. Borm, D. Robbins, S. Haubold, T. Kuhlbusch, H. Fissan, K. Donaldson, et al., The potential risks of nanomaterials: a review carried out for ECETOC, Part Fibre Toxicol. 3 (2006) 11. Available from: https://doi.org/10.1186/1743-8977-3-11.

[47] A. Behera, P. Mallick, Chapter 20 — Application of nanofibers in aerospace industryISBN: 978-0-12-819904-6in: B. Han, S. Sharma, T.A. Nguyen, L. Longbiao, K. Subrahmanya Bhat (Eds.), Fiber-Reinforced Nanocomposites: Fundamentals and Applications, Elsevier, 2020. Available from: https://doi.org/10.1016/B978-0-12-819904-6.00020-7.

[48] Z. Liu, M. Frasconi, J. Lei, Z.J. Brown, Z. Zhu, D. Cao, et al., Selective isolation of gold facilitated by second-sphere coordination with α-cyclodextrin, Nat. Commun. 14 (4) (2013) 1855. Available from: https://doi.org/10.1038/ncomms2891.

[49] P. Pati, S.P. McGinnis, P. Vikesland, Waste not want not: life cycle implications of gold recovery and recycling from nanowaste, Environ. Sci. Nano 3 (2016) 1133—1143. Available from: https://doi.org/10.1039/C6EN00181E.

[50] A. Behera, B. Swain, D.K. Sahoo, Chapter 16 — Fiber reinforced ceramic matrix nanocomposites. ISBN: 978-0-12-819904-6Available from: https://doi.org/10.1016/B978-0-12-819904-6.00016-5in: B. Han, S. Sharma, T.A. Nguyen, L. Longbiao, K. Subrahmanya Bhat (Eds.), Fiber-Reinforced Nanocomposites: Fundamentals and Applications, Elsevier, 2020, pp. 359—368.

[51] J. Zou, Y. Dai, K. Pan, B. Jiang, C. Tian, G. Tian, et al., Recovery of silicon from sewage sludge for production of high-purity nano-SiO_2, Chemosphere 90 (8) (2013) 2332−2339. Available from: https://doi.org/10.1016/j.chemosphere.2012.10.087.

[52] B.S. Kauldhar, S.K. Yadav, Turning waste to wealth: a direct process for recovery of nano-silica and lignin from paddy straw agro-waste, J. Clean. Prod. 194 (2018) 158−166. Available from: https://doi.org/10.1016/j.jclepro.2018.05.136.

[53] G. Sapra, V. Chaudhary, P. Kumar, P. Sharma, A. Saini, et al., Recovery of silica nanoparticles from waste PV modules, Materials Today: Proceedings 45 (4) (2021) 3863−3868. Available from: https://doi.org/10.1016/j.matpr.2020.06.093.

[54] A. Mohajerani, L. Burnett, J.V. Smith, H. Kurmus, J. Milas, A. Arulrajah, et al., Nanoparticles in construction materials and other applications, and implications of nanoparticle use, Materials (Basel) 12 (19) (2019) 3052. Available from: https://doi.org/10.3390/ma12193052.

[55] Z. Zhuang, X. Xu, Y. Wang, Y. Wang, F. Huang, Z. Lin, Treatment of nanowaste via fast crystal growth: with recycling of nano-SnO_2 from electroplating sludge as a study case, J. Hazard. Mater. 211−212 (2012) 414−419. Available from: https://doi.org/10.1016/j.jhazmat.2011.09.036.

[56] P. Dwivedi, P.K. Mishra, M.K. Mondal, N. Srivastav, Non-biodegradable polymeric waste pyrolysis for energy recovery, Hiliyon 5 (8) (2019) e02198. Available from: https://doi.org/10.1016/j.heliyon.2019.e02198.

[57] C. Zhuo, Y.A. Levendis, Upcycling waste plastics into carbon nanomaterials: a review, J. Appl. Polym. Sci. 131 (4) (2014). Available from: https://doi.org/10.1002/app.39931.

[58] L. Muchova, E. Bakker, P. Rem, Precious metals in municipal solid waste incineration bottom ash, Water Air Soil Pollut: Focus 9 (2009) 107−116. Available from: https://doi.org/10.1007/s11267-008-9191-9.

[59] Y. Wu, Q. Fang, X. Yi, G. Liu, R.-W. Li, Recovery of gold from hydrometallurgical leaching solution of electronic waste via spontaneous reduction by polyaniline, Prog. Nat. Sci. Mater. Int. 27 (4) (2017) 514−519. Available from: https://doi.org/10.1016/j.pnsc.2017.06.009.

[60] D.D.L. Chung, 7 – Carbon-matrix composites, in: D.D.L. Chung (Ed.), Carbon Composites, second ed., Butterworth-Heinemann, 2017, pp. 387−466. Available from: https://doi.org/10.1016/B978-0-12-804459-9.00007-5.

[61] I. Mansuri, R. Khanna, V. Sahajwalla, Recycling carbonaceous industrial/commercial waste as a carbon resource in iron and steelmaking, Steel Res. Int. 88 (6) (2017) 1600333. Available from: https://doi.org/10.1002/srin.201600333.

[62] R. Nagarajan, S. Thirumalaisamy, E. Lakshumanan, Impact of leachate on groundwater pollution due to non-engineered municipal solid waste landfill sites of erode city, Tamil Nadu, India, Iran. J. Environ. Health Sci. Eng. 9 (1) (2012) 35. Available from: https://doi.org/10.1186/1735-2746-9-35.

[63] M.J. Hollamby, J. Eastoe, A. Chemelli, O. Glatter, S. Rogers, R.K. Heenan, et al., Separation and purification of nanoparticles in a single step, Langmuir 26 (10) (2010) 6989−6994. Available from: https://doi.org/10.1021/la904225k.

[64] C.O. Dimkpa, P.S. Bindraban, Fortification of micronutrients for efficient agronomic production: a review, Agron. Sustain. Dev. 36 (2016) 7. Available from: https://doi.org/10.1007/s13593-015-0346-6.

[65] S. Bolisetty, M. Peydayesh, R. Mezzenga, Sustainable technologies for water purification from heavy metals: review and analysis, Chem. Soc. Rev. 48 (2019) 463−487. Available from: https://doi.org/10.1039/C8CS00493E.

[66] J.F. Liu, J. Sun, G. Jiang, Use of cloud point extraction for removal of nanosized copper oxide from wastewater, Chin. Sci. Bull. 55 (2010) 346−349. Available from: https://doi.org/10.1007/s11434-009-0695-0.

[67] E. Lahtinen, L. Kivijärvi, R. Tatikonda, A. Väisänen, K. Rissanen, M. Haukka, Selective recovery of gold from electronic waste using 3D-printed scavenger, ACS Omega 2 (10) (2017) 7299−7304. Available from: https://doi.org/10.1021/acsomega.7b01215.

[68] I. Ojea-Jiménez, F.M. Romero, N.G. Bastús, V. Puntes, Small gold nanoparticles synthesized with sodium citrate and heavy water: insights into the reaction mechanism, J. Phys. Chem. C 114 (4) (2010) 1800−1804. Available from: https://doi.org/10.1021/jp9091305.

CHAPTER 17

Recycled nanomaterials for construction and building materials

P.O. Awoyera, C.O. Nwankwo and D.P. Babagbale
Department of Civil Engineering, Covenant University, Ota, Nigeria

17.1 Introduction

Sustainability in construction calls for reduction in the use of virgin materials and repurposing of waste materials in concrete production. This leads to the repurposing of micro-sized materials such as fly ash (FA), plastics, glass, agricultural wastes, silica fume, granulated ground blast furnace slag, and so on as cement substitutes in concrete production. Pozzolanic reaction is the dominant mechanism in play in dealing with supplementary cementitious materials (SCMs). One of the precursors to this pozzolanic reaction is a product of the cement hydration reaction; therefore the pozzolanic reaction is a much slower reaction that largely contributes to concrete strength at later stages. One of the most effective ways of improving the early strengths of pozzolanic cements is by the introduction of nanomaterials, typically in the range of 10^{-9} m.

17.2 Application of nanomaterials in construction

Today, the production of construction materials such as concrete and asphalt is centered on the crucial factor of durability so as to increase the useful life of construction projects and decrease economic costs. As a result, commercially available nanomaterials such as nanosilica (SiO_2), nanoiron (Fe_2O_3), nanotitanium (TiO_2), nanoalumina (Al_2O_3), nano-calcium carbonate ($CaCO_3$), nanolime, and many more have been used to enhance the properties of concrete [1]. These nanomaterials can be additional sources of silica and alumina, act as nucleation seeds for calcium-silicate-hydrate (C-S-H), or act as inert fillers within the cement matrix [2,3]. Like any other emerging technology, the use of nanomaterials is limited by inadequate knowledge and standardization of these materials [4]. These nanomaterials are typically more effective when used in smaller quantities; in larger quantities they tend to agglomerate because of their high Van der Waal forces [5]. Though nanoparticles are used in very low percentages (usually on the order of 1%–3%), commercially available nanomaterials are relatively expensive. Martins et al. [6] reported that nano-TiO_2 accounted for 96% of the material cost in a concrete mix containing FA without accounting for the

---

Nanomaterials Recycling
DOI: https://doi.org/10.1016/B978-0-323-90982-2.00016-0

© 2022 Elsevier Inc.
All rights reserved.

extra safety measures needed in handing the nanomaterial. The high cost of nanomaterials has moved researchers to consider other materials with similar physical and chemical properties but from different sources, such as specific waste materials [7].

The typical pozzolans used in concrete have shown better performance when smaller particle sizes are used [8]. Kim et al. [9] investigated how the size of FA can affect its performance in cement mortars. It was found that reducing the size of normal FA particles by one-twentieth gave a 28-compressive strength percentage increase of 24%. It was concluded that the smaller the particle size, the better the activation of the pozzolanic reaction, owing to the increased surface area of the particles. The high surface areas of these particles accelerate the pozzolanic reaction, and the particles fill voids within the cement matrix just like commercial nanomaterials [10].

There has been a rise in construction and demolition (C&D) wastes ending up in landfills, easily attributable to the high rate of urban development and the growing demand for housing and transportation. In China these wastes accounted for 30%—40% of total solid wastes generated in 2012 [11]; in Europe they made up about 25%—30% of the total solid waste stream in 2019 [12]; in Singapore they accounted for 59% in 2011 [13]; and the total was as high as 80% in the United Arab Emirates in 2010 [14].

These wastes are also sources of recycled nanomaterials that can be used in construction. These wastes are typically repurposed as backfill material or further processed to get recycled. During this processing, large quantities of reactive particles smaller than 200 nm are produced [15].

Even asphaltic concrete can benefit from nanomaterials. Modification of asphalt is done to improve its mechanical properties, such as aging resistance, adhesion, friction properties, durability, adhesion, and oxidation resistance. The microstructure is the sole determinant of a material's macro properties; thus nano-modified asphalt offers significant improvement over the fundamental material properties and is superior to other asphalt modification methods [16].

17.3 Nanomaterials in concrete design and development

Although there is limited research on the study of nanomaterials and their inclusion in construction materials, their effects and future prospects in concrete design and development have gradually come to light. The current studies on the workability, microstructure, hydration, and performance of concrete, asphalt, and other building applications developed with nanomaterials are summarized and presented here.

17.3.1 Concrete

Several nanomaterials have been used by researchers in a bid to enhance concrete's fresh or hardened properties or for other environmental reasons, such as a management option for pollution.

For one, Luo et al. [11] researched the use of demolition waste and observed that using demolition waste nanomaterials resulted in varied material performance, owing to the different sources of the demolition waste. These nanomaterials are much smaller with larger surface areas; hence they can fill smaller pores, have greater packing densities, and encourage more pozzolanic reactions compared to the typical micromineral admixtures. They also use lesser doses. The typical nanoparticles that were reviewed were nanosilica (NS), nano-$CaCO_3$, and nano-TiO_2, which also decompose organic particles in the mix.

Zhang et al. [15] studied the effect in concrete of recycled powder (RP), which is obtained from the processing of C&D wastes and observed that RP is a rich source of C-S-H and so could serve as C-S-H seeds, ultimately enhancing cementitious materials' mechanical properties. C-S-H seeds enhance the cement hydration process and increase early strength development [17,18], but the temperature, pH, and moisture content affect their efficiency. These factors along with their high cost have limited their use in the construction industry. Similar to C-S-H seeds, RP could be used as this gel; this thought comes with the challenge of preparing the nanoparticles from RP because the finer the particle, the stronger is the nucleation effect. The typical dry grinding of the C&D materials typically sets the upper limit of particle size to 2 μm, but this has been brought down to 0.1–0.6 μm when wet grinding is used [19–22]. With RP comes the problem of impurities too. The authors got RP from hardened cement paste and used wet grinding, producing wet ground recycled powder (WGRP). It was noted that via the wet grinding process, most of components in the RP, including the impurities, could be crushed as ultrafine particles, thus allowing the WGRP to exhibit the nucleation effect and filling effect efficiently. This resultant nano-RP with a median particle size (D_{50}) of 249 nm helped to increase the early strength and overall strength of the mortar, and it reduced the chloride ion permeability and porosity, enhancing the durability of the mix. It was suggested that this nano-RP be used as an accelerator in precast concrete.

Jones et al. [23] investigated the use of ultrafine fly ash (UFFA). Postprocessing of FA was done to refine the pore size with the added advantage of removing carbonaceous and clay residues. The UFFA was separated from the parent FA material by using a particle air-cyclonic separation technique and was defined by the author as particles less than 10 μm in diameter. The ultrafine FA had a 23% increase in Al_2O_3 content and a 44% increase in SiO_2; CaO, and Fe_2O_3 were reduced by 4.5%–9.9%, respectively. There was also a 5% reduction in the loss on ignition, resulting not from unburned carbon but from the presence of other components. The study revealed how the size change of FA can affect the strength and workability of concrete. Mortar flow speed significantly increased with the addition of UFFA compared to the parent FA because of the cenospherical nature of the UFFA particles. Though agglomerations were noticed in the ultrafine materials due to electrostatic attraction, they were easily dispersed in water.

In an attempt to curtail the harmful effects of the cement industry, much higher than usual proportions of cement were replaced with FA in what is referred to as high-volume fly ash (HVFA). Shaikh and Supit [5] conducted research on the mechanical properties of HVFA concrete containing $CaCO_3$ nanoparticles. SCMs such as FA are usually limited to 15%—20% replacement of ordinary Portland cement (OPC) in practice. Beyond this percentage there is typically a drop in the strength of concrete. Nanomaterials have slowly been gaining popularity as means of improving the strength of concrete. The results of the study showed that the HVFA concretes containing 1% $CaCO_3$ nanoparticles had reasonably higher compressive strength, lower volume of permeable voids, lower porosity, higher resistance to water sorptivity, chloride permeability, and chloride ion diffusivity than the counterpart HVFA concretes and even OPC concrete. It has also been found that the addition of 1% $CaCO_3$ nanoparticles improves the microstructure by forming additional calcium silicate hydrate (CSH) gels and decreases the calcium hydroxide and calcium silicates of HVFA concretes. The addition of $CaCO_3$ nanoparticles not only led to much denser microstructure in the HVFA matrix but also changed the formation of the hydration products, hence contributed to improvement of the early-age compressive strength and durability properties of HVFA concretes.

Supit et al. [24] also conducted a study on UFFA, which explored both cement mortars and HVFA mortars containing varying proportions of FA that were prepared and used as control mortars. UFFA was used to replace cement at percentages of up to 15%. The results revealed that the cement mortars with 8% UFFA of cement replacement exhibited higher compressive strength at 7—28 days than control mortars. There was also great improvement in compressive strength of HVFA mortars, particularly at early age. In fact, chloride-induced corrosion, water sorptivity, volume of permeable voids, chloride ion penetration, chloride diffusivity, and porosity of HVFA concretes were all improved. and in most cases the HVFA concrete containing 32% FA and 8% UFFA exhibited superior durability properties in comparison to ordinary concrete containing 100% cement [25].

Choi et al. [8] explored how FA of different sizes affected the strength of mortars. It was found that the smaller the particle size, the stronger the mortar.

Kim et al. [9] corroborated this in their paper, which examined FA of different sizes and their effects on the strength development of mortars. Reducing the size of normal FA particles by one-twentieth gave a 28-compressive strength percentage increase of about 24%. The smaller particle sizes hastened the rate of the pozzolanic reaction because of the increased surface area. Samples with nano–fly ash (NFA) with an average particle size of 655 nm had enhanced durability performance. He concluded that the smaller the binder size, the better activation of the pozzolanic reaction at early stages, owing to the densifying of the microstructure.

Alrulraj and Carmichael [26] observed as high as a 46% compressive strength increase with a 10% replacement of coarse aggregate with NFA when compared to the reference concrete. More so, workability of the concrete with NFA was found to be up to 90% higher than that of conventional concrete at the same replacement percentage. The results indicated that workability and strength increased proportionally with increasing replacement percentages of NFA; a 30% replacement yielded as much as a 60% strength increase and a whopping 140% increase in workability when compared to the control mortar.

Martins et al. [6] conducted an experimental investigation of nano-TiO_2 and FA—based high-performing concrete and found that the nano-TiO_2 particles also minimized the strength loss associated with the use of FA but only at 28 days.

Arshad et al. [27] looked into the effect of different sizes of nano-rice husk ash (nRHA) on the compressive strength of concrete. In the study, a 10% cement replacement with the nRHA, which is about 91% SiO_2, negatively affected the compressive strength. However, it was observed that the smaller the particle size, the better the mortar properties up to a certain level. The grinding time was as long as 63 hours to achieve a 65-nm particle size.

Ramezanianpour et al. [28] researched and discovered that RHA shows the highest pozzolanic activity among other plant residue. Using nRHA in concrete can reduce the porosity of the matrix, modify (densify) the pore structure and reduce permeability thus reducing the ingress of aggressive chloride agents. nRHA has high reactivity, owing to its noncrystalline silica [6—8]. The nanoparticles accelerated the hydration and pozzolanic reactions by acting as cement hydrate crystallization centers. The nRHA that was used had an SiO_2 content of 83.74% with a mean particle size of 35.86 nm after being ground for about 13 hours. The author used an alkali to ensure the nanoparticles were effectively dispersed in the matrix. Increasing the percentage of nRHA from 2.5% to 5.0% and then to 7.5% showed increasing 28-day and 90-day compressive strengths. Early strength for 5% nRHA and 7.5% nRHA was reduced, but the 2.5% nRHA maintained a comparable strength value to the control concrete. Electrical resistivity improved with increasing nRHA content both at early and later ages, and the nRHA particles effectively filled the microstructure voids, densifying the matrix and resulting in reduced capillary absorption and increased chloride resistivity.

Pathak and Tiwari [29] studied the effects of nano-zinc oxide on concrete and noted that there was an increase in the concrete's compressive strength and durability but observed a reduction in its properties above 1% replacement.

Seifan et al. [30] did a comprehensive review of the use of NS and microsilica (MS) materials in mortar. They found that the high surface area of the nanoparticles enhances their reactivity and promoted cement hydration, making it much more beneficial than the MS. The results showed that using NS with average particle size 20 nm and MS with an average particle size of 2—10 μm increased the packing densities of mortars by filling their pores, resulting in a stronger material and one with reduced permeability. However, the author noted that the addition of the nanoparticles adversely affected the flowability of the mortar. This is because the addition of the nanoparticles to the cement created a demand for increased water content to ensure that the workability of the fresh mixture is retained [31]. On the other hand, MS did not affect the workability compared to NS. The author concluded that 5% cement replacement with NS is the optimum concentration to meet higher mechanical properties and performance.

Garg et al. [10] worked with MS of 0.2 μm and NS of 40 nm and studied their effects on mortar when they were used to replace cement at different percentages. He expressed that the nanoparticles typically had a twofold purpose: enhancing the pozzolanic reaction and filling pores within the mix. The setting time was reduced for the mixes containing varying proportions of NS and for the mixes containing a blend of NS and MS, owing to the enhanced hydration that led to early hardening. The mixes with just MS were found to have longer setting times. An increase in compressive and split tensile strength was obtained by increasing the content of MS and NS from 5.0% to 15% and from 0.5% to 1.0%, respectively, along with a slight reduction in strength afterwards. This decrease in strength was because of the reduction in homogeneity of the cement matrix at the higher content of MS, while the decrease in strength at the higher content of NS was attributed to the agglomeration of nanoparticles at higher content in the cement matrix. SEM-EDX analysis confirmed the loss of pozzolanic activity of MS and NS at higher content due to agglomeration and friction among particles.

Wang et al. [31] carried out investigation on the synergistic effects of NS and FA on properties of cement-based composites. The results revealed that the strength growth rate of mortar with FA replacement of cement had an increase as high as 30% when infused with NS. This shows that NS accelerates not only the hydration of cement particles, but also hydration of FA at 7 days. Thus more gels are generated, which increases the compressive strength. The influence of FA fineness on mortar strength was also investigated. It was observed that the compressive strengths of FA mortars were significantly improved at all ages by mixing NS of moderate fineness (540 m^2/kg).

Song et al. [32] found that the addition of NS and carbon nanotubes (CNTs) generally improved the interfacial tensile and shear strengths of recycled aggregate concrete for dosage not exceeding 2.0−0.5 wt.%, respectively. The maximum tensile and shear strength improvements were 51%−53%, respectively, both obtained from use of CNTs.

Carmichael and Arulraj [4] determined the impact strength of concrete in which cement is replaced with different nanomaterials [nanocement (NC), NFA, NS, and NS fume (NSF)] at replacement percentages of up to 50%. Their experiment revealed that for each of the nanomaterials, the concrete impact strength increased by as much as 50% with cement replacement of up to 50%. It was also found that replacement of cement with the nanomaterials increased the ductility index at a replacement level of not more than 40%.

17.3.2 Asphalt concrete

Modification of asphalt is done to improve its mechanical properties, such as aging resistance, adhesion, friction properties, durability, adhesion, and oxidation resistance. The evolution of asphalt material science has led to improving these properties on a nano-scale, since the microstructure of materials is the determinant of their macro properties, thus bringing about the emergence of the field of asphalt nanomaterials science in 2006 [16]. Though research papers have used various nanoclays, NS, zycosoil, and nanoirons [33–35] to modify asphalt, this is an emerging field with limited studies done [35]. There are even fewer studies of the use of recycled nanomaterials.

Saha et al. [34] attempted to use zycosoil to treat stone aggregates and bitumen used for a bituminous mix. Zycosoil has an estimated molecule size of 3–6 nm, and these nanomolecules contain alkoxy groups of silane. They chemically react with hydroxyl (OH) groups present in pavement materials and forms siloxane bonds (Si—O—Si), which are known to be one of the strongest natural bonds. This chemical reaction makes pavement water-repellent. The Marshall quotient, which is used as a measure of the permanent deformation, shear stress, and rutting of the mix, was found to be 1.25 times higher than the quotient of the bituminous mix prepared by using plane stone when 0.5% of zycosoil was added with bitumen during preparation. Similarly, the tensile strength ratio values obtained from the zycosoil-infused asphalt mixture were significantly greater than those of the normal asphalt mix, owing to the formation of hydrophobic strong siloxane bonds. Therefore the use of zycosoil with both aggregate and binder can improve the stability of mix by 30% as well as remarkably reducing the construction cost of the bituminous macadam by 3%.

Sarsam and Al-Shujairy [36] carried out experiments to assess the fatigue life of reclaimed asphalt concrete recycled with nanomaterial additives. The nanomaterials used, silica fumes and FA, drastically increased the fatigue life of the unconditioned mix and conditioned mix at strain levels of 250 and 750 $\mu\varepsilon$ with increases as high as 600% when the recycled asphalts were treated with soft asphalt and silica fumes at the unconditioned 250 $\mu\varepsilon$ strain level.

The results of the research by Crucho et al. [35] showed that the modifications of asphalt concrete with NS and nanoclay improved the aging resistance of the mixture, indicating enhanced durability.

Yao and Zou [33] evaluated the mechanical properties of asphalt modified with micro- and nanomaterials. The nanomaterials used (nonmodified nanoclay, polymodified nanoclay, nanomer, carbon microfiber, and NS materials) enhanced the rut factor and dynamic modulus of the modified asphalt mixtures.

Fang et al. [16] presented a review discussing various nanomaterials (e.g., nanobentonite, nano-layered silicate [37–40], nano-SiO_2, nano-TiO_2, nano-$CaCO_3$, montmorillonite, nanoclays [41,42], Fe_3O_4 [43]) and their effects on the modification of asphalt.

The best method for asphalt modification was found to be mother liquor melting, in which the nanomaterials are typically introduced into the asphalt through a solvent that is compatible with both the base asphalt and the nanomaterial. This ensures the uniform distribution of the nanomaterial and the subsequent stable storage of the modified asphalt. However, this dispersion technique is quite complicated. From all the literature reviewed, it is apparent that nanomaterials play a positive role in enhancing the properties of asphaltic concrete, such as resistance to cracking, elasticity, and aging resistance.

17.3.3 Other building applications

Zoriyeh et al. [44] studied the effects of nanomaterials (in this case, nanoclay) on the engineering properties of a high-plasticity building soil. The results of the experiments showed that an increase in the percentage of nanoclay increased the liquid limit and plastic limit and decreased the plasticity index, owing to the high water absorptive property of the nanoclay. As the ratio of nanomaterials increases, the optimum water content and dry unit weight decrease. The nanoclay reduced the pores by filling the gap between the particles and caused the particles to bond to each other, thus achieving a tighter soil structure. That caused the soils with nanoclay to be harder when compared to original soils. An increase in the percentage of nanoclay caused an increase in unconfined compressive strength test results relative to the original soil, as much as 110% at the end of 7 days. Nanoclay has a great effect on the engineering properties of the soil, owing to its high specific surface area, and even a small percentage of this material is sufficient to achieve better results.

Kulanthaivel et al. [45] presented a study on the stabilizing effects of NS and white cement at varying proportions on clay soil. The study found that the nanomaterials improved the performance of the clay soil with the optimum dosage of NS alone and white cement alone added to soil being 7% by weight of soil. The combination of NS with the white cement optimum dosage was established as 2% NS and 3% white cement by weight of soil. These dosages increased the unconfined compressive strength of the soils and their California Bearing Ratio, thus ensuring enhanced durability.

In addition, the permeability was reduced by 45% when the soil was treated with 2% NS and 3% white cement.

17.3.4 Health implications

The use of nanomaterials is accompanied by risks to workers during dosing. Nanomaterials are also expensive, and this limits their use. The handling of nanoparticles has to be done safely and responsibly because mishandling can endanger humans, animals, and the environment. Nanoparticles are much smaller than average admixtures, which makes them more prone to inhalation, and inhaling any of these nanoparticles, such as NS, nanoalumina, nanotitania, nano-iron oxide, and nanoclay particles, can lead to health complications, such as silicosis, lipid peroxidation, and even lung cancer [46].

Nanoparticles volatilize easily. NS in particular had a volatilization greater than 42 wt.%, one of the highest of concrete materials [46]. In open dosage this causes uncertainty in achieving accurate doses and, even more seriously, engenders environmental contamination due to the copious amounts of breathable nanomaterials that are released into the atmosphere and can be detrimental to human life.

Thus for accuracy and safety, closed dosing systems (which are usually used for materials with high fluidity) rather than open dosage should be used. Alternatively, the nanoparticles could be incorporated in a liquid vector, usually an aqueous phase, although this would translate to higher cost implications in using nanoparticles.

Nanoparticles can be easily breathed in or absorbed through the skin [47,48] and can freely circulate through the bloodstream and penetrate organs and tissues. There is still little knowledge about the toxicology of nanoparticles, how they affect life, and the most effective health surveillance systems. Necessary precautions have to be taken, such as the wearing of masks and gloves [6] and other forms of respiratory protection when handling nanomaterials.

17.4 Conclusion

Using more durable concrete can reduce maintenance and durability costs and greatly extend the service life of buildings. To reduce the quantity of cement in construction using nanoparticles alone is not a sustainable solution, since they can be used only in small percentages. However, the advantage of nanoparticles can be seen when used in conjunction with other SCMs. The durability of concrete can be enhanced by using SCMs such as FA, and the nanomaterials can be used to counter the slow strength development associated with the use of these SCMs. As such, the negative environmental effects of cement can be curtailed to an extent, owing to the facilitation of higher amounts of SCMs to be used without the usual accompanying strength loss.

17.5 Recommendations

Even dispersion of the nanoparticles within the mix is a big issue, as high nanoparticle content causes agglomeration, so the authors recommend that the particles be introduced in a solution of some kind. Without even dispersion of the nanoparticles, pockets could be formed that would impede the formation of C-S-H, resulting in a more porous matrix. Rather than using NS in dry powder form, colloidal NS is recommended, which is NS in aqueous suspension form and consists of amorphous hydroxylated silica nanoparticles with particle size in the range of $1-500$ nm. The particles of colloidal NS show less segregation and better dispersion in cement mortar and are considered to facilitate the production of CSH gel with high stiffness [10,49].

Along the lines of health and safety, because this is a developing technology, the long-term effects of nanomaterials in cementitious materials need to be more closely examined. Also, in terms of the handling of nanomaterials, necessary caution must be maintained during their production, dosage, and disposal. Furthermore, exposure prevention measures should be established and systematized to avoid the risk of developing ailments linked to nanomaterials exposure.

Production difficulties, inadequate knowledge, and lack of standardization hinder the use of nanomaterials in construction. Therefore the next step in nanoengineering would be to institute and regulate nanomaterials production, dosage, supply, use, disposal, and research and development to further close the many gaps that are seen in this emerging field.

References

[1] N. Lovecchio, F. Shaikh, M. Rosano, R. Ceravolo, W. Biswas, Environmental assessment of supplementary cementitious materials and engineered nanomaterials concrete, AIMS Environ. Sci. 7 (1) (2020) 13−30. Available from: https://doi.org/10.3934/environsci.2020002.

[2] K. Sobolev, I. Flores, L.M. Torres-Martinez, P.L. Valdez, E. Zarazua, E.L. Cuellar, Engineering of SiO_2 nanoparticles for optimal performance in nano cement-based materials, Nanotechnol. Constr. 3 (2009) 139−148.

[3] J. Sun, Z. Xu, W. Li, X. Shen, Effect of nano-SiO_2 on the early hydration of alite-sulphoaluminate cement, Nanomaterials 7 (5) (2017) 102.

[4] M. Jemimah Carmichael, G.P. Arulraj, Impact resistance of concrete with nano materials, Mater. Today Proc. (2020). Available from: https://doi.org/10.1016/j.matpr.2020.05.635.

[5] F.U.A. Shaikh, S.W.M. Supit, Mechanical and durability properties of high volume fly ash (HVFA) concrete containing calcium carbonate ($CaCO_3$) nanoparticles, Constr. Build. Mater. 70 (2014) 309−321. Available from: https://doi.org/10.1016/j.conbuildmat.2014.07.099.

[6] T. Martins, F.P. Torgal, S. Miraldo, J.B. Aguiar, J. Carlos, An experimental investigation on nano-TiO_2 and fly ash based high performing concrete, Indian Concr. J. 90 (2016) 23−31.

[7] H.M. Hamada, G.A. Jokhio, F.M. Yahaya, A.M. Humada, Applications of nano palm oil fuel ash and nano fly ash in concrete, IOP Conf. Ser. Mater. Sci. Eng. 342 (2018) 012068.

[8] S.J. Choi, S.S. Lee, P.J.M. Monteiro, Effect of fly ash fineness on temperature rise, setting, and strength development of mortar, J. Mater. Civ. Eng. 24 (5) (2012) 499−505.

[9] S.S. Kim, R. Doug Hooton, T.-J. Cho, J.-B. Lee, Comparison of innovative nano fly ash with conventional fly ash and nano silica, Can. J. Civ. Eng. 41 (5) (2014) 396−402. Available from: https://doi.org/10.1139/cjce-2012-0419.

[10] R. Garg, R. Garg, M. Bansal, Y. Aggarwal, Experimental study on strength and microstructure of mortar in presence of micro and nano-silica, Mater. Today Proc. 43 (Part 2) (2021) 769−777. Available from: https://doi.org/10.1016/j.matpr.2020.06.167.

[11] Z. Luo, et al., Current progress on nanotechnology application in recycled aggregate concrete, J. Sustain. Cem. Mater. 8 (2) (2019) 79−96. Available from: https://doi.org/10.1080/21650373.2018.1519644.

[12] European Commission. Waste and recycling. Available from: <http://ec.europa.eu/environment/waste/construction_demolition.htm> (accessed 06.01.21).

[13] A. Giannis, M.J. Chen, K. Yin, H.H. Tong, A. Veksha, Application of system dynamics modelling for evaluation of different recycling scenarios in Singapore, J. Mater. Cycles Waste Manag 19 (2017) 1177–1185. Available from: https://doi.org/10.1007/s10163-016-0503-2.

[14] S. Rogers, Battling construction waste and winning: lessons from UAE, Proc. Inst. Civ. Eng. Eng. 164 (2011) 41–48.

[15] J. Zhang, H. Tan, X. He, R. Zhao, J. Yang, Y. Su, Nano particles prepared from hardened cement paste by wet grinding and its utilization as an accelerator in Portland cement, J. Clean. Prod. 283 (2020) 124632. Available from: https://doi.org/10.1016/j.jclepro.2020.124632.

[16] C. Fang, R. Yu, S. Liu, Y. Li, Nanomaterials applied in asphalt modification: a review, J. Mater. Sci. Technol. 29 (7) (2013) 589–594. Available from: https://doi.org/10.1016/j.jmst.2013.04.008.

[17] C. Xu, H. Li, X. Yang, Effect and characterization of the nucleation C-S-H seed on the reactivity of granulated blast furnace slag powder, Constr. Build. Mater. 238 (2020) 117726. Available from: https://doi.org/10.1016/j.conbuildmat.2019.117726.

[18] P. Zhang, F.H. Wittmann, M. Vogel, H.S. Müller, T. Zhao, Influence of freeze-thaw cycles on capillary absorption and chloride penetration into concrete, Cem. Concr. Res. 100 (2017) 60–67. Available from: https://doi.org/10.1016/j.cemconres.2017.05.018.

[19] N. Kotake, M. Kuboki, S. Kiya, Y. Kanda, Influence of dry and wet grinding conditions on fineness and shape of particle size distribution of product in a ball mill, Adv. Powder Technol. 22 (1) (2011) 86–92. Available from: https://doi.org/10.1016/j.apt.2010.03.015.

[20] Y. Liu, Z. Qin, B. Chen, Experimental research on magnesium phosphate cements modified by red mud, Constr. Build. Mater. 231 (2020) 117131. Available from: https://doi.org/10.1016/j.conbuildmat.2019.117131.

[21] H. Tan, M. Li, X. He, Y. Su, J. Zhang, H. Pan, et al., Preparation for micro-lithium slag via wet grinding and its application as accelerator in Portland cement, J. Clean. Prod. 250 (2019) 119528. Available from: https://doi.org/10.1016/j.jclepro.2019.119528.

[22] D. Wang, Q. Wang, J. Xue, Reuse of hazardous electrolytic manganese residue: detailed leaching characterization and novel application as a cementitious material, Resour. Conserv. Recycl. 154 (2020) 104645. Available from: https://doi.org/10.1016/j.resconrec.2019.104645.

[23] M.R. Jones, A. McCarthy, A.P.P.G. Booth, Characteristics of the ultrafine component of fly ash, Fuel 85 (16) (2006) 2250–2259. Available from: https://doi.org/10.1016/j.fuel.2006.01.028.

[24] S.W.M. Supit, F.U.A. Shaikh, P.K. Sarker, Effect of ultrafine fly ash on mechanical properties of high volume fly ash mortar, Constr. Build. Mater. 51 (2014) 278–286. Available from: https://doi.org/10.1016/j.conbuildmat.2013.11.002.

[25] F.U.A. Shaikh, S.W.M. Supit, Compressive strength and durability properties of high volume fly ash (HVFA) concretes containing ultrafine fly ash (UFFA), Constr. Build. Mater. 82 (2015) 192–205. Available from: https://doi.org/10.1016/j.conbuildmat.2015.02.068.

[26] P.G. Alruraj, M. Jemimah Carmichael, Effect of nano-flyash on strength of concrete, Int. J. Comput. Civ. Struct. Eng. 2 (2) (2011) 475–482.

[27] M.F. Arshad, et al., Effect of nano black rice husk ash on the chemical and physical properties of porous concrete pavement, J. Southwest Jiaotong Univ. 53 (5) (2018). Available from: https://doi.org/10.3969/j.issn.0258-2724.2018.015.

[28] A.A. Ramezanianpour, M. Balapour, E. Hajibadeh, Effect of nano rice husk ash against penetration of chloride ions in mortars, AUT J. Civ. Eng. 2 (1) (2018) 92–107. Available from: https://doi.org/10.22060/ajce.2017.12387.5203.

[29] A. Pathak, P.A. Tiwari, Effect of zinc oxide nanoparticle on compressive, Int. J. Res. Appl. Sci. Eng. Technol. 5 (8) (2017) 683–687.

[30] M. Seifan, S. Mendoza, A. Berenjian, Mechanical properties and durability performance of fly ash based mortar containing nano- and micro-silica additives, Constr. Build. Mater. 252 (2020) 119121. Available from: https://doi.org/10.1016/j.conbuildmat.2020.119121.

[31] J. Wang, M. Liu, Y. Wang, Z. Zhou, D. Xu, P. Du, et al., Synergistic effects of nano-silica and fly ash on properties of cement-based composites, Constr. Build. Mater. 262 (2020) 120737. Available from: https://doi.org/10.1016/j.conbuildmat.2020.120737.

[32] X. Song, C. Li, D. Chen, X. Gu, Interfacial mechanical properties of recycled aggregate concrete reinforced by nano-materials, Constr. Build. Mater. 270 (7) (2020) 121446. Available from: https://doi.org/10.1016/j.conbuildmat.2020.121446.

[33] H. Yao, Z. You, Effectiveness of micro- and nanomaterials in asphalt mixtures through dynamic modulus and rutting tests, J. Nanomater. 2016 (2016) 2645250. Available from: https://doi.org/10.1155/2016/2645250.

[34] A. Saha, B. Singh, S. Biswas, Effect of nano-materials on asphalt concrete mixes, a case study, Eur. Transp. (2017). no. 65.

[35] J.M.L. Crucho, J.M. Coelho das Neves, S.D. Capitão, L.G. de Picado-Santosa, Evaluation of the durability of asphalt concrete modified with nanomaterials using the TEAGE aging method, Constr. Build. Mater. 214 (2019) 178−186.

[36] S.I. Sarsam, A.M. Al-Shujairy, Assessing fatigue life of reclaimed asphalt concrete recycled with nanomaterial additives, Int. J. Adv. Mater. Res. 1 (1) (2015) 1−7.

[37] L.F. Ran, W.R. Huang, B.H. Zhu, Mechanism study on nanometer bentonite modified asphalt, J. Chongqing Jiaotong Univ. (Nat. Sci.) (1)(2008) 73−76 (in Chinese).

[38] J.Y. Yu, L. Wang, X. Zeng, S.P. Wu, B. Li, Effect of montmorillonite on properties of styrene−butadiene−styrene copolymer modified bitumen, Polym. Eng. Sci. 47 (2007) 1289−1295.

[39] J.Y. Yu, X. Zeng, S.P. Wu, L. Wang, G. Liu, Preparation and properties of montmorillonite modified asphalts, Mater. Sci. Eng. A 447 (2007) 233−238.

[40] G. Liu, S.P. Wu, M.V.D. Ven, J.Y. Yu, A. Molenaar, Influence of sodium and organo-montmorillonites on the properties of bitumen, Appl. Clay Sci. 49 (2010) 69−73.

[41] S.S. Galooyak, B. Dabir, A.E. Nazarbeygi, A. Moeini, Rheological properties and storage stability of bitumen/SBS/montmorillonite composites, Constr. Build. Mater. 24 (2010) 300−307.

[42] S.G. Jahromi, A. Khodaii, Effects of nanoclay on rheological properties of bitumen binder, Constr. Build. Mater. 23 (8) (2009) 2894−2904.

[43] J.S. Zhang, Z. Li, M.T. Li, J. Xu, W.J. Yin, L. Liu, Highway 8 (2005) 142−146 (in Chinese).

[44] H. Zoriyeh, S. Erdem, E. Gürbüz, I. Bozbey, Nano-clay modified high plasticity soil as a building material: micro-structure linked engineering properties and 3D digital crack analysis, J. Build. Eng. 27 (2020) 101005. Available from: https://doi.org/10.1016/j.jobe.2019.101005.

[45] P. Kulanthaivel, B. Soundara, S. Velmurugan, V. Naveenraj, Experimental investigation on stabilization of clay soil using nano-materials and white cement, Mater. Today Proc. 45 (2021) 507−511. Available from: https://doi.org/10.1016/j.matpr.2020.02.107.

[46] M. Torres-carrasco, J.J. Reinosa, M.A. De Rubia, E. Reyes, F.A. Peralta, J.F. Fernández, Critical aspects in the handling of reactive silica in cementitious materials: effectiveness of rice husk ash vs nano-silica in mortar dosage, Constr. Build. Mater. 223 (2019) 360−367. Available from: https://doi.org/10.1016/j.conbuildmat.2019.07.023.

[47] T. Forbe, M. García, E. Gonzalez, Potencial risks of nanoparticles, Food Sci. Technol. 31 (2011) 835−842.

[48] M. Crosera, M. Bovenzi, G. Maina, G. Adami, C. Zanette, C. Florio, et al., Nanoparticle dermal absorption and toxicity: a review of the literature, Occup. Environ. Heal. 82 (2009) 1043−1055.

[49] M. Ltifi, A. Guefrech, P. Mounanga, A. Khelidj, Experimental study of the effect of addition of nano-silica on the behaviour of cement mortars, Proc. Eng. 10 (2011) 900−905.

CHAPTER 18

Nanomaterials recycling in industrial applications

Marjan Hezarkhani[1], Abdulmounem Alchekh Wis[1], Yusuf Menceloglu[1,2] and Burcu Saner Okan[1]

[1]Sabanci University Integrated Manufacturing Technologies Research and Application Center & Composite Technologies Center of Excellence, Teknopark Istanbul, Istanbul, Turkey
[2]Faculty of Engineering and Natural Sciences, Sabanci University, Istanbul, Turkey

Abbreviations

AAS	Atomic absorption spectrometry
AES	Atomic emission spectrometry
AFM	Atomic force microscopy
AF4	Asymmetric flow field flow fractionation
APS	Aerodynamic particle sizer
BET	Brunauer-Emmett-Teller theory
CE	Capillary electrophoresis
CLSM	Confocal laser scanning microscopy
CPC	Condensation particle counter
DLS	Dynamic light scattering
EDB	Electronic diffusion battery
EDX	Energy-dispersive X-ray spectroscopy
EELS	Electron energy loss spectroscopy
ELPI	Electrical low-pressure impactor
FFF	Field-flow fractionation
FMPS	Fast mobility particle sizer
FS	Fluorescence spectroscopy
GC−MS	Gas chromatography−mass spectrometry
HAADF-STEM	Scanning transmission electron microscopy with a high-angle annular dark-field detector
HDC	Hydrodynamic chromatography
HPLC	High-performance liquid chromatography
HR-TEM	High-resolution transmission electron microscopy
ICP-Ms	Inductively coupled plasma mass spectrometry
ICP-OES	Inductively coupled plasma atomic emission spectroscopy
MALS	Multiangle light scattering
NTA	Nanoparticle tracking analysis
SAED	Selected area electron diffraction
SAXS	Small-angle X-ray scattering
SEC	Size-exclusion chromatography
SEM	Scanning electron microscopy
SIMS	Secondary ion mass spectrometry

Nanomaterials Recycling
DOI: https://doi.org/10.1016/B978-0-323-90982-2.00017-2

© 2022 Elsevier Inc.
All rights reserved.

SMPS	Scanning mobility particle sizer
SP-ICP-MS	Single-particle inductively coupled plasma mass spectrometry
STM	Scanning tunneling microscope
TEM	Transmission electron microscopy
TD-GC—MS	Thermal desorption gas chromatography—mass spectrometry
UV/Vis	Ultraviolet-visible spectrophotometry
XPS	X-ray photoelectron spectroscopy

18.1 Introduction

In recent years, utilizing sustainable resources and recycling waste materials in the industrial applications have gained significant attention from the scientific and engineering communities [1]. In recent years, nanomaterials have shown great potential for commercial applications in catalysts [2], electronics [3], energy stockpiling [4], biomedicine [5], and so forth [6]. Moreover, large-scale production of different functional nanomaterials, ranging from zero-dimensional to two-dimensional applications, is also noted for their possible industrialization [7,8]. Sustaining the high quality of nanomaterials requires the use of expensive raw materials and reagents, which significantly increases the cost of nano-scale materials [9]. Even though these materials are finding use in ever more applications at an industrial scale, owing to their extraordinary properties as compared to their macro-level counterparts, very little effort has been put into the management of nanomaterial waste. For instance, an adsorption property has a significant application at an industrial scale, while desorption is essential in processing, synthesis, and manufacturing of the high-cost materials. To augment the efficiency of the adsorption processes, which involve removing dyes, phenols, heavy metals, metalloids, and organic contaminants from aqueous solutions in industry, it is essential to use nanomaterials. In the meantime, it is critical to utilize an economically abundant nanomaterial with the ability to be regenerated by continuous sequences of adsorption and desorption. The cost and dependence of such procedures in industry can be reduced significantly by usage of sustainable resource of nanomaterials that possesses a desorption feature. Therefore choosing an appropriate elutant is essential for a successful desorption technique, and it directly relies on the adsorption mechanism and the sort of material [10]. In other words, the elutant must be cheap, inert with respect to the nanomaterials, efficient, and ecofriendly. Various methods, such as magnetic recycling of nanocatalysts and recovery of magnetic nanomaterials [11], have been developed to achieve recycling and reuse of nanowaste in an efficient and cost-effective manner. However, the main priority of all these approaches has been to develop the methodology of recycling. High cost, high toxicity, and resource limitations are the main reasons behind these efforts [12]. Almost none of these investigations have focused on understanding the efficiency of these nanomaterial-recycling paths, which is the key to a reliable and sustainable economy. Therefore more comprehensive research is needed to determine a feasible methodology for the renewal of nanomaterials from waste sources [13]. There have been numerous attempts to recover constituents from thermoset and thermoplastic or biobased composites to provide sustainable solutions. However, there are still limitations for scaling up the developed technologies in the laboratories and increasing the yield of the recycling process, especially for nanomaterials. This chapter provides a brief explanation related to sustainable nanoparticle-recycling issues, regulations, and standards in addition to reusing different types of recycled nanoparticles such as graphene and metals for their possible industrial development.

18.2 Current problems in nanowaste sustainability and its management

Nanotechnology is a growing field that has attracted researchers because of the unique properties and applications of nanomaterials, such as high surface area, high interactivity, and potential to cross cell membranes. Despite the wide interest in nanomaterials and their potential benefits, they still have some negative aspects, such as their high cost and especially not being environmentally friendly, owing to their greater toxicity. Therefore the effect of this modern science on the surrounding environment must be considered in terms of both sustainability and waste management [14,15].

A new concept has emerged from nanotechnology, termed nanowaste, which deals with waste nanoparticles with at least one dimension in the range of $1-100$ nm [16]. There are various sources of nanowaste, but the most productive sources are the cosmetics industry and the personal care industry, which produce more than 50% of all nanowaste [17]. The main issue of nanotechnology is the handling nanowaste materials after reusing them, since they are particles with extremely small sizes, which make them difficult to monitor and track. Meanwhile, because this technology is a very recent topic, there is not enough knowledge about their waste treatment and the implications for the environment [18,19]. Environmental issues with nanowaste are getting worse compared to bulk materials as a result of various classifications, such as synthetic and natural, inorganic and organic, spheres and clusters, wires, nanofibers, plates, and thin films. The actual problem is that each type of nanomaterial behaves differently; therefore nanomaterials may potentially bond with pollutants because of their shape, large quantum effects, large surface area, biological reactivity, and especially small size, like that of cellular components [20,21] Hence it is important to pass legislation that oblige these industries to control and safely dispose of their nanowaste.

18.2.1 Regulations and standards for nanowaste management

The simulation of use, recycling, or end-of-life processes of nanomaterial-based products has been conducted to minimize environmental and toxicological risks for cradle-to-grave monitoring in keeping with United States Environmental Protection Agency regulations. Industrial waste disposal scenarios using a life cycle assessment approach have been developing that assess the global environmental impacts of a process or a product. There are very few policies to monitor and regulate the disposal of nanowastes. These policies mostly mention the safe techniques of usage of nanomaterials related to their identification or classification according to their size, function, and source. There are only two regulations related to this area, the first of which is ISO/TS 80004. This policy describes the applications and special vocabulary of nanomaterials and introduces legislation and uniform standards. The other policy, ISO/TR 13121:2011, which is more specific and accurate than the previous one, accurately describes the process of identifying, treating, reporting potential risks, taking measures, and developing the use of nanomaterials for the sake of human safety (consumers and workers) and the environment. In addition, this policy proposes several methods that enable companies to be more transparent and subject to accountability in how to manage this type of waste. There is currently a big gap between these policies and their implementation, since there is no control procedure or regulation to monitor the massive production currently taking place, and there is no knowledge or reliable prediction of the effects of this nanowaste in the future [22].

The huge difference in the properties and shapes of nanomaterials is the reason for the lack of legislations and laws in the field of nanowaste recycling. Each nanomaterial is distinguished by its characteristics, reactivity, bonding, and exposure risks and thus relative recycling methods must be specially developed for each nanomaterial type. Hence identification of recovered nanomaterials according to their physicochemical properties is an essential aspect to be considered. In the following subsection of this chapter, nanoparticles are classified on the basis of their production method, characterization, sizes, crystal morphologies, and toxicity levels.

18.2.2 Identification of recovered waste nanoparticles by characterization techniques

The difference in manufacturing methods of nanomaterials leads to major obstacles in legislation, as was mentioned above, each type of nanomaterial has a different manufacturing technique, crystal morphologies and sizes, and different toxicity levels [22]. The size classification, mass concentration, and particle number concentration are factors that explain the difference between nanomaterials, such as colloids, nanoparticles, and ions, and may determine the level of toxicity of each type by relying on a list of analytical tools. These factors are listed in Table 18.1, which also shows the analytical devices that can be used to identify these factors. It should be noted that these analytical methods are used only for identifying factors and not for treating nanowaste [22–24].

Table 18.1 Key parameters for the characterization of nanomaterials in the environment and proposed analytical methods.

Characterizing parameter	Analytical techniques
Particle size	AFM, APSa, CE, CLSM, CPCa, DMAa, DLS, ELPIa, FFF, FMPSa, FS, HDC, NTA, MALS, SAXS, SEC, SEM, SMPSa, SP-ICP-Ms, TEM, XRD
Elemental composition	EDX, GC—Ms, ICP-Ms, SPICP-Ms, XPS
Particle size distribution	AFM, APSa, CE, CLSM, CPCa, DMAa, DLS, ELPIa, FFF, FMPSa, FS, HDC, NTA, MALS, SAXS, SEC, SEM, SMPSa, SP-ICP-Ms, TEM, XRD
Surface charge	AFM, BET, zeta potential by DLS, XPS
Concentration related to particle number or mass	FS, GC—Ms, MALS, ICP-Ms, SPICP-Ms, UV/vis
Shape	AFM, CLSM, FFF, SEM, TEM
Surface speciation	AFM, fluorescence labeling, SIMS, STM, XPS
Surface area	AFM, BET, SEM, TEM, XPS
Structure, crystallinity	HR-TEM, SAED, SAXS, XRD
Surface functionality	EELS, FTIR, Raman, XPS

Source: Reproduced with permission from Copyright Elsevier F. Part, G. Zecha, T. Causon, E.-K. Sinner, M. Huber-Humer, Current limitations and challenges in nanowaste detection, characterisation and monitoring, Waste Manage, 43 (2015) 407—420.

Although there are some techniques for treating nanowaste, there is still obstacle that is the low effectiveness of materials recovered. In many of these approaches, the number of recovered materials does not exceed 0%—40%, and this is a very small number that can question the effectiveness of these methods. Finding new effective techniques to improve the recovery yield of the nanowaste materials is the future target for researchers. Some of the major obstacles to reaching such a target are fast growth in nanomaterial application and lack of sufficient data on the quality and quantity of nanomaterials in products. Moreover, the different geographical sources that nanowaste is disposed and the ability of these materials to penetrate or interact with the surrounding environment are of crucial concern for management of their recovery procedure.

Overall, based on the preceding discussions, one can comprehend that the following issues must be meticulously understood regarding the current situation of nanowaste materials and their variety. The initial step would be establishment of international policies and regulations for treatment of nanowaste. The second stage will consist of development of efficient recycling methodologies for nanowaste materials based on their physical aspects such as quantity and size. So far, the fate of these waste materials is unknown. Therefore finding effective ways to recycle or reuse nano-scale waste is yet to be uncovered. The following section describes some of the available methodologies in this regard.

18.3 Recovery of sustainable metals and inorganic nanoparticles from waste for catalytic and magnetic properties

Recycling of sustainable nanomaterials is a goal that is almost a dream. The presented instance here opens a wide door in this field and brings us a step closer to making this dream a reality. The acquisition of nanomaterials with high efficiency and sustainable catalytic properties is the focus of the following study. One of the important properties of nanomaterials with metal-based nature is their catalytic performability. Inorganic and metallic nanoparticles are widely used in various industrial fields. There are some studies for the recovery of metal nanoparticles from waste materials. For instance, Li et al. [2] recovered Humi-Fe_3O_4 nanoparticles strengthen with coal-derived humic acid to use these compounds to absorb toxic substances such as nitrophenol. The results showed an exceptional ability of this catalyst for three times reusing without affecting their distinctive properties and be able to absorb the toxic p-nitrophenol and reduction it to p-aminophenol upto 90%—92% after 6 h. In addition to that, these materials are also magnetically recoverable, and this opens the doors of great industrial applications in the future. A schematic showing reusing techniques is shown in Fig. 18.1.

In a related context, high purity nano-α-Fe_2O_3 (96.89%) with different sizes and shapes are manufactured from acid pickling waste by using nonionic surfactant polyethylene glycol polymer. The different sizes and shapes of nano-α-Fe_2O_3 showed well-crystallized and an interesting diversity possessed various characteristics, which would lead them potential application in pigment, catalysis, gas sensor industries, and nonlinear optics from the point of practical application. The nano-α-Fe_2O_3 catalyst could be recycled at least three times without any appreciable loss of its catalytic activity. Fig. 18.2 represented different morphology of produced nano-α-Fe_2O_3 in varies pH medium [25].

In another work, Biswas et al. [11] demonstrated that good quality magnetic iron oxide or iron hydroxide nanomaterials recycled from scrap iron that was accumulated from iron factories or at blacksmiths. The synthesis route of magnetic iron oxide nanoparticles from scrap iron is represented in Fig. 18.3 [11].

Growth in consumption of batteries is expected to continue, owing to their low maintenance, versatility, low cost, and high requirements for electronic devices. However, the recovery rate for various groups of batteries has been reported to be very low. For instance, the recovery rate of NiCd, lead-acid, and Li primary batteries has been reported to be 4.4%—5.5%; that of zinc-carbon and alkaline batteries has been reported to be below 1.7% [26]. With regard to recycling important metals, there is a trend to recycle nanozinc and lithium by recovering used batteries, which are large and important sources of these metals. Zhan et al. reproduced highly efficient photocatalyst nano-ZnO particles by using high-temperature under vacuum evaporation and controlling the oxidation reaction from waste zinc manganese batteries. The synthesis pathway of nanozinc particles is represented in Fig. 18.4 [27]. In another study, cobalt and nickel

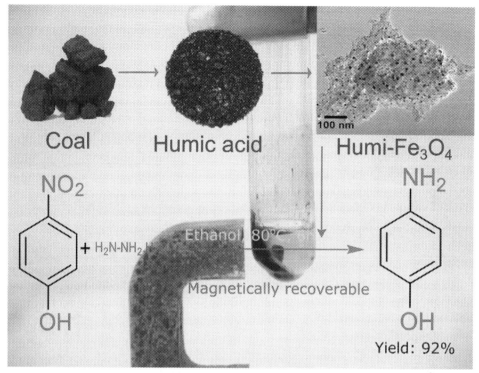

Figure 18.1 Schematic representation of humi-Fe$_3$O$_4$ nanoparticles recycling. *Reproduced from T. Das, G. Kalita, P.J. Bora, D. Prajapati, G. Baishya, B.K. Saikia, Humi-Fe$_3$O$_4$ nanocomposites from low-quality coal with amazing catalytic performance in reduction of nitrophenols, J. Environ. Chem. Eng. 5 (2) (2017) 1855–1865. Copyright Elsevier.*

nanoparticles were recovered from waste lithium-ion batteries that are widely used in power banks for mobile phones, by using cleaner and energy-saving technology via step-by-step reduction under vacuum. The nickel particles were reduced at a lower temperature vacuum reduction and then were separated by magnetic separation. After magnetic separation, cobalt compounds were secondarily reduced from residue compounds at a higher-temperature vacuum reduction. The chemical process of the step-by-step reduction of separation electrode materials is represented in Fig. 18.5. [28]

In addition to the catalytic and magnetic performances of zinc nanoparticles, dye removal from polluted air and water could be mentioned for potential application of zinc-based material. There is an approach for successfully synthesized ZnO porous nanosheet material from spent zinc-carbon batteries through a simple precipitation and calcination technique under very low temperatures. The product was used as a photocatalyst for the photodegradation of methylene blue dye removal from an aqueous medium. This recycled Zn-based material could be utilized in the water treatment industry [26].

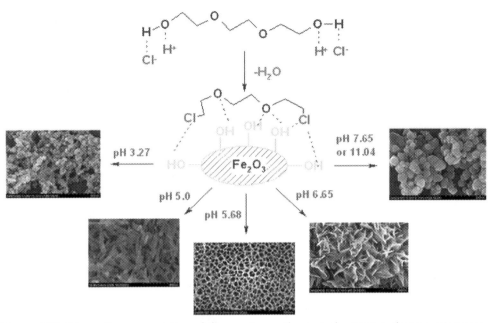

Figure 18.2 Schematic representation of the reaction pathway and scanning electron microscopy images of morphological structure of produced nano-α-Fe$_2$O$_3$ particles in different pH medium. Reproduced from X. Li, X. You, B. Lu, X. Wu, J. Zhao, Q. Cai, Reclamation of acid pickling waste: preparation of nano α-Fe$_2$O$_3$ and its catalytic performance, Ind. Eng. Chem. Res. 53 (52) (2014) 20085–20091. Copyright American Chemical Society.

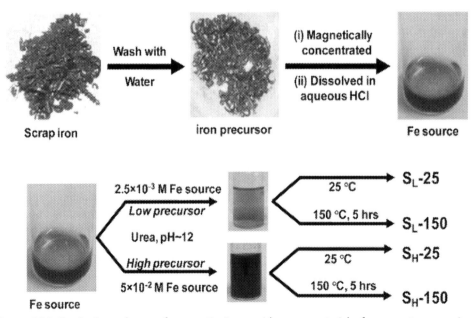

Figure 18.3 Synthetic pathway of magnetic iron oxide nanomaterials from waste scrap iron. Reproduced from A. Biswas, A.K. Patra, S. Sarkar, D. Das, D. Chattopadhyay, S. De, Synthesis of highly magnetic iron oxide nanomaterials from waste iron by one-step approach, Colloids Surf. A Physicochem. Eng. Asp. 589 (2020) 124420. Copyright Elsevier.

Nanomaterials recycling in industrial applications 383

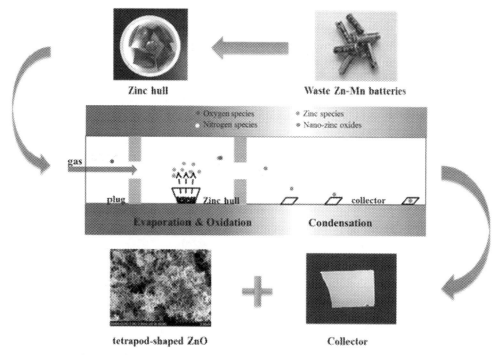

Figure 18.4 Scheme of synthesis of nano-ZnO particles from waste zinc manganese batteries. *Reproduced with permission from L. Zhan, O. Li, Z. Xu, Preparing nano-zinc oxide with high-added-value from waste zinc manganese battery by vacuum evaporation and oxygen-control oxidation, J. Clean. Prod. 251 (2020) 119691. Copyright Elsevier.*

Since different types of very expensive inorganic materials such as iron, zinc, nickel, and cobalt are widely used in the nano industry, recycling and management of sustainable nanomaterials could be a manner for decreasing their high cost and negative impact on the environment. Recently, the possibility of recycling nanometals and reusing them in the industry for their magnetic and catalytic properties have been taken into consideration in the field of rerepairing electronic sensors, such as batteries and supercapacitors.

18.4 Utilization of recycled nano-scale and micron-scale reinforcements in composite applications

One of the widely used applications for nanomaterials recycled from waste sources is for manufacturing thermoset and thermoplastic composites. Reinforced composites are rich sources of nanomaterials that could be recycled. Recycling sustainable reinforced material could be followed by remelting and remolding, thermal and chemical treatment, and mechanical recycling. There are various types of nanodepot composite materials based on their precursors, such as thermoset, thermoplastic polymers, and

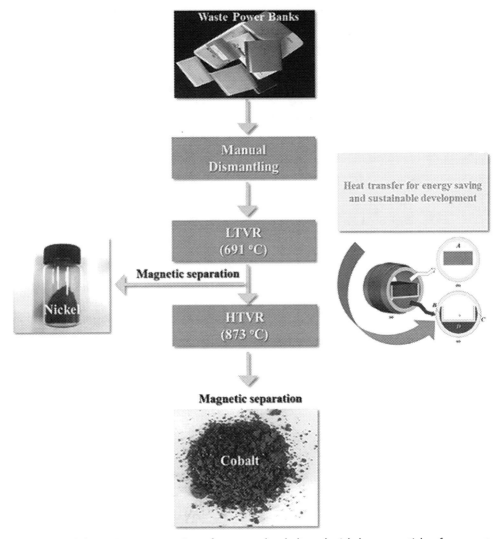

Figure 18.5 Schematic representation of recovered cobalt and nickel nanoparticles from waste power banks. *Reproduced with permission from Z. Huang, J. Zhu, R. Qiu, J. Ruan, R. Qiu, A cleaner and energy-saving technology of vacuum step-by-step reduction for recovering cobalt and nickel from spent lithium-ion batteries, J. Clean. Prod. 229 (2019) 1148–1157. Copyright Elsevier.*

resins, and production techniques. Current and future environmental legislation and waste management require all composite materials to be properly recovered from end-of-life (EOL) products, such as wind turbines, automobiles, and aircraft, and reused. The structural recycling of different types of composite material is represented in Fig. 18.6 [29].

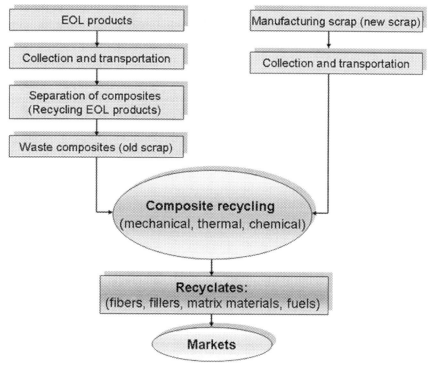

Figure 18.6 Structure of recycling system for composite materials. *Reproduced with permission from Y. Yang, R. Boom, B. Irion, D.-J. van Heerden, P. Kuiper, H. de Wit, Recycling of composite materials, Chem. Eng. Process. Process Intensif. 51 (2012) 53–68. Copyright Elsevier.*

There have been some studies of the utilization of recycled nanoparticles in production of reinforced composite material for different industrial applications, such as aircraft, automotive, and membrane development. For instance, there is a study about reinforcing recycled zinc oxide from spent batteries and reusing it in uncured thermosetting epoxy resins reinforced composite by a rotary molding system followed by ZnO deposition in uncured epoxy resins. The manufacturing molding setup is represented in Fig. 18.7. Usage of 30 wt.% recycled ZnO particles in reinforced composite showed enhancement in composite glass transition temperature as 1.4%, hardness as 82.3%, and stiffness as 19.2%. This approach is an easily applicable method for the production of other reinforced polymer networks, as demonstrated with the commercial epoxy ZnO composites [30]. Moreover, the recycled ZnO particles from waste batteries could be acceptable nanomaterial for the production of zinc-coated cured epoxy resins in nanocomposite systems for hydrophobic anticorrosion and hydrophobic properties [31].

One of the most essential sources for nanomaterials is nanoparticle deposited fibrous structural exists as composite materials. A green and sustainable strategy is introduced to recycle and repeatedly used nanotellurium from waste ultrathin

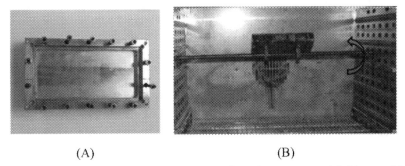

Figure 18.7 Rotary mold for manufacturing ZnO reinforced composites. (A) Disassembled rotary mold. (B) Rotary mold working in the furnace. *Reproduced with permission from I. Lorero, M. Campo, G. Del Rosario, F.A. López, S.G. Prolongo, New manufacturing process of composites reinforced with ZnO nanoparticles recycled from alkaline batteries, Polym. (Basel) 12 (7) (2020) 17–21. doi: 10.3390/polym12071619. Licensed under the Creative Commons Attribution 4.0 License (CC-BY 4.0).*

tellurium nanowires [32]. Besides, Okan et al. [33] produced lightweight graphene nanoplatelet by cost-effective recycling technique of waste tire from automotive. Graphene is a broadly favored carbon-based material, owing to its high electrical conductivity and fast electron mobility, which is widely utilized in fabricating electrospun and sprayed graphene reinforced polymeric nanofibers, spheres, and foams. For instance, a three-dimensional sea urchin–shaped composite electrode was attained by the incorporation of graphene oxide in the presence of PAN polymer by using core-shell electrospinning technology [34]. In another study, Pt-supported graphene-based structural composite electrodes are fabricated with the use of graphene nanoplatelet that is recycled from waste tires by using core-shell electrospinning and electrospraying technique with tailored morphology [35]. In another study, Okan et al. [36] produced polypropylene nanocomposites reinforced by waste tire–derived graphene nanoplatelets via a melt-mixing method and tailored crystallization and mechanical characteristics with controlled polymer grade. Recent advances show that recycling nanoparticles from waste-reinforced composites and reusing them as valuable products could reduce environmental pollution and high-cost composite development. However, there are still gaps that need to be improving in this field of study.

18.5 The growth of nanomaterials from waste plastics by an upcycling process

Inexpensive feedstocks yielded from plastic chemical recycling enable a new pathway to manufacture some value-added products. This process is then termed upcycling process, as the quality and value of the final products are upgraded. Upcycling is a sustainable method of reusing and helps to keep the environment from being degraded while converting the waste materials into value-added products. It is possible to synthesis carbon nanotubes (CNTs), graphene, or other carbonaceous from plastic waste materials [37,38]. Some attempts have been made to grow carbon structures on organically modified montomorillonite and zeolites. Gong et al. [39] used waste polypropylene (PP) as raw materials catalyzed by organically modified montomorillonite (OMMT) to realize highly yielding graphene flakes (GFs). As described in the original article, the uniform mixture containing PP ($\sim$89 wt.%), talcum ($\sim$11 wt.%), and OMMT was placed in a crucible. The mixture was heated upto 700°C for 15 min to obtain the carbonized char, which was then immersed in hydrogen fluoride (HF) and HNO_3 after cooling. Impurities (e.g., montomorillonite (MMT) and talcum additives) were dissolved by HF, and amorphous carbon was oxidized by HNO_3. GFs were obtained after repeatedly centrifuging and isolating from the solution. In another work, Cui et al. [40] represented a solid-state chemical vapor deposition method for converting the plastic wastes to valuable graphene foils with high quality, fewer defects, and very high electrical conductivity, much higher than that of common free-standing graphene films treated at ultrahigh temperatures. The structure of this flexible GF is proper and functional for producing an anode for the fabrication of foldable lithium-ion batteries. The strategy of generating high-quality free-standing GF from plastic wastes indeed provides a trash-to-treasure way for graphene production and practical applications. Moreover, Jiang et al. [41] mixed melted polypropylene with a catalyst and then heated the mixture to a temperature of 830°C to obtain CNTs. Compound mixtures of PP, OMMT, and nickel were combusted to synthesize MWCNTs. In this process, different nickel compounds acted as catalysts to synthesize MWCNTs in the presence of OMMT. The combustion temperature and the types and contents of the nickel catalysts and OMMT had effects on the yield of MWCNTs. In this method the heat gradually diffused from the surface to the core of the samples. Meanwhile, the degradation products of PP diffused in a reverse direction of heat diffusion, and a part of them were catalyzed to form MWCNTs and hydrogen due to the presence of the degraded OMMT and Ni catalyst. Choi et al. [42] successfully converted the "noncarbonizable" linear low-density polyethylene (LLDPE) into an ordered carbon by using thermal oxidation as a degradation pathway for plastics. Also, the conversion yield of this transformation reached 50%. The graphitized LLDPE-based carbon has an exceptional electrical performance, which makes it an applicable material for lithium-ion battery fabrication. In another work, Song et al. [43] investigated the effects of both OMMT and Ni_2O_3 on the carbonization of PP during pyrolysis. The catalyst H-MMT (montmorillonite, which contains hydrogen protons) has an important influence on the degradation and carbonization behavior of PP and also the formation of multiwalled CNTs (MWCNTs). The higher yields of MWCNTs obtained by combination of OMMT and the nickel catalyst during pyrolysis due to the acidic sites upon the H-MMT layers and the metallic Ni formed in situ from the reduction of Ni_2O_3 during the transformation of PP into MWCNTs. The presence of carbenium ions as intermediates in the catalytic degradation of PP promotes the formation of

MWCNTs from the degradation products as carbon sources, especially those with higher carbon content. Liu et al. [44] provided a potential way to convert waste plastics into carbon nanomaterials and hydrogen by demonstrating a newly developed process to continuously convert PP to hydrogen and carbon materials. HZSM-5 zeolite and NiO were used as catalysts in a screw kiln reactor for degrading PP and a moving-bed reactor for decomposing the pyrolysis gas separately. The highest yield of MWCNTs was obtained at the decomposition temperature of 700°C. The graphitization degree and thermal stability of the MWCNTs synthesized at a high decomposition temperature were higher than those at low decomposition temperatures. Bajad et al. [45] studied the synthesis of CNTs from waste plastic by a combustion technique and using Ni/Mo/MgO as a catalyst. The amount of obtained carbon product measured the catalytic activity of the three components Ni, Mo, and MgO. The activity of the components is observed to be interdependent and the component Ni is found to be more effective. To conclude, the existence of carbon chains within polymer seeds attracts attention to the ability of these plastic wastes to convert to some more valuable materials from carbon families. In other words, solid plastic waste is one of the prominent types of waste that can be used as the precursor for the synthesis of graphene nanosheets and carbon nanostructures. The waste plastics can therefore provide a carbon source for carbon-based value-added products.

18.6 Recovery and reuse of metal nanoparticles from waste electronic components

The electronics industry provides a diversity of products and parts, and electronic waste or e-waste is one of emerging problems, owing to high usage of electronic products. Therefore waste disposal and recycling scenarios have been developing to minimize the environmental impacts of e-waste. Electroplating is a process that creates a metal coating on a solid substrate through the reduction of cations of that metal through a direct electric current [46]. Electroplating is widely used in the industry and decorative arts to improve surface qualities. It may also be used to manufacture metal plates with complex shapes. It is also used to purify metals such as copper [38]. The term "electroplating" may also be used occasionally for processes that use an electric current to achieve oxidation of anions onto a solid substrate, as in the formation of silver chloride on silver wire to make silver/silver chloride electrodes [47].

This method was designed as a strategy for recycling SnO_2 from tin plate electroplating sludge that mainly contains amorphous Sn and Fe compounds. The procedure for this method is divided into six sections: reagents, analysis of original electroplating sludge, treatment of sludge mineralization, acid treatment of sludge after mineralization, experimental range experiment, and finally characterization. High purity of 90% of nano-SnO_2 powder was obtained from this method. A summary of the entire process is provided in Fig. 18.8. In short, the mentioned technique provides a relatively

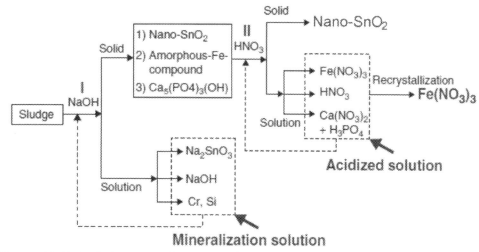

Figure 18.8 Method of recycling of nano-SnO₂ from electroplating sludge. *Reproduced with permission from Z. Zhuang, X. Xu, Y. Wang, Y. Wang, F. Huang, Z. Lin, Treatment of nanowaste via fast crystal growth: With recycling of nano-SnO₂ from electroplating sludge as a study case, J. Hazard. Mater. 211 (2012) 414–419. Copyright Elsevier.*

inexpensive method to extract nano- SnO₂ from slag of electronic parts and to use these recycled materials in electronic parts applications [48].

Recovering important metals that are used in electronics and communication applications is the focus of this section. The methods mentioned in this section gain importance owing to the possibility of using them in many applications, such as restoration of optical sensors and electronic chip applications.

18.7 Selective recovery of metal nanoparticles by using α-cyclodextrin

It was revealed that in addition to contamination from nanowastes, many technologies use resource-limited materials, such as rare earth elements and precious metals (e.g., gold and silver), that should be recovered. One purchase of reducing high-cost nano-size elements is to be recovered and reused continuously. Pati et al. [49] developed a laboratory-scale method to recover nano-scale gold. This method could provide a solution to the nanotechnology problems. In this method the gold was recovered at least in the shape of nanoparticles. In this process, alpha-cyclodextrin (α-CD) is used to facilitate the formation of the host-guest inclusion complex that includes the coordination of the second domain of $[K(OH_2)_6]^+$ and $[AuBr_4]$, which is used to recover the gold. Then the recovered metal is used to produce new nanoparticles. The results of the life cycle assessment indicate that the recovery and recycling of gold nanoparticles can significantly reduce the environmental impact of the synthesis of gold nanoparticles [49].

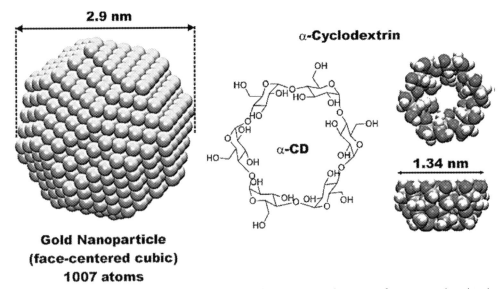

Figure 18.9 A quasispherical gold nanoparticle with an average diameter of 2.9 nm and molecular structure of alpha-cyclodextrin (α-CD) and its top and side views [55]. *Reproduced with permission from M.V. Slavgorodska, A.V. Kyrychenko, Binding preference of α-cyclodextrin onto gold nanoparticle, Nanosistemi Nanomateriali Nanotehnologii, 17(1) (2019) 133–144.*

The method is characterized by high recovery yields of around 90%. Therefore this method is significantly superior to other methods of treating nanowaste, whose efficiency ranges from 0% to 40%. Because mercury and cyanide are not used in the α-CD gold recovery method, this method is distinguished from the traditional methods and with the selective ability of gold recovery with high efficiency [50]. In summary, this research provides a relatively inexpensive solution to recover the gold used as nanomaterial in many applications and reuse nanogold in industrial applications.

The spontaneous assembly of a one-dimensional supramolecular complex with an extended $\{[K(OH_2)_6][AuBr_4](\alpha\text{-CD})_2\}_n$ chain superstructure formed during the rapid coreciptation of α-CD and KAuBr$_4$ in water. The method was highly selective for gold salt. The study hypothesized that a perfect match in molecular recognition between α-CD and [AuBr4] leads to a near-axial orientation of the ion with respect to the α-CD channel. This study could be a good start for a green method to recover gold from gold-bearing raw materials, especially with α-CD an inexpensive and environmentally benign carbohydrate [51]. Fig. 18.9 represents quasispherical gold nanoparticles and α-CD with the specific particle sizes.

18.8 Nanomaterials for environmental cleanup applications

Environmental cleanup projects are of global and local interest, involving the cleanup of landfills, oil fields, and industrial sites that can contain and leak pesticides, oils, heavy metals, and other dangerous chemicals. There are some affected resources, such as aquifers and groundwater, resulting in risks to surrounding populations [52]. The origin of microplastics is known to be the disintegration of larger plastics into fragments, microfibers; sources include beauty products and tires, to name a few. In one study, Jian et al. [53] mixed CNTs with manganese to form hollow nanostructures called nanocoils. The manganese inside the nanocoils generates free radicals, which are short-lived, highly reactive oxygen molecules that attack the microplastics and cause them to fragment into smaller pieces. Eventually, these microplastic fragments are converted to carbon dioxide gas and water. Nanocoils are provided to convert 50% of microplastics to carbon dioxide and water. When these microplastic fragments were added to a growing algae colony, the algae grew faster than colonies without the fragments.

In another study related to biobased waste issues, Patwa et al. [54] use mucus from common resources as a biomaterial to catch the nanoparticles, particularly for the accumulation of gold and quantitative point nanoparticles in the mucus coming from natural species, for example, jellyfish. Removing nanoparticles with jellyfish, secreted mucus can remove specific nanoparticles <50 nm in diameter of fluid colloidal suspensions at room temperature. This treatment was reprimanded for the alleged effect of bioaccumulation of nanoparticles. In short, this study demonstrated the ability of natural materials to get rid of nanowastes. This study opens the door to major industrial applications in this field [54]. Consequently, this section explains the use of natural materials in the recycling of nanowaste. These methods open applications in the fields of ocean cleaning, water purification, and liquid waste treatment applications (Fig. 18.10).

18.9 Conclusions and potential outlook

The nanowaste resulting from the development of nanotechnology has become a real problem that needs solutions, and researchers are obligated to find new ways to deal with this problem. However, owing to the lack of information, legislation, and the complexity of the problem, especially at this time, research and evaluation must be conducted periodically and continuously for nanowaste and its safe disposal and toxicity.

As the journey of a thousand miles begins with a single step, determining the effects of nanowaste on the environment and people is the first step in this field, as it is important to determine the current and future sources of nanowaste, their quantities, where they will be disposed of, and in what form. On the same level, the characteristics of nanowastes and their differences must be determined from the beginning of production to the stage of safe disposal or recycling. Fortunately, this matter has brought increasing interest from scientists in recent years to solve this problem. In this chapter, some examples of solutions presented in the field of recycling or reuse of

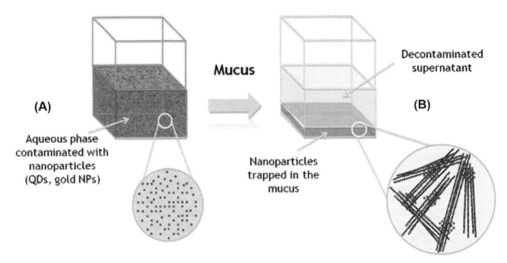

Figure 18.10 Schematic representation of the removal of nanoparticles from an aqueous suspension using mucus secreted by jellyfish. *Reproduced from A. Patwa et al., Accumulation of nanoparticles in "jellyfish" mucus: a bio-inspired route to decontamination of nano-waste, Sci. Rep. 5 (2015) 11387. Copyright 2015 Springer Nature.*

nanowaste are discussed and presented, with a discussion of important aspects in improving the field of industrial applications of nanowaste. These aspects include assessment of the problem of nanoparticles, challenges of recycling of nanoparticles, assessment of toxicological risks, current legislative frameworks, and treatment techniques.

In this chapter, important technologies were discussed that represent promising areas in the field of nanowaste recycling. Examples were chosen for various fields. The example of the catalytic performance of inorganic nanoparticles was chosen, which represents the field of methods for recovering nanoparticles and reuse in high-efficiency electronic chips and sensors. The example of removing dyes by using recycled nanomaterials was chosen as an example in the field of removing dyes, toxic substances, and harmful pharmaceutical substances from waste. In composites, the instance that was chosen was recycled zinc oxide that was reused as a reinforcement in composites. This field is wide, as nanomaterials were reused in composites as strength and durability reinforcement. The example of nano-SnO_2 was chosen as an example in the field of extracting nanomaterials from electronic parts. The example of nano-gold recovery was chosen as an example of the recovery of precious metals using highly efficient materials and methods. Finally, an environmentally friendly method was chosen in which nanomaterials were collected by using jellyfish mucus. This is an example of the new environmental methods in the field of recycling nanowaste.

Consequently, this chapter has shown the path of the pioneers in the field of recycling nanomaterials, which will develop rapidly and remarkably in the near future. The ambition of the sustainability assessment highlights the environmental and cost benefits due to producing nanomaterials from industrial wastes, such as tires and plastics, instead of virgin sources. Another challenge is to build and assess several potential end-of-life scenarios in order to demonstrate the potential environmental and economic benefits of the recycling or reuse of waste nanomaterials instead of their disposal after use and to demonstrate the contribution of the newly developed technology to resources conservation and circular economy.

References

[1] A.K. Mohanty, S. Vivekanandhan, J.-M. Pin, M. Misra, Composites from renewable and sustainable resources: challenges and innovations, Science 362 (6414) (2018) 536−542.

[2] T. Das, G. Kalita, P.J. Bora, D. Prajapati, G. Baishya, B.K. Saikia, Humi-Fe_3O_4 nanocomposites from low-quality coal with amazing catalytic performance in reduction of nitrophenols, J. Environ. Chem. Eng. 5 (2) (2017) 1855−1865.

[3] B. Li, S. Ye, I.E. Stewart, S. Alvarez, B.J. Wiley, Synthesis and purification of silver nanowires to make conducting films with a transmittance of 99%, Nano Lett. 15 (10) (2015) 6722−6726.

[4] L. Mai, X. Tian, X. Xu, L. Chang, L. Xu, Nanowire electrodes for electrochemical energy storage devices, Chem. Rev. 114 (23) (2014) 11828−11862.

[5] B. Tian, C.M. Lieber, Nanowired bioelectric interfaces: focus review, Chem. Rev. 119 (15) (2019) 9136−9152.

[6] N. Sharma, H. Ojha, A. Bharadwaj, D.P. Pathak, R.K. Sharma, Preparation and catalytic applications of nanomaterials: a review, Rsc Adv. 5 (66) (2015) 53381−53403.

[7] L. Zhang, et al., Continuous and scalable production of well-controlled noble-metal nanocrystals in milliliter-sized droplet reactors, Nano Lett. 14 (11) (2014) 6626−6631.

[8] K.R. Paton, et al., Scalable production of large quantities of defect-free few-layer graphene by shear exfoliation in liquids, Nat. Mater. 13 (6) (2014) 624−630.

[9] J.-L. Wang, J.-W. Liu, S.-H. Yu, Recycling valuable elements from the chemical synthesis process of nanomaterials: a sustainable view, ACS Mater. Lett. 1 (5) (2019) 541−548.

[10] N. Das, Recovery of precious metals through biosorption—a review, Hydrometallurgy 103 (1−4) (2010) 180−189.

[11] A. Biswas, A.K. Patra, S. Sarkar, D. Das, D. Chattopadhyay, S. De, Synthesis of highly magnetic iron oxide nanomaterials from waste iron by one-step approach, Colloids Surf. A Physicochem. Eng. Asp. 589 (2020) 124420.

[12] S. Sharifi, S. Behzadi, S. Laurent, M.L. Forrest, P. Stroeve, M. Mahmoudi, Toxicity of nanomaterials, Chem. Soc. Rev. 41 (6) (2012) 2323−2343.

[13] G. Bystrzejewska-Piotrowska, J. Golimowski, P.L. Urban, Nanoparticles: their potential toxicity, waste and environmental management, Waste Manage. 29 (9) (2009) 2587−2595.

[14] B.J. Marquis, S.A. Love, K.L. Braun, C.L. Haynes, Analytical methods to assess nanoparticle toxicity, Analyst 134 (3) (2009) 425−439.

[15] S.A. Love, M.A. Maurer-Jones, J.W. Thompson, Y.-S. Lin, C.L. Haynes, Assessing nanoparticle toxicity, Annu. Rev. Anal. Chem. 5 (2012) 181−205.

[16] A. Caballero-Guzman, T. Sun, B. Nowack, Flows of engineered nanomaterials through the recycling process in Switzerland, Waste Manage. 36 (2015) 33−43.

[17] N. Musee, Nanowastes and the environment: potential new waste management paradigm, Environ. Int. 37 (1) (2011) 112−128.

[18] H.U. Sverdrup, K.V. Ragnarsdottir, D. Koca, An assessment of metal supply sustainability as an input to policy: security of supply extraction rates, stocks-in-use, recycling, and risk of scarcity, J. Clean. Prod. 140 (2017) 359–372.

[19] S. Zhang, Y. Ding, B. Liu, C. Chang, Supply and demand of some critical metals and present status of their recycling in WEEE, Waste Manage. 65 (2017) 113–127.

[20] T. Faunce, B. Kolodziejczyk, Nanowaste: need for disposal and recycling standards, G20 Insights, Policy Era Agenda, 2030, 2017.

[21] B. Mrowiec, Risk of nanowastes, Inżynieria i Ochr. Środowiska 19 (4) (2016) 469–478.

[22] F. Part, G. Zecha, T. Causon, E.-K. Sinner, M. Huber-Humer, Current limitations and challenges in nanowaste detection, characterisation and monitoring, Waste Manage. 43 (2015) 407–420.

[23] S. Bandyopadhyay, J.R. Peralta-Videa, J.L. Gardea-Torresdey, Advanced analytical techniques for the measurement of nanomaterials in food and agricultural samples: a review, Environ. Eng. Sci. 30 (3) (2013) 118–125.

[24] V.K. Sharma, R.A. Yngard, Y. Lin, Silver nanoparticles: green synthesis and their antimicrobial activities, Adv. Colloid Interface Sci. 145 (1–2) (2009) 83–96.

[25] X. Li, X. You, B. Lu, X. Wu, J. Zhao, Q. Cai, Reclamation of acid pickling waste: preparation of nano α-Fe_2O_3 and its catalytic performance, Ind. Eng. Chem. Res. 53 (52) (2014) 20085–20091.

[26] S. Maroufi, R.K. Nekouei, M. Assefi, V. Sahajwalla, Waste-cleaning waste: synthesis of ZnO porous nano-sheets from batteries for dye degradation, Environ. Sci. Pollut. Res. 25 (28) (2018) 28594–28600.

[27] L. Zhan, O. Li, Z. Xu, Preparing nano-zinc oxide with high-added-value from waste zinc manganese battery by vacuum evaporation and oxygen-control oxidation, J. Clean. Prod. 251 (2020) 119691.

[28] Z. Huang, J. Zhu, R. Qiu, J. Ruan, R. Qiu, A cleaner and energy-saving technology of vacuum step-by-step reduction for recovering cobalt and nickel from spent lithium-ion batteries, J. Clean. Prod. 229 (2019) 1148–1157.

[29] Y. Yang, R. Boom, B. Irion, D.-J. van Heerden, P. Kuiper, H. de Wit, Recycling of composite materials, Chem. Eng. Process. Process Intensif. 51 (2012) 53–68.

[30] I. Lorero, M. Campo, G. Del Rosario, F.A. López, S.G. Prolongo, New manufacturing process of composites reinforced with ZnO nanoparticles recycled from alkaline batteries, Polym. (Basel) 12 (7) (2020) 17–21. Available from: https://doi.org/10.3390/polym12071619.

[31] S. Ammar, K. Ramesh, B. Vengadaesvaran, S. Ramesh, A.K. Arof, Amelioration of anticorrosion and hydrophobic properties of epoxy/PDMS composite coatings containing nano ZnO particles, Prog. Org. Coat. 92 (2016) 54–65.

[32] J.-L. Wang, et al., Recycling nanowire templates for multiplex templating synthesis: a green and sustainable strategy, Chem. Eur. J. 21 (13) (2015) 4935–4939.

[33] B.S. Okan, Y. Mencelo\uglu, B.G. Ozunlu, Y.E. Yagci, Graphene from waste tire by recycling technique for cost-effective and light-weight automotive plastic part production, AIP Conf. Proc. 2205 (1) (2020) 20046.

[34] L. Haghighi Poudeh, I. Letofsky-Papst, F.Ç. Cebeci, Y. Menceloglu, M. Yildiz, B. Saner Okan, Facile synthesis of single-and multi-layer graphene/Mn_3O_4 integrated 3D Urchin-shaped hybrid composite electrodes by core-shell electrospinning, ChemNanoMat 5 (6) (2019) 792–801.

[35] L. Haghighi Poudeh, D. Cakiroglu, F.C. Cebeci, M. Yildiz, Y.Z. Menceloglu, B. Saner Okan, Design of Pt-supported 1D and 3D multilayer graphene-based structural composite electrodes with controlled morphology by core–shell electrospinning/electrospraying, ACS Omega 3 (6) (2018) 6400–6410.

[36] J.S.M. Zanjani, L.H. Poudeh, B.G. Ozunlu, Y.E. Yagci, Y. Menceloglu, B. Saner Okan, Development of waste tire-derived graphene reinforced polypropylene nanocomposites with controlled polymer grade, crystallization and mechanical characteristics via melt-mixing, Polym. Int. 69 (9) (2020) 771–779.

[37] M.P. Aji, et al., Polymer carbon dots from plastics waste upcycling, Environ. Nanotechnol. Monit. Manag. 9 (2018) 136–140.
[38] C. Zhuo, Y.A. Levendis, Upcycling waste plastics into carbon nanomaterials: a review, J. Appl. Polym. Sci. 131 (4) (2014) 1–14. Available from: https://doi.org/10.1002/app.39931.
[39] J. Gong, et al., Upcycling waste polypropylene into graphene flakes on organically modified montmorillonite, Ind. Eng. Chem. Res. 53 (11) (2014) 4173–4181. Available from: https://doi.org/10.1021/ie4043246.
[40] L. Cui, X. Wang, N. Chen, B. Ji, L. Qu, Trash to treasure: converting plastic waste into a useful graphene foil, Nanoscale 9 (26) (2017) 9089–9094. Available from: https://doi.org/10.1039/c7nr03580b.
[41] Z. Jiang, R. Song, W. Bi, J. Lu, T. Tang, Polypropylene as a carbon source for the synthesis of multi-walled carbon nanotubes via catalytic combustion, Carbon N Y 45 (2) (2007) 449–458. Available from: https://doi.org/10.1016/j.carbon.2006.08.012.
[42] D. Choi, D. Jang, H.I. Joh, E. Reichmanis, S. Lee, High performance graphitic carbon from waste polyethylene: thermal oxidation as a stabilization pathway revisited, Chem. Mater. 29 (21) (2017) 9518–9527. Available from: https://doi.org/10.1021/acs.chemmater.7b03737.
[43] R. Song, et al., The combined catalytic action of solid acids with nickel for the transformation of polypropylene into carbon nanotubes by pyrolysis, Chem. A Eur. J. 13 (11) (2007) 3234–3240. Available from: https://doi.org/10.1002/chem.200601018.
[44] J. Liu, Z. Jiang, H. Yu, T. Tang, Catalytic pyrolysis of polypropylene to synthesize carbon nanotubes and hydrogen through a two-stage process, Polym. Degrad. Stab. 96 (10) (2011) 1711–1719.
[45] G. Bajad, V. Guguloth, R.P. Vijayakumar, S. Bose, Conversion of plastic waste into CNTs using Ni/Mo/MgO catalyst—an optimization approach by mixture experiment, Fuller. Nanotub. Carbon Nanostruct. 24 (2) (2016) 162–169.
[46] P.T. Williams, Hydrogen and carbon nanotubes from pyrolysis-catalysis of waste plastics: a review, Waste Biomass Valor. 12 (1) (2021) 1–28. Available from: https://doi.org/10.1007/s12649-020-01054-w.
[47] L. Gal-Or, I. Silberman, R. Chaim, Electrolytic ZrO_2 coatings: I. Electrochemical aspects, J. Electrochem. Soc. 138 (7) (1991) 1939.
[48] Z. Zhuang, X. Xu, Y. Wang, Y. Wang, F. Huang, Z. Lin, Treatment of nanowaste via fast crystal growth: With recycling of nano-SnO_2 from electroplating sludge as a study case, J. Hazard. Mater. 211 (2012) 414–419.
[49] P. Pati, S. McGinnis, P.J. Vikesland, Waste not want not: life cycle implications of gold recovery and recycling from nanowaste, Environ. Sci. Nano 3 (5) (2016) 1133–1143.
[50] Z. Liu, A. Samanta, J. Lei, J. Sun, Y. Wang, J.F. Stoddart, Cation-dependent gold recovery with α-cyclodextrin facilitated by second-sphere coordination, J. Am. Chem. Soc. 138 (36) (2016) 11643–11653.
[51] V. Oestreicher, C.S. Garcia, G.J.A.A. Soler Illia, P.C. Angelome, Gold recycling at laboratory scale: from nanowaste to nanospheres, ChemSusChem 12 (21) (2019) 4882–4888. Available from: https://doi.org/10.1002/cssc.201901488.
[52] S.S. Patil, U.U. Shedbalkar, A. Truskewycz, B.A. Chopade, A.S. Ball, Nanoparticles for environmental clean-up: a review of potential risks and emerging solutions, Environ. Technol. Innov. 5 (2016) 10–21.
[53] J. Kang, L. Zhou, X. Duan, H. Sun, Z. Ao, S. Wang, Degradation of cosmetic microplastics via functionalized carbon nanosprings, Matter 1 (3) (2019) 745–758. Available from: https://doi.org/10.1016/j.matt.2019.06.004.
[54] A. Patwa, et al., Accumulation of nanoparticles in "jellyfish" mucus: a bio-inspired route to decontamination of nano-waste, Sci. Rep. 5 (2015) 11387.
[55] M.V. Slavgorodska, A.V. Kyrychenko, Binding preference of α-cyclodextrin onto gold nanoparticle, Nanosistemi Nanomateriali Nanotehnologii 17(1) (2019) 133–144.

Index

Note: Page numbers followed by "*f*" and "*t*" refer to figures and tables, respectively.

A

Acrylic acid (AA), 225–226, 232–237
Adsorption, 240–242
Alpha-cyclodextrin (α-CD), 192–193, 389
    selective recovery of metal nanoparticles by using, 389–390
Alzheimer's disease, 64
Antisolvent technique by using CO_2, 160–162
Applications in nanomaterials recycling, 232–242
    adsorption, 240–242
    aqueous two-phase system (ATPS), 232–237
    catalysts, 238–240
Aqueous dispersion techniques, 162
Aqueous two-phase system (ATPS), 232–237
Arsenic, 179
Asphalt concrete, 369–370
Atomic force microscopy (AFM), 309–312
Attenuated total reflection (ATR), 297
Auger electron spectroscopy (AES), 303–305, 304*f*

B

Basic ionic liquids (B-ILs), 272
Biocidal products regulations, 80
Bioionic liquids (Bio-ILs), 272
Biological treatment of nanowastes, 134
Bragg diffraction, 293*f*
Bragg's law, 293–294
Buckminsterfullerene, 253
Building applications, 370
1-Butyl-3-methylimidazolium chloride, 277–278
Butyl methacrylate (BMA), 232–237

C

Cadmium, 179–181
Caenorhabditis elegans, 130
Carbon-based nanomaterials, 23, 38–40, 150–151, 182–183
Carbon-based nanoparticles, 253
Carbon black, 39
Carbon fiber, 346
Carbon nanodots, 23
Carbon nanofiber, 39
Carbon nanotubes (CNTs), 11–12, 23, 39, 182–183, 194–195, 253
    recycled, 326–327
Carboxylate, 225–226
Carboxyl group, 225–226
Catalysts, 238–240
Cell nanotoxicity, mechanism of, 134*f*
Cellulose, 328–329
Cellulose nanofibers (CNFs), 199, 328–330
Centrifugation/solvent evaporation technique, 163
Ceramic-based nanomaterials, 151–152
Ceramic nanoparticles, 253
Challenges in nanowaste management, 24–25, 140–141
Characterization techniques, identification of recovered waste nanoparticles by, 378–379
Chemical hygiene plan (CHP), 186
Chemical oxygen demand (COD), 29
Chemical properties of nanomaterials, 250*t*
Chemical recycling processes, 190–193
Chemicals (hazard information and packaging for supply) regulations, 75–76
    classification and labeling of individual substance, 76
    safety data sheet requirements, 75
Chemical vapor deposition (CVD), 38–39
Chiral ILs, 272
Chlorinated furans, 30
Chromium, 181
Citrus maximus, 201
Classification, Labelling, and Packaging (CLP), 96–98
Classification of nanoparticles, 126*f*, 252–254
    carbon-based nanoparticles, 253
    ceramic nanoparticles, 253
    metal nanoparticles, 253
    polymeric nanoparticles, 254
        lipid-based nanoparticles, 254
    semiconductor nanoparticles, 253

398 Index

Classification of nanowastes, 4—7, 25—27, 26f, 42—45, 112—114, 126t, 183—186, 184f
  on the basis of risk factors, 6t
Cloud point extraction (CPE), 150
  nanoparticle recovery by, 219
Coagulation technique, 198
Colloidal solvent technique, 163
  recycling nanoparticles by employing, 219
Competent Authorities for REACH and Classification and Labelling (CARACAL), 96, 99
Competent authority (CA), 74
Compounded annual growth rate (CAGR), 37
Concrete, 364—368
  nanomaterials in, 200f
Construction, application of nanomaterials in, 363—364
  asphalt concrete, 369—370
  building applications, 370
  concrete, 364—368
  health implications, 370—371
Construction and demolition (C&D) wastes, 364—365
Construction and destruction (C&D) scrap, 32
Construction industries' nanowaste, 347—348
Consumer products, uses of nanomaterials in, 258
Consumer Product Safety Commission (CPSC), 94
Contamination of nanowaste, sources and routes of, 7—8
Controlling exposure of nanomaterial, 67—71
  elimination/substitution, 67
  engineering controls, 67—68
  personal protective equipment (PPE), 69—71
    eye protection, 69
    hand protection, 69
    protective clothing, 70
    respiratory protection, 70—71
  safe laboratory work practices, 68—69
Control of major accident hazards (COMAH) regulations, 73—74
Control of substances hazardous to health (COSHH) regulations, 76—79
  assessment of hazards and exposure, 77
  control of exposure, 78
  health surveillance, 78—79
  instruction and training, 79
  issues under COSHH regulation, 79

  monitoring exposure, 78
  prevention/control of exposure, 77
  prevention of exposure, 77
  risk management, 79
Copper-containing waste etchants (CCWEs), 218—219
Copper-tin NPs, 194
Core-shell NP, 231—232
Cosmetics industry, nanomaterials in, 258
Council for Science, Technology and Innovation (CSTI), 100
Critical solution temperature (CST), 226—227

D

Dangerous substances and explosive atmospheres regulations, 79
Dangers of nanotechnology, 264—265
Department of Toxic Substances Control (DTSC), 94
Dermal exposure, 65
Dichloroethylene (DCE), 153
Differential magnetic catch and release (DMCR), 159
Diffuse reflectance spectroscopy, 297
N,N-Dimethylaminoethyl methacrylate (DMAEMA), 225—226, 232—237
Disposal and recycling of nanomaterials in waste, 262
Disposal of nanoparticle waste, 188
Disposal of nanowaste, 115—116
Dynamic sources, 48—51

E

Ecotoxicology, 259—260
Electrodeposition deposition and electrokinetic process, 195
Electronic industries' nanowaste, 346
Electron probe microanalysis, 312—314
Electroplating, 388
End-of-life single-wall carbon nanotubes (EOL-SWCNTs), 299
Energetic ionic liquids (E-ILs), 272
Energy-harvesting industries' nanowaste, 347
Engineered nanomaterials (ENMs), 29—32, 84
Environment, impact of nanowastes on, 11—13, 131—133
Environment, recovering nanomaterials from, 211—212

Environmental cleanup applications, nanomaterials for, 390–391
Environmental hazards, 66
Environmental Protection Agency (EPA), 15–16
Ethical, legal, and societal issues (ELSI), 90
1-Ethyl-3-methylimidazolium acetate, 278
1-Ethyl-3-methylimidazolium tetrafluoroborate, 269–270
Ethyl acrylic acid (EAA), 225–226
Ethylene oxide–propylene oxide (EOPO), 232–237
European Chemicals Agency (ECHA), 96
European Commission (EC), 15–16
European Observatory for Nanomaterials (EUON), 97
Existing substances regulation (ESR), 80
Exposure routes, 64–65
  dermal exposure, 65
  ingestion, 65
  inhalation, 64–65
Eye protection, 69

F

Federal Insecticide, Fungicide, and Rodenticide Act (FIFRA), 93–94
Ferromagnetic core-shell nanoparticles, synthesis of, 216, 216f
Fiber-type NMs, 155–156
Filtering facepiece respirators, 70
Fire and explosion hazards, 66
Forms of nanowaste, 112f
Fourier transform infrared spectroscopy, 296–298
Fullerenes, 23, 39–40, 253
Fundamental of nanoparticles, 251–252

G

Gastrointestinal tract, 251
Generation of nanowaste, 131
Glucose reduction process, 199
Gold nanoparticles, 41
Graphene, 23, 192
Green synthesis of nanomaterials, 128f
Group Assessing Already Registered Nanomaterials (GAARN), 96

H

Half- or full-face respirators, 70–71
Hand protection, 69

Health, impact of nanowaste on, 133–134
Health and Safety Executive (HSE), 73–75
Healthcare industries' nanowaste, 345–346
Health hazards, 63–65
  exposure routes, 64–65
    dermal exposure, 65
    ingestion, 65
    inhalation, 64–65
Health implications, 370–371
Heracleum persicum, 154–155
High-gradient magnetic separation (HGMS) technique, 159
High-volume fly ash (HVFA), 366
Hydrogel, 230–231
Hydrophilic ILs, 273–275
Hydrophobic ILs, 273
Hydroxyapatite (HAP) component, 294–296
Hyperbranched polyethylenimine (HPEI), 230, 240

I

Incidental nanowaste, 16–17
Incineration of waste that contains nanomaterials, 28–33
  legislative framework, 33
  nanowaste management problems and issues, 32–33
  nanowaste treatment
    in landfills, 30–31
    in waste incineration plants (WIPs), 29–30
    in waste treatment plants, 28–29
  recycling of waste containing nanomaterials, 31–32
Industrial applications, nanomaterials recycling in, 376–377
  α-cyclodextrin, selective recovery of metal nanoparticles by using, 389–390
  environmental cleanup applications, nanomaterials for, 390–391
  identification of recovered waste nanoparticles by characterization techniques, 378–379
  regulations and standards for nanowaste management, 377–378
  sustainable metals and inorganic nanoparticles' recovery from waste for catalytic and magnetic properties, 380–383
  upcycling process, growth of nanomaterials from waste plastics by, 386–388

Industrial applications, nanomaterials recycling in (*Continued*)
    utilization of recycled nano-scale and micron-scale reinforcements in composite applications, 383–386
    waste electronic components, recovery and reuse of metal nanoparticles from, 388–389
Industrial scale up applications of nanomaterials recycling, 341–343
    future perspectives, 358
    market scenario of recycled nanomaterials, 343–344
    nanowaste generation from various industries and practices, 344–349
        construction industries' nanowaste, 347–348
        electronic industries' nanowaste, 346
        energy-harvesting industries' nanowaste, 347
        medical industries' nanowaste, 345–346
    nanowaste recycling processes, 353–358
        chemical recycling, 354–355
        landfilling treatment, 356
        mechanical recycling, 353–354
        microemulsion process, 356–357
        other methods of recycling, 357–358
        thermal processing, 355–356
    potential nanobyproduct materials, their recovery, and recycling, 349–353
        carbon-based nanobyproducts, 349–350
        metal nanobyproducts, 350–351
        metal oxide and nonmetal oxide nanobyproducts, 351–353
        polymer nanobyproducts, 353
Ingestion, 65
Inhalation, 64–65
Inorganic nanomaterials, 40–42
    metallic nanoparticles, 41
    metal oxide nanoparticles, 42
Interagency Nanotechnology Working Group (IWGN), 21
Internal reflectance spectroscopy, 297
International law on nanomaterials, 94
International Uniform Chemical Information Database (IUCLID), 96
International Union for Conservation of Nature (IUCN), 15–16
Ionic liquids (ILs), 269–270
    applications of, for recycling, 276–278
    recycling, 273–276

    scope of, 270–271
    synthesis of, 271–272
    types of, 272, 273*f*
Iron, 41

K

Korea Research Institute of Standards and Science (KRISS), 101

L

Landfills, nanowaste treatment in, 30–31
Layer-by-layer assembling, 199
Lead, 179
Lead poisoning, 179
Legislation
    for recipients and users of chemicals, 73
    for regulatory authority, 73–75
        exposure assessment in NONS, 75
    for suppliers, 72–73
Lignocellulose nanofibers (LCNF), 329–330
    recycled, 328–330
Lipid-based nanoparticles, 254
Lower critical solution temperature (LCST), 226–227, 240–242
Lungs, 251

M

Magnetic flocculants (MFs), 160
Magnetic graphene oxide (M-GO), 158–159
Magnetic separation technique, 157–160
Maleic anhydride, 225–226
Maleic anhydride grafted polypropylene (MAPP) coupling agent, 330–331
Management of nanomaterial wastes, 125
    biological treatment of nanowastes, 134
    challenges in nanowaste management, 140–141
    difficulties and concerns about, 28
    generation of nanowaste, 131
    impact of nanowaste on health, 133–134
    impact of nanowaste on the environment, 131–133
    recycling of nanowaste, 135–140
    regulations and standards for, 377–378
    synthesis of nanomaterials, 127
    toxicity of nanomaterials and their release to the environment, 127–130
    types of nanomaterials and nanowaste, 125
Manufactured nanomaterials, safe handling of, 61–62

controlling exposure, 67–71
  elimination or substitution, 67
  engineering controls, 67–68
  personal protective equipment (PPE), 69–71
  safe laboratory work practices, 68–69
environmental hazards, 66
fire and explosion hazards, 66
health hazards, 63–65
  dermal exposure, 65
  ingestion, 65
  inhalation, 64–65
precautionary measures, 63
  organizational measures, 63
  personal measures, 63
  technical measures, 63
precautionary principles, 62
regulations, 72–76
  chemicals (hazard information and packaging for supply) regulations, 75–76
  recipients, legislation for, 73
  regulatory authority, legislation for, 73–75
  suppliers, legislation for, 72–73
risk evaluation, 66–67
spills, 71–72
storage, 71
waste handling, 71
workplace risk management, 76–80
  biocidal products regulations, 80
  control of major accident hazards regulations, 80
  control of substances hazardous to health (COSHH) regulations, 76–79
  dangerous substances and explosive atmospheres regulations, 79
  existing substances regulation (ESR), 80
Manufacturing of NMs, 252
Market scenario of recycled nanomaterials, 343–344
Material safety data sheet (MSDS), 89, 186
Mechanical properties of recycled nanomaterials, 324–331
  recycled carbon nanotubes, 326–327
  recycled lignocellulose nanofibers, 328–330
  recycled nanoalumina, 330–331
  recycled nano-$CaCO_3$, 324–325
  recycled nanoclay, 324
  recycled nanosilver, 326
  recycled nano-SiO_2, 327–328

recycled nano-zinc oxide, 326
Medical industries' nanowaste, 345–346
Metal-based nanomaterials, 152–155
Metallic ionic liquids (M-ILs), 272
Metallic nanoparticles, 41, 276–277
Metallic silver, 41
Metal nanoparticles, 253
Metal oxide nanoparticles, 42
Methacrylic acid (MAA), 225–226
N,N'-Methylene-bis-acrylamide (MBA), 240
Micelle, 228–229
Microbiological process, 197–198
Microemulsion process, 197
  nanoparticle recovery using, 218–219, 219f
Microwave irradiation (MWI), 271
Multi-walled carbon nanotubes (MWCNTs), 150, 194–195, 318, 386–388

N

Nanoalumina, recycled, 330–331
Nanoarsenic waste risk, 179
Nano-$CaCO_3$
  recycled, 324–325
Nanocadmium waste risk, 179–181
Nanochromium waste risk, 181
Nanoclay, recycled, 324
Nanoelectromechanical systems (NEMS), 200
Nano inventiveness, 91
Nanolead waste risk, 179
Nanomaterial-reinforced composite materials, 155–157
Nanomaterials, 21–22
  in concrete design and development, 364–371
  in construction, 363–364
Nanometer, 249
Nanoplatinum waste risk, 182
Nanopollution, 4, 4f, 17
Nanoporous materials and membrane separation, 198
Nano regulations, 93–94, 105
Nano safety, 91, 100–101, 105
Nanosafety and Ethics Strategic Plan, 102
Nanoscience, 254–255
Nanosilica, 347–348
Nanosilver, recycled, 326
Nanosilver waste risk, 181–182
Nano-SiO_2
  recycled, 327–328

Nanosocialism, 90–91
Nanotechnology, 3, 9–10
  scope of, 254–255
Nanotechnology Development Promotion Act
    (NDPA), 100–101
Nanotoxicity, causes of, 133f
Nanowastes, 3–4, 21–23, 37–38, 344–349
  biological treatment of, 134
  carbon-based nanomaterials, 23
  challenge of, 24–25
  challenges in the management of, 140–141
  classification of, 4–7, 25–27, 26f, 42–45, 43t,
    112–114, 113f, 126t, 183–186
    on the basis of risk factors, 6t
  from construction industries, 347–348
  difficulties and concerns about management of,
    28
  disposal of, 115–116
  from electronic industries, 346
  from energy-harvesting industries, 347
  forms of, 112f
  future perspectives and challenges, 16–17
  generation of, 131
  impact on environment, 11–13, 131–133
  impact on health, 133–134
  incineration of waste that contains
    nanomaterials, 28–33
    landfills, nanowaste treatment in, 30–31
    legislative framework, 33
    nanowaste management problems and issues,
      32–33
    recycling of waste containing nanomaterials,
      31–32
    waste incineration plants (WIPs), nanowaste
      treatment in, 29–30
    waste treatment plants, nanowaste treatment
      in, 28–29
  from medical industries, 345–346
  nanomaterials, types of, 38–42
    carbon-based nanomaterials, 38–40
    inorganic nanomaterials, 40–42
    organic nanomaterials, 40
  production, 4f, 5f
  production pathway of, 22f
  prospective concerns around, 28–33
  recycling of, 116–119, 135–140
  regulatory bodies for generation and
    management of, 15–16

  silver nanoparticles, 23–24
  sources and routes of contamination of, 7–8
  sources of, 45–51
    dynamic sources, 48–51
    miscellaneous sources, 51
    stationary sources, 47–48
  titanium dioxide nanoparticles, 28
  toxic effects of, 9–11
  treatment strategies of, 13–14, 14f, 15t
Nanowaste management problems and issues,
    32–33
Nanowaste recycling, various processes for,
    188–199
  chemical processes, 190–193
  coagulation technique, 198
  electrodeposition deposition and electrokinetic
    process, 195
  glucose reduction process, 199
  layer-by-layer assembling, 199
  microbiological process, 197–198
  microemulsion process, 197
  nanoporous materials and membrane separation,
    198
  physical processes, 189–190
  sludge treatment process, 196–197
  thermal processes, 194–195
Nano-zinc oxide, recycled, 326
Nanozinc waste risk, 181
National Nanotechnology Initiative (NNI), 90, 94
Neural disease, 64
Neutral ionic liquids (N-ILs), 272
4-Nitrophenol (4-NP), 153–154
Notification of new substances (NONS), 74–75

O

Octadecylamine (ODA), 194
Organically modified montomorillonite (OMMT),
    386–388
Organic nanomaterials, 40
Organic solvent, 215
Organisation for Economic Cooperation and
    Development (OECD), 15–16
Organizational measures, 63

P

Parkinson's disease, 64
1-Pentyl-3-methyl-imidazolium bromide,
    269–270

Personal measures, 63
Personal protective equipment (PPE), 69–71
  eye protection, 69
  hand protection, 69
  protective clothing, 70
  respiratory protection, 70–71
    filtering facepiece respirators, 70
    half- or full-face respirators, 70–71
Phenylboronic acid (PBA), 225–226
pH/thermal responsive materials, 224–226
  applications in nanomaterials recycling, 232–242
    adsorption, 240–242
    aqueous two-phase system (ATPS), 232–237
    catalysts, 238–240
  stimuli-responsive nanostructures, 227–232
    core-shell NP, 231–232
    hydrogel, 230–231
    micelle, 228–229
    polymer brush, 230
    vesicle, 229
  thermoresponsive materials, 226–227
Physical properties of nanomaterials, 250t
Physical recycling processes, 189–190
Piaractus mesopotamicus, 127–129
Poly(2-acrylamido-2-methylpropane sulfonic acid), 225–226
Poly[(2-diisopropylamino)ethyl methacrylate] (PDPA), 230
Poly(4-styrenesulfonic acid), 225–226
Polyacrylamide (PAAm), 227
Poly(acrylamide-co-butyl methacrylate), 227
Poly(acrylic acid) (PAA), 227
Polyamide 6 (PA6), 155–156
Polycyclic aromatic hydrocarbons (PAH), 23
Polyethylene (PE), 194–195, 307–308
Polyethylene terephthalate (PET), 194–195
Polyionic liquids (P-ILs), 269
Polymer brush, 230
Polymeric nanoparticles, 40, 254
  lipid-based nanoparticles, 254
Polymer ILs (P-ILs), 272
Polymers polypropylene (PP), 155–156
Poly(N-isopropylacrylamide) (PNIPAM), 227–231, 239–240
Polyolefin, 194–195
Polypropylene (PP), 192, 194–195, 386–388
Polyvinyl alcohol (PVA), 164, 194–195

Pomelo (*Citrus maximus*), 201
Potential nanopollutants, 4
Precautionary measures, 63
  organizational measures, 63
  personal measures, 63
  technical measures, 63
Precautionary principles, 62
Printed circuit boards (PCBs), 197
Prochilodus lineatus, 129
Production pathway of nanowastes, 22f
Products, recovering nanomaterials from, 212
Prospective concerns around nanowastes, 28–33
Protective clothing, 70
Protic ionic liquids (Pr-ILs), 272

R

Raman spectroscopy, 298–299
REACH Implementation Project on Nanomaterials (RIPoN), 96
Recipients, legislation for, 73
Recyclable nanomaterials, 224–225, 232, 233t
  and recycling methods, 166t
Recycling definitions, 318–324
  economic aspects, 323–324
  environmental aspects, 323
Recycling of nanomaterials, 175–183, 255–256
  benefits of nanomaterials recycling, 202
  classification of nanowaste, 183–186, 184f
  disposal of nanoparticle waste, 188
  humans and environment, nanomaterials posing risks to, 179–183
  limitations of nanomaterials recycling, 202
  nanowaste recycling, various processes for, 188–199
    chemical processes, 190–193
    coagulation technique, 198
    electrodeposition deposition and electrokinetic process, 195
    glucose reduction process, 199
    layer-by-layer assembling, 199
    microbiological process, 197–198
    microemulsion process, 197
    nanoporous materials and membrane separation, 198
    physical processes, 189–190
    sludge treatment process, 196–197
    thermal processes, 194–195
  recycling of nanocomposites, 202

404 Index

Recycling of nanomaterials (*Continued*)
  safety guidelines for handling nanoparticles,
    186–188
  various nanowaste recycling products, 199–201
    low-cost sensors for energy storage
      applications, 201
    nanomaterials applied in suspensions,
      200–201
    nanomaterials in concrete production,
      199–200
Recycling of nanowaste, 116–119, 135–140
Recycling of waste containing nanomaterials,
  31–32
Recycling operations, nanomaterials in, 263
Recycling procedures, nanowaste streams with,
  320*t*
REFINE project, 104
Registration, Evaluation, Authorization and
  Restriction of Chemicals (REACH),
  96–98
Registration and Evaluation of Chemicals in the
  Republic of Korea (Korea REACH),
  100–101
Regulations, 72–76
  chemicals (hazard information and packaging for
    supply) regulations, 75–76
    classification and labeling of an individual
      substance, 76
    safety data sheet requirements, 75
  legislation for recipients and users of chemicals,
    73
  legislation for regulatory authority, 73–75
    exposure assessment in NONS, 75
  legislation for suppliers, 72–73
Regulatory authority, legislation for, 73–75
  exposure assessment in notification of new
    substances (NONS), 75
Regulatory bodies for nanowaste generation and
  management, 15–16
Respiratory protection, 70–71
  filtering facepiece respirators, 70
  half- or full-face respirators, 70–71
Restriction of Hazardous Substances Directive
  (RoHS), 96
Risk evaluation of nanomaterial, 66–67
Risks and benefits of nanomaterials, 180*t*, 261*t*
Risks related to nanomaterials in waste, 262–263
Ryegrass biomass, 178–179

S

Safety and global regulations for application of
  nanomaterials, 83–87
  approaches of democratic governance to
    nanotechnology, 89–91
  arguments against regulation of nanomaterials,
    95
  future perspectives, 105
  international law on nanomaterials, 94
  nano inventiveness, 91
  response from governments all over the world,
    92–105
    application of nanotechnology in Thailand,
      101–102
    Canadian policy on nanotechnology, 99–100
    European Union, 95–99
    Japanese nano policy, 100
    regulation of nanomaterials for clinical
      application, 103–105
    response from advocacy groups, 102–103
    South Korean policy on nanotechnology,
      100–101
    technical aspects of nanomaterials, 103
    United Kingdom, 95
    United States, 92–94
  risks management for environment and health
    safety, 88–89
Safety guidelines for handling nanoparticles,
  186–188
Scanning electron microscopy, 306–308
Scherer's formula, 293–294
Science, technology and innovation (STI), 100
Scientific Committee on Consumer Products
  (SCCP), 96
Scientific Committee on Emerging and Newly
  Identified Health Risks (SCENIHR), 97
Semiconductor nanoparticles, 253
Sewage sludge, 196
Silica nanoparticles (SNPs), 42, 307–308
Silicon dioxide, 200
Silver, 181–182
Silver nanoparticles, 11–13, 23–24
Single-walled carbon nanotubes (SWCNTs), 150,
  299, 318
Skin, 251
Sludge treatment process, 196–197
Solid electrolyte interphase (SEI) layer, 299
Solvent evaporation method, 214–215

classification of, 215
new modifications of, 215
recycling of nanomaterials by, 216—217
Solvent extraction method, 215, 218f
recycling of waste by, 218, 218f
types of, 215
Sources of nanowastes, 45—51
dynamic sources, 48—51
miscellaneous sources, 51
stationary sources, 47—48
Spills, 71—72
Spodoptera frugiperda, 130
Sports industry, nanomaterials in, 259
Standards, nanomaterials recycling, 249—251
categories of wastage of nanomaterials, 259
classification of nanoparticles, 252—254
carbon-based nanoparticles, 253
ceramic nanoparticles, 253
lipid-based nanoparticles, 254
metal nanoparticles, 253
polymeric nanoparticles, 254
semiconductor nanoparticles, 253
dangers of nanotechnology, 264—265
disposal and recycling of nanomaterials in waste, 262
fundamental of nanoparticles, 251—252
importance of recycling in waste management, 256—257
nanomaterials and the industries they are used in, 258—259
nanowaste ecotoxicology and treatment, 259—260
recycling of nanomaterials, 255—256
recycling operations, nanomaterials in, 263
risks related to nanomaterials in waste, 262—263
scope of nanotechnology, 254—255
uses of nanomaterials, 258
in consumer products, 258
waste generated during production, 260—262
Stationary sources, 47—48
Stimuli-responsive materials, fabrication of, 228f
Stimuli-responsive nanostructures, 227—232
core-shell NP, 231—232
hydrogel, 230—231
micelle, 228—229
polymer brush, 230
vesicle, 229
Stokes radiation, 298—299

Storage, 71
Supercritical antisolvent precipitation (SAP), 161—162
Supplementary cementitious materials (SCMs), 363
Suppliers, legislation for, 72—73
Supported ionic liquids (S-ILs), 272
Sustainable metals and inorganic nanoparticles' recovery from waste for catalytic and magnetic properties, 380—383
Synthesis of nanomaterials, 127

T

Technical measures, 63
Techniques used for the recovery of nanomaterials from wastes, 157—165
antisolvent technique by using CO_2, 160—162
aqueous dispersion techniques, 162
centrifugation/solvent evaporation technique, 163
colloidal solvent technique, 163
magnetic separation technique, 157—160
Techniques used to study physiochemical properties of recycled nanomaterials, 291—296
atomic force microscopy (AFM), 309—312
Auger electron spectroscopy (AES), 303—305, 304f
electron probe microanalysis, 312—314
Fourier transform infrared spectroscopy, 296—298
Raman spectroscopy, 298—299
scanning electron microscopy, 306—308
transmission electron microscopy (TEM), 308—309
X-ray diffraction (XRD) method, 291—296
X-ray fluorescence (XRF) spectroscopy, 305—306
X-ray photoelectron spectroscopy (XPS), 299—302
Telurium nanowires, 185—186
Textile industry, nanomaterials in, 133f
Thermal technique, 194—195
Thermoresponsive materials, 226—227
Thymbra spicata, 153—154
Tin plate electroplating sludge, 352—353
Titania, 42
Titania nanoparticles, 42
Titanium dioxide, 42

406 Index

Titanium dioxide nanoparticles, 28, 259
Toxic effects of nanowastes, 9–11
Toxicity of nanomaterials, 127–130
Toxic Substances Control Act (TSCA), 93–94
Transmission electron microscopy (TEM),
    308–309
Treatment strategies, of nanowaste, 13–14, 14f, 15t
Trichloroethylene (TCE), 153
Triton X-114-based cloud point extraction
    method, 150, 150f
Types of nanomaterials, 38–42, 125
    carbon-based nanomaterials, 38–40
    inorganic nanomaterials, 40–42
        metallic nanoparticles, 41
        metal oxide nanoparticles, 42
    organic nanomaterials, 40
Types of nanomaterial wastes, 149–157
    carbon-based nanomaterials, 150–151
    ceramic-based nanomaterials, 151–152
    metal-based nanomaterials, 152–155
    nanomaterial-reinforced composite materials,
        155–157

U
Ultrafine fly ash (UFFA), 365–366
Ultrafine particles, 109
Upcycling process, growth of nanomaterials from
    waste plastics by, 386–388
Upper critical solution temperature (UCST),
    226–227
Uses of nanomaterials, 258
    in consumer products, 258

V
Vacuum membrane distillation technique,
    275–276
Vesicle, 229

W
Wastage of nanomaterials, categories of, 259
Waste containing nanomaterials (WCNM), 210
Waste electronic components, recovery and reuse
    of metal nanoparticles from, 388–389

Waste generated during production, 260–262
Waste handling, 71
Waste incineration plants (WIPs), nanowaste
    treatment in, 29–30
Waste management
    importance of recycling in, 256–257
    recycling in, 209–210
        classification of wastes, 210
Waste nanoparticles, environmental applications of,
    213t
Waste printed circuit boards (WPCB), 194–195,
    197
Wastes, classification of, 210
Waste treatment plants (WTPs), 28
    nanowaste treatment in, 28–29
Workplace risk management, 76–80
    biocidal products regulations, 80
    control of major accident hazards regulations, 80
    control of substances hazardous to health
        (COSHH) regulations, 76–79
        assessment of hazards and exposure, 77
        control of exposure, 78
        health surveillance, 78–79
        instruction and training, 79
        issues under COSHH regulation, 79
        monitoring exposure, 78
        prevention/control of exposure, 77
        prevention of exposure, 77
        risk management, 79
    dangerous substances and explosive atmospheres
        regulations, 79
    existing substances regulation (ESR), 80

X
X-ray diffraction (XRD) method, 291–296
X-ray fluorescence (XRF) spectroscopy, 305–306
X-ray photoelectron spectroscopy (XPS),
    299–302

Z
Zinc, 181
Zinc nanoparticles, 41
Zinc oxide (ZnO) nanoparticles, 42

Printed in the United States
by Baker & Taylor Publisher Services